Advanced Textbooks in Control and Signal Processing

For further volumes:
http://www.springer.com/series/4045

Seddik Bacha • Iulian Munteanu
Antoneta Iuliana Bratcu

Power Electronic Converters Modeling and Control

with Case Studies

Seddik Bacha
Grenoble Electrical Engineering Laboratory
Saint Martin D'Heres, France

Iulian Munteanu
Control Systems Department
Grenoble Image Speech Signal Control Systems Laboratory
Saint-Martin d'Hères, France

Antoneta Iuliana Bratcu
Control Systems Department
Grenoble Image Speech Signal Control Systems Laboratory
Saint-Martin d'Hères, France

ISSN 1439-2232
ISBN 978-1-4471-5477-8 ISBN 978-1-4471-5478-5 (eBook)
DOI 10.1007/978-1-4471-5478-5
Springer London Heidelberg New York Dordrecht

Library of Congress Control Number: 2013952983

Printed on acid-free paper

Springer is part of Springer Science+Business Media (www.springer.com)

I tmurt-iw, tamurt idurar
I yizmawen-is

To our families

Foreword

In my 30-year experience as researcher in the field of power converters, I have witnessed how the interest of the industry in the power electronics area and the corresponding applications of the power converters have had an exponential growth. In this sense, currently we can find power converter applications such as mobile devices, battery chargers and advanced lighting systems, transportation systems (electric or hybrid cars, trains or airplanes), energy storage systems, integration of renewable energy sources and power quality and power distribution systems.

In general, we can say that these are just a few examples where power converters have been of paramount importance for the development of new markets with the goal of improving the system performance in terms of efficiency, robustness, versatility, reduced volume, low maintenance and cost. Today power converters are ubiquitous and have penetrated most of the strategic sectors of the modern industries.

Power converters development requires from a concurrent expertise on many areas, like semiconductors, circuits design, advanced mathematics, modeling and control of the converters among others. The new markets frequently have to deal with nonlinear loads, with a complex behavior and consequently requiring from the designer a deep knowledge of advanced techniques to achieve the system stability, controllability and new performances that are required to be met in a very competitive market. The challenges of this new market can only be overcome by a deep knowledge of the converters, where a mathematical model is fundamental to achieve a precise control. In this sense, I think that this book is an important, in time, contribution to help engineers to achieve these goals. The book gives a global view of the available modeling and control design techniques, which can be very helpful to both the novice and the expert. In this sense, the book can be considered as self-contained, starting with the most basic techniques moving to the most advanced ones and presenting numerous application examples that help in the clarification of complex concepts.

This work represents an advanced textbook that covers most of the aspects of power converters modeling, as well as the most widely used control approaches, selected upon their already proven effectiveness.

The book offers a teaching perspective *ex nihilo*, beginning from the basics of electricity laws and switches' behavior and arriving at obtaining dynamical models of converters ready to be used for control purposes. It also provides the reader with the tools for designing various types of control structures for a wide range of switching converters (with both DC and AC stages).

Presentation of the theoretical approach and then of a pragmatic one in case of each modeling and control method is a particular feature of the authors' pedagogical vision. Each chapter – except for the introductory chapters – contains at least a case study illustrating the concepts dealt with within the chapter.

The book's main audience is composed of the master students, but it remains open for all specialists in the field, from both academia and industry. The book is divided into two parts, respectively dedicated to modeling and control of power electronic converters.

The first part begins with an introductory chapter on the modeling topics. The switched (topological) model – based on physical description by differential equations and on the classical assumption of perfect switches – is described in Chap. 3. This model succeeds in capturing the time-varying nature of the system; it can be used for building other models (e.g., averaged or sampled-data models) or directly for simulation purpose and/or electromagnetic-compatibility analysis (e.g., switching harmonics). It can also be employed in sliding-mode control laws design. The classical (state-space) averaged model for large- and small-signal behavior of DC-DC converters is studied and its limitations are assessed within Chap. 4. Knowing the limitations of this classical model, two alternatives are explored: first, the generalized averaged modeling, which is extended to high-order harmonic dynamics (i.e., to converters having also AC power stages), second, the reduced-order modeling based upon mode separation, which is suitable for describing the discontinuous-conduction operation or for modeling order reduction in order to mitigate the complexity (Chaps. 5 and 6 respectively).

The second part of the book exploits the results presented in the first part, that is, it shows how the different models can be used for control purpose. The basics of linear and nonlinear control approaches aspects have been presented after recalling the prerequisites in Chaps. 7 and 10 respectively.

The use of the linear control approach is separately detailed for the DC-DC power converters and for the converters having AC stages. Whereas for the DC-DC converters the control design relies mainly upon PI and lead-lag control in Chap. 8, for converters integrating AC stages some more sophisticated approaches – like use of *dq* or combined *dq*-stationary frameworks or of resonant controllers – are necessary, like detailed in Chap. 9.

The nonlinear control applied to power electronic converters is relatively new in the field (beginning of 1990s). The electrical engineers are not familiar with these aspects for several reasons, among which the first and most important one is the difficult apprehension of the dedicated tools. The authors have strived to provide an intuitive way to achieve such kind of control laws. However, the theoretical aspects have been developed in parallel to sustain this intuitive approach. The examples and case studies illustrate each of the control methods.

The nonlinear control approaches have been divided into two families: continuous and discontinuous. The class of continuous nonlinear control methods is represented by the feedback-linearization control, dealt with in Chap. 11; and the energy-based control laws, stabilizing control and passivity-based control, which are exposed in Chap. 12. Combinations of these two classes are obviously possible. The second class consists in the variable-structure control, also known as sliding-mode control; it is amply presented in Chap. 13. This kind of control is widely used in power electronics for its intrinsic robustness. Its limitations are mainly due to structural limits, internal dynamics and arbitrary switching frequency, issues that are extensively addressed in the book.

As a conclusion of the above analysis, the textbook *Power Electronic Converters Modeling and Control: With Case Studies* proposes a collection of concepts, organized in a synergic manner such that to ease comprehension of the control design. The book's contribution goes towards completing the already existing literature by offering a useful integration of control techniques, worthy to be read, understood and employed in the most various applications.

Sevilla
February 2013

Leopoldo García Franquelo

Series Editors' Foreword

The topics of control engineering and signal processing continue to flourish and develop. In common with general scientific investigation, new ideas, concepts and interpretations emerge quite spontaneously, and these are then discussed, used, discarded or subsumed into the prevailing subject paradigm. Sometimes these innovative concepts coalesce into a new sub-discipline within the broad subject tapestry of control and signal processing. This preliminary battle between old and new usually takes place at conferences, through the Internet and in the journals of the discipline. After a little more maturity has been acquired by the new concepts, archival publication as a scientific or engineering monograph may occur.

A new concept in control and signal processing is known to have arrived when sufficient material has evolved for the topic to be taught as a specialised tutorial workshop or as a course to undergraduate, graduate or industrial engineers. *Advanced Textbooks in Control and Signal Processing* are designed as a vehicle for the systematic presentation of course material for both popular and innovative topics in the discipline. It is hoped that prospective authors will welcome the opportunity to publish a structured and systematic presentation of some of the newer emerging control and signal processing technologies in the textbook series.

The problem of modelling and the control of power electronic systems is that they have a circuit topology that includes continuous time elements such as resistors, inductors, capacitors and voltage and current sources that are interfaced with electronic devices like diodes, and electronic switches typically thyristors, transistors, and MOSFETs. This leads to system types that involve both continuous time and discontinuous discrete time switching system behaviour. As with most branches of technology, the desire to use the power of computer based simulation technology leads firstly to the development of sets of mathematical models; however, these devices are used in the control of other systems and there are already classical control solutions in existence. Consequently, it is a short step to use the new models to assess and further develop classical control solutions, and then proceed to investigate the use of advanced control methods in the designs; however, a key challenge is to introduce more analytical and computer-based approaches without losing sight of the real-world applications and the practical limitations and constraints that arise. This advanced-course textbook *Power Electronics Converters Modeling and Control: With Case Studies* by Seddik Bacha, Iulian Munteanu and Antoneta I. Bratcu achieves these objectives extremely well.

The textbook is structured into two parts:

- Part I, *Modelling*, comprises five chapters and, beginning from that simplest of questions, "What is a model?", passes through four chapters devoted to switched models, classical averaged models, equivalent averaged generator models and, finally, generalised averaged models. The model approaches use the state-space model formalism that has many practical and pedagogical advantages, not the least of which is the straightforward step to constructing MATLAB®/Simulink® model simulations.
- Part II, *Control*, comprises seven chapters in total. The opening chapter of this part presents a general overview of control in power electronics. There are then two chapters on approaches to linear system control methods. The second of these linear control chapters looks specifically at DC–AC and AC–DC power converter control. There then follow four chapters based on more advanced control and nonlinear methods. As with the linear control chapters, this group of chapters begins with a general overview of relevant mathematical methods and the remaining three chapters tackle specific nonlinear control approaches: feedback linearization, energy-based methods, and variable-structure (sliding-mode) control designs respectively.

A notable feature of the book is the frequent use in every chapter of examples and case-study material. Throughout the textbook there are continual references to what is practical, and the advantages and disadvantages of the modelling and control methods described. For the student reader, each essential study chapter is provided with problems where the first few problems have solutions given, and then the student is invited to tackle some problems without solutions.

The authors of this textbook have worked together for over a decade or so, and have much relevant experience in power electronics and related subjects. Professor Bacha has been teaching and researching this subject area since 1990. Most importantly, he has taught the material of this advanced course textbook at Master's level for a number of years. Doctors Munteanu and Bratcu have worked within the field of control engineering and have an interest in wind energy systems. Indeed, they co-authored (with N-A. Cutululis and E. Ceangă) the excellent *Advances in Industrial Control* series monograph *Optimal Control of Wind Energy Systems* (ISBN 978-1-84800-079-7, 2008) on this very topic.

Readers might like to complement this fine addition to the *Advanced Textbooks in Control and Signal Processing* series with a research monograph on power electronics that is published in the *Advances in Industrial Control* series entitled *Dynamics and Control of Switched Electronic Systems: Advanced Perspectives for Modelling, Simulation and Control of Power Converters* and edited by Francesco Vasca and Luigi Iannelli (ISBN 978-1-4471-2884-7, 2012).

Industrial Control Centre
Glasgow, Scotland, UK
February 2013

M.J. Grimble
M.A. Johnson

Preface

Modern power electronics has evolved into a new era of electrical energy processing. In this context, power electronic controlled systems have become indispensable to the proper operation of power systems. Control systems theory and signal processing have become, in the last decades, vectors of research and technological innovation in the field of power electronics. Following this trend, this textbook applies control systems theory to the field of power electronics and is intended to be a reference for students and professionals working in the power systems field. It provides the reader with the tools for obtaining various models and control structures for a wide range of switching converters (with both DC and AC stages). The subject covers not only linear control techniques that use the ubiquitous proportional–integral controller, devised as early as the 1980s, but also more modern nonlinear continuous or variable-structure control.

The textbook *Power Electronic Converters Modeling and Control: With Case Studies* originates from the course "Modeling and control of power electronic structures", taught by Professor Seddik Bacha to bachelor engineers and masters in electrical engineering at Grenoble Institute of Technology and Joseph Fourier University in France since 1994. Its content has been enriched by topics and case studies issued from research work and theses developed at Grenoble Electrical Engineering Laboratory in France in switching converters and renewable energy conversion control. Its writing has begun following the encouragements of Professor Jean-Paul Hautier, former General Administrator of École Nationale Supérieure des Arts et Métiers in France.

Like its main topics, the presentation style of this textbook places it at the intersection of power electronics, control systems and signal processing, partially covering some industrial electronics areas. The spirit of the writing assumes that students possess basic knowledge within the aforementioned disciplines. Within the book, each problem is approached in both theoretical and practical fashion, employing illustrative examples. Case studies issue from close-to-real-world problems and are treated in a most complete way. Simulations and comments therein are placed so as to allow insight into what concerns switching converter–control structure closed-loop operation.

Effort has been made to synthesize information from a quite well-developed domain possessing a rich bibliography, merge key terms, achieve the case studies and to unify visions, notations and styles to obtain a presentation both power engineers and control engineers can comprehend.

The discourse of this book was heavily influenced by the experience Dr. Iulian Munteanu and Dr. Antoneta Iuliana Bratcu had as students and co-workers of Professor Emil Ceangă from Dunărea de Jos University of Galaţi in Romania. We appreciate his helpful suggestions, which have inspired the pedagogical presentation of many control approaches within this book.

We thank Professor Leopoldo García Franquelo from University of Sevilla in Spain for evaluating our work and endorsing this textbook. We also thank Professor Jean-Pierre Rognon from Grenoble Institute of Technology in France for his helpful comments and suggestions, which improved the quality of this text.

Grenoble, France Seddik Bacha
January 2013 Iulian Munteanu
Antoneta Iuliana Bratcu

Contents

Chapter 1
Introduction

This introduction aims at placing the present textbook within the interdisciplinary topics related to power electronics and control systems. It offers an overview of power electronic converter roles and objectives in the context of their use as power processing elements of power systems and emphasizes the crucial role of their control. Then, the needs of analysis through modeling and simulation of switching converters are assessed. Finally, the scope and organization of the book are stated and its contents outlined.

1.1 Role and Objectives of Power Electronic Converters in Power Systems

The domain of power electronics concerns electrical power processing by electronic devices which have a controllable behavior. The central idea is the use of power electronic (switching) converters for controlling the electric energy flows within power structures, the general aim being output power conditioning with respect to a certain application. This goal obviously determines the way in which raw input power must be processed. Its implementation results in a control structure that yields the corresponding control input that effectively acts on the converter, thereby modifying its behavior (Erikson and Maksimović 2001).

This association, between power structures and converters, has led to a new electrical power context in which the former have become more diversified, more flexible and more efficient. This situation has been accelerated by the powerful combination of microprocessor-based control devices and high-quality switching devices and by the significant improvement of power handling capabilities and of output power quality (Bose 2001).

Various basic functions may be performed by power electronic converters. DC-DC converters are fed with DC voltage and they also output DC voltage with different value and possibly different polarity than the input voltage. DC-AC converters – or

S. Bacha et al., *Power Electronic Converters Modeling and Control: with Case Studies*, Advanced Textbooks in Control and Signal Processing, DOI 10.1007/978-1-4471-5478-5_1,

inverters – are devices that transform a DC voltage into a bipolar-wave AC voltage with variable magnitude and frequency. AC-DC converters rectify an AC voltage, outputting a unipolar voltage with a significant DC component. In these devices both the input current waveform and output voltage DC value may be controlled. The power flow through all these systems may be reversible – that is, the converters may be bidirectional – thus interchanging the role between the input and output ports (Mohan et al. 2002). Also, corresponding to actual application needs, electric power filtering may be justified and insulation between input and output ground references may be added to these converters.

Power electronic converters may be found in a plethora of applications such as electrical machine motion control, switched-mode power supplies (SMPS), lighting drives, energy storage, distributed power generation, active power filters, flexible AC transmission systems (FACTS), renewable energy conversion and vehicular and embedded technology.

In all these systems control of power converters is ubiquitous and it is responsible for the overall application's proper operation (Kassakian et al. 1991). Control goals that issue from the converter role may include a multitude of functional objectives, thus leading to a somewhat complex control structure that must not adversely affect converter power efficiency and output power quality. Converter control structures should work in a most polluted (noisy) environment due to the hard switching and high-frequency modulation (Tan et al. 2011). The intrinsic nonlinearities, limitations, parameter and load variations (the latter being stochastic in most of the cases) do not render easy the operation of the control structures (Sira-Ramírez and Silva-Ortigoza 2006). Yet, optimal operation of power converters is always achieved through control.

In this context, control of power electronic converters represents a very important issue for their proper operation, requiring special effort of analysis, selection of the most suitable set of specifications and choice of the most adequate control design method in each application case.

1.2 Requirements of Modeling, Simulation and Control of Power Electronic Converters

The design of a switching converter is a nontrivial task due to multiple objectives that must be fulfilled. Cost, gauge, power efficiency, power quality and reliability of the overall structure must be considered when one is defining the objective to be optimized in the design process. Good behavior in terms of operation and energy efficiency depends on choosing the appropriate topology and component types, on sizing in terms of voltage and current handling capacities and on choosing the switching frequency. Voltage and current filters are crucial for the power quality and converter time response. Choice and design of gate drivers, including modulation (e.g., PWM) stage, galvanic insulation, etc., affect the accuracy of control input delivery. Insertion of sensors adds complexity to the power conversion structure and negatively affects its reliability.

For a given set of specifications the design engineer must perform the above mentioned actions but also take into account the presence and influence of controllers and control loops in the overall power converter operation. This usually increases the number of iterations in the design process.

To conclude, the analysis and design of power electronic converters presents significant challenges (Maksimović et al. 2001). These may be alleviated by using modeling and simulation of power electronic converters and their associated control structures, which helps the design engineer better understand converter operation. With this knowledge, the designer may anticipate if the circuit performance will meet the imposed specifications for various changes in operating conditions.

As the circuit is switched at high frequency, the amount of computing power needed to capture the circuit behavior by simulation may be important. In this respect, the type and accuracy of the circuit model are critical in simulations and in computer-aided design. An oversimplified modeling may fail to render the correct converter behavior; conversely, a complex model may lead to impractically slow simulations.

Modeling is also an important step in converter control design. Conventional control approaches always use some form of model in order to manipulate the low-frequency (averaged) behavior of the converter so that it complies with the set of dynamical specifications (Sun and Grotstollen 1992; Blasko and Kaura 1997). Models used for control purposes may be different (and most often simpler) than those used for circuit design or simulation.

Control objectives are set according to the converter role in a certain application and, together with the control approach, determine the legitimacy of using one model or another. In general, good design has to do with the output power quality, which must fulfill certain standards. For example, in the case of switched-mode power supplies the control aims at feeding a DC load with constant DC voltage (in the sense that voltage variations must be contained within certain limits around the rated value) irrespective of load value. In the case of rectifiers one may impose a double goal: to regulate the output voltage while extracting a controllable reactive power. Stand-alone inverters must output a voltage wave with constant voltage and frequency irrespective of load conditions. Power quality is addressed in the active power filters, where the control goal is high-order harmonics reduction, while the power balance between the power structure elements is maintained (Kannan and Al-Haddad 2012). In the control of renewable energy conversion systems, the balance between the produced power and the delivered power has to be ensured in spite of primary resource variations (Teodorescu et al. 2011). Grid-connected applications may also require one impose the delivery of a certain amount of reactive power to the power inverters or FACTS (thyristor-controlled reactors, static compensators, etc.) (Bacha et al. 2011). This list is far from exhaustive and may be continued with torque tracking in AC machine drives (Kazmierkowsky et al. 2011), load matching in induction heating, etc.

1.3 Scope and Structure of the Book

This textbook is oriented towards presenting some classical approaches to modeling and control of power electronic converters having both AC and DC power stages. Starting from converter physical laws, the modeling in this book is oriented to control purposes and covers both averaged and switched (exact) models. Far from being exhaustive, this book concerns only basic and well-established approaches in the relevant literature. Both linear and nonlinear control methods are explored, mostly in their analog form.

This book opens the way to exploring advanced control approaches in the power electronics field. It is dedicated mainly to bachelor and master students, but may also be useful to undergraduate students. Researchers and engineers that work at the intersection of the fields of power electronics, industrial electronics and control systems may benefit from insights presented in this textbook. Prerequisites the reader should master include both electrical circuits theory (Bird 2010) and signals and systems (Oppenheim et al. 1997). A good basis in power electronic circuits (Mohan et al. 2002), industrial electronics (Wilamowsky and Irwin 2011) and control systems theory (d'Azzo et al. 2003; Dorf and Bishop 2008) is also recommended before reading the present textbook.

The remainder of this book is organized in 12 chapters. The first part comprises five chapters that cover the topics of power electronic converter modeling: switched models, classical and general averaged models, and reduced-order models. The second part of the book groups together the last seven chapters, which explore power electronics control approaches. Some of these are based on averaged models – like linear control of DC-DC and AC-DC (or DC-AC) converters, feedback linearization control and stabilizing and passivity-based control – while others, like the sliding-mode approach, use switched models.

Each chapter contains illustrative examples, at least one case study that explains in detail the application of the respective modeling or control method, one or several problems with solutions and a set of proposed problems, which the reader is invited to solve. MATLAB®-Simulink® software is extensively used throughout the textbook to support numerical simulations presented and discussed within examples and case studies. Each chapter ends with a list of the most significant references for the topic discussed.

References

Bacha S, Frey D, Lepelleter E, Caire R (2011) Power electronics in the future distribution grid. In: Hadjsaid N, Sabonnadiere JC (eds) Electrical distribution networks. Wiley/ISTE, Hoboken/London, pp 416–438

Bird J (2010) Electrical circuit theory and technology, 4th edn. Elsevier, Oxford

Blasko V, Kaura V (1997) A new mathematical model and control of a three-phase AC-DC voltage source converter. IEEE Trans Power Electron 12(1):116–123

Bose BK (2001) Modern power electronics and AC drives. Prentice-Hall, Upper Saddle River
d'Azzo JJ, Houpis CH, Sheldon SN (2003) Linear control system analysis and design with MATLAB, 5th edn. Marcel-Dekker, New York
Dorf RC, Bishop RH (2008) Modern control systems, 11th edn. Pearson Prentice-Hall, Upper Saddle River
Erikson RW, Maksimović D (2001) Fundamentals of power electronics, 2nd edn. Kluwer, Dordrecht
Kannan HY, Al-Haddad K (2012) Three-phase current-injection rectifiers. Ind Electron Mag 6(3):24–40
Kassakian JG, Schlecht MF, Verghese GC (1991) Principles of power electronics. Addison-Wesley, Reading
Kazmierkowsky MP, Franquelo LG, Rodriguez J, Perez MA, Leon JI (2011) High-performance motor drives. Ind Electron Mag 5(3):6–26
Maksimović D, Stanković AM, Thottuvelil VJ, Verghese GC (2001) Modeling and simulation of power electronic converters. Proc IEEE 89(6):898–912
Mohan N, Undeland TM, Robbins WP (2002) Power electronics: converters, applications and design, 3rd edn. Wiley, Hoboken
Oppenheim AV, Willsky AS, Hamid S (1997) Signals and systems, 2nd edn. Prentice-Hall, Upper Saddle River
Sira-Ramirez H, Silva-Ortigoza R (2006) Control design techniques in power electronics devices. Springer, London
Sun J, Grotstollen H (1992) Averaged modeling of switching power converters: reformulation and theoretical basis. In: Proceedings of the IEEE Power Electronics Specialists Conference – PESC 1992. Toledo, Spain, pp 1166–1172
Tan S-C, Lai Y-M, Tse C-K (2011) Sliding mode control of switching power converters: techniques and implementation. CRC Press/Taylor & Francis Group, Boca Raton
Teodorescu R, Liserre M, Rodriguez P (2011) Grid converters for photovoltaic and wind power systems. Wiley, Chichester
Wilamowsky BM, Irwin JD (2011) Fundamentals of industrial electronics. CRC Press/Taylor & Francis Group, Boca Raton

Part I
Modeling of Power Electronic Converters

Chapter 2
Introduction to Power Electronic Converters Modeling

This chapter deals with a brief presentation of the main modeling aspects of power electronic converters. The chapter overviews modeling basics, provides useful hints about the main modeling methodologies, accompanied by some illustrative examples, and suggests some possible uses of models.

2.1 Models

2.1.1 What Is a Model?

Modeling of a phenomenon or process is based on its observation and relies upon capturing into an approximate, but sufficiently comprehensive, representation, its most significant features from the point of view of a given application. Modeling requires generalization in the sense that the studied phenomenon must be regarded in the context of similar phenomena so common features may be extracted.

Generally speaking, there are two main modeling approaches: one that uses *black-box models*, based on the process behavior observation of its response to some known input signals, and one based on the known information about the system to be modeled (i.e., representation centered on the behavior laws). The latter approach is employed not only to model physical processes, but also biological, economics or even social systems. Mixing between the two approaches is also encountered, leading to the so-called *gray-box models*.

The interest of this textbook is on power electronic converter modeling using the "information" approach. This means that model representations will be made using the available physical knowledge about the considered converter. In general, physical knowledge about system results in mathematical description of mass and energy conservation laws. Thus, energy accumulation variations within the system are described by so-called *state variables*. In the particular case of power converters, information is embodied in Kirchhoff's laws of the converter circuit, Ohm's laws for the various loads and, finally, in the states of various solid-state switches.

S. Bacha et al., *Power Electronic Converters Modeling and Control: with Case Studies*, Advanced Textbooks in Control and Signal Processing, DOI 10.1007/978-1-4471-5478-5_2,

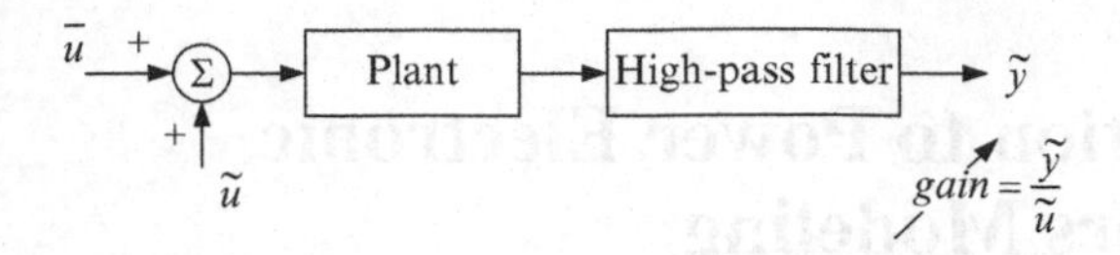

Fig. 2.1 Basic idea of linear identification approach, where $\overline{u}$ and $\widetilde{u}$ are input's low-frequency and high-frequency components, respectively

2.1.2 Scope of Modeling

Next within this textbook one seeks to obtain quasi-general power electronic converter dynamical models to simulate converter dynamic behavior and to construct various control laws. Steady-state converter behavior (static models) can also be obtained, either by zeroing the time derivatives in the dynamical models in the case of DC variables, or by zeroing the derivative of both magnitude and phase in the case of AC variables.

Concerning simulation, a plethora of software renders power converter time-domain behavior in a very precise and reliable way (see, for example, SPICE®, SABER®, MATLAB®). With these programs, however, simulation results are not general. For instance, even if they provide various time waveforms of the internal variables, they do not give direct information about the converter modes. As a consequence, obtaining a model necessary for control purposes cannot be done by using these software packages, at least not directly. Certainly, it is possible to identify a power electronic converter based on the input-output variables evolution file obtained by simulation and to elaborate a frequency-domain model. But, since the quasi-totality of power electronic converters is naturally a nonlinear or linear time-varying system and any linear input-output model depends on the operating point, the information acquired has limited validity.

Figure 2.1 shows how the linearized system looks. Another important drawback of the linear identification approach is that system identification in the frequency domain will be limited at half the switching frequency, according to Shannon's theorem.

An analytical model based on knowledge of the circuit's physical behavior is needed for control purposes. According to the intended use, various levels of modeling can be considered. Model choice also relies on the following criteria:

- the required dynamic or steady-state accuracy;
- whether internal, input or output variables should explicitly appear in the model;
- an acceptable level of complexity;
- the domain of definition.

These requirements are not totally synergistic and often are antagonistic, necessitating an optimal choice. For example, the accuracy of plant replication increases with model complexity.

2.2 Model Types

One can specify some simplifying assumptions, sufficiently accurate so as not to affect the validity of the models to be implemented:

1. Switches are considered 'perfect' in the sense that they behave as a zero-value resistance during conduction (the so-called ON state) and as an infinite-value resistance when the switch is turned off (the so-called OFF state). Also, the switching time is infinitely short.
2. Generators are considered 'perfect' (for example, they provide infinite short-circuit power in the case of voltage sources).
3. Passive elements are considered linear and invariant.

If the two first modeling assumptions are easy to understand, the third deserves more attention. Let us consider as an example a nonlinear inductance whose value depends both on time and on the current $i(t)$ passing through it. Inductance voltage is given by the following equation:

$$v(t) = \frac{d}{dt}(L(i,t) \cdot i(t))$$

Developing this equation gives a nontrivial expression:

$$v(t) = \left(\frac{\partial L(i,t)}{\partial t} + \frac{\partial L(i,t)}{\partial i} \cdot \frac{di(t)}{dt} \right) \cdot i(t) + L(i,t) \cdot \frac{di(t)}{dt}. \tag{2.1}$$

Being too complicated, Eq. (2.1) is practically unusable in modeling. Moreover, this complexity is not justified, as the first term is typically not important in most of applications.

These remarks reasonably justify the adoption of the above assumptions, which however do not essentially affect the modeling methodology. Obviously, in order to increase modeling accuracy, one can successively add detail to an initial, simplified model. For example, models of circuit elements can be enriched by taking into account the dissipative elements (internal resistance of a power source, winding resistance of a coil, etc.). Figure 2.2 shows how the model of a diode can be enriched.

Single-pole–single-throw (SPST) switches can be implemented in various power electronic devices (Erikson and Maksimović 2001). Examples of their schematics are represented in Fig. 2.3. The single-quadrant SPST symbols (e.g., diode and transistor) and their ideal characteristics are presented in Fig. 2.3a, b. Two-quadrant voltage-bidirectional SPSTs can be of many types. Their common features – interesting from the modeling and control viewpoint – have been unified into a unique schematic that will be used throughout this book, as exemplified in Fig. 2.3c.

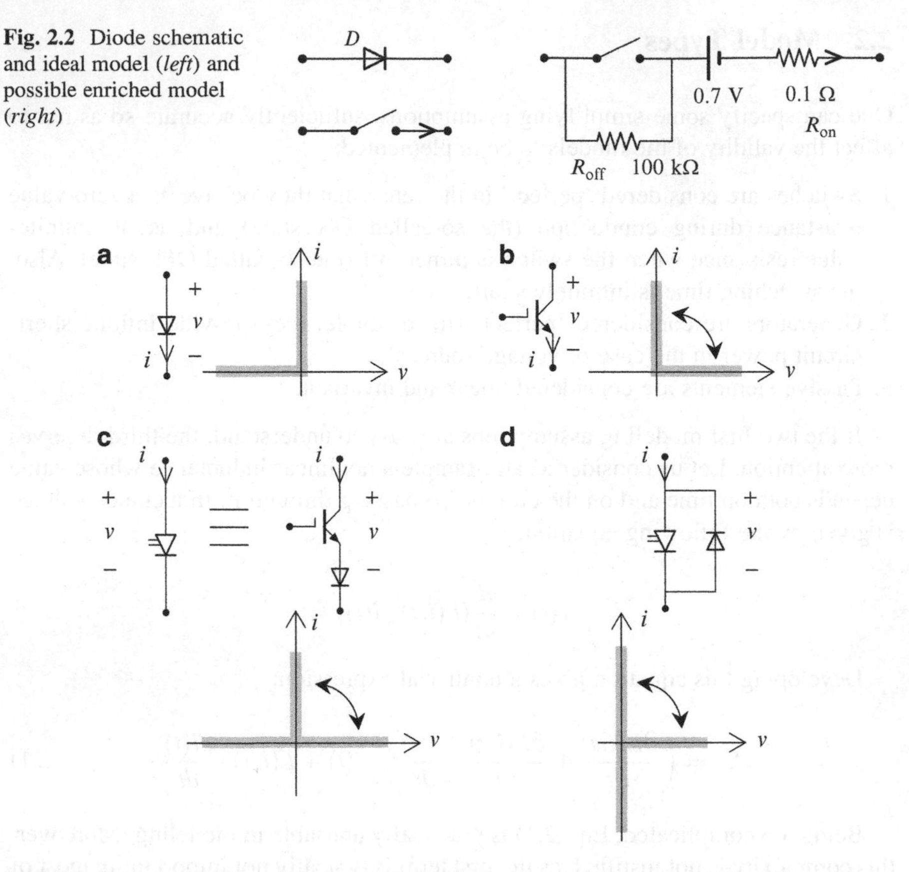

Fig. 2.2 Diode schematic and ideal model (*left*) and possible enriched model (*right*)

Fig. 2.3 Various SPST switches and their ideal characteristics: **(a)** diode; **(b)** transistor (BJT/IGBT); **(c)** two-quadrant voltage-bidirectional SPST; **(d)** two-quadrant current-bidirectional SPST (Erikson and Maksimović 2001)

2.2.1 Switched Models

The switched model is the less elaborated model of the converter, meaning it describes the electrical equations for each circuit configuration. It is sometimes named the "*exact*" model because, under the previously stated assumptions, it describes the converter behavior exactly. The reader may refer to Kassakian et al. (1991) or to Erikson and Maksimović (2001) to get an overview of specific converter switched models and their analysis.

Let us consider the example of the buck converter in Fig. 2.4, where the switch is driven by signal $u(t)$, called the *switching function* (Fig. 2.4a). Let us consider that $u(t)$ is periodic, having T as switching period and α as duty ratio:

$$u(t) = \begin{cases} 1, & 0 \le t < \alpha T \\ 0, & \alpha T \le t < T \end{cases}, \quad u(t-T) = u(t) \forall t.$$

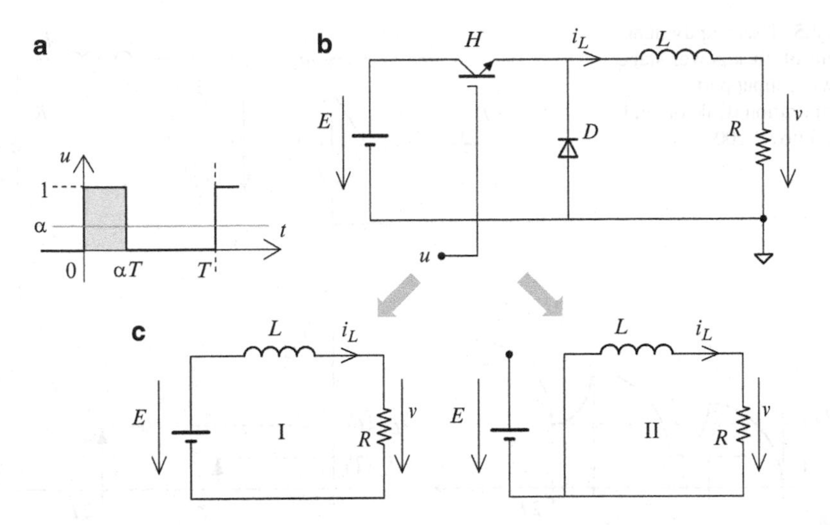

Fig. 2.4 Example of a buck power stage: (**a**) signal driving switch H; (**b**) converter diagram; (**c**) various configurations of the converter: configuration **I** – between time 0 and $\alpha \cdot T$, configuration **II** – between time $\alpha \cdot T$ and T

One can easily note that α represents the average value of $u(t)$.

According to the state of the switch H, the circuit takes the configuration **I** (switch turned on) and configuration **II** (switch turned off) in Fig. 2.4.

Configuration **I** corresponds to a time t (modulo T) between 0 and $\alpha \cdot T$; the system behavior is given by the first equation from (2.2). Configuration **II** corresponds to a time t between $\alpha \cdot T$ and T. One may observe that the circuit has in fact two switching devices, as the diode also naturally turns on and off (Sira-Ramírez and Silva-Ortigoza 2006). Hence, its governing equations are:

$$\begin{cases} E = L\dfrac{di_L}{dt} + Ri_L \\ 0 = L\dfrac{di_L}{dt} + Ri_L, \end{cases} \tag{2.2}$$

where $v = R \cdot i_L$. An elegant way to describe this behavior is to use the switching function u and to compress the representation (2.2) into the form below:

$$E \cdot u(t) = L\frac{di_L}{dt} + Ri_L. \tag{2.3}$$

Function u takes the values 1 (switch turned on) and 0 (switch turned off) according to the configuration. This leads to the equivalent electrical diagram presented in Fig. 2.5, which will from now on be called the *exact equivalent circuit*.

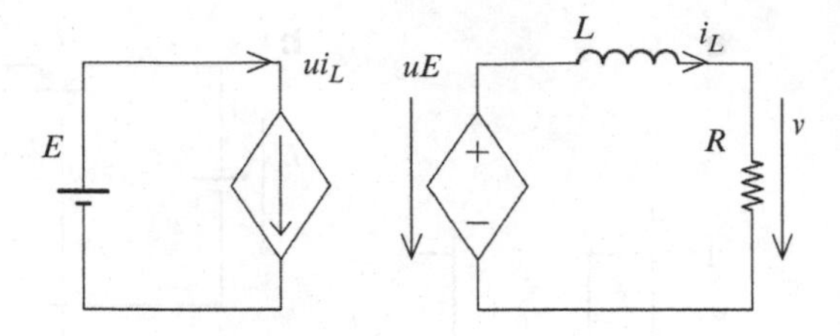

Fig. 2.5 Exact equivalent circuit of buck power stage showing input port configuration (Erikson and Maksimović 2001)

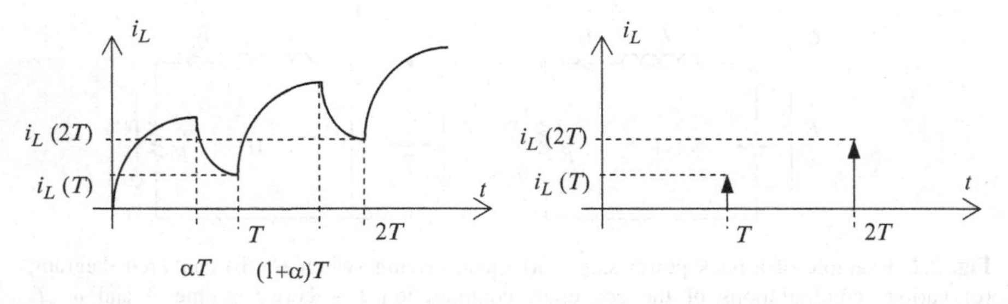

Fig. 2.6 Buck converter sampling and sampled-data model

2.2.2 Sampled-Data Models

A sampled-data model is a model that provides information about the system state in a periodic manner. In the present case, it is a representation sampled not at the switching moments but at each complete operating period (Verghese and Stanković 2001). In the buck converter case detailed in Fig. 2.4, the system switches between two circuit configurations and the inductor current time evolution can be like the one presented in Fig. 2.6. If one considers the current values at each switching period T, a recurrent equation is obtained as below:

$$i_L((k+1)T) = \left(i_L(kT) - \frac{E}{R}\right) \cdot e^{-\frac{R}{L}T} + \frac{E}{R} \cdot e^{-\frac{R}{L}(1-\alpha)T}. \tag{2.4}$$

Equation (2.4) can be put into the more general matrix form:

$$\mathbf{x}_{k+1} = \Phi(\mathbf{x}_k, \mathbf{u}_k, \mathbf{p}_k), \tag{2.5}$$

where **x**, **u** and **p** are the state, control input and disturbance vectors, respectively. Model (2.5) gives the system state at each sampling period but does not provide any information about the concerned variables between the two sampling points. By virtue of its discrete-time description, this model may be useful in digital control of converters (Maksimović et al. 2001).

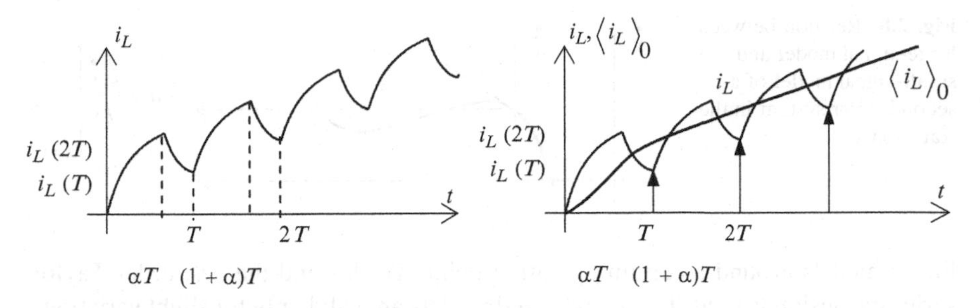

Fig. 2.7 Behaviors of switched and averaged models

2.2.3 Averaged Models

As their names show, these models replicate an average behavior of the system state. This average does not remain constant for a time period comparable with the system time constant but changes as the system is excited. It is computed on a time window of width T, which is sufficiently small in relation to the system dynamics. This window must be regarded as sliding on the time axis, and so it will be called *sliding average* (*moving average* in statistics).

In the case of a chopper inductor current, sliding average is expressed as

$$\langle i_L \rangle_0(t) = \frac{1}{T} \cdot \int_{t-T}^{t} i_L d\tau. \tag{2.6}$$

The averaging of the exact model (2.3) gives

$$\frac{d}{dt} \langle i_L \rangle_0 = -\frac{R}{L} \langle i_L \rangle_0 + \alpha \cdot \frac{E}{L}. \tag{2.7}$$

Model (2.7) is the averaged model of the circuit in Fig. 2.4. Its time evolution is shown in Fig. 2.7.

One can remark that the averaged model is less accurate than the sampled-data model at the precise sampling times; conversely, this approach provides information between the sampling points. A thorough analysis of averaged models for specific power stages has been performed by Kislovsky et al. (1991) and by Erikson and Maksimović (2001).

2.2.4 Large-Signal and Small-Signal Models

Converter dynamic behavior is nonlinear with few exceptions. Sometimes, in order to perform a modal analysis or to build linear control laws, it is necessary to develop

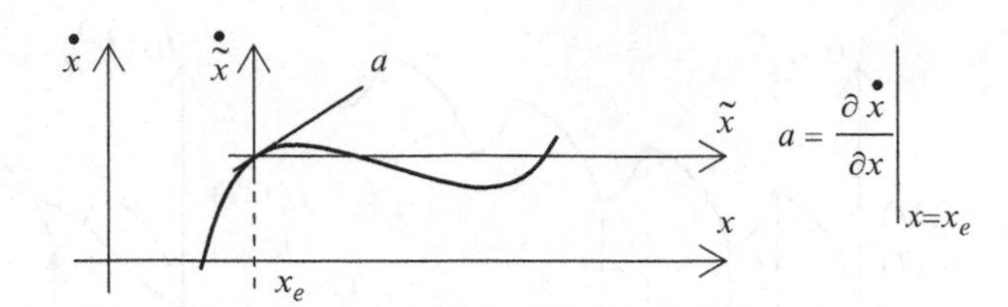

Fig. 2.8 Relation between large-signal model and small-signal model of a second-order system in the state space

linear models around a certain operating point. To this end the first-order Taylor series expansion is used. These linearized models are valid only for slight variations around the considered operating point. This is why they are called *small-signal models*, also known as *tangent linear models*. Conversely, the initial models, valid on the entire definition range, are called *large-signal models* by the power electronics community.

Figure 2.8 suggests the relation between the large-signal model and the small-signal one in the intuitive case of a second-order system state-space trajectory. If the large-signal model is linear, then it is identical to the small-signal model. This is a quite seldom encountered situation – an example is the ideal buck converter case for constant load value. This approach is similar to both averaged modeling and sampled-data modeling.

Let us consider the general case of a continuous nonlinear system:

$$\begin{cases} \dfrac{d}{dt}\mathbf{x} = f(\mathbf{x}(t), \mathbf{u}(t)) \\ \mathbf{y} = h(\mathbf{x}(t), \mathbf{u}(t)), \end{cases} \tag{2.8}$$

where $\mathbf{x}$, $\mathbf{u}$ and $\mathbf{y}$ are the state, input and output vectors.

2.2.4.1 Obtaining Steady-State Models

By zeroing the derivatives one obtains the steady-state input-output characteristic, i.e., the locus of the system's equilibrium points, denoted with the subscript e and represented by a generally nonlinear curve in the input-output plane:

$$\mathbf{y}_e = g(\mathbf{u}_e).$$

2.2.4.2 Building Small-Signal Models

Let us now consider the small variations $\widetilde{\mathbf{x}} = \mathbf{x} - \mathbf{x}_e$, $\widetilde{\mathbf{u}} = \mathbf{u} - \mathbf{u}_e$ and $\widetilde{\mathbf{y}} = \mathbf{y} - \mathbf{y}_e$ around a given equilibrium point $\mathbf{y}_e$, established in response to input $\mathbf{u}_e$. Thus, the linearized system around the specified equilibrium point is written as

$$\begin{cases} \dot{\widetilde{\mathbf{x}}} = \mathbf{A} \cdot \widetilde{\mathbf{x}} + \mathbf{B} \cdot \widetilde{\mathbf{u}} \\ \widetilde{\mathbf{y}} = \mathbf{C} \cdot \widetilde{\mathbf{x}} + \mathbf{D} \cdot \widetilde{\mathbf{u}}, \end{cases} \tag{2.9}$$

with

$$\begin{cases} \mathbf{A} = \left(\dfrac{\partial f(\mathbf{x}, \mathbf{u})}{\partial \mathbf{x}} \right)_{\mathbf{x}_e, \mathbf{u}_e} & \mathbf{B} = \left(\dfrac{\partial f(\mathbf{x}, \mathbf{u})}{\partial \mathbf{u}} \right)_{\mathbf{x}_e, \mathbf{u}_e} \\ \mathbf{C} = \left(\dfrac{\partial h(\mathbf{x}, \mathbf{u})}{\partial \mathbf{x}} \right)_{\mathbf{x}_e, \mathbf{u}_e} & \mathbf{D} = \left(\dfrac{\partial h(\mathbf{x}, \mathbf{u})}{\partial \mathbf{u}} \right)_{\mathbf{x}_e, \mathbf{u}_e} \end{cases} . \tag{2.10}$$

In the case of bilinear systems, i.e., when nonlinearity is given by a product between two state variables or between a state variable and an input variable, another approach is used, as follows.

In the model given by (2.8) one introduces the above defined small variations. Further, some simplifications are made by:

- neglecting the products of variations corresponding to higher-than-second-order terms in the Taylor series expansion;
- simplifying terms corresponding to $\mathbf{x} = 0$.

The resulting model is described by the same matrices as those corresponding to the linearized model (2.9), i.e., those given by (2.10).

Example. The example below concerns a bilinear system. The two above-presented approaches will be employed in order to obtain the small-signal model.

$$\begin{cases} \dot{x_1} = 2x_1x_2 - x_2u \\ \dot{x_2} = x_1 + x_2 \\ y = x_1^2 + u. \end{cases} \tag{2.11}$$

First, one searches the equilibrium points of the system (2.11) for an input $u = u_e$. This is done by zeroing the x_1 and x_2 derivatives. By solving the resulting algebraic system two solutions can be found. The first one is trivial, namely $x_{1e} = x_{2e} = 0$, which gives $y_e = u_e$. The other is $x_{1e} = u_e/2$ and $x_{2e} = -u_e/2$, which gives $y_e = 3u_e/4$. Second, the matrices of the linearized model must be obtained. Two methods may be employed as follows:

- *First method*: Using relations (2.10) one obtains matrices **A**, **B**, **C** and **D**:

$$\begin{cases} \mathbf{A} = \begin{bmatrix} \dfrac{\partial f_1(x,u)}{\partial x_1} & \dfrac{\partial f_1(x,u)}{\partial x_2} \\ \dfrac{\partial f_2(x,u)}{\partial x_1} & \dfrac{\partial f_2(x,u)}{\partial x_2} \end{bmatrix}_{x_{1e}, x_{2e}, u_e} = \begin{bmatrix} -u_e & 0 \\ 1 & 1 \end{bmatrix}, \\ \mathbf{B} = \begin{bmatrix} \dfrac{u_e}{2} & 0 \end{bmatrix}^T, \quad \mathbf{C} = [u_e \quad 0], \quad \mathbf{D} = 1. \end{cases}$$

- *Second method*: One considers the small-signal variations $\widetilde{\mathbf{x}} = \mathbf{x} - \mathbf{x}_e$, where $\mathbf{x}^T = [x_1 \quad x_2]^T$, $\mathbf{x}_e^T = [x_{1e} \quad x_{2e}]^T$, $\widetilde{u} = u - u_e$ and $\widetilde{y} = y - y_e$ which are substituted in equation set (2.11); hence,

$$\begin{cases} \dot{\widetilde{x_1}} + \dot{x_{1e}} = 2(\widetilde{x_1} + x_{1e})(\widetilde{x_2} + x_{2e}) - (\widetilde{x_2} + x_{2e})(\widetilde{u} + u_e) \\ \dot{\widetilde{x_2}} + \dot{x_{2e}} = (\widetilde{x_1} + x_{1e}) + (\widetilde{x_2} + x_{2e}) \\ y + y_e = (\widetilde{x_1} + x_{1e})^2 + (\widetilde{u} + u_e). \end{cases} \tag{2.12}$$

At the steady-state point, the system is described by the relations below:

$$\begin{cases} 0 = 2x_{1e}x_{2e} - x_{2e}u_e \\ 0 = x_{1e} + x_{2e} \\ y_e = x_{1e}^2 + u_e. \end{cases} \tag{2.13}$$

Next, one develops the products. By neglecting the products of variations $\widetilde{x_1} \cdot \widetilde{x_2}, \widetilde{x_1}^2$ and $\widetilde{x_2} \cdot \widetilde{u}$, knowing that $\dot{x_{1e}} = \dot{x_{2e}} = 0$ and taking into account (2.13), one consequently obtains

$$\begin{cases} \dot{\widetilde{x_1}} = 2x_{1e}\widetilde{x_1} + (2x_{1e} - u_e)\widetilde{x_2} - x_{2e}\widetilde{u} \\ \dot{\widetilde{x_2}} = \widetilde{x_1} + \widetilde{x_2} \\ \widetilde{y} = 2x_{1e}\widetilde{x_1} + \widetilde{u}. \end{cases} \tag{2.14}$$

From (2.13) one computes the steady-state values of the state variables as functions of u_e: $x_{1e} = u_e/2$ and $x_{2e} = -u_e/2$. By replacing these values, the system (2.14) becomes:

$$\begin{cases} \dot{\widetilde{x_1}} = -u_e\widetilde{x_1} + \dfrac{u_e}{2}\widetilde{u} \\ \dot{\widetilde{x_2}} = \widetilde{x_1} + \widetilde{x_2} \\ \widetilde{y} = u_e\widetilde{x_1} + \widetilde{u}, \end{cases} \tag{2.15}$$

which corresponds to the matrices **A**, **B**, **C** and **D** computed using the first method.

The small-signal model may also be expressed in the frequency domain, i.e., as a transfer function, which can be computed based upon the above deduced matrices:

$$H(s) \triangleq \frac{\widetilde{Y}(s)}{\widetilde{U}(s)} = \mathbf{C}(s\mathbf{I} - \mathbf{A})^{-1}\mathbf{B} + \mathbf{D},$$

where $\widetilde{U}(s)$ and $\widetilde{Y}(s)$ are the Laplace transforms of scalar time signals $\widetilde{u}$ and $\widetilde{y}$, respectivcly. The same final expression of $H(s)$ can be reached if the Laplace transform is applied to Eq. (2.15). By denoting with $\widetilde{X_1}(s)$ and $\widetilde{X_2}(s)$ the Laplace transforms of the small-signal model state variables, one obtains

$$\frac{\widetilde{X_1}(s)}{\widetilde{U}(s)} = \frac{u_e/2}{s + u_e}, \frac{\widetilde{X_2}(s)}{\widetilde{X_1}(s)} = \frac{1}{s - 1}, \widetilde{Y}(s) = u_e X_1(s) + U(s),$$

allowing us to remark that the state $\widetilde{x_2}$ is unobservable – because it does not appear in the expression of the output – and also unstable.

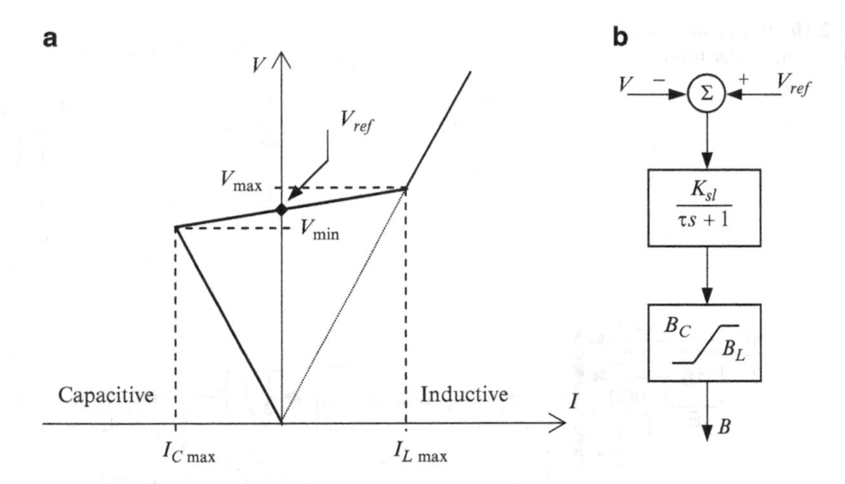

Fig. 2.9 (**a**) Static regulation characteristic of a SVC (Watanabe et al. 2011); (**b**) elementary dynamic model of a SVC

2.2.5 Behavioral Models

It is common to use more or less elaborate black-box-type models to replicate the steady-state or dynamic input-output characteristic of *flexible AC transmission systems* (FACTS). Such models are called *behavioral models*.

The simplest models are the static ones, as illustrated for example by the case of the *static VAR compensator* (SVC) from Fig. 2.9a. Here one can note a regulation zone described by the following relation:

$$V = V_{ref} + X_{sl}I,$$

where I is the current exchanged by the SVC and the utility grid, V is the voltage at the point of coupling, V_{ref} is the voltage reference value and X_{sl} is the slope of the regulation characteristic. As a matter of fact, FACTS are modeled by means of a reactance or susceptance depending on the desired voltage.

A simple (first-order) dynamical model can be built based upon the static curve from Fig. 2.9a; it can be seen in Fig. 2.9b. The term K_{sl} is the inverse of reactance X_{sl}, τ is the time constant of the system and B_C and B_L are the susceptances corresponding to the SVC's capacitor and inductance, respectively. The model output is the susceptance relative to the error $V_{ref} - V$. Other more or less complicated models are recommended by CIGRE (1995) and IEEE (1993), which rely upon the same principle, matching the input-output characteristics of FACTS.

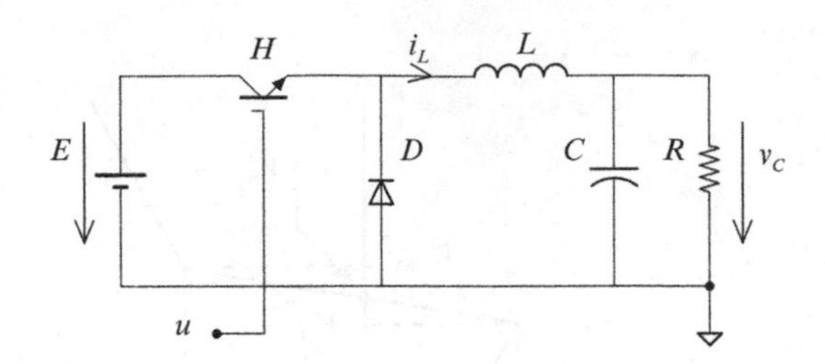

Fig. 2.10 Buck converter with second-order filter

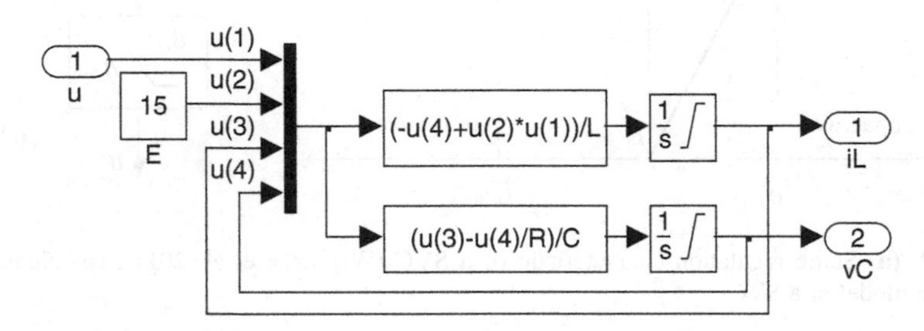

Fig. 2.11 Example of MATLAB®-Simulink® implementation of exact model of a buck converter

2.2.6 *Examples*

Let us consider the case of a buck converter as shown in Fig. 2.10. Via the topological analysis of Fig. 2.4 and Eq. (2.3), the governing equations are

$$\begin{cases} L\dfrac{di_L}{dt} = -Ri_L + Eu(t) \\ C\dfrac{dv_C}{dt} = i_L - \dfrac{v_C}{R}. \end{cases} \tag{2.16}$$

Model (2.16) can express the exact (switched) behavior of the system, as well as its averaged behavior, depending on whether the input signal u is represented by a switched, discontinuous-time-derivative time function or by its continuous-time-derivative average, respectively.

One can attempt a comparison between the different models by means of simulation diagrams; an example of implementation using the Simulink® library is presented in Fig. 2.11. State variables i_L and v_C are the outputs, switching function u and voltage E are the independent inputs. Note also the state variables low limitation required by modeling the unidirectional nature of the converter.

Figure 2.12 allows a comparison between the switched and averaged time evolutions of the two state variables of the system in Fig. 2.11 during the start-up regime, which consists in feeding the system with a duty ratio step. Note the underdamped response due to the unusual choice of L and C, the inductor current limitation and the slight difference in capacitor voltage in the two models.

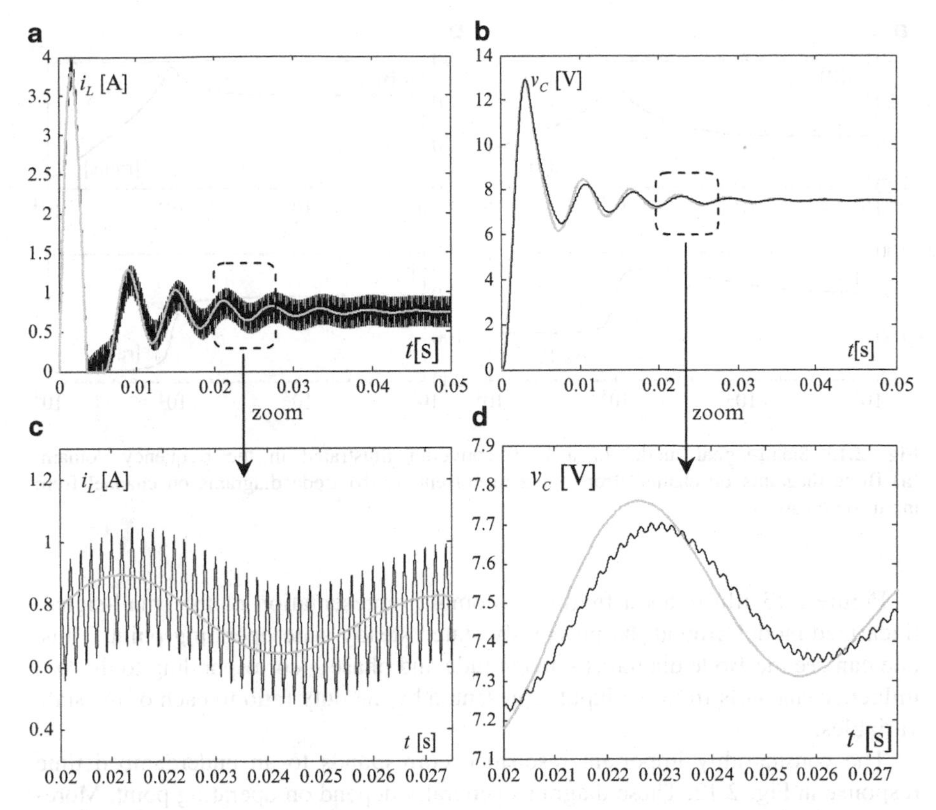

Fig. 2.12 Switched versus averaged model of a buck converter illustrated by time evolution of state variables at start: (**a**) current i_L evolution, switched (*black*) and averaged (*gray*); (**b**) voltage v_C evolution, switched (*black*) and averaged (*gray*)

As is valid for the entire operating range, model (2.16) is a large-signal one. However, one may be interested in capturing the behavior around a typical steady-state operating point (u_e, i_{Le}, v_{Ce}). In the usual operating range, the form expressed by Eq. (2.14) is linear; this form also characterizes small-signal behavior around the given operating point. The latter belongs to the input-output steady-state model that can be obtained by zeroing derivatives in (2.16):

$$\begin{cases} i_{Le} = \dfrac{E \cdot u_e}{R} \\ v_{Ce} = E \cdot u_e, \end{cases}$$

where the second equation shows the converter's static behavior: the output voltage is obtained by stepping down the input voltage proportionally with the steady-state duty ratio, u_e.

Small-signal analysis may result in a transfer-function-type description of the system. In this case, these transfers are defined between the chosen control input (duty ratio) and each of the states.

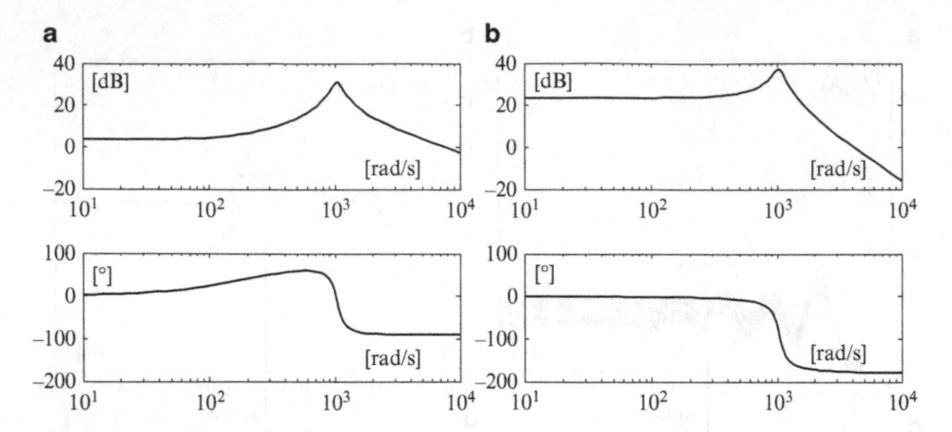

Fig. 2.13 Small-signal model of a buck converter illustrated in the frequency domain: (**a**) Bode diagrams on channel from input to current i_L; (**b**) Bode diagrams on channel from input to voltage v_C

Figure 2.13 illustrates a frequency-domain representation of the small-signal, linearized model around the previously stated steady-state operating point. Thus, one can see the Bode diagrams – magnitude and phase – corresponding to the two influence channels from the input represented by the duty ratio to each of the state variables.

The considerably important resonance corresponds to an underdamped time response in Fig. 2.12. These diagrams generally depend on operating point. Moreover, load resistance, which is a parameter in transfer functions, may vary. Therefore, the converter control based on this model should be robust enough to handle all these uncertainties.

2.3 Use of Models

2.3.1 Relations Between Various Types of Models

Let us consider a control purpose and a certain control approach to fulfill this purpose.

According to the control law type and to the imposed closed-loop performances, a certain type of modeling is appropriate. Figure 2.14 shows there are two main modeling branches, which lead to discrete-time and continuous-time models, respectively. Many ways of performing conversions between various models are available. One should note that some transformations are more accurate than others.

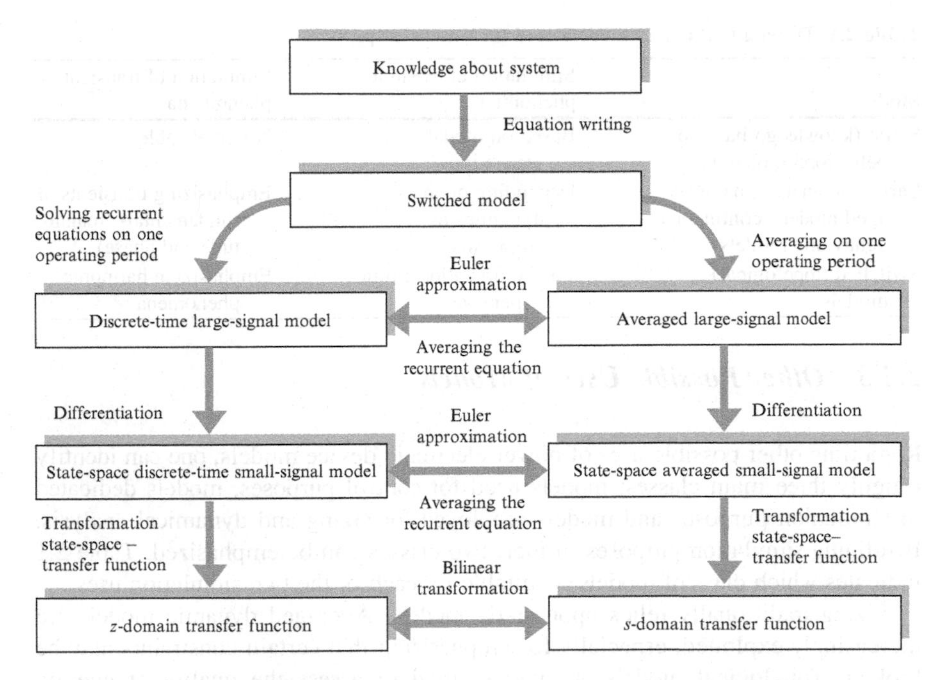

Fig. 2.14 Relations between different types of models (Bacha and Etxeberria 2006)

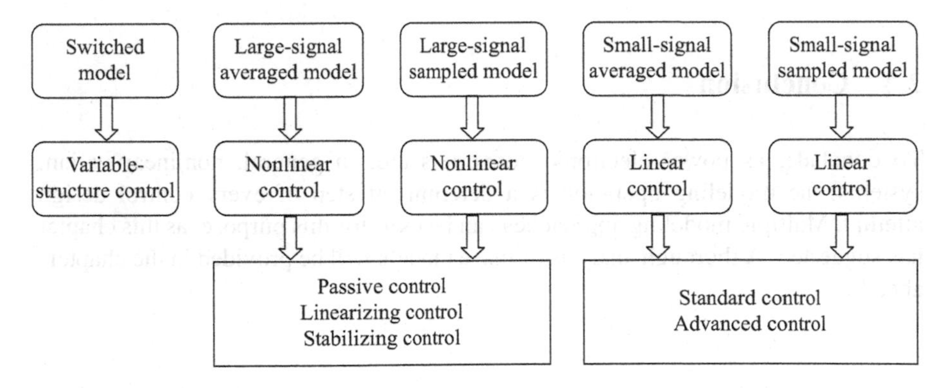

Fig. 2.15 Relations between models and control laws

2.3.2 *Relations Between Modeling and Control*

Figure 2.15, showing certain relations between control laws and associated models, is not exhaustive. Indeed, it is possible to build a variable-structure control law using a continuous large-signal model even though this is not the natural approach. However, there are some difficult cases like, for example, tuning a continuous-time proportional–integral controller using a large-signal nonlinear model.

Table 2.1 Different classes of models used for simulation purposes

Model	Simulation of dynamic phenomena	Simulation of transient phenomena
Static (knowledge-based or behavioral models)	Based on modal separation	Not applicable
Large- or small-signal averaged models, continuous behavioral models	Depending on the dynamics to emphasize	Emphasizing transients of fundamentals (magnitude and phase)
Switched (topological) models	Too computational-time expensive	Emphasizing harmonic phenomena

2.3.3 Other Possible Uses of Models

Regarding other possible uses of power electronic device models, one can identify roughly three main classes: models used for control purposes, models dedicated to simulation purposes and models employed for sizing and dynamical analysis. If refining simulation purposes further, two classes can be emphasized. Table 2.1 indicates which class of models is suitable for each of the two simulation uses.

Sizing traditionally relies upon static models. Averaged dynamic models are increasingly exploited, especially for emphasizing that certain constraints may be broken. Topological models are widely used to assess the quality of energy, especially those related to harmonic spectra and corresponding filtering.

2.4 Conclusion

To conclude, as power electronic converters are, in general, nonlinear variant systems, the modeling operation is a determinant step of every control design attempt. Multiple modeling approaches can be used for this purpose, as this chapter has suggested. A thorough analysis of these models will be provided in the chapters ahead.

References

Bacha S, Etxeberria I (2006) Modeling elements (in French: Éléments de modélisation). In: Crappe M (ed) Exploiting of electrical power grids by means of power electronics systems (in French: L'exploitation des réseaux d'énergie électrique avec l'électronique de puissance). Hermès Lavoisier, Paris, pp 121–139

CIGRE Groupe d'action 38.02.08 (1995) Tools for long-term dynamical simulation (in French: Outils de simulation de la dynamique à long terme). Electra 163:150–166

Erikson RW, Maksimović D (2001) Fundamentals of power electronics, 2nd edn. Kluwer, Dordrecht

IEEE Special Stability Control Working Group (1993) Static Var compensator for power flow and dynamic performance simulation. In: Proceedings of the IEEE-PES winter meeting, Columbus, 31 January–5 February 1993

Kassakian JG, Schlecht MF, Verghese GC (1991) Principles of power electronics. Addison-Wesley, Reading

Kislovsky AS, Redl R, Sokal NO (1991) Dynamic analysis of switching-mode DC/DC converters. Van Nostrand Reinhold, New York

Maksimović D, Stanković AM, Thottuvelil VJ, Verghese GC (2001) Modeling and simulation of power electronic converters. Proc IEEE 89(6):898–912

Sira-Ramirez H, Silva-Ortigoza R (2006) Control design techniques in power electronics devices. Springer, London

Verghese GC, Stanković AM (2001) Introduction to power electronic converters and models. In: Banerjee S, Verghese GC (eds) Nonlinear phenomena in power electronics: attractors, bifurcations, chaos and nonlinear control. IEEE Press, Piscataway, pp 25–37

Watanabe EH, Aredes M, Barbosa PG, De Araujo Lima FK, Da Silva Dias RF, Santos G (2011) Flexible AC transmission systems. In: Rashid MH (ed) Power electronics handbook, 3rd edn. Elsevier, Burlington, pp 851–880

Chapter 3
Switched Model

This chapter focuses on methodologies for obtaining the so-called switched model. This model describes basic low-frequency dynamics, as given by the energy accumulation variations, and it captures the switching dynamics of power electronic converters as well.

The switched model is a useful analysis tool, which emphasizes the presence of the external control action. In its bilinear form this model can be directly used for simulation and control design purposes. The switched model offers a starting point for obtaining other types of models, such as averaged or reduced-order models.

The chapter first states the mathematical framework, then provides a general modeling methodology. Some illustrative examples and a case study complete the presentation of switched-model-related topics. Problems with solutions, as well as proposed problems, can be found at the end of chapter.

3.1 Mathematical Modeling

3.1.1 General Mathematical Framework

Because of its multiple combinations of switch states, a power electronic converter exhibits a periodically repeated sequence of possible configurations during its operation time interval, also called *switching period*. Each such configuration represents in fact a unique circuit containing sources and passive elements, which can be mathematically described by a set of differential equations.

Under the assumptions already stated in Chap. 2, a generic power electronic converter, represented in Fig. 3.1 which switches between N distinct configurations,

S. Bacha et al., *Power Electronic Converters Modeling and Control: with Case Studies*, Advanced Textbooks in Control and Signal Processing, DOI 10.1007/978-1-4471-5478-5_3,

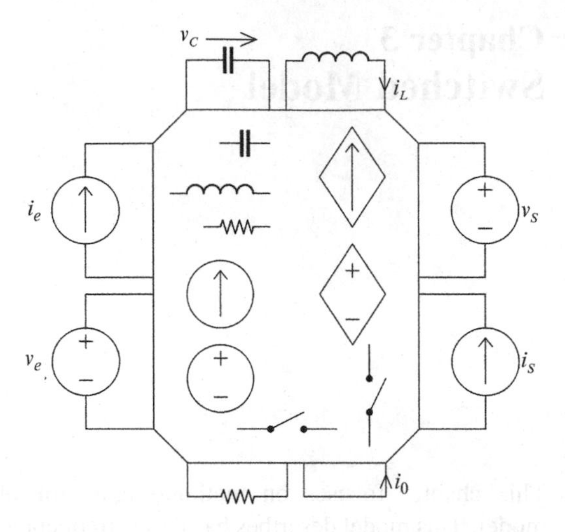

Fig. 3.1 Conventional symbols for power electronic converter representation (Kassakian et al. 1991)

is described as a dynamical system (Tymerski et al. 1989; Sun and Grotstollen 1992; Maksimović et al. 2001):

$$\frac{d}{dt}\mathbf{x}(t) = \mathbf{A}_i \cdot \mathbf{x}(t) + \mathbf{B}_i \cdot \mathbf{e}(t), t_i \leq t \leq t_{i+1}, \tag{3.1}$$

with

$$\sum_{i=1}^{N} (t_i - t_{i-1}) = T,$$

where T is the switching time, t_i are different time points defining the switching between N configurations, $\mathbf{A}_i$ and $\mathbf{B}_i$ are the $n \times n$ state matrix and $n \times p$ input matrix respectively, corresponding to configuration i, $\mathbf{x}(t)$ is the n-length state vector and $\mathbf{e}(t)$ is the p-length vector of the independent sources of the system. Note that in Eq. (3.1) the control input does not appear explicitly.

A more compact form of (3.1) is

$$\frac{d}{dt}\mathbf{x}(t) = \sum_{i=1}^{N} (\mathbf{A}_i\mathbf{x}(t) + \mathbf{B}_i\mathbf{e}(t)) \cdot h_i, \tag{3.2}$$

where h_i are respective *validation functions* associated to configurations. These functions take values 1 or 0 depending on whether their respective configurations are activated or not.

Comments on power electronic converter classification
In general, power electronic converters are classified according to the following criteria:

- conversion mode: DC-DC, DC-AC, etc.;

- innermost control type (at switching level): pulse width modulation (PWM), hysteresis control, sliding-mode control, current-programmed control, etc.;
- operating regime: either natural or forced switching, either continuous or discontinuous conduction, etc.

A classification – different from the above mentioned in the sense that it is focused on control aspects – is proposed next.

Let $\mathbf{h} = [h_1 \cdots h_N]^T$ be the vector containing the various validation functions under the general representation (3.2). Based on this formulation, a classification approach is given by Krein et al. (1990) and rediscussed by Sun and Grotstollen (1992). The method is based on the functional dependencies of the elements in vector **h**; thus, three classes emerge:

- **h** does not depend on the state **x**;
- **h** depends on both the state **x** and the time;
- **h** depends only on the state **x**.

Note that the dependence of $\mathbf{h}(\mathbf{x})$ can either be implicit – because of the circuit operational manner – or explicit, due to the presence of state feedback.

The first class of functions in the above list is characterized by the fact that switching is controlled by a function depending solely on time, $\mathbf{h} = \mathbf{H}(t)$, or external actions. The corresponding converters are simple to model and analyze. For example, this is the case of buck converters with continuous conduction controlled at variable duty ratio.

As regards the second class, some functions h_i may only depend on time, whereas others may depend on the system state. For example, the buck converter operating in discontinuous conduction and a thyristor-based rectifier belong to this class. Finally, the third class is illustrated by a diode-bridge rectifier connected to the grid or by a current-controlled buck converter.

One can note that a given configuration may belong to different classes according to the manner in which it operates.

It appears that analysis and control methods will be more suited to one of the above mentioned families rather than to another one. Thus, for example, as shown later in this book, the classical averaged model is a simple tool, easy to use for converters belonging to the first family, but it is unsuitable for the other two classes. The same holds for variable-structure control and associated sliding modes.

In conclusion, to the best of our knowledge, at present there is no systematic approach able to provide a general methodology allowing a uniform analysis of converter behavior irrespective of its operation. In this respect, the action of exact modeling providing the switched model of a given studied converter is even more important.

3.1.2 Bilinear Form

The bilinear form provides a more compact representation of a power electronic converter switched model, while showing the control inputs. Instead of using validation functions for describing N configurations by means of N models, it is possible to condense the information in a single, unified model fed with p binary functions, denoted by u_k and named *switching functions*. The number p is determined as the smallest integer satisfying the relation $2^p \geq N$.

The bilinear form of the switched model is expressed by the general equation:

$$\dot{\mathbf{x}} = \mathbf{A}\mathbf{x} + \sum_{k=1}^{p} (\mathbf{B}_k\mathbf{x} + \mathbf{b}_k) \cdot u_k + \mathbf{d}, \tag{3.3}$$

where, for every k from 1 to p, $\mathbf{B}_k$ are $n \times n$ matrices, $\mathbf{b}_k$ are n-length column vectors and $\mathbf{d}$ also are n-length column vectors. Equation (3.3) shows explicitly the control input vector $\mathbf{u} = [u_1 \quad u_2 \cdots u_p]^T$. Note the presence of products between state variables and control inputs, which gives the bilinear feature of the model. Next, one will see that every power electronic converter can be modeled by Eq. (3.3), in which particularizations are made. For example, the buck converter has $\mathbf{B}_k = 0$, leading to a linear model; in the case of boost converter $\mathbf{b}_k = 0$. An advantage of model (3.3) is the fact that the small-signal model can be easily obtained from it. Employing the procedure detailed in Sect. 2.2.4 from Chap. 2, the following relations hold:

$$\dot{\tilde{\mathbf{x}}} = \widetilde{\mathbf{A}} \cdot \tilde{\mathbf{x}} + \widetilde{\mathbf{B}} \cdot \tilde{\mathbf{u}},$$

with $\tilde{\mathbf{x}}$ being the vector of state variables' variations around the steady-state point $\mathbf{x}_e$, established in response to vector $\mathbf{u}_e = [u_{1e} \quad u_{2e} \cdots u_{pe}]^T$, where $\tilde{\mathbf{u}} = \mathbf{u} - \mathbf{u}_e$. State matrix $\widetilde{\mathbf{A}}$ and input matrix $\widetilde{\mathbf{B}}$ have the forms

$$\begin{cases} \widetilde{\mathbf{A}} = \mathbf{A} + \sum_{k=1}^{p} \mathbf{B}_k u_{ke} \\ \widetilde{\mathbf{B}} = \sum_{k=1}^{p} (\mathbf{B}_k \mathbf{x}_e + \mathbf{b}_k). \end{cases} \tag{3.4}$$

Model (3.4) can be used directly in designing linear control laws.

3.2 Modeling Methodology

3.2.1 *Basic Assumptions. State Variables*

The switched model is built under the assumptions listed in Sect. 2.2 from Chap. 2. As there is no supplementary approximation, the switched model is also called the *exact model*.

Modeling methodology is based upon classical techniques of systems theory (Cellier et al. 1996). One can note that within a switching subinterval, i.e., during the time interval when the control input is constant, the system evolves continuously in time, that is, it is characterized by variables having finite derivatives. This means that during the actual time subinterval the system can be described by a set of ordinary differential equations obeying the energy conservation laws particular to the circuit configuration (Sanders 1993; van Dijk et al. 1995).

As usual, state variables are chosen to reflect the variation of energy accumulation. In the particular case of power electronic converters, these are the currents passing through inductors, the capacitor voltages and/or linear/nonlinear combinations of both. These variables meet Dirichlet's conditions because energy variations naturally occur in a continuous manner.

If the number of capacitors is denoted by n_C and the number of inductances is n_L, then system order n verifies the relation $n \leq n_C + n_L$.

3.2.2 General Algorithm

Some a priori knowledge about converter operation is indispensable in deriving the switched model. A preliminary analysis is necessary, aimed at one or more of the following:

- checking different waveforms of the studied converter;
- providing the sequence of configurations taken during a switching time interval;
- giving the set of equations that mathematically describe each configuration (an example of how this can be done automatically can be found in Merdassi et al. 2010).

Nowadays, commercial computer-aided design software products are available to assist the user in performing the above actions.

The manner of obtaining a switched model is not unique. Irrespective of the method employed, the following procedure is general and can be applied in most of applications.

Remark. In the case of only two circuit configurations, the two respective validation functions are complementary, i.e., $h_1 = 1 - h_2$. Hence, one can reasonably take $u = h_1 \in \{0;1\}$ in the case of converters and $u = 2h_1 - 1 \in \{-1; 1\}$ in the case of DC-AC converters. In cases with more than two configurations there is no general rule that applies.

For cases where the number of possible configurations is quite large, directly applying Algorithm 3.1 requires significant time and effort. An alternative procedure may be based on identifying so-called *switched* (or else intermediary) *variables* and reaching the bilinear form more easily. The steps are detailed in Algorithm 3.2.

As an example, in the case of DC-DC converters the switches are implemented by means of transistors and diodes, connected in the so-called *switch network* (Erikson and Maksimović 2001). Depending on their position in the converter circuit, both of these elements can switch both current and voltage, as shown in Fig. 3.2.

Algorithm 3.1.

Obtaining the switched model of a given power electronic converter based upon analysis of all possible configurations

#1. Either collect the waveforms and corresponding circuit configurations or simply study the waveforms provided by an adequate software package.
#2. Choose the state variables, by either directly taking capacitor voltages and inductor currents or by taking an appropriate combination of these variables.
#3. (a) Write the expressions of the derivatives of the above state variables for each configuration.
(b) Identify the transition conditions between configurations and write the model in a compact manner by making the control variables (switching functions) appear explicitly.
#4. Make an equivalent topological (or exact) diagram of the studied converter where the different coupling terms appear. This step is optional.

Algorithm 3.2.

Deducing the switched model of a given power electronic converter based upon identifying the switched variables

#1. Choose the state variables by either directly taking capacitor voltages and inductor currents or by taking an appropriate combination of these variables.
#2. Following the initial analysis, identify the switched variables; e.g., these can be the transistor voltages and the diode currents, or vice versa. Write their mathematical expressions depending on states of switches (on/off) and on state variables.
#3. Write Kirchhoff's voltage laws for expressing the derivatives of inductor currents and Kirchhoff's current laws for obtaining the derivatives of capacitor voltages.
#4. Introduce the switching functions **u** and write the switched variables as functions of **u**.
#5. Replace the switched variables in the state-space equations developed at step #3. Obtain the bilinear form.

3.2.3 Examples

Example 1. Power electronic converter for induction-based heating
Let us consider the power electronic converter given in Fig. 3.3. It is composed of a diode rectifier, an LR filter with inductance L_f and resistance r and a current source

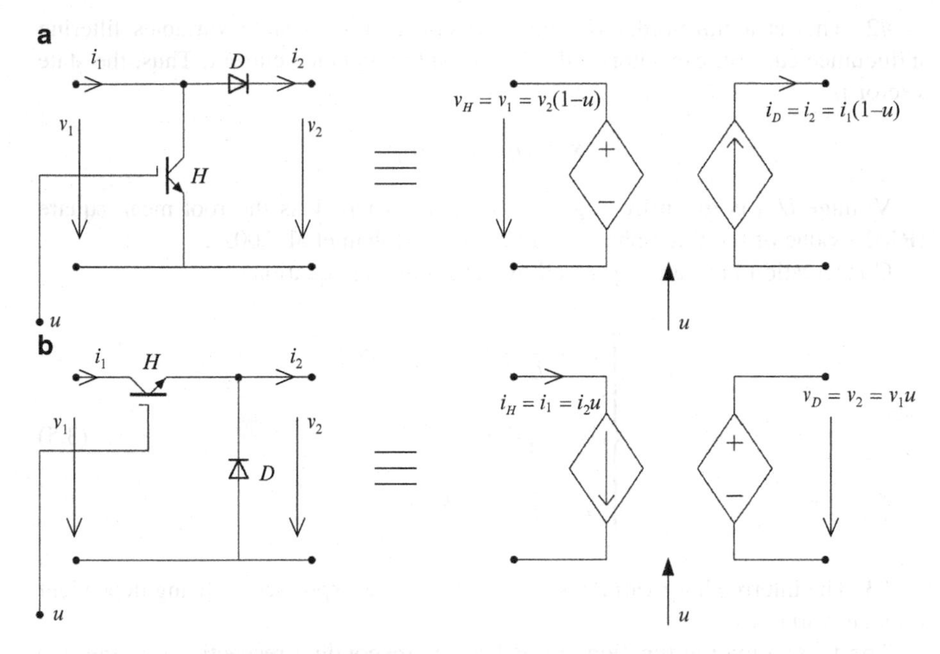

Fig. 3.2 Example of switch network showing switched variables: (**a**) boost converter case, (**b**) buck converter case

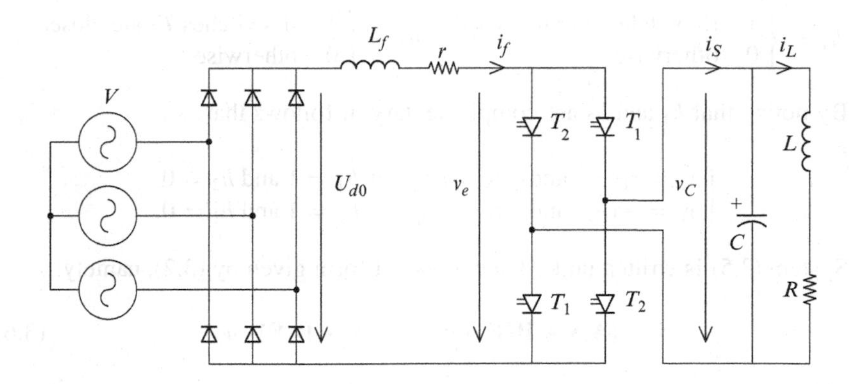

Fig. 3.3 Power electronic converter for induction-based heating

inverter supplying a resonant circuit composed of a capacitor C and an induction-based heater of inductance L and resistance R.

The current source inverter operates in full wave; it is conceived such that switches T_1 and switches T_2 operate complementarily (turning on T_1 determines T_2 being turned off and inversely).

In order to obtain the exact (switched) model of this structure by using Algorithm 3.1 presented in Sect. 3.2.2, one can proceed as follows.

Given sufficient initial information about the process, one can begin directly with the second step of Algorithm 3.1.

#2. This is a third-order dynamic system having as state variables filtering inductance current, capacitor voltage and load inductance current. Thus, the state vector is

$$\mathbf{x} = \begin{bmatrix} i_f & v_C & i_L \end{bmatrix}^T.$$

Voltage U_{d0} is given by $U_{d0} = 3\sqrt{6}/\pi \cdot V$, where V is the root mean square (RMS) value of the three-phase grid voltages (Mohan et al. 2002).

Conservation laws are expressed by the following equations:

$$\begin{cases} \dot{i_f} = \dfrac{1}{L_f}\left(U_{d0} - v_e - ri_f\right) \\ \dot{v_C} = \dfrac{1}{C}(i_S - i_L) \\ \dot{i_L} = \dfrac{1}{L}(v_C - Ri_L). \end{cases} \tag{3.5}$$

#3. The intermediary variables v_e and i_S must be expressed as being dependent on state variables.

The two validation functions h_1 and h_2, corresponding respectively to the two configurations taken by the current source inverter, are defined as follows:

$$h_1 = \begin{cases} 1 & \text{if switches } T_1 \text{ are closed} \\ 0 & \text{otherwise,} \end{cases} \qquad h_2 = \begin{cases} 1 & \text{if switches } T_2 \text{ are closed} \\ 0 & \text{otherwise.} \end{cases}$$

By noting that h_1 and h_2 are complementary, it follows that

$$\begin{cases} v_e = v_C & \text{and} \quad i_S = i_f & \text{if } h_1 = 1 \text{ and } h_2 = 0 \\ v_e = -v_C & \text{and} \quad i_S = -i_f & \text{if } h_2 = 1 \text{ and } h_1 = 0. \end{cases}$$

System (3.5) is written under the equivalent form given by (3.2), namely:

$$\dot{\mathbf{x}} = (\mathbf{A}_1\mathbf{x} + \mathbf{B}_1\mathbf{E}) \cdot h_1 + (\mathbf{A}_2\mathbf{x} + \mathbf{B}_2\mathbf{E}) \cdot h_2, \tag{3.6}$$

where

$$\begin{cases} \mathbf{A}_1 = \begin{bmatrix} -\dfrac{r}{L_f} & -\dfrac{1}{L_f} & 0 \\ \dfrac{1}{C} & 0 & -\dfrac{1}{C} \\ 0 & \dfrac{1}{L} & -\dfrac{R}{L} \end{bmatrix}, \quad \mathbf{A}_2 = \begin{bmatrix} -\dfrac{r}{L_f} & +\dfrac{1}{L_f} & 0 \\ -\dfrac{1}{C} & 0 & -\dfrac{1}{C} \\ 0 & \dfrac{1}{L} & -\dfrac{R}{L} \end{bmatrix}, \\ \mathbf{B}_1 = \mathbf{B}_2 = \begin{bmatrix} \dfrac{1}{L_f} & 0 & 0 \end{bmatrix}^T, \qquad \mathbf{E} = U_{d0}. \end{cases} \tag{3.7}$$

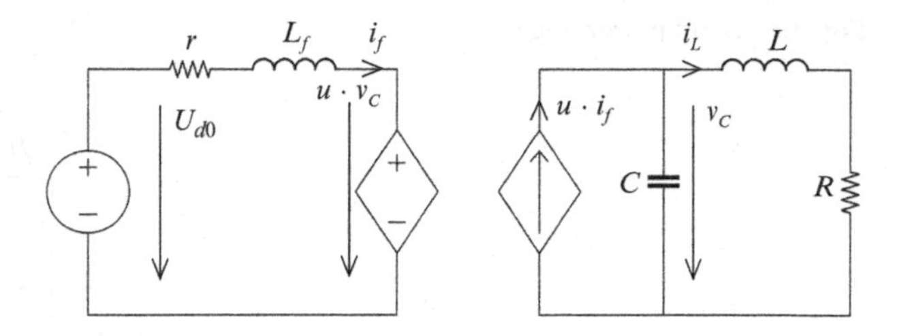

Fig. 3.4 Equivalent circuit of the switched model of converter presented in Fig. 3.2

Representation (3.6) must be condensed into the bilinear form by using a single switching function, which is sufficient in order to represent two configurations ($p = 1$ is the largest integer satisfying $2^p \geq N = 2$).

If the switching function u is chosen such that

$$u = 2h_1 - 1,$$

then a unique representation yields

$$\dot{\mathbf{x}} = \mathbf{A} \cdot \mathbf{x} + \mathbf{B} \cdot \mathbf{x} \cdot u + \mathbf{d}, \tag{3.8}$$

where function u takes values within the discrete set $\{-1; 1\}$ and

$$\mathbf{A} = \frac{\mathbf{A}_1 + \mathbf{A}_2}{2}, \mathbf{B} = \frac{\mathbf{A}_1 - \mathbf{A}_2}{2}, \mathbf{d} = \mathbf{B}_1 U_{d0}.$$

Once the above expressions have been established, corresponding matrices are given by Eq. (3.9).

One can note that the topological model obtained as Eqs. (3.8) and (3.9) is bilinear, i.e., it contains products of form $\mathbf{x} \cdot u$; this is the case of the quasi-totality of power electronic converters.

The equivalent circuit translating exactly the switched model (3.8) is given in Fig. 3.4, where several features can be identified:

- the obtained circuit is simplified in relation to the original converter;
- the obtained circuit reveals the coupling terms by means of so-called *coupled sources*, which are represented by diamonds.

$$\mathbf{A} = \begin{bmatrix} -\dfrac{r}{L_f} & 0 & 0 \\ 0 & 0 & -\dfrac{1}{C} \\ 0 & \dfrac{1}{L} & -\dfrac{R}{L} \end{bmatrix}, \quad \mathbf{B} = \begin{bmatrix} 0 & -\dfrac{1}{L_f} & 0 \\ \dfrac{1}{C} & 0 & 0 \\ 0 & 0 & 0 \end{bmatrix}, \quad \mathbf{d} = \begin{bmatrix} \dfrac{U_{d0}}{L_f} \\ 0 \\ 0 \end{bmatrix}. \tag{3.9}$$

For experienced users, the equivalent circuit in Fig. 3.4 can be directly obtained by making use of certain rules of transformation that allow transposition of switch operation of the original circuit (Fig. 3.3) into operation of coupled sources.

Fig. 3.5 Boost power stage

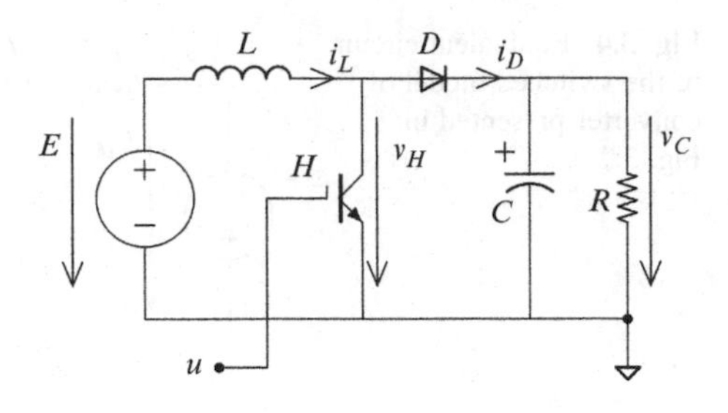

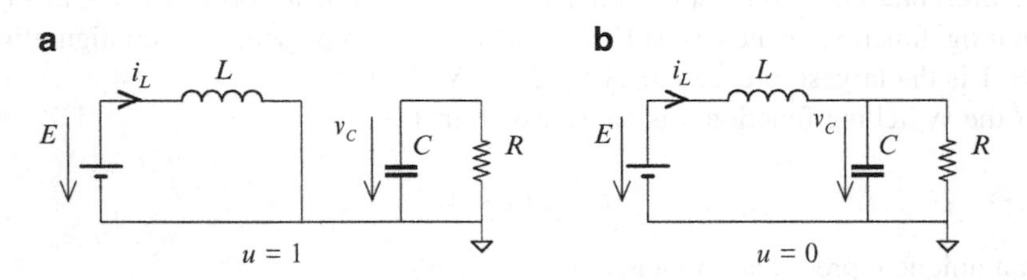

Fig. 3.6 The two possible configurations of circuit in Fig. 3.4

Example 2. Boost DC-DC converter
The example of an ideal boost converter (Fig. 3.5) was chosen to illustrate how the switched model may be found in the case of continuous-conduction mode (ccm). A discussion concerning the case of discontinuous conduction is also presented.

Case of continuous-conduction mode
As stated earlier, one can obtain the switched model of any power electronic converter and its bilinear form in two ways:

(a) by listing all its possible configurations and finding a general structure which leads to the bilinear form, according to Algorithm 3.1;
(b) by applying the method of emphasizing the variables exhibiting switched-time evolution, as described in Algorithm 3.2.

Each of these methods will be detailed next in the case of the boost converter operating in continuous-conduction mode.

(a) Using the list of all possible configurations
According to preliminary analysis, the boost converter operating in continuous conduction can take two configurations, as shown in Fig. 3.6: cases (a) and (b) correspond to switch H being turned on ($h_1 = 1$) and turned off ($h_2 = 1$) respectively. As this is a DC-DC case having two configurations, switching function u can be taken such that $u = h_1 = 1 - h_2$, as discussed in Sect. 3.2.2.

The state variables are the inductor current i_L and the capacitor voltage v_C. The state-space equations corresponding to the two circuit configurations are listed below.

$$u = 1 : \begin{cases} \dot{i_L} = E/L \\ \dot{v_C} = -v_C/(RC), \end{cases} \quad u = 0 : \begin{cases} \dot{i_L} = E/L - v_C/L \\ \dot{v_C} = i_L/C - v_C/(RC). \end{cases} \tag{3.10}$$

Equations from (3.10) can be condensed into a single form by employing Eq. (3.2), in which validation functions h_1 and h_2 are expressed using switching function u, namely $h_1 = u$ and $h_2 = 1 - u$:

$$\begin{cases} \dot{i_L} = \dfrac{E}{L}u + \dfrac{E - v_C}{L}(1-u) \\ \dot{v_C} = -\dfrac{v_C}{RC}u + \left(\dfrac{i_L}{C} - \dfrac{v_C}{RC}\right)(1-u), \end{cases}$$

from which one can derive

$$\begin{cases} \dot{i_L} = -(1-u)v_C/L + E/L \\ \dot{v_C} = (1-u)i_L/C - v_C/(RC), \end{cases} \tag{3.11}$$

which can be used directly for simulation purposes. Equation (3.11) allows one to obtain the bilinear form:

$$\begin{bmatrix} \dot{i_L} \\ \dot{v_C} \end{bmatrix} = \underbrace{\begin{bmatrix} 0 & -1/L \\ 1/C & -1/(RC) \end{bmatrix}}_{\mathbf{A}} \cdot \begin{bmatrix} i_L \\ v_C \end{bmatrix} + \underbrace{\begin{bmatrix} 0 & 1/L \\ -1/C & 0 \end{bmatrix}}_{\mathbf{B}} \cdot \begin{bmatrix} i_L \\ v_C \end{bmatrix} \cdot u + \underbrace{\begin{bmatrix} E/L \\ 0 \end{bmatrix}}_{\mathbf{d}}, \tag{3.12}$$

with $\mathbf{b} = [0 \quad 0]^T$ (see Eq. (3.3)).

(b) Identifying the switched variables

The first step – choosing the state variables – is the same as in the previous procedure; i_L and v_C are the state variables. In the case presented, the switched variables are the transistor voltage v_H and the diode current i_D which can be written as functions of state variables (see Fig. 3.5):

$$v_H = \begin{cases} 0 & \text{if } H \text{ is turned on} \\ v_C & \text{if } H \text{ is turned off} \end{cases} \quad \text{and} \quad i_D = \begin{cases} 0 & \text{if } H \text{ is turned on} \\ i_L & \text{if } H \text{ is turned off.} \end{cases}$$

The equations defining the circuit behavior are obtained by applying Kirchhoff's voltage law for expressing di_L/dt and Kirchhoff's current law for dv_C/dt (see Fig. 3.6). That is,

$$\begin{cases} L \cdot \dot{i_L} = E - v_H \\ C \cdot \dot{v_C} = i_D - v_C/R. \end{cases} \tag{3.13}$$

Next, the switched variables must be expressed as depending on a suitably defined switching function. If the switching function is introduced as

$$u = \begin{cases} 1 & \text{if } H \text{ is turned on} \\ 0 & \text{if } H \text{ is turned off,} \end{cases}$$

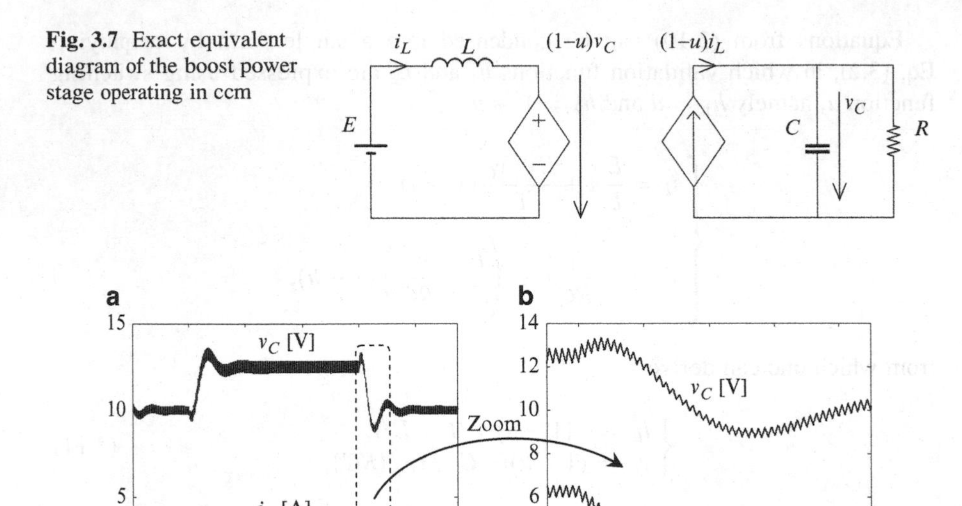

Fig. 3.7 Exact equivalent diagram of the boost power stage operating in ccm

Fig. 3.8 Dynamic behavior of the switched model of the ideal boost converter

then the switched variables are written as:

$$\begin{cases} v_H = v_C(1-u) \\ i_D = i_L(1-u). \end{cases} \tag{3.14}$$

By substituting Eq. (3.14) into Eq. (3.13), one obtains

$$\begin{cases} \dot{i_L} = -(1-u)\cdot v_C/L + E/L \\ \dot{v_C} = (1-u)\cdot i_L/C - v_C/(RC), \end{cases}$$

which is identical to Eq. (3.11) and from which one can derive the bilinear form expressed by (3.12).

Irrespective of the method used in deducing the bilinear form, the exact equivalent circuit of the boost converter in ccm results from Eq. (3.11) and it is presented in Fig. 3.7.

The two dependent sources define a coupling between the circuit's input and output. The behavior is like that of an ideal DC transformer with the variable ratio controlled by an external action.

The system represented by Eq. (3.12) can be simulated using dedicated software. Figure 3.8 shows the behavior of the ideal boost power stage at step-variation of duty ratio as simulated in Simulink®. To this end, the system is fed by a switched function $u(t)$, obtained by PWM modulation of the duty ratio. Note the ripple in the state variables due to the switching of the circuit Eq. (3.10) with the modulation

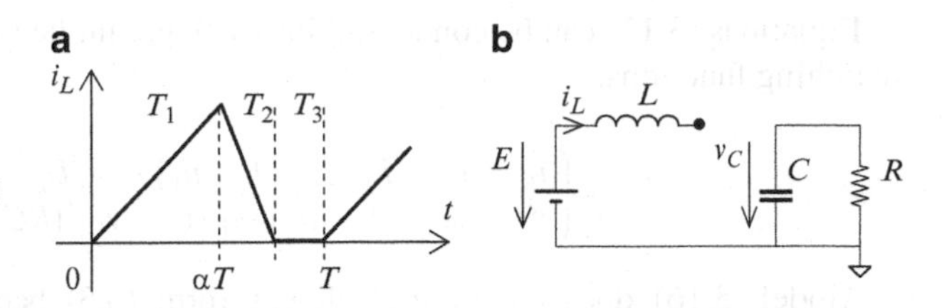

Fig. 3.9 Boost converter operating in dcm: (**a**) inductor current evolution; (**b**) circuit configuration corresponding to dcm ($i_L = 0$, time subinterval T_3)

frequency. Also, the time response is variable with the operating point (v_C, i_L), indicating that the system is nonlinear. For the same reason, the ripple has variable magnitude, depending on the operating point. Non-minimum-phase behavior can also be observed.

Case of discontinuous-conduction mode

Now consider that the boost power stage operates in discontinuous-conduction mode (dcm). This means that the switching period is large enough with respect to inductance to allow the inductor current to become zero (see Fig. 3.9a). This occurs when the load current value determines a mean value for the inductor current, smaller than its ripple (Vorpérian 1990; Sun et al. 1998).

An analysis of the circuit for this case shows that within a switching period there are three time subintervals, each corresponding to a circuit configuration. Intervals T_1 and T_2 are associated with the circuit topologies from Figs. 3.6a, b, respectively. In the third subinterval, T_3, the converter remains in the configuration shown in Fig. 3.8b, but current i_L is interrupted. State equations describing the three circuit configurations are given by Eq. (3.15).

$$\boxed{T_1}\quad \begin{cases} \dot{i_L} = E/L \\ \dot{v_C} = -v_C/(RC), \end{cases} \qquad \boxed{T_2}\quad \begin{cases} \dot{i_L} = E/L - v_C/L \\ \dot{v_C} = i_L/C - v_C/(RC), \end{cases} \qquad \boxed{T_3}\quad \begin{cases} \dot{i_L} = i_L = 0 \\ \dot{v_C} = -v_C/(RC). \end{cases} \tag{3.15}$$

As the number of switching functions p must obey the relation $2^p \geq N = 3$ (see Sect. 3.1.2), a second switching function must be introduced in order to obtain a unified switched model.

To conclude, the first switching function is an external independent action, whereas the second switching function depends on an internal state, i_L. They can be defined as follows:

$$u_1 = \begin{cases} 1 & \text{if } H \text{ is turned on} \\ 0 & \text{if } H \text{ is turned off} \end{cases} \quad \text{and } u_2 = \frac{1 + \operatorname{sgn}(i_L)}{2},$$

where

$$\operatorname{sgn}(i_L) = \begin{cases} 1 & \text{if } i_L > 0 \text{ (ccm)} \\ -1 & \text{if } i_L \leq 0 \text{ (dcm)}. \end{cases}$$

Equations (3.15) can be condensed into a single unified model showing the two switching functions:

$$\begin{cases} \dot{i_L} = u_2 \cdot (E/L - (1 - u_1) \cdot v_C/L) \\ \dot{v_C} = u_2 \cdot (1 - u_1) \cdot i_L/C - v_C/(RC). \end{cases} \tag{3.16}$$

Model (3.16) does not fit the bilinear form (3.3) because here products of switching functions ($u_1 \cdot u_2$) appear (*trilinear form*). Moreover, this form cannot be directly used for control purposes.

3.3 Case Study: Three-Phase Voltage-Source Converter as Rectifier

Let us consider the example of the three-phase four-wire voltage-source converter (VSC) used as rectifier, whose electrical diagram is presented in Fig. 3.10.

This figure already reveals the switching functions u_1, u_2 and u_3, defined as

$$u_k = \begin{cases} 1 & \text{if switch } H_k \text{ is turned on} \\ 0 & \text{if switch } H_k \text{ is turned off} \end{cases} \quad k = 1, 2, 3.$$

The switched model of this converter and its bilinear form will next be obtained by means of the two algorithms already presented:

(a) according to Algorithm 3.1, i.e., by listing all its possible configurations and finding a general written form which leads to the bilinear form;
(b) or by applying the method of emphasizing the switched variables, as described in Algorithm 3.2.

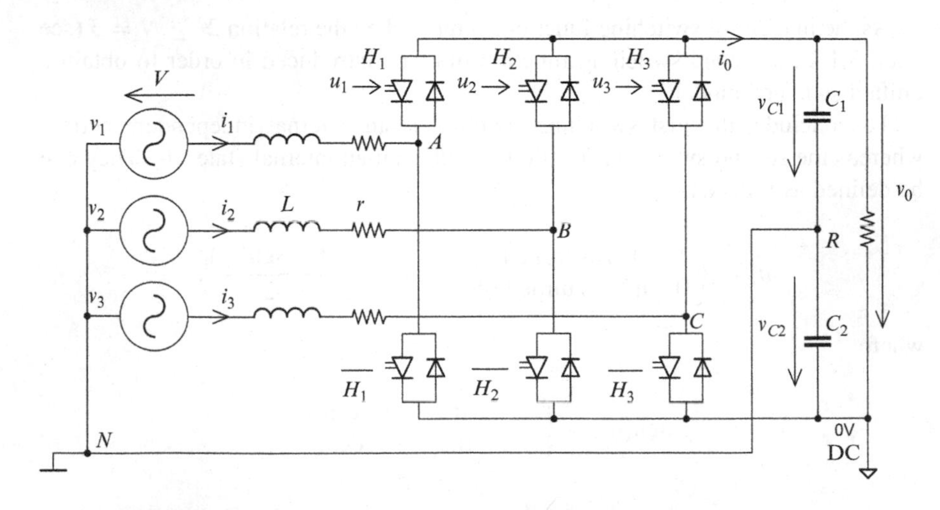

Fig. 3.10 Electrical diagram of three-phase four-wire VSC used as rectifier

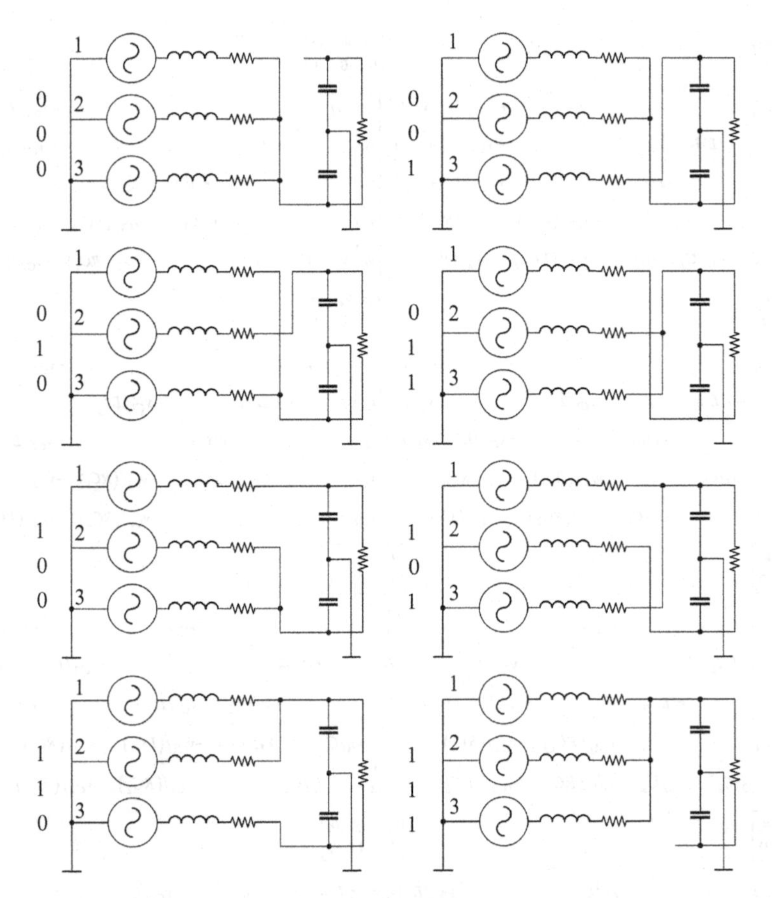

Fig. 3.11 All eight possible configurations taken by three-phase AC-DC converter, respectively corresponding to binary combinations of switching functions u_1, u_2 and u_3

(a) Listing of all possible configurations
The converter can take eight possible configurations, respectively corresponding to the 2^3 binary combinations of the $p = 3$ switching functions defined as $u_k \in \{0;1\}$, $k = 1, 2, 3$; these configurations are shown in Fig. 3.11.

The converter is described by five state variables composing the state vector $\mathbf{x} = [i_1 \quad i_2 \quad i_3 \quad v_{C1} \quad v_{C2}]^T$. The governing equations of each of the eight topologies represented in Fig. 3.11 can be found in Fig. 3.12.

These equations allow one to deduce a general rule for writing the derivatives of the state variables, depending upon the switching functions, as presented in (3.17). This relation can directly and easily be used for simulation and control purposes; it also allows one to obtain the equivalent circuit diagram shown in Fig. 3.13.

u_1	u_2	u_3
0	0	0

$$\begin{cases} \dot{i}_1 = -r/L\cdot i_1 + v_{C2}/L + v_1/L \\ \dot{i}_2 = -r/L\cdot i_2 + v_{C2}/L + v_2/L \\ \dot{i}_3 = -r/L\cdot i_3 + v_{C2}/L + v_3/L \\ \dot{v}_{C1} = -v_{C1}/(RC_1) - v_{C2}/(RC_1) \\ \dot{v}_{C2} = -i_1/C_2 - i_2/C_2 - i_3/C_2 - v_{C1}/(RC_2) - v_{C2}/(RC_2) \end{cases}$$

u_1	u_2	u_3
0	0	1

$$\begin{cases} \dot{i}_1 = -r/L\cdot i_1 + v_{C2}/L + v_1/L \\ \dot{i}_2 = -r/L\cdot i_2 + v_{C2}/L + v_2/L \\ \dot{i}_3 = -r/L\cdot i_3 - v_{C1}/L + v_3/L \\ \dot{v}_{C1} = i_3/C_1 - v_{C1}/(RC_1) - v_{C2}/(RC_1) \\ \dot{v}_{C2} = -i_1/C_2 - i_2/C_2 - v_{C1}/(RC_2) - v_{C2}/(RC_2) \end{cases}$$

u_1	u_2	u_3
0	1	0

$$\begin{cases} \dot{i}_1 = -r/L\cdot i_1 + v_{C2}/L + v_1/L \\ \dot{i}_2 = -r/L\cdot i_2 - v_{C1}/L + v_2/L \\ \dot{i}_3 = -r/L\cdot i_3 + v_{C2}/L + v_3/L \\ \dot{v}_{C1} = i_2/C_1 - v_{C1}/(RC_1) - v_{C2}/(RC_1) \\ \dot{v}_{C2} = -i_1/C_2 - i_3/C_2 - v_{C1}/(RC_2) - v_{C2}/(RC_2) \end{cases}$$

u_1	u_2	u_3
0	1	1

$$\begin{cases} \dot{i}_1 = -r/L\cdot i_1 + v_{C2}/L + v_1/L \\ \dot{i}_2 = -r/L\cdot i_2 - v_{C1}/L + v_2/L \\ \dot{i}_3 = -r/L\cdot i_3 + v_{C2}/L + v_3/L \\ \dot{v}_{C1} = i_2/C_1 + i_3/C_1 - v_{C1}/(RC_1) - v_{C2}/(RC_1) \\ \dot{v}_{C2} = -i_1/C_2 - v_{C1}/(RC_2) - v_{C2}/(RC_2) \end{cases}$$

u_1	u_2	u_3
1	0	0

$$\begin{cases} \dot{i}_1 = -r/L\cdot i_1 - v_{C1}/L + v_1/L \\ \dot{i}_2 = -r/L\cdot i_2 + v_{C2}/L + v_2/L \\ \dot{i}_3 = -r/L\cdot i_3 + v_{C2}/L + v_3/L \\ \dot{v}_{C1} = i_1/C_1 - v_{C1}/(RC_1) - v_{C2}/(RC_1) \\ \dot{v}_{C2} = -i_2/C_2 - i_3/C_2 - v_{C1}/(RC_2) - v_{C2}/(RC_2) \end{cases}$$

u_1	u_2	u_3
1	0	1

$$\begin{cases} \dot{i}_1 = -r/L\cdot i_1 - v_{C1}/L + v_1/L \\ \dot{i}_2 = -r/L\cdot i_2 + v_{C2}/L + v_2/L \\ \dot{i}_3 = -r/L\cdot i_3 - v_{C1}/L + v_3/L \\ \dot{v}_{C1} = i_1/C_1 + i_3/C_1 - v_{C1}/(RC_1) - v_{C2}/(RC_1) \\ \dot{v}_{C2} = -i_2/C_2 - v_{C1}/(RC_2) - v_{C2}/(RC_2) \end{cases}$$

u_1	u_2	u_3
1	1	0

$$\begin{cases} \dot{i}_1 = -r/L\cdot i_1 - v_{C1}/L + v_1/L \\ \dot{i}_2 = -r/L\cdot i_2 - v_{C1}/L + v_2/L \\ \dot{i}_3 = -r/L\cdot i_3 + v_{C2}/L + v_3/L \\ \dot{v}_{C1} = i_1/C_1 + i_2/C_1 - v_{C1}/(RC_1) - v_{C2}/(RC_1) \\ \dot{v}_{C2} = -i_3/C_2 - v_{C1}/(RC_2) - v_{C2}/(RC_2) \end{cases}$$

u_1	u_2	u_3
1	1	1

$$\begin{cases} \dot{i}_1 = -r/L\cdot i_1 - v_{C1}/L + v_1/L \\ \dot{i}_2 = -r/L\cdot i_2 - v_{C1}/L + v_2/L \\ \dot{i}_3 = -r/L\cdot i_3 - v_{C1}/L + v_3/L \\ \dot{v}_{C1} = i_1/C_1 + i_2/C_1 + i_3/C_1 - v_{C1}/(RC_1) - v_{C2}/(RC_1) \\ \dot{v}_{C2} = -v_{C1}/(RC_2) - v_{C2}/(RC_2) \end{cases}$$

Fig. 3.12 State-space equations corresponding to each of the eight possible configurations of the VSC listed in Fig. 3.11

$$\begin{cases} \dot{i_k} = -r/L\cdot i_k - u_k\cdot v_{C1}/L + (1-u_k)\cdot v_{C2}/L + v_k/L \\ \dot{v_{C1}} = 1/C_1\cdot \sum_{k=1}^{3} u_k\cdot i_k - v_{C1}/(RC_1) - v_{C2}/(RC_1) \\ \dot{v_{C2}} = -1/C_2\cdot \sum_{k=1}^{3}(1-u_k)\cdot i_k - v_{C1}/(RC_2) - v_{C2}/(RC_2) \end{cases} \quad k = 1,2,3. \tag{3.17}$$

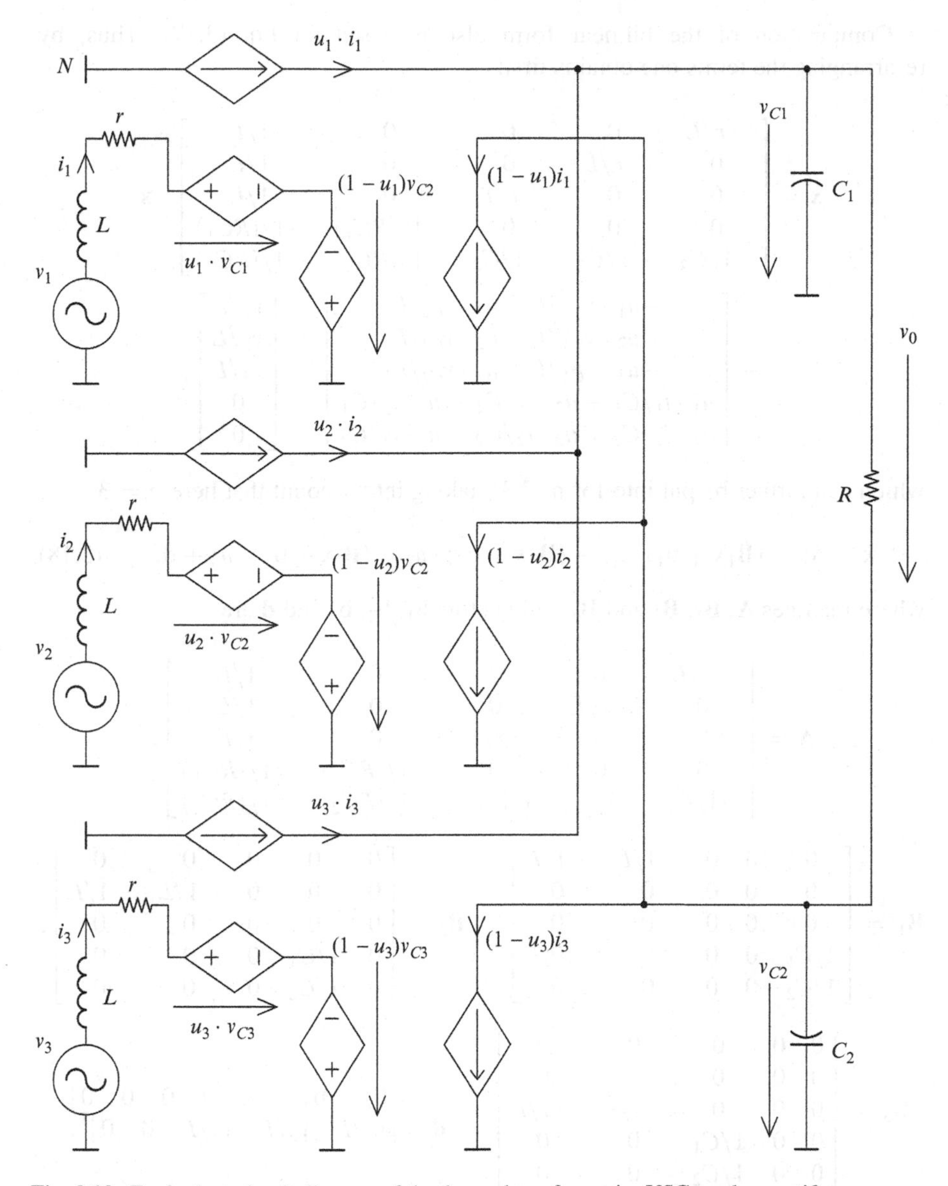

Fig. 3.13 Equivalent circuit diagram of the three-phase four-wire VSC used as rectifier

Computation of the bilinear form also is based on Eq. (3.17). Thus, by re-arranging the terms one obtains first:

$$\dot{\mathbf{x}} = \begin{bmatrix} -r/L & 0 & 0 & 0 & 1/L \\ 0 & -r/L & 0 & 0 & 1/L \\ 0 & 0 & -r/L & 0 & 1/L \\ 0 & 0 & 0 & -1/(RC_1) & -1/(RC_1) \\ -1/C_2 & -1/C_2 & -1/C_2 & -1/(RC_2) & -1/(RC_2) \end{bmatrix} \cdot \mathbf{x}$$
$$+ \begin{bmatrix} -u_1 \cdot v_{C1}/L - u_1 \cdot v_{C2}/L \\ -u_2 \cdot v_{C1}/L - u_2 \cdot v_{C2}/L \\ -u_3 \cdot v_{C1}/L - u_3 \cdot v_{C2}/L \\ u_1 \cdot i_1/C_1 + u_2 \cdot i_2/C_1 + u_3 \cdot i_3/C_1 \\ u_1 \cdot i_1/C_2 + u_2 \cdot i_2/C_2 + u_3 \cdot i_3/C_2 \end{bmatrix} + \begin{bmatrix} v_1/L \\ v_2/L \\ v_3/L \\ 0 \\ 0 \end{bmatrix},$$

which can further be put into form (3.3), taking into account that here $p = 3$:

$$\dot{\mathbf{x}} = \mathbf{A}\mathbf{x} + (\mathbf{B}_1\mathbf{x} + \mathbf{b}_1) \cdot u_1 + (\mathbf{B}_2\mathbf{x} + \mathbf{b}_2) \cdot u_2 + (\mathbf{B}_3\mathbf{x} + \mathbf{b}_3) \cdot u_3 + \mathbf{d}, \tag{3.18}$$

where matrices $\mathbf{A}$, $\mathbf{B}_1$, $\mathbf{B}_2$ and $\mathbf{B}_3$ and vectors $\mathbf{b}_1$, $\mathbf{b}_2$, $\mathbf{b}_3$ and $\mathbf{d}$ are

$$\mathbf{A} = \begin{bmatrix} -r/L & 0 & 0 & 0 & 1/L \\ 0 & -r/L & 0 & 0 & 1/L \\ 0 & 0 & -r/L & 0 & 1/L \\ 0 & 0 & 0 & -1/(RC_1) & -1/(RC_1) \\ -1/C_2 & -1/C_2 & -1/C_2 & -1/(RC_2) & -1/(RC_2) \end{bmatrix},$$

$$\mathbf{B}_1 = \begin{bmatrix} 0 & 0 & 0 & -1/L & -1/L \\ 0 & 0 & 0 & 0 & 0 \\ 0 & 0 & 0 & 0 & 0 \\ 1/C_1 & 0 & 0 & 0 & 0 \\ 1/C_2 & 0 & 0 & 0 & 0 \end{bmatrix}, \quad \mathbf{B}_2 = \begin{bmatrix} 0 & 0 & 0 & 0 & 0 \\ 0 & 0 & 0 & -1/L & -1/L \\ 0 & 0 & 0 & 0 & 0 \\ 0 & 1/C_1 & 0 & 0 & 0 \\ 0 & 1/C_2 & 0 & 0 & 0 \end{bmatrix},$$

$$\mathbf{B}_3 = \begin{bmatrix} 0 & 0 & 0 & 0 & 0 \\ 0 & 0 & 0 & 0 & 0 \\ 0 & 0 & 0 & -1/L & -1/L \\ 0 & 0 & 1/C_1 & 0 & 0 \\ 0 & 0 & 1/C_2 & 0 & 0 \end{bmatrix}, \quad \begin{array}{l} \mathbf{b}_1 = \mathbf{b}_2 = \mathbf{b}_3 = [0 \ \ 0 \ \ 0 \ \ 0 \ \ 0]^T \\ \mathbf{d} = [v_1/L \ \ v_2/L \ \ v_3/L \ \ 0 \ \ 0]^T. \end{array}$$

(b) Identifying switched variables

The state variables have already been identified: $\mathbf{x} = [i_1 \ \ i_2 \ \ i_3 \ \ v_{C1} \ \ v_{C2}]^T$. As regards the switched variables, these are five, namely:

- the voltages e_1, e_2 and e_3 of the points A, B and C (Fig. 3.9) considered with respect to the neutral N;
- the output currents feeding the RC filter, i_{01} and i_{02}.

As a consequence, the following relations hold:

$$e_k = \begin{cases} v_{C1}, & H_k \text{ is turned on} \\ -v_{C2}, & H_k \text{ is turned off} \end{cases} \quad k = 1, 2, 3.$$

In AC systems one may prefer to use bipolar switching functions defined as $u_k{}^* \in \{-1, 1\}$, $k = 1, 2, 3$. In this way the switched variables can be expressed as depending on the state variables:

$$\begin{cases} e_k = v_{C1} \cdot \dfrac{1+u_k^*}{2} - v_{C2} \cdot \dfrac{1-u_k^*}{2}, \quad k = 1, 2, 3 \\ i_{01} = i_1 \cdot \dfrac{1+u_1^*}{2} + i_2 \cdot \dfrac{1+u_2^*}{2} + i_3 \cdot \dfrac{1+u_3^*}{2} \\ i_{02} = -i_1 \cdot \dfrac{1-u_1^*}{2} - i_2 \cdot \dfrac{1-u_2^*}{2} - i_3 \cdot \dfrac{1-u_3^*}{2}, \end{cases}$$

from which the state equations can further be obtained as

$$\begin{cases} L\dot{i_k} = -r \cdot i_k - \dfrac{1+u_k}{2} \cdot v_{C1} + \dfrac{1-u_k}{2} \cdot v_{C2} + v_k \\ C_1\dot{v_{C1}} = \displaystyle\sum_{k=1}^{3} \dfrac{1+u_k}{2} \cdot i_k - \dfrac{v_{C1}}{R} - \dfrac{v_{C2}}{R} \\ C_2\dot{v_{C2}} = -\displaystyle\sum_{k=1}^{3} \dfrac{1-u_K}{2} \cdot i_k - \dfrac{v_{C1}}{R} - \dfrac{v_{C2}}{R} \end{cases} \quad k = 1, 2, 3. \tag{3.19}$$

Note that Eqs. (3.17) and (3.19) are equivalent provided that the change of variable $u_k = (1 + u_k{}^*)/2$ has been made.

MATLAB®-Simulink® numerical simulation results

The system described by Eq. (3.18) has been implemented in Simulink®. The three-phase voltage system $\{v_1, v_2, v_3\}$ has constant frequency and magnitude. The system has been fed by a three-phase set of duty ratios α_1, α_2, α_3 having the same frequency and phase as the voltage system $\{v_1, v_2, v_3\}$ and a constant magnitude. In order to obtain the switching functions $u_k(t)$, $k = 1, 2, 3$, three PWM modulators have been used.

Figure 3.14 shows system behavior (state variable evolutions) at start up in two cases: a and b when the AC voltage system is balanced, and c and d when the magnitude of v_3 is 20 % smaller than the magnitude of the other two voltages. The first part of time evolutions shows the dynamical behavior of the system (the rise time, quite reduced resonance, etc.). Note that the two capacitor voltages' instantaneous values are different in the second case and the steady-state magnitudes of currents are different, also.

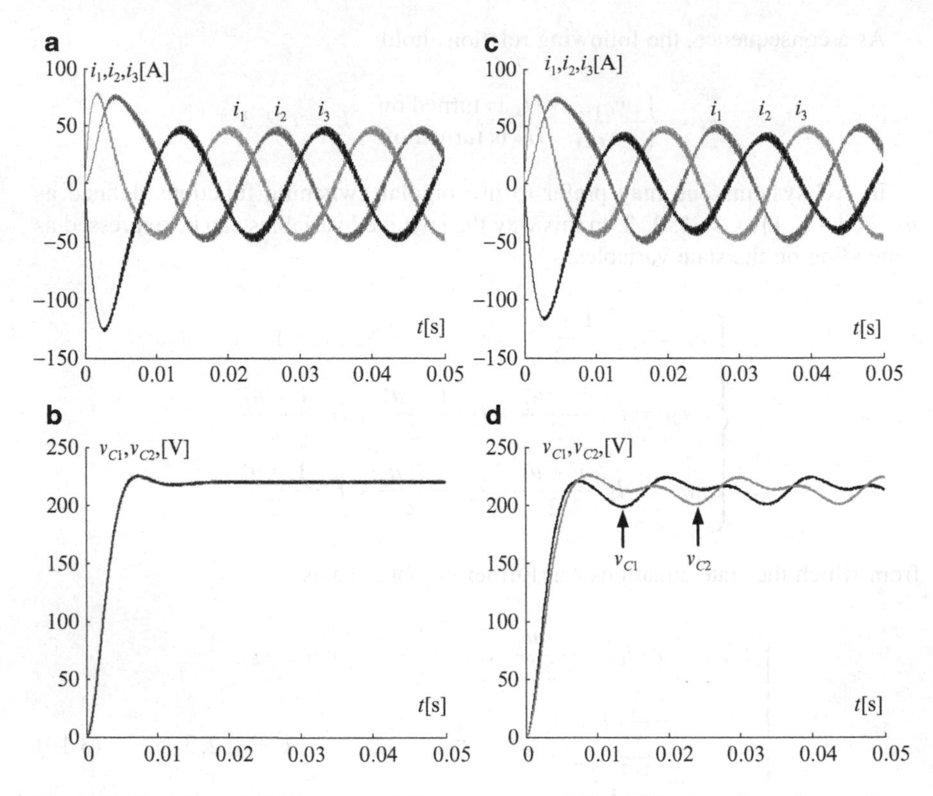

Fig. 3.14 Evolution of VSC state variables at turn on: (**a**) and (**b**) when three-phase voltage system is balanced; (**c**) and (**d**) when three-phase voltages are unbalanced (magnitude of v_3 is 20 % smaller)

Figures 3.15 represent zooms of Figs. 3.14c, d, respectively. The ripple due to switching is quite reduced (due to inductances and capacitor filtering effects) and irregular, depending on the system operating point.

The choice of state vector $\mathbf{x}$ is not unique: combinations of capacitor voltages and inductor currents can be taken as state variables as discussed above. Let us make a different choice. For example, one can consider as state variables the three inductor currents, the sum of the capacitor voltages, $v_0 = v_{C1} + v_{C2}$, and the difference of these voltages, $\Delta v_C = v_{C1} - v_{C2}$. This allows one to use the state feedback to regulate the circuit output without changing the system order. Moreover, one state gives information about the system imbalance.

Differences between the balanced case and the unbalanced one, as reflected in the time evolutions of the two newly introduced voltages at start up, can be seen in Fig. 3.16. Results are predictable because, given that the neutral is connected, a homopolar voltage nonzero component determines the presence of a nonzero homopolar current, which, at its turn, induces imbalance between the two capacitor voltages v_{C1} and v_{C2}.

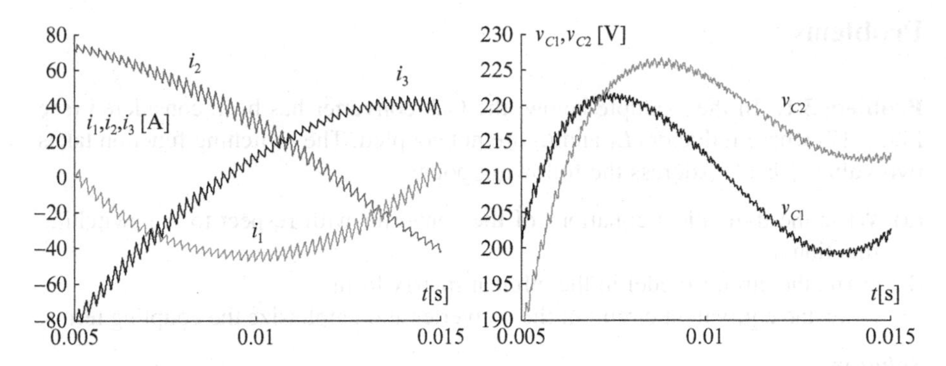

Fig. 3.15 Zooms on the evolutions of state variables in the unbalanced case

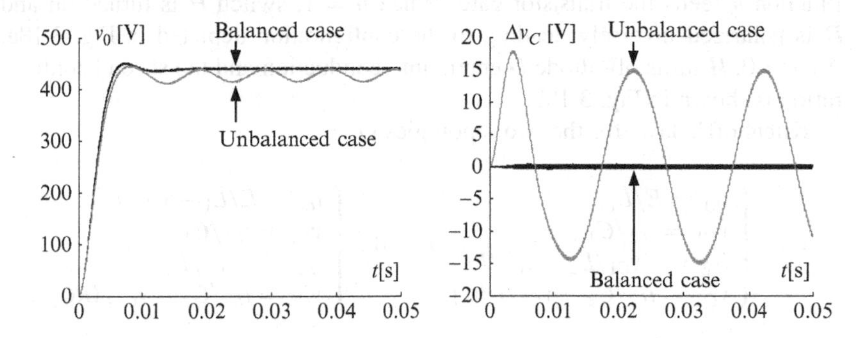

Fig. 3.16 Evolutions of newly introduced state variables at system start up in both balanced and unbalanced case

3.4 Conclusion

The notions of switched models and their corresponding equivalent diagrams are basic and will serve in the development of approaches in the rest of the book.

These models will prove valuable for the following tasks, among others:

- MATLAB®-Simulink® simulation of converter dynamical behavior, provided ease of implementation;
- building other types of models that result from the switched models (for example, averaged model, large-signal model, small-signal model, etc.);
- designing nonlinear control laws, e.g., sliding-mode control (Sanders and Verghese 1992; Malesani et al. 1995).

The reader is invited to solve the problems in the next section. Solutions for the first two are given.

Problems

Problem 3.1. In the example below, the Ćuk converter has been considered (see Fig. 3.17) where inductors L_1 and L_2 are not coupled. The switching function takes two values $\{0; 1\}$. Address the following points.

(a) Write the dynamical equations of the converter with respect to the switching function u.
(b) Write the circuit model in the bilinear matrix form.
(c) Draw the equivalent circuit of the converter and emphasize the coupling terms.

Solution.

(a) Let us take the current sense and voltage polarity as indicated in Fig. 3.17. Transistor H and diode D form the switching network; a single switching function u feeds the transistor gate. When $u = 1$, switch H is turned on and D is polarized inversely, leading to the configuration depicted in Fig. 3.18a. As $u = 0$, H turns off, diode D enters into conduction and the second configuration is shown in Fig. 3.18b.

Kirchhoff's laws for the two topologies give

$$u = 1: \begin{cases} \dot{i_{L1}} = E/L_1 \\ \dot{v_{C1}} = i_{L2}/C_1 \\ \dot{i_{L2}} = -v_{C1}/L_2 - v_{C2}/L_2 \\ \dot{v_{C2}} = i_{L2}/C_2 - v_{C2}/(RC), \end{cases} \qquad u = 0: \begin{cases} \dot{i_{L1}} = E/L_1 - v_{C1}/L_1 \\ \dot{v_{C1}} = i_{L1}/C_1 \\ \dot{i_{L2}} = -v_{C2}/L_2 \\ \dot{v_{C2}} = i_{L2}/C_2 - v_{C2}/(RC_2). \end{cases}$$

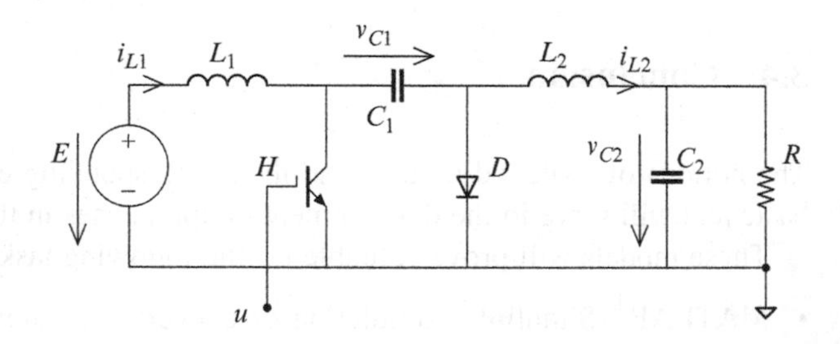

Fig. 3.17 Electrical circuit of Ćuk DC-DC converter

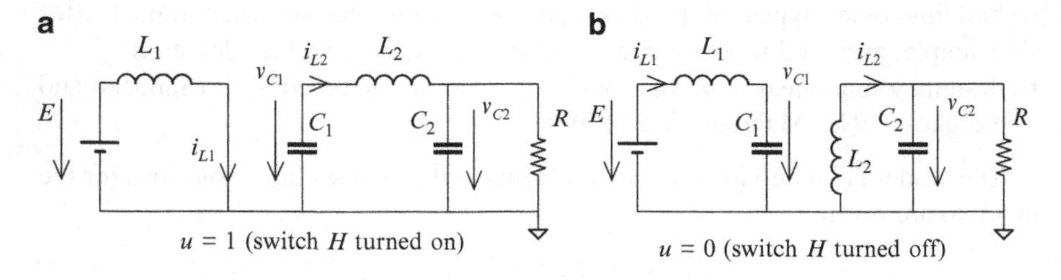

Fig. 3.18 The two possible configurations of the Cúk converter

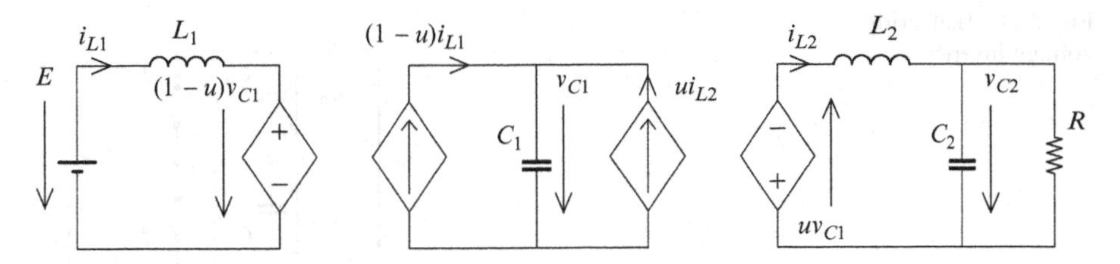

Fig. 3.19 Equivalent circuit of switched model of Cúk converter, presented in Fig. 3.17

By multiplying the two sets of equations with u and $(1–u)$, respectively, and summing, one obtains

$$\begin{cases} \dot{i_{L1}} = -(1-u)\cdot v_{C1}/L_1 + E/L_1 \\ \dot{v_{C1}} = (1-u)\cdot i_{L1}/C_1 + u\cdot i_{L2}/C_1 \\ \dot{i_{L2}} = -u\cdot v_{C1}/L_2 - v_{C2}/L_2 \\ \dot{v_{C2}} = i_{L2}/C_2 - v_{C2}/(RC_2). \end{cases} \tag{3.20}$$

(b) By denoting by $\mathbf{x} = [i_{L1} \quad v_{C1} \quad i_{L2} \quad v_{C2}]^T$ the state vector (each of its components describes an energy accumulation), the set of Eq. (3.20) can be rewritten in matrix form as

$$\dot{\mathbf{x}} = \begin{bmatrix} 0 & -1/L_1 & 0 & 0 \\ 1/C_1 & 0 & 0 & 0 \\ 0 & 0 & 0 & -1/L_2 \\ 0 & 1/C_2 & 0 & -1/(RC_2) \end{bmatrix} \cdot \mathbf{x} + \begin{bmatrix} u\cdot v_{C1}/L_1 \\ u\cdot i_{L1}/C_1 + u\cdot i_{L2}/C_1 \\ -u\cdot v_{C1}/L_2 \\ 0 \end{bmatrix} + \begin{bmatrix} E/L_1 \\ 0 \\ 0 \\ 0 \end{bmatrix}.$$

Note that there is a single switching function, therefore $p = 1$. By processing the second term of the above matrix relation, one obtains

$$\dot{\mathbf{x}} = \mathbf{A}\cdot\mathbf{x} + (\mathbf{B}\cdot\mathbf{x} + \mathbf{b})\cdot u + \mathbf{d},$$

with

$$\begin{cases} \mathbf{A} = \begin{bmatrix} 0 & -1/L_1 & 0 & 0 \\ 1/C_1 & 0 & 0 & 0 \\ 0 & 0 & 0 & -1/L_2 \\ 0 & 1/C_2 & 0 & -1/(RC_2) \end{bmatrix}, & \mathbf{B} = \begin{bmatrix} 0 & 1/L_1 & 0 & 0 \\ 1/C_1 & 0 & 1/C_1 & 0 \\ 0 & -1/L_2 & 0 & 0 \\ 0 & 0 & 0 & 0 \end{bmatrix}, \\ \mathbf{b} = [0 \quad 0 \quad 0 \quad 0]^T, & \mathbf{d} = [E/L_1 \quad 0 \quad 0 \quad 0]^T. \end{cases}$$

Equations (3.20) lead to the equivalent circuit in Fig. 3.19. The first equation of (3.20) gives the first circuit, where the dependent voltage source is a function of the inductor current in the second circuit. The second equation gives the second circuit. Note that the two dependent current sources are coupled with variables from the other two circuits (the inductor current from the first circuit and the capacitor voltage from the third one). These coupled sources can be seen as variable-ratio DC ideal transformers. The third circuit is the image of the last two equations in (3.19).

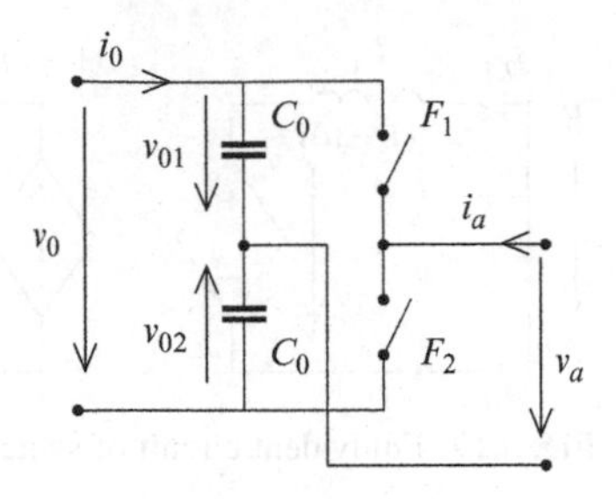

Fig. 3.20 Half-bridge voltage inverter

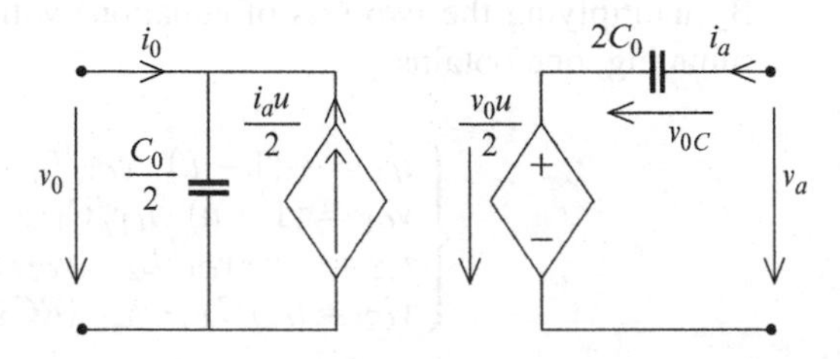

Fig. 3.21 Equivalent topologic diagram of inverter from Fig. 3.20

Problem 3.2. Let consider the half-bridge voltage inverter (current rectifier) of Fig. 3.20. Switching function u is defined to take the value 1 if switch F_1 is turned on and the value -1 if switch F_2 is turned on. Zero indices concern continuous variables, whereas indices a concern alternative variables.

(a) Write the system equations by making appear the switching function u.
(b) Prove that the switched model of the system may be represented by the equivalent circuit from Fig. 3.21.

Solution. The evolution of the filtering capacitor voltages is given by

$$\begin{cases} C_0 \cdot \dot{v_{01}} = i_0 + \dfrac{1+u}{2} \cdot i_a \\ C_0 \cdot \dot{v_{02}} = -i_0 + \dfrac{1-u}{2} \cdot i_a. \end{cases}$$

The expression of the alternative voltage is $v_a = \frac{1+u}{2} \cdot v_{01} + \frac{1-u}{2} \cdot v_{02}$. Taking $v_{0c} = (v_{01} + v_{02})/2$ and knowing that $v_0 = v_{01} - v_{02}$ one obtains

$$\begin{cases} v_a = v_{0c} + \dfrac{v_0}{2} \cdot u \\ 2C_0 \cdot \dot{v_{0c}} = i_a \\ \dfrac{C_0}{2} \cdot \dot{v_0} = i_0 + \dfrac{i_a}{2} \cdot u. \end{cases} \tag{3.21}$$

Equation (3.21) allows the equivalent circuit of Fig. 3.21 to be built as representation of the switched model. This proof is useful for solving Problem 3.3.

The following problems are left to the reader to solve.

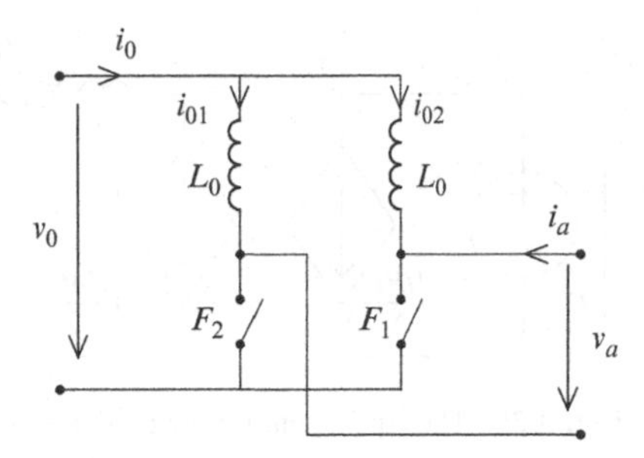

Fig. 3.22 Half-bridge current source inverter

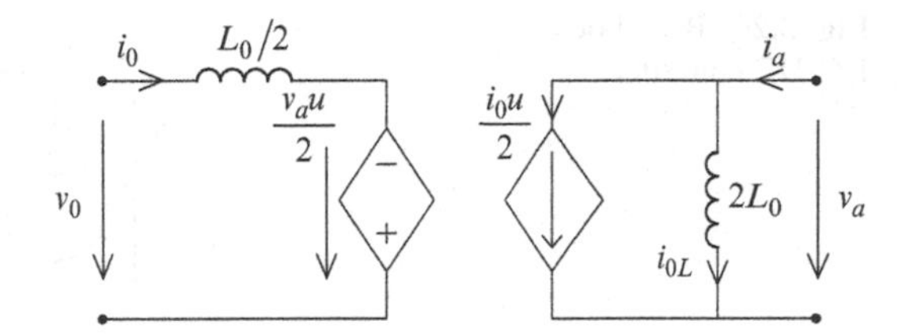

Fig. 3.23 Equivalent topological diagram of current source inverter from Fig. 3.22

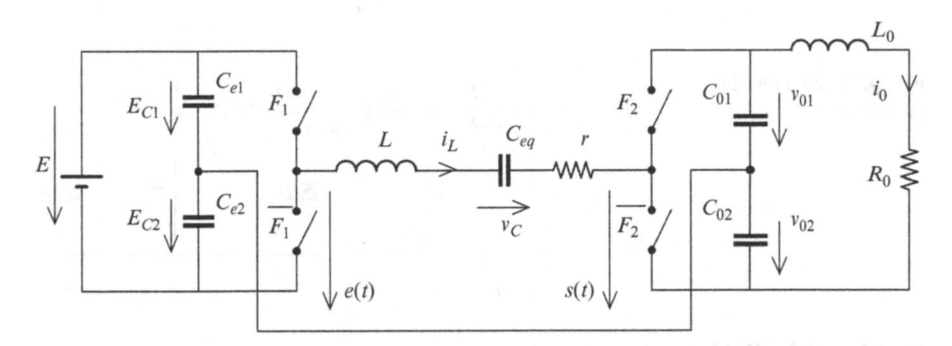

Fig. 3.24 Series-resonance voltage power supply based on capacitive half-bridges

Problem 3.3. Current-source inverter
Let us consider the half-bridge current source inverter presented in Fig. 3.22. Prove that the equivalent diagram shown in Fig. 3.23 corresponds to the circuit given in Fig. 3.22.

Problem 3.4. Series-resonance power supply
Let us consider the converter given in Fig. 3.24. Switching functions u_1 and u_2 are such that

$$u_1 = \begin{cases} 1 & \text{if } F_1 \text{ is closed} \\ -1 & \text{if } F_1 \text{ is open.} \end{cases}$$

(a) Write the observable state model of the system, i.e., of appropriate order, and prove that for this model the equivalent circuit is the one shown in Fig. 3.25.
(b) Find out the value of capacitor C_{eq} from the equivalent circuit.

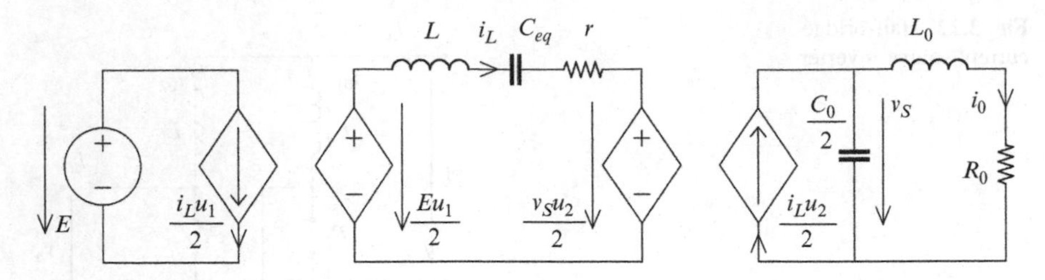

Fig. 3.25 The equivalent topologic diagram of the converter in Fig. 3.24

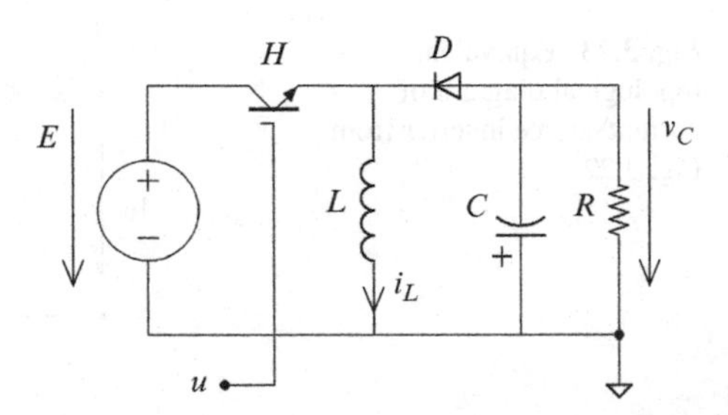

Fig. 3.26 Buck-boost DC-DC converter

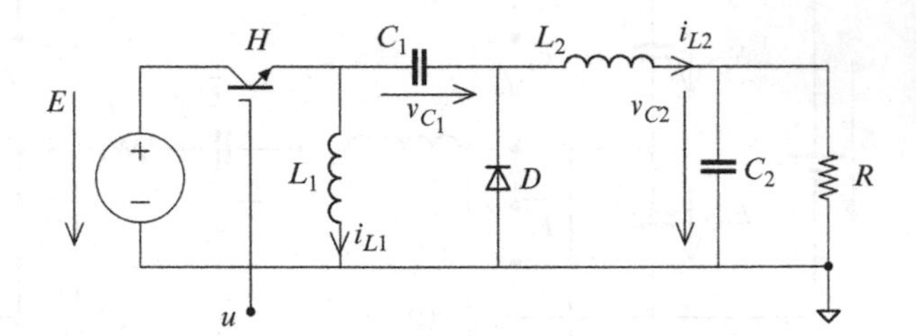

Fig. 3.27 Zeta DC-DC converter

Problem 3.5. Buck-boost power stage
Consider the DC-DC converter given in Fig. 3.26.

(a) Establish the switched model of converter in Fig. 3.26 by considering a switching function that takes its values in the set {0;1}, namely, value 1 corresponds to switch H being turned on and value 0 to the same switch being turned off.
(b) Identify the variables switched by power switches H and D.
(c) Draw the equivalent diagram.

Problem 3.6. Zeta DC-DC power stage

(a) Deduce the switched model of the DC-DC converter from Fig. 3.27 by considering a switching function that takes value 1 when switch H is turned on and value 0 when the same switch is turned off. Consider the voltage polarity and current sense as indicated in Fig. 3.27.
(b) Identify the variables switched by power switches H and D.
(c) Draw the equivalent diagram.

References

Cellier F, Elmqvist H, Otter M (1996) Modeling from physical principles. In: Levine WS (ed) The control handbook. CRC Press/IEEE Press, Boca Raton, pp 99–107

Erikson RW, Maksimović D (2001) Fundamentals of power electronics, 2nd edn. Kluwer, Dordrecht, The Netherlands

Kassakian JG, Schlecht MF, Verghese GC (1991) Principles of power electronics. Addison-Wesley, Reading, Massachusetts

Krein PT, Bentsman J, Bass RM, Lesieutre B (1990) On the use of averaging for the analysis of power electronic systems. IEEE Trans Power Electron 5(2):182–190

Maksimović D, Stanković AM, Thottuvelil VJ, Verghese GC (2001) Modeling and simulation of power electronic converters. Proc IEEE 89(6):898–912

Malesani L, Rossetto L, Spiazzi G, Tenti P (1995) Performance optimization of Cúk converters by sliding-mode control. IEEE Trans Power Electron 10(3):302–309

Merdassi A, Gerbaud L, Bacha S (2010) Automatic generation of average models for power electronics systems in VHDL-AMS and modelica modelling languages. HyperSci J Model Simul Syst 1(3):176–186

Mohan N, Undeland TM, Robbins WP (2002) Power electronics: converters, applications and design, 3rd edn. Wiley, Hoboken

Sanders SR (1993) On limit cycles and the describing function method in periodically switched circuits. IEEE Trans Circuit Syst 40(9):564–572

Sanders SR, Verghese GC (1992) Lyapunov-based control for switched power converters. IEEE Trans Power Electron 7(1):17–24

Sun J, Grotstollen H (1992) Averaged modelling of switching power converters: reformulation and theoretical basis. In: Proceedings of the IEEE Power Electronics Specialists Conference – PESC 1992, Toledo, pp 1165–1172

Sun J, Mitchell DM, Greuel ME, Krein PT, Bass RM (1998) Modeling of PWM converters in discontinuous conduction mode – a reexamination. In: Proceedings of the 29th annual IEEE Power Electronics Specialists Conference – PESC 1998, Fukuoka, Japan, vol 1, pp 615–622

Tymerski R, Vorpérian V, Lee FCY, Baumann WT (1989) Nonlinear modelling of the PWM switch. IEEE Trans Power Electron 4(2):225–233

van Dijk E, Spruijt HJN, O'Sullivan DM, Klaassens JB (1995) PWM-switch modeling of DC-DC converters. IEEE Trans Power Electron 10(6):659–665

Vorpérian V (1990) Simplified analysis of PWM converters using model of PWM switch. Part II: discontinuous conduction mode. IEEE Trans Aerosp Electron Syst 26(3):497–505

Chapter 4
Classical Averaged Model

This chapter deals with methodologies of obtaining the so-called averaged model, which focuses on capturing the low-frequency behavior of power electronic converters while neglecting high-frequency variations due to circuit switching. This appears to be a natural action, as every converter employs filters in order to limit the ripple of various variables. The result is a continuous-time model, one which is easier to handle by classical analysis and control formalisms.

This chapter is organized as follows. It starts by presenting the basics of averaging methodology and states some theoretical fundamentals. Then the methodology of obtaining small-signal and large-signal averaged models and their equivalent averaged diagrams are given. The error introduced by averaging, computed with respect to the exact sampled-data model, is also analyzed. A case study will serve at illustrating the various approaches. The chapter ends with some problems and their solutions and some proposed problems.

4.1 Introduction

In the previous chapter it was shown that electrical circuits containing static converters can be mathematically described by a cyclic set of state equations, corresponding to different electrical configurations listed along the entire converter operation cycle. The switched model is the product of such an analysis and it is particularly suitable for designing nonlinear control laws, such as variable-structure or hysteresis control.

However, in most control applications it is the low-frequency behavior that is interesting. In this context, the various high-frequency switching phenomena are parasitic and must be neglected. When certain control laws (e.g., linear control) need to be implemented, the designer must transform the original discontinuous model in a continuous invariant model that provides the best representation of the system macroscopic behavior. The obtained model should be easy to employ; to this end, the averaging method is strongly recommended.

S. Bacha et al., *Power Electronic Converters Modeling and Control: with Case Studies*, Advanced Textbooks in Control and Signal Processing, DOI 10.1007/978-1-4471-5478-5_4,

Because of its undoubted utility, this kind of model – named the *averaged model* – has been studied since early 1970s, either by circuit averaging or state-space averaging (Wester and Middlebrook 1973; Middlebrook and Ćuk 1976) or by averaged equivalent electrical diagram analysis (Pérard et al. 1979).

The utility of the averaged model for simulation purposes, using dedicated software products such as SPICE®, SABER®, MATLAB®, has also been largely proved (Sanders and Verghese 1991; Ben-Yakoov 1993; Vuthchhay and Bunlaksananusorn 2008). This kind of model is useful for analytically expressing the essential dynamical behavior of power electronic circuits, both in continuous-time (Middlebrook 1988; Rim et al. 1988; Lehman and Bass 1996) and discrete-time domain (Maksimović and Zane 2007). Averaging techniques can also be used for modeling the inner current control loop of converters (Verghese et al. 1989; Rodriguez and Chen 1991; Tymerski and Li 1993).

4.2 Definitions and Basics

Some fundamental notions and terms are given in order to describe the averaged dynamic behavior of a power electronic (switched) circuit.

4.2.1 Sliding Average

Let us consider the signal from Fig. 4.1, which is not mandatory cyclic. If signal $f(t)$ is averaged on a time window of width T, which is moving along the time axis, one obtains the expression of the so-called *sliding average* (or *local average*) (Maksimović et al. 2001), as given by Eq. (4.1).

$$\langle f(t)\rangle_0(t) = \frac{1}{T}\cdot\int_{t-T}^{t} f(\tau)d\tau, \tag{4.1}$$

Unlike the classical average of the signal $f(t)$, the term $\langle f(t)\rangle_0$ is time-dependent because the associated time window changes its position on the time axis. However, if the signal $f(t)$ is periodic and reaches its steady-state regime, the moving average becomes identical with the classical average.

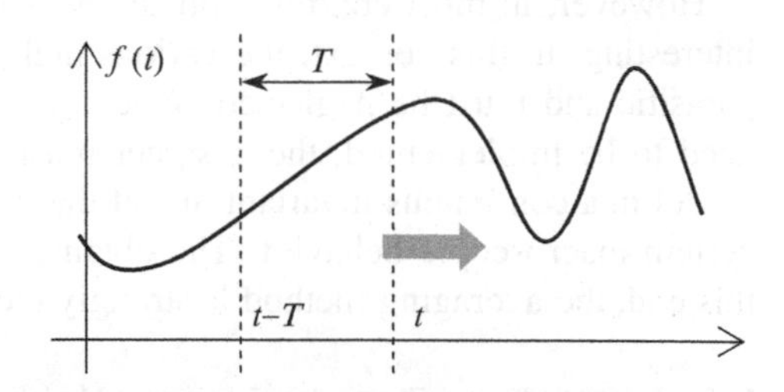

Fig. 4.1 Moving average illustration

A fundamental property of the moving average is that the time derivative of a signal sliding average is the sliding average of the signal time derivative, that is, the time derivation and the sliding averaging operation are commutative:

$$\frac{d}{dt}\langle f(t)\rangle_0(t) = \left\langle \frac{d}{dt}f(t)\right\rangle_0(t). \tag{4.2}$$

In order to give a brief justification of Eq. (4.2), let us denote by F the primitive of function f; therefore, $F(\tau) = \int f(\tau)d\tau + C$, where C is a suitably chosen constant, and $\frac{d}{d\tau}F(\tau) = f(\tau)$. One obtains successively

$$\begin{aligned}\frac{d}{dt}\langle f(t)\rangle_0(t) &= \frac{d}{dt}\left(\frac{1}{T}\cdot\int_{t-T}^{t} f(\tau)d\tau\right) = \frac{d}{dt}\left(\frac{1}{T}\cdot(F(t)-F(t-T))\right)\\ &= \frac{1}{T}\cdot\left(\frac{d}{dt}F(t)-\frac{d}{dt}F(t-T)\right) = \frac{1}{T}\cdot(f(t)-f(t-T)).\end{aligned} \tag{4.3}$$

On the other hand, Definition 4.1 of the sliding average allows the form

$$\left\langle \frac{d}{dt}f(t)\right\rangle_0(t) = \frac{1}{T}\cdot\int_{t-T}^{t}\frac{d}{d\tau}f(\tau)\,d\tau = \frac{1}{T}\cdot f(\tau)|_{t-T}^{t} = \frac{1}{T}\cdot(f(t)-f(t-T)). \tag{4.4}$$

From Eqs. (4.3) and (4.4) it follows that relation (4.2) is proved.

Next, for sake of simplicity "average" will be used to denote the "sliding average". Note also that, because the sliding average is not fixed, but dependent on time, the average time evolution can be replicated in real time.

4.2.2 State Variable Average

Let us consider a state variable x, whose dynamic is given by

$$\frac{d}{dt}x(t) = f(x,u,t), \tag{4.5}$$

with f being a generally nonlinear function. Its average is given by Eq. (4.1). Therefore, by applying the property expressed by Eq. (4.2) one obtains:

$$\frac{d}{dt}\langle x\rangle_0(t) = \langle f(x,u,t)\rangle_0(t). \tag{4.6}$$

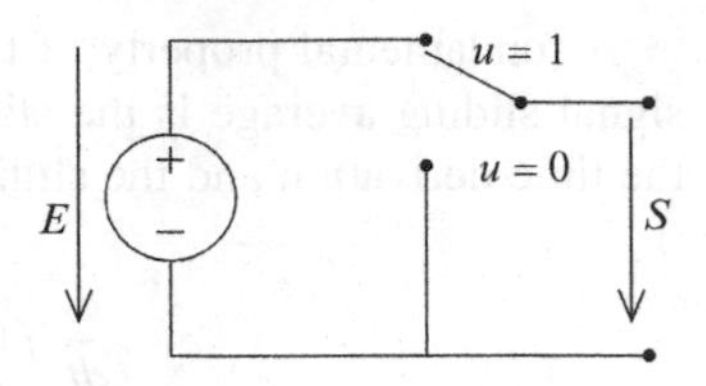

Fig. 4.2 On-off switch

4.2.3 *Average of a Switch*

Generally speaking, a switch changes its output variable between several states; in this text an on-off switch will be approached as shown in Fig. 4.2.

The switch in Fig. 4.2 has two states, given by the switching function u, and outputs a variable S, which takes the value $E \cdot u$. Without employing any simplification its average value is

$$\langle S \rangle_0 = \langle E \cdot u \rangle_0, \tag{4.7}$$

When this expression is developed, two cases can occur:

- E is a constant value and one obtains an exact relation:

$$\langle S \rangle_0 = \langle E \rangle_0 \cdot \langle u \rangle_0 = E \cdot \langle u \rangle_0; \tag{4.8}$$

- E is variable, S being given by an approximation:

$$\langle S \rangle_0 \approx \langle E \rangle_0 \cdot \langle u \rangle_0. \tag{4.9}$$

Equation (4.9) is justified as one of the two variables E or u is close to its average value (the small-ripple assumption). If voltage E is supposed constant, it will be equal with its average. One can identify here an instance of the averaged model principle: to assimilate (by approximation) the average of a product by the product of averages. In other words, the product and averaging operators are commutative under certain conditions.

4.2.4 *Complete Power Electronic Circuit Average*

The configuration corresponding to a circuit's average behavior is called *circuit average*. Adopting the invariance conditions of passive circuit elements, as assumed in the previous chapter, the well-known relations linking voltages $v(t)$ and currents $i(t)$ in a passive circuit element are as follows:

- for an inductor L: $v(t) = L \cdot \frac{d}{dt} i(t)$;
- for a capacitor C: $i(t) = C \cdot \frac{d}{dt} v(t)$;
- for a resistor R: $v(t) = R \cdot i(t)$.

According to Eq. (4.2), the derivative of the average is the average of the derivative, implying that the relations between currents and voltages are the same

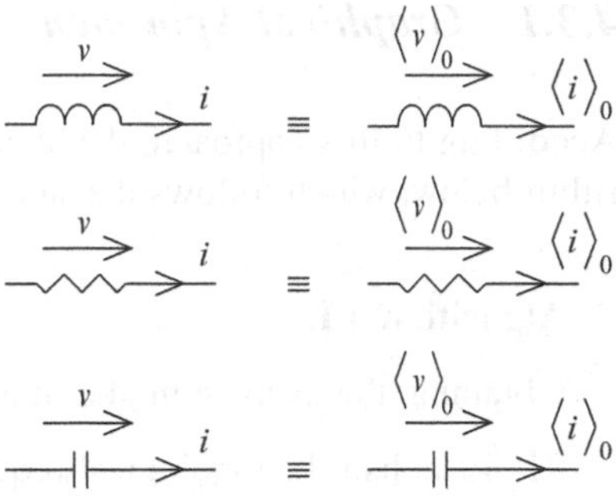

Fig. 4.3 Average of elementary passive circuits

as those linking their averages. Hence, the passive elements R, L and C remain unchanged after the circuit averaging operation (see Fig. 4.3). In this way, another property of averaging is emphasized: the configuration of a passive circuit remains unchanged by averaging.

In order to obtain the averaged model of a more complex circuit, one that also includes switches apart from its passive elements, one can use the previously presented properties, namely preservation of circuit configuration by replacement of the variables by their averages and replacement of the products of variables by the products of averages. An example clarifies this approach, as follows.

Let us consider the boost circuit represented by its topological diagram given in Fig. 4.4a, having switching function u defined as usual. The average scheme preserves the circuit configuration. The state variables i_L and v_C are replaced by their averages. The next step consists in developing the products (couplings) $\langle v_C(1-u)\rangle_0$ and $\langle i_L(1-u)\rangle_0$, an operation which gives the approximate relations:

$$\begin{cases} \langle v_C(1-u)\rangle_0 \approx \langle v_C\rangle_0 \cdot \langle (1-u)\rangle_0 = \langle v_C\rangle_0 \cdot (1-\alpha) \\ \langle i_L(1-u)\rangle_0 \approx \langle i_L\rangle_0 \cdot \langle (1-u)\rangle_0 = \langle i_L\rangle_0 \cdot (1-\alpha), \end{cases} \tag{4.10}$$

where α denotes the duty ratio corresponding to averaging the switching signal u. If the voltage E is supposed constant, it will be identical with its average. The diagram in Fig. 4.4c represents the averaged scheme of the boost converter.

Note that, as long as one does not make approximations concerning the average products, the averaged diagram is identical to the exact diagram. The approximations in Eq. (4.10) are even more valid if the current i_L and the voltage v_C are filtered appropriately, thus making them sufficiently close to their average values.

4.3 Methodology of Averaging

As seen in the above considered example, the averaged model can be directly built by employing the topological (exact) diagram. Then, it is sufficient to use the electrical circuit laws for establishing the analytical averaged model. The problem can also be solved analytically. Two averaging approaches emerge – the graphical and the analytical– as described below.

4.3.1 Graphical Approach

According to this approach, the averaged models can be obtained using the algorithm below, which follows the steps in Fig. 4.4 and uses Eq. (4.10).

Algorithm 4.1.

Obtaining the average model of a power electronic converter

#1. Establish the topological (equivalent circuit) diagram where the various coupling terms are emphasized.
#2. Preserve the diagram structure and replace the variables by their averages.
#3. Develop the coupling terms by approximating the product average by the product of averages.
#4. Deduce the averaged model equations based upon the obtained diagram.

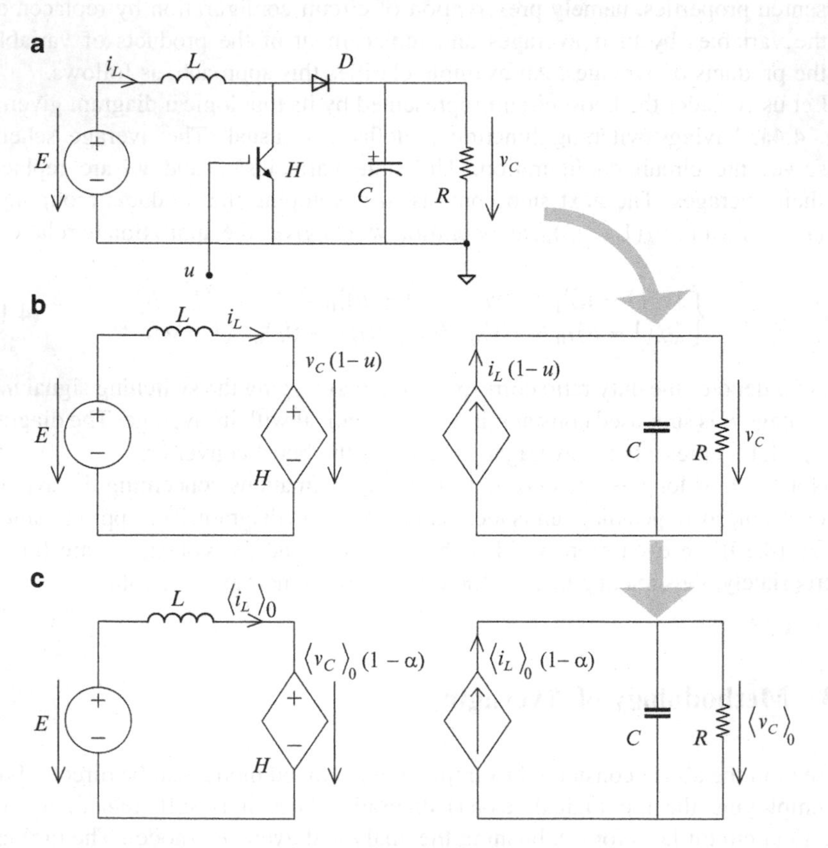

Fig. 4.4 Phases of circuit averaging (boost power stage)

4.3.2 Analytical Approach

The analytical approach uses topological model equations. The converter can be described in a general manner by Eq. (3.2) from Chap. 3:

$$\dot{\mathbf{x}} = \sum_{i=1}^{N} (\mathbf{A}_i\mathbf{x} + \mathbf{B}_i\mathbf{e}) \cdot h_i, \tag{4.11}$$

i.e., as a linear system switching between N configurations, where the matrix pairs $(\mathbf{A}_i,\mathbf{B}_i)$ denote the state model of configuration i, h_i are the enable functions and vector $\mathbf{e}$ denotes the free sources vector. Applying the average operator to Eq. (4.11) and taking into account relation (4.2) one obtains

$$\frac{d\langle\mathbf{x}\rangle_0}{dt} = \left\langle \sum_{i=1}^{N} (\mathbf{A}_i\mathbf{x} + \mathbf{B}_i\mathbf{e}) \cdot h_i \right\rangle_0,$$

which can be rewritten by taking into account the linearity of the averaging operation:

$$\frac{d\langle\mathbf{x}\rangle_0}{dt} = \left\langle \sum_{i=1}^{N} (\mathbf{A}_i h_i) \cdot \mathbf{x} + \sum_{i=1}^{N} (\mathbf{B}_i h_i) \cdot \mathbf{e} \right\rangle_0.$$

After development and approximation one obtains

$$\frac{d\langle\mathbf{x}\rangle_0}{dt} \approx \left\langle \sum_{i=1}^{N} (\mathbf{A}_i h_i) \right\rangle_0 \cdot \langle\mathbf{x}\rangle_0 + \left\langle \sum_{i=1}^{N} (\mathbf{B}_i h_i) \right\rangle_0 \cdot \langle\mathbf{e}\rangle_0.$$

By introducing the notation

$$\mathbf{A}_m = \left\langle \sum_{i=1}^{N} (\mathbf{A}_i h_i) \right\rangle_0, \mathbf{B}_m = \left\langle \sum_{i=1}^{N} (\mathbf{B}_i h_i) \right\rangle_0, \tag{4.12}$$

the averaged model of the power electronic circuit can be written as

$$\frac{d\langle\mathbf{x}\rangle_0}{dt} = \mathbf{A}_m \cdot \langle\mathbf{x}\rangle_0 + \mathbf{B}_m \cdot \langle\mathbf{e}\rangle_0. \tag{4.13}$$

Note that matrices $\mathbf{A}_m$ and $\mathbf{B}_m$ are not the state and input matrices, respectively. Matrices $\mathbf{A}_m$ and $\mathbf{B}_m$ are dependent on the state $\mathbf{x}$ and on the control input, which does not appear explicitly here.

A similar result can be obtained starting from the bilinear form, detailed in Eq. (3.3) from Sect. 3.1.2 in Chap. 3:

$$\dot{\mathbf{x}} = \mathbf{A}\mathbf{x} + \sum_{k=1}^{p} (\mathbf{B}_k\mathbf{x} + \mathbf{b}_k) \cdot u_k + \mathbf{d}, \tag{4.14}$$

which makes the control input vector $\mathbf{u} = [u_1\ u_2\ \cdots\ u_p]^T$ appear explicitly as being composed of p switching functions, where p is the smallest integer satisfying the relation $2^p \geq N$. In (4.14), for every k from 1 to p, $\mathbf{B}_k$ are square matrices of the same dimension as matrix $\mathbf{A}$, and $\mathbf{b}_k$ and $\mathbf{d}$ are column vectors of the same dimension. Applying averaging to relation (4.14) and supposing that the average of a product can be approximated by the product of averages and that matrices $\mathbf{A}$, $\mathbf{B}_k$ and $\mathbf{b}_k$ are invariant, gives the following:

$$\frac{d\langle\mathbf{x}\rangle_0}{dt} = \mathbf{A} \cdot \langle\mathbf{x}\rangle_0 + \sum_{k=1}^{p} \left(\mathbf{B}_k \cdot \langle\mathbf{x}\rangle_0 + \mathbf{b}_k\right) \cdot \alpha_k + \mathbf{d}, \tag{4.15}$$

where $\alpha_k = \langle u_k \rangle_0$ is the duty ratio of switching function u_k for each k from 1 to p.

4.4 Analysis of Averaging Errors

The goal here is to perform an analysis of errors introduced by averaging. The error between the average of a given switched model state-space solution and the averaged model state-space solution (on the same time window) is easy to put into light by numerical simulation, like shown in Fig. 4.5, where the error signal has been denoted by $\varepsilon(t)$. An estimation of this kind of error can also be performed analytically by referring to the sampled-data model introduced in Sect. 2.2.2 from Chap. 2 (Pérard et al. 1979).

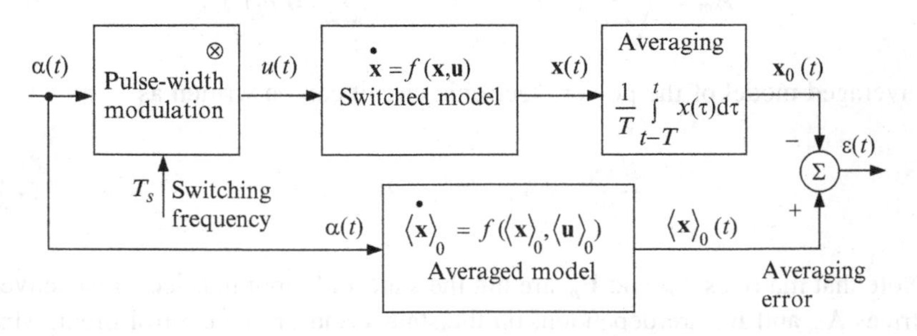

Fig. 4.5 Averaged model output vs. average of switched model output

4.4.1 Exact Sampled-Data Model

Let us consider a linear system that switches between two configurations, α being the duty ratio; it can be described by the following set of differential equations:

$$\frac{d\mathbf{x}}{dt} = \begin{cases} \mathbf{A}_1 \cdot \mathbf{x} + \mathbf{B}_1 \cdot \mathbf{e} \ \text{ for } t \in [kT, (k+\alpha)T) \\ \mathbf{A}_2 \cdot \mathbf{x} + \mathbf{B}_2 \cdot \mathbf{e} \ \text{ for } t \in [(k+\alpha)T, (k+1)T), \end{cases} \tag{4.16}$$

which is valid for each switching period $[kT, (k+1)T)$, where $k \in \mathbb{N}$. It is proposed that system (4.16) be solved by integrating each configuration and postulating the state variables' continuity. Let us introduce the general notation of the transition matrix $\Phi(t) = \exp(\mathbf{A} \cdot t)$. The transition matrices associated with the two configurations are $\Phi_1(t) = \exp(\mathbf{A}_1 \cdot t)$ and $\Phi_2(t) = \exp(\mathbf{A}_2 \cdot t)$.

For the first configuration one can write

$$\mathbf{x}[(k+\alpha)T] = \Phi_1(\alpha T) \cdot \mathbf{x}(kT) + \int_0^{\alpha T} \Phi_1(\tau) \cdot \mathbf{B}_1 \mathbf{e}\, d\tau,$$

which gives, after integration

$$\mathbf{x}[(k+\alpha)T] = \Phi_1(\alpha T) \cdot \mathbf{x}(kT) + \mathbf{A}_1{}^{-1} \cdot (\Phi_1(\alpha T) - \mathbf{I}_n) \cdot \mathbf{B}_1 \mathbf{e}, \tag{4.17}$$

where $\mathbf{I}_n$ is the identity matrix of the same dimension as $\mathbf{A}$. For the second configuration the result is given directly (algebra is the same):

$$\mathbf{x}[(k+1)T] = \Phi_2(\alpha T) \cdot \mathbf{x}[(k+\alpha)T] + \mathbf{A}_2{}^{-1} \cdot (\Phi_2[(1-\alpha)T] - \mathbf{I}_n) \cdot \mathbf{B}_2 \mathbf{e}. \tag{4.18}$$

By introducing the value of $\mathbf{x}[(k+\alpha)T]$, given in Eq. (4.17), into Eq. (4.18) one obtains a recurrent equation of the form

$$\mathbf{x}_{k+1} = \Phi(\mathbf{x}_k, \alpha_k, T). \tag{4.19}$$

Equations (4.18) and (4.19) are forms of the *exact sampled-data model*. Equation (4.19) is very complex and difficult to manipulate. In order to avoid cumbersome mathematical developments certain systems having simple representations will be taken into account. This can be obtained from Eq. (4.18) by employing an adequate change of variable.

$$\frac{d\mathbf{x}}{dt} = \begin{cases} \mathbf{A}_1 \cdot \mathbf{x} \ \text{ for } t \in [kT, (k+\alpha)T) \\ \mathbf{A}_2 \cdot \mathbf{x} \ \text{ for } t \in [(k+\alpha)T, (k+1)T). \end{cases} \tag{4.20}$$

The system of Eq. (4.20) corresponds to the dynamic depicted in Fig. 4.6.

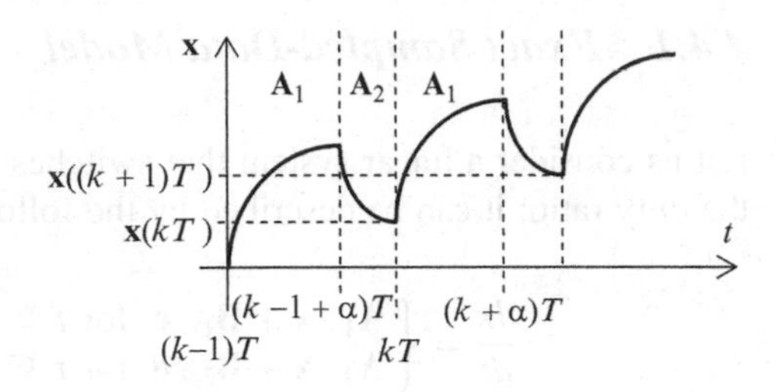

Fig. 4.6 Dynamic of system described by Eq. (4.20)

Integrating the first equation of (4.20), one obtains

$$\mathbf{x}[(k+\alpha)T] = \Phi_1(\alpha T)\cdot \mathbf{x}(kT), \tag{4.21}$$

whereas for the second configuration the integration gives

$$\mathbf{x}[(k+1)T] = \Phi_2[(1-\alpha)T]\cdot \mathbf{x}[(k+\alpha)T]. \tag{4.22}$$

By substituting Eq. (4.21) into Eq. (4.22), one obtains

$$\mathbf{x}[(k+1)T] = \Phi_2[(1-\alpha)T]\cdot \Phi_1(\alpha T)\cdot \mathbf{x}(kT), \tag{4.23}$$

which is the output of the switched model at switching instants in a recurrent form.

Like the model described by Eq. (4.19), the model in Eq. (4.23) is the sampled-data topological model. The computation of matrices Φ_1 and Φ_2 can be simplified more or less satisfactorily, by their first-order expansions, respectively

$$\begin{cases} \Phi_1(\alpha T) \approx \mathbf{I} + \mathbf{A}_1\cdot \alpha T \\ \Phi_2[(1-\alpha)T] \approx \mathbf{I} + \mathbf{A}_2\cdot (1-\alpha)T, \end{cases} \tag{4.24}$$

where $\mathbf{I}$ is the identity matrix of the same dimension as $\mathbf{A}_1$ and $\mathbf{A}_2$. Introducing the simplified expressions (4.24) into Eq. (4.23) yields the *first-order approximated sampled-data model*.

4.4.2 Relation Between Exact Sampled-Data Model and Exact Averaged Model

Note that Eq. (4.23), providing the solution of the switched model, contains matrix products. In the general case a product of matrices is not commutative, i.e., the following relation generally holds between the two state-space matrices: $\mathbf{A}_1\cdot\mathbf{A}_2 \neq \mathbf{A}_2\cdot\mathbf{A}_1$.

Case of commutative matrices

In this subsection the exceptions where $\mathbf{A}_1\cdot\mathbf{A}_2 = \mathbf{A}_2\cdot\mathbf{A}_1$ are addressed. In such cases the matrix exponentials are also switching, i.e., $\Phi_1\cdot\Phi_2 = \Phi_2\cdot\Phi_1$. Therefore, the following relation holds:

$$\exp(\mathbf{A}_1)\cdot\exp(\mathbf{A}_2) = \exp(\mathbf{A}_2)\cdot\exp(\mathbf{A}_1) = \exp(\mathbf{A}_1+\mathbf{A}_2). \tag{4.25}$$

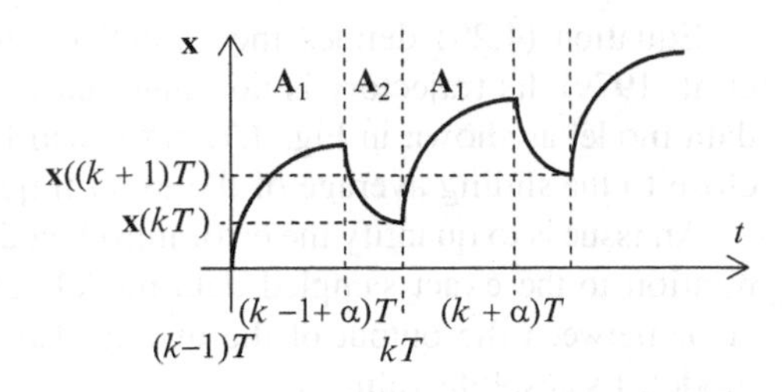

Fig. 4.7 Exact behavior and exact averaged model

This very interesting result, if applied to the exact switched model (4.23), gives

$$\mathbf{x}[(k+1)T] = \Phi_m(\alpha, T) \cdot \mathbf{x}(kT), \tag{4.26}$$

where for matrix Φ_m one obtains successively

$$\begin{aligned}\Phi_m &= \Phi_2[(1-\alpha)T] \cdot \Phi_1(\alpha T) = \Phi_1(\alpha T) \cdot \Phi_2[(1-\alpha)T] \\ &= \exp\{\mathbf{A}_1 \cdot \alpha T\} \cdot \exp\{\mathbf{A}_2 \cdot (1-\alpha)T\} = \exp\{[\mathbf{A}_1 \cdot \alpha + \mathbf{A}_2 \cdot (1-\alpha)] \cdot T\}.\end{aligned}$$

The latter result can be expressed more synthetically as

$$\Phi_m = \exp(\mathbf{A}_m \cdot T), \tag{4.27}$$

where

$$\mathbf{A}_m = \mathbf{A}_1 \cdot \alpha + \mathbf{A}_2 \cdot (1-\alpha)$$

is the state matrix of the averaged model.

Equation (4.27) represents the solution of the system (averaged model)

$$\frac{d}{dt}\langle \mathbf{x} \rangle_0 = (\mathbf{A}_1 \cdot \alpha + \mathbf{A}_2 \cdot (1-\alpha)) \cdot \langle \mathbf{x} \rangle_0. \tag{4.28}$$

In addition, at sampling moments it holds that $\mathbf{x}[(k+1)T] = \langle \mathbf{x} \rangle_0[(k+1)T]$. This is why the model expressed by Eq. (4.28) is called the *exact averaged model* (Pérard et al. 1979); its dynamic is represented in Fig. 4.7.

General case
Unfortunately, the assumption of matrices being commutative does not hold in the quasi-totality of power converters. In the general case, the matrix product between $\mathbf{A}_1$ and $\mathbf{A}_2$ is not commutative; hence, their exponentials are not commutative:

$$\exp(\mathbf{A}_1) \cdot \exp(\mathbf{A}_2) \neq \exp(\mathbf{A}_2) \cdot \exp(\mathbf{A}_1).$$

Therefore, relation (4.28) becomes

$$\frac{d}{dt}\langle \mathbf{x} \rangle_0 \approx \mathbf{A}_m \cdot \langle \mathbf{x} \rangle_0 = [\mathbf{A}_1 \cdot \alpha + \mathbf{A}_2 \cdot (1-\alpha)] \cdot \langle \mathbf{x} \rangle_0. \tag{4.29}$$

Equation (4.29) defines the so-called *approximated averaged model* (Pérard et al. 1979). Its trajectory is no longer passing through the points of the sampled-data model as shown in Fig. 4.7, but it will be an averaged trajectory more or less close to the sliding average of the exact trajectory.

An issue is to quantify the error introduced by the approximation in this model in relation to the exact sampled-data model. This represents an upper bound of the error between the output of the averaged model and the average of the switched model. Its absolute value is

$$\mathbf{Err} = \Phi_m(\alpha, T) - \Phi_2[(1-\alpha)T] \cdot \Phi_1(\alpha T). \tag{4.30}$$

The complete computation of error expressed by (4.30) is not trivial and can only be done in numerical form. For the sake of simplicity a second-order approximation of the matrix exponential is employed,

$$\exp(\mathbf{A}t) \approx \mathbf{I} + \mathbf{A}t + \frac{\mathbf{A}^2 t^2}{2},$$

in order to express the matrix Φ_m and the product $\Phi_2[(1-\alpha)T] \cdot \Phi_1(\alpha T)$. One obtains successively:

$$\begin{aligned}
\Phi_m(\alpha, T) &\approx \mathbf{I} + \mathbf{A}_m T + \frac{\mathbf{A}_m^2 T^2}{2} = \mathbf{I} + [\mathbf{A}_1\alpha + \mathbf{A}_2(1-\alpha)]T + \frac{[\mathbf{A}_1\alpha + \mathbf{A}_2(1-\alpha)]^2 T^2}{2} \\
&= \mathbf{I} + [\mathbf{A}_1\alpha + \mathbf{A}_2(1-\alpha)]T + \frac{\mathbf{A}_1^2\alpha^2}{2} + \frac{\mathbf{A}_2^2(1-\alpha)^2}{2} T^2 \\
&\quad + (\mathbf{A}_1\mathbf{A}_2 + \mathbf{A}_2\mathbf{A}_1)\frac{\alpha(1-\alpha)}{2} T^2,
\end{aligned}$$

$$\begin{aligned}
\Phi_2[(1-\alpha)T] \cdot \Phi_1(\alpha T) &\approx \mathbf{I} + [\mathbf{A}_1\alpha + \mathbf{A}_2(1-\alpha)]T + \frac{[\mathbf{A}_1\alpha + \mathbf{A}_2(1-\alpha)]^2 T^2}{2} \\
&= \mathbf{I} + [\mathbf{A}_1\alpha + \mathbf{A}_2(1-\alpha)]T + \frac{\mathbf{A}_1^2\alpha^2}{2} + \frac{\mathbf{A}_2^2(1-\alpha)^2}{2} T^2 \\
&\quad + \mathbf{A}_1\mathbf{A}_2 \frac{\alpha(1-\alpha)}{2} T^2.
\end{aligned}$$

One can remark that the matrix error **Err** expressed by Eq. (4.30) between the second-order developments is reduced to the matrix **E** as follows:

$$\mathbf{Err} \approx \mathbf{E} = (\mathbf{A}_1\mathbf{A}_2 - \mathbf{A}_2\mathbf{A}_1)\frac{\alpha(1-\alpha)}{2} T^2. \tag{4.31}$$

Equation (4.31) shows that if matrices $\mathbf{A}_1$ and $\mathbf{A}_2$ are commutative then the sampled-data and averaged models are confused, as shown in the previous paragraph. In the general case, the smaller the norm of the error matrix in relation to the norms of the other state matrices weighted by their associated enabling times within a switching period, αT and $(1 - \alpha)T$, respectively, the more precise the approximate model is. This further requires that time T be small – i.e., the converter to operate at high frequency – and that the duty ratio α be close to one or zero. If these assumptions hold, this leads to smaller ripple of the state variables. To conclude, the averaged model can be seen as an "ideal", operating at infinite frequency, whereas the switched model, operating at finite frequency, exhibits, besides the variables' ripples, an average different from the averaged model solution.

Discussions on this subject will be detailed in the example presented in Sect. 4.5.3.

4.5 Small-Signal Averaged Model

The large-signal averaged model of a power electronic converter is in general – with very few exceptions – a nonlinear model. If a modal analysis or linear controller design is envisaged, a linear representation of the converter model – in the state space or in the frequency domain – is needed.

4.5.1 Continuous Small-Signal Averaged Model

There are two ways of obtaining the continuous small-signal averaged model, namely:

- either by starting from a previously obtained small-signal sampled-data model using the adequate transform;
- or by basing it upon the state-space representation of the large-signal averaged model.

The second manner is more "prudent" because it is direct and without a supplementary simplification degree. Therefore one proceeds to the use of a Taylor series development limited to the first order around the equilibrium (steady-state or quiescent) operating points chosen upon the method described in Sect. 2.2.4 of Chap. 2.

In order to establish the small-signal model in the frequency domain, one takes as a starting point its averaged state-space representation (in variations):

$$\begin{cases} \dot{\widetilde{\mathbf{x}}} = \mathbf{A}\widetilde{\mathbf{x}} + \mathbf{B}\widetilde{\mathbf{u}} \\ \widetilde{\mathbf{y}} = \mathbf{C}\widetilde{\mathbf{x}} + \mathbf{D}\widetilde{\mathbf{u}}, \end{cases}$$

with $\widetilde{\mathbf{x}}$ and $\widetilde{\mathbf{y}}$ being the state vector and output vector, respectively, and $\mathbf{A}$, $\mathbf{B}$, $\mathbf{C}$ and $\mathbf{D}$ the matrices of the averaged state-space model. The transfer matrix is computed as

$$\mathbf{H}(s) = \mathbf{C}(s\mathbf{I} - \mathbf{A})^{-1}\mathbf{B} + \mathbf{D}, \tag{4.32}$$

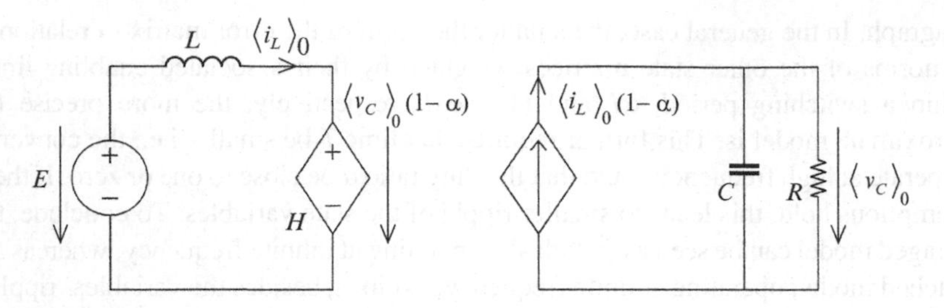

Fig. 4.8 Boost power stage averaged diagram

where **I** is the identity matrix of appropriate dimension. Note that if matrix **D** is nonzero, the transfer function relative degree is zero. Its poles are the eigenvalues of the state matrix **A**. In the multivariable case, Eq. (4.32) corresponds to a transfer matrix, i.e., a matrix of transfer functions. The transfer matrix contains the transfer functions from all exogenous variables to the system outputs, so it provides important information for control design. Software environments dedicated to linear system analysis can directly compute the transfer function/matrix corresponding to a given linear state-space representation.

Alternatively, by using the superposition principle in the previous state-space representation, one may obtain the transfer function of the channel of interest by zeroing the input variations on the other channels and expressing the output variable as a function of the single remaining input. An equivalent circuit diagram that can ease the computation by circuit transformations can thus be drawn; this is called an *operational* approach.

4.5.2 Sampled-Data Small-Signal Model

The sampled-data small-signal model can be established in different ways:

- either by starting from the large-signal sampled-data model of the recursive equation, which takes into account an entire operation time slot by making a differentiation analogous to the one employed for the continuous model (Brown and Middlebrook 1981);
- or by starting from the small-signal averaged model in either state-space or frequency representation.

Obviously, one must take care when choosing the sampling frequency in order to avoid the spectrum aliasing.

4.5.3 Example

Let us consider the boost converter already analyzed in this chapter in Sect. 4.2.4, whose averaged diagram is recalled in Fig. 4.8. This diagram allows the description of the averaged model of the chopper.

Let us consider the following notations for the averages of the state variables:

$$x_1 = \langle i_L \rangle_0 \text{ and } x_2 = \langle v_C \rangle_0.$$

The large-signal averaged model is given by

$$\begin{cases} \dot{x}_1 = \dfrac{1}{L}(E - x_2(1-\alpha)) \\ \dot{x}_1 = \dfrac{1}{C}x_1(1-\alpha) - \dfrac{x_2}{RC}. \end{cases}$$

The equilibrium point is computed by zeroing the derivatives in the previous relations:

$$\begin{cases} x_{1e} = \dfrac{E}{(1-\alpha_e)^2 R} \\ x_{2e} = \dfrac{E}{(1-\alpha_e)}, \end{cases}$$

where α_e is the duty ratio corresponding to the equilibrium point. These relations give the static behavior of the ideal boost power stage.

Differentiation of the large-signal model will be performed around the equilibrium point in order to extract the small-signal model. Let "$\widetilde{\cdot}$" denote the small variations around the equilibrium point. Consider for simplicity that the circuit supply E and the load R are constant. The other variables in the system may be written around the equilibrium point as $\alpha = \alpha_e + \widetilde{\alpha}$, $x_1 = x_{1e} + \widetilde{x_1}$ and $x_2 = x_{2e} + \widetilde{x_2}$. The state-space model can be rewritten as

$$\begin{cases} L\dot{\widetilde{x_1}} = E - x_{2e} - \widetilde{x_2} + x_{2e} \cdot \alpha_e + x_{2e} \cdot \widetilde{\alpha} + \alpha_e \cdot \widetilde{x_2} + \widetilde{x_2} \cdot \widetilde{\alpha} \\ C\dot{\widetilde{x_2}} = x_{1e} + \widetilde{x_1} - x_{1e} \cdot \alpha_e - x_{1e} \cdot \widetilde{\alpha} - \alpha_e \cdot \widetilde{x_1} - \widetilde{x_2}/R - \widetilde{x_1} \cdot \widetilde{\alpha}. \end{cases}$$

Note that in these two equations the last terms are very small with respect to the others and will be neglected. Moreover, in the equilibrium point the derivatives of the large-signal averaged model are zero:

$$\begin{cases} E - x_{2e}(1-\alpha_e) = 0 \\ x_{1e}(1-\alpha_e) - x_{2e}/R = 0, \end{cases}$$

leading to some simplification in the small-signal model:

$$\begin{cases} L\dot{\widetilde{x_1}} = x_{2e} \cdot \widetilde{\alpha} - (1-\alpha_e) \cdot \widetilde{x_2} \\ C\dot{\widetilde{x_2}} = (1-\alpha_e) \cdot \widetilde{x_1} - x_{1e} \cdot \widetilde{\alpha} - \widetilde{x_2}/R. \end{cases}$$

Further, by replacing the previously obtained steady-state variables x_{1e} and x_{2e}, one obtains that

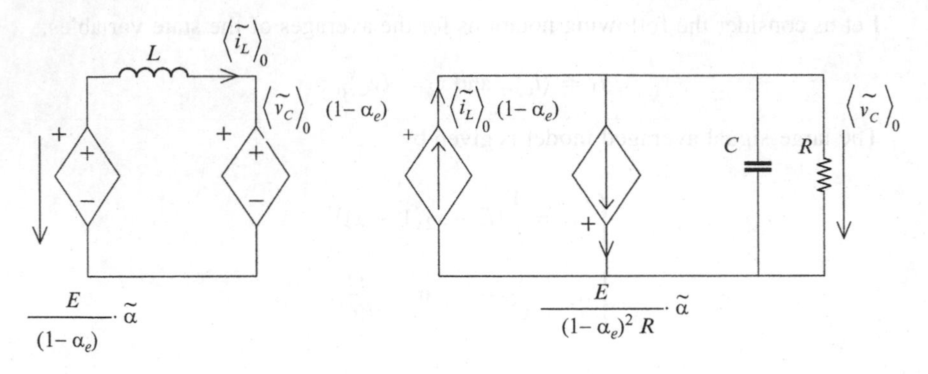

Fig. 4.9 Small-signal equivalent diagram of boost circuit

$$\dot{\widetilde{\mathbf{x}}} = \begin{bmatrix} 0 & -\dfrac{(1-\alpha_e)}{L} \\ \dfrac{(1-\alpha_e)}{C} & \dfrac{1}{RC} \end{bmatrix} \widetilde{\mathbf{x}} + \begin{bmatrix} \dfrac{E}{L(1-\alpha_e)} \\ -\dfrac{E}{(1-\alpha_e)^2 RC} \end{bmatrix} \widetilde{\alpha}. \tag{4.33}$$

The equivalent electrical diagram corresponding to the small-signal model in Eq. (4.33) is presented in Fig. 4.9. Equations (4.33) and Fig. 4.9 enable time-domain or frequency-domain dynamical behavior analysis of the converter and represent a base for its control design.

Remark. In obtaining the small-signal model one can also introduce the variation of the input voltage E and load resistance R, which stand for disturbance inputs. In this way, the influence of the disturbances over the state variables can be assessed in the context of a control design approach.

Note that the coupling element has a fixed transfer ratio, as α_e is constant (it represents the duty ratio's steady-state value). Therefore, it may be seen as an ideal transformer with ratio $(1-\alpha_e) : 1$. The diagram in Fig. 4.9 may be drawn in a different manner, allowing operational computation of the small-signal transfer functions (Erikson and Maksimović 2001) – see Fig. 4.10. In this figure $\alpha_e' = 1 - \alpha_e$.

Further, the voltage source and the inductor can be pushed through the transformer in order to obtain the circuit from Fig. 4.11 and to establish a direct input–output relation. The transfer function from the control input $\widetilde{\alpha}$ to the output variable $\langle \widetilde{v_C} \rangle_0$ is computed by writing Kirchhoff's equations in the circuit from Fig. 4.11:

$$\frac{x_{2e}}{\alpha_e'} \cdot \widetilde{\alpha} = \frac{sL}{\alpha_e'^2} \cdot \left(x_{1e} \cdot \widetilde{\alpha} + \langle \widetilde{v_C} \rangle_0 \cdot \frac{sCR+1}{R} \right) + \langle \widetilde{v_C} \rangle_0$$

or

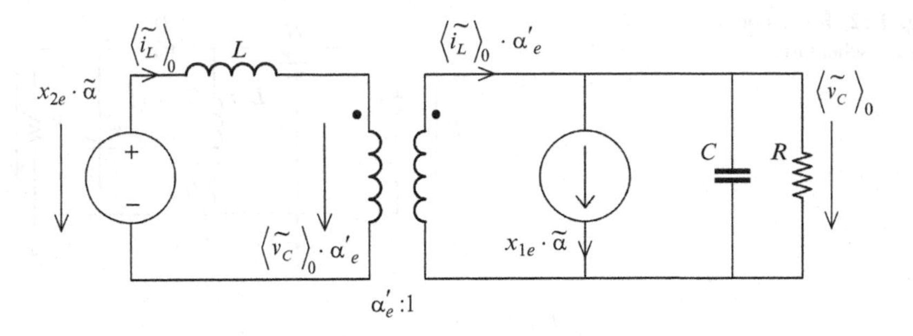

Fig. 4.10 Small-signal equivalent diagram of boost circuit redrawn

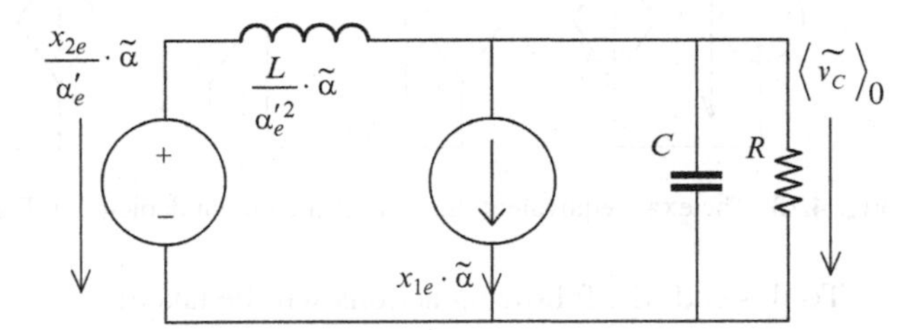

Fig. 4.11 Manipulation of left-side circuit elements: push through the transformer

$$\left(x_{2e}\cdot\alpha'_2 - sL\cdot x_{1e}\right)\cdot\tilde{\alpha} = \left(s^2LC + s\frac{L}{R} + \alpha_e'^2\right)\cdot\langle\tilde{v_C}\rangle_0.$$

This finally leads to the control-to-output transfer function:

$$H_{\alpha\to v_C}(s) = \frac{\langle\tilde{v_C}\rangle_0}{\tilde{\alpha}} = \frac{x_{2e}}{\alpha'_e}\cdot\frac{1 - s\dfrac{L\cdot x_{1e}}{x_{2e}\cdot\alpha'_e}}{s^2\dfrac{LC}{\alpha_e'^2} + s\dfrac{L}{R\alpha_e'^2} + 1},$$

whose elements (gain, time constants, etc.) depend on the steady-state operating point. Also, a right-half-complex-plane zero can be identified, which corresponds to nonminimum-phase behavior.

4.6 Case Study: Buck-Boost Converter

The buck-boost power stage from Fig. 4.12 has the following parameters: $C = 100$ μF, $L = 0.5$ mH, $R = 15$ Ω and $E = 100$ V. The input voltage E and the load resistance R are supposed constant. The averaging time window corresponds to switching frequency 100 kHz, i.e., $T = 10$ μs. The scope is to perform a large-signal time-domain analysis and a small-signal analysis in the frequency domain.

Fig. 4.12 Buck-boost circuit schematic

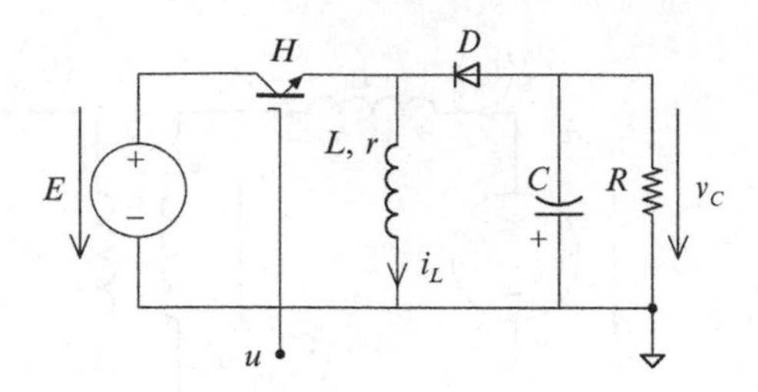

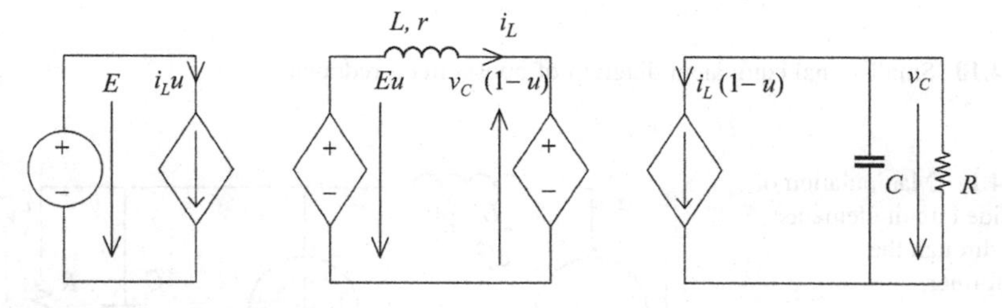

Fig. 4.13 The exact equivalent diagram of the circuit depicted in Fig. 4.12

To this end, the following actions will be taken:

(a) establish the switched model;
(b) build the approximate averaged model;
(c) deduce the error matrix **E** and quantize it;
(d) for a given α_e, compute the equilibrium point of the system;
(e) show that the large-signal averaged model is bilinear and deduce the state-space small-signal model;
(f) deduce the transfer function of the system having duty ratio variations as input and capacitor voltage v_C as output;
(g) show that the system has an unstable zero and analyze its effect in open loop.

Next, the above listed actions will be detailed, illustrated where relevant by numerical simulation results obtained in MATLAB®-Simulink®.

(a) Obtaining the switched model
Let u be the switching function (control input) acting on the converter. It takes the value 1 if the switch H is turned on and 0 if the switch is turned off. In order to render this matrix regular a small resistance r is added in series with the inductance L.

The exact model can be written as

$$\begin{cases} \dot{i}_L = \dfrac{1}{L}[Eu + v_C(1-u) - ri_L] \\ \dot{v}_C = \dfrac{1}{C}\left[-i_L(1-u) - \dfrac{v_C}{R}\right]. \end{cases} \tag{4.34}$$

The exact diagram corresponding to Eq. (4.34) is given in Fig. 4.13.

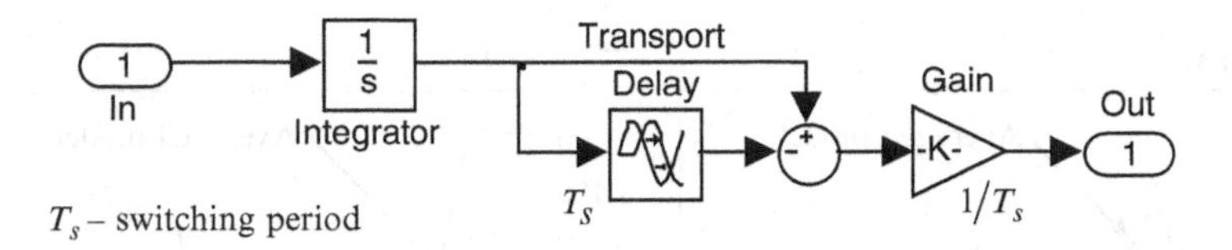

Fig. 4.14 Simulink® block diagram implementing the average of a signal on a given time window

(b) Obtaining the approximate averaged model

The exact averaged model can be deduced from the exact equivalent diagram (Fig. 4.13) by replacing the state variables i_L and v_C and the coupling terms $i_L \cdot (1-u)$, $v_C \cdot (1-u)$ and $E \cdot u$ by their sliding averages, without further development. When writing the coupling terms one can make the approximations

$$\begin{cases} \langle i_L(1-u)\rangle_0 \approx \langle i_L\rangle_0 \cdot (1-\alpha) \\ \langle v_C(1-u)\rangle_0 \approx \langle v_C\rangle_0 \cdot (1-\alpha). \end{cases}$$

Note that if voltage E is constant, one can write without approximation that $\langle Eu\rangle_0 = E\alpha$. The approximate averaged model results finally as

$$\begin{cases} \langle \dot{i_L}\rangle_0 = \dfrac{1}{L}\left[E\alpha + \langle v_C\rangle_0(1-\alpha) - r\langle i_L\rangle_0\right] \\ \langle \dot{v_C}\rangle_0 = \dfrac{1}{C}\left[-\langle i_L\rangle_0(1-\alpha) - \dfrac{\langle v_C\rangle_0}{R}\right]. \end{cases} \tag{4.35}$$

Note that model (4.35) has the same form as model (4.34); the difference is that the former contains averaged variables instead of switched ones. Model (4.35) is easy to implement in MATLAB®-Simulink®; Fig. 4.14 shows how to compute the average of a signal on a given time window by using Simulink® blocks.

Figure 4.15 presents a comparison between the switched and the averaged behavior of the converter in Fig. 4.12 (see Fig. 4.5); the time evolutions of state variables – inductor current i_L and capacitor voltage v_C – are shown. The switched time evolution in Fig. 4.15 was obtained by feeding model (4.34) with a PWM-modulated duty ratio signal $u(t)$, whereas the averaged time evolution resulted in response to feeding the same model with the average value of $u(t)$, computed according to Eq. (4.1) (see diagram in Fig. 4.14).

Figure 4.16 shows the averaged model response at duty ratio α's step variations. In this figure one can note that system behavior depends on operating point. The system's dynamic characteristics can be assessed: settle time, overshoot, etc. Remarks regarding the static behavior can also be made: the converter can perform as a voltage inverter, as a boost when the duty ratio is larger than 0.5 (losses taken into account), etc. Elements useful for design purposes can result from analyzing such time evolutions.

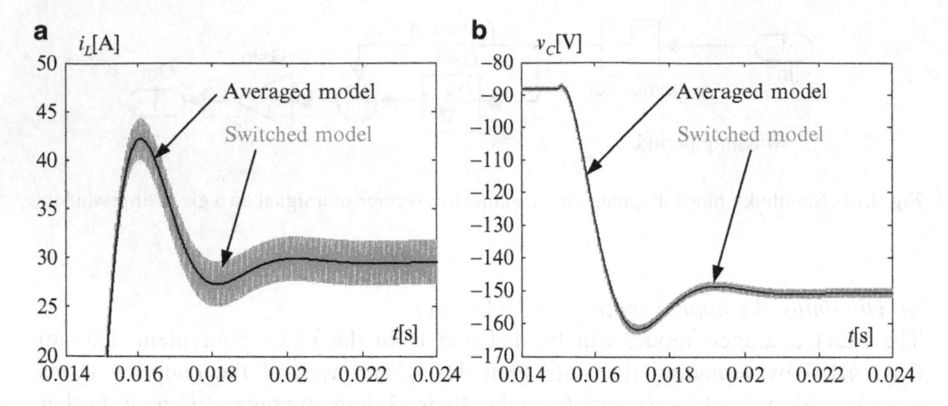

Fig. 4.15 Time evolution of state variables: averaged vs. switched model

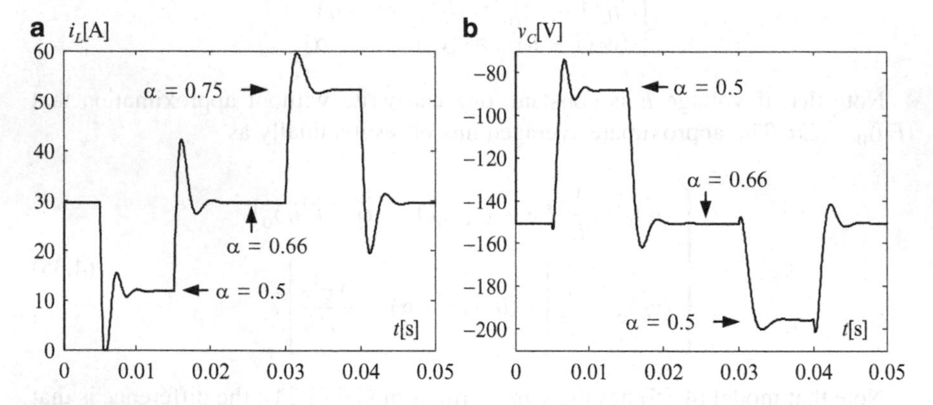

Fig. 4.16 Averaged model response at step variations of the duty ratio α

(c) Analysis of error between the exact sample-data model and the approximated averaged model

According to Eq. (4.31), the second-order error matrix **E** is computed as

$$\mathbf{E} = (\mathbf{A}_1 \cdot \mathbf{A}_2 - \mathbf{A}_2 \cdot \mathbf{A}_1)\frac{\alpha(1-\alpha)}{2}T^2,$$

where state matrices $\mathbf{A}_1$ and $\mathbf{A}_2$ correspond to the circuit configurations having switch H turned on and turned off, respectively, i.e., for $u = 1$ and $u = 0$:

$$\mathbf{A}_1 = \begin{bmatrix} -\dfrac{r}{L} & 0 \\ 0 & -\dfrac{1}{RC} \end{bmatrix}, \mathbf{A}_2 = \begin{bmatrix} -\dfrac{r}{L} & \dfrac{1}{L} \\ -\dfrac{1}{C} & -\dfrac{1}{RC} \end{bmatrix}.$$

Algebraic calculus gives that

$$\mathbf{E} = \frac{\alpha(1-\alpha)}{2}T^2 \cdot \begin{bmatrix} 0 & -\dfrac{r}{L^2} + \dfrac{1}{RLC} \\ \dfrac{1}{RC^2} - \dfrac{r}{LC} & 0 \end{bmatrix}.$$

Provided numerical data, by taking $\alpha = 0.6$ and neglecting r one obtains:

$$\mathbf{A}_1 = \begin{bmatrix} 0 & 0 \\ 0 & -10^4/1.5 \end{bmatrix}, \mathbf{A}_2 = \begin{bmatrix} 0 & 10^4/5 \\ -10^4 & -10^4/1.5 \end{bmatrix},$$

$$\mathbf{E} = \begin{bmatrix} 0 & 0.16 \cdot 10^{-4} \\ 0.8 \cdot 10^{-4} & 0 \end{bmatrix}.$$

Using matrix norm computation (largest singular value) as provided by MATLAB®, one obtains the following values: $\|\mathbf{E}\| = 8 \cdot 10^{-5}$, $\|\mathbf{A}_1\| \approx 666.67$, $\|\mathbf{A}_2\| \approx 10^4$. The ratio between the norms of matrices $\mathbf{A}_1$ and $\mathbf{A}_2$, weighted by their respective enabling times, and the norm of matrix $\mathbf{E}$ is of the order of 10^{-2}. If switching frequency is reduced to one-tenth its value, i.e., 10 kHz, then the ratios increase by ten. If one significantly reduces this frequency, e.g., by taking it 1 kHz instead of 100 kHz, then the ratios get close to 1, which is critical.

(d) Averaging error analysis

Averaging error is defined as the difference between two time signals, the output of the averaged model and the average, in the usual sense, of the output of the switched model (see time signal $\boldsymbol{\varepsilon}(t) = \langle \mathbf{x} \rangle_0(t) - \mathbf{x}_0(t)$ in Fig. 4.5).

Numerical simulations in MATLAB®-Simulink® environment allow the averaging error to be highlighted in the time evolutions of both state variables, i_L and v_C. The relation of these evolutions with the output of the switched model can also be analyzed.

Figure 4.17 shows the dependence of averaging error on switching frequency for the two state variables of the converter – current (Fig. 4.17a) and voltage (Fig. 4.17b) – in both dynamical and steady-state regimes following a step variation of the duty ratio α.

Figure 4.18 contains zoomed plots of the time evolutions illustrated in Fig. 4.16, or more precisely, those corresponding to the dynamical regime occurring when a duty ratio step from $\alpha = 0.75$ to $\alpha = 0.66$ has been applied. As expected intuitively, one can see that averaging error decreases as switching frequency increases (Figs. 4.18(a) vs. (b) and (c) vs. (d)).

(e) Obtaining the steady-state model

Equilibrium points are obtained by zeroing the derivatives of the system described by Eq. (4.35). One obtains

$$\begin{cases} \langle i_L \rangle_{0e} = \dfrac{\alpha_e}{(1-\alpha_e)^2 R} E \\ \langle v_C \rangle_{0e} = -\dfrac{\alpha_e}{(1-\alpha_e)} E. \end{cases} \tag{4.36}$$

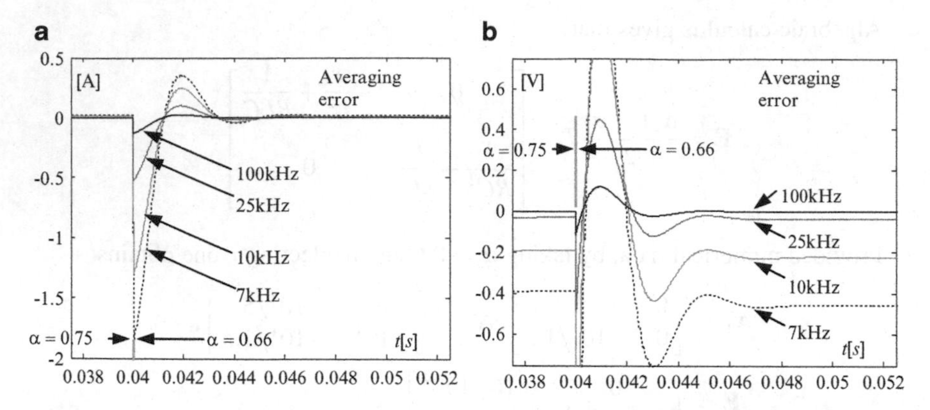

Fig. 4.17 Illustration of averaging error on dynamical regimes of state variables of a buck-boost converter: (**a**) inductor current; (**b**) capacitor voltage

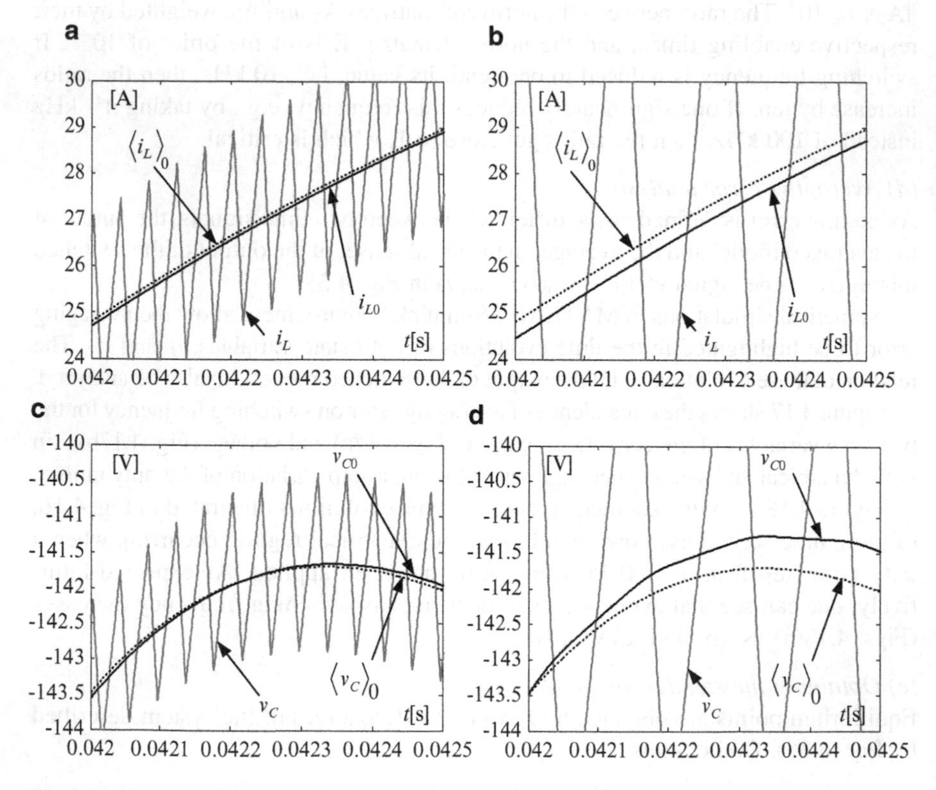

Fig. 4.18 Dynamical regime of buck-boost converter state variables in response to step variations of duty ratio – switched evolution, average of switched evolution and averaged model output: (**a**) inductor current for switching frequency 25 kHz; (**b**) inductor current for switching frequency 7 kHz; (**c**) capacitor voltage for switching frequency 25 kHz; (**d**) capacitor voltage for switching frequency 7 kHz

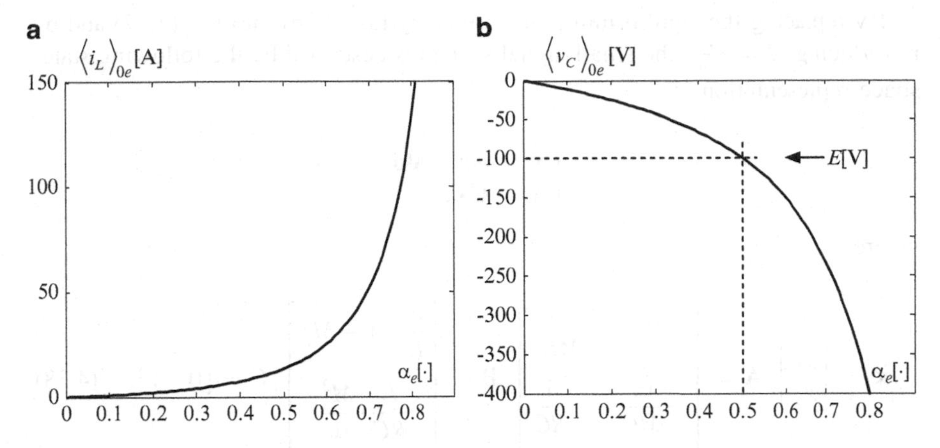

Fig. 4.19 Steady-state characteristics of buck-boost converter

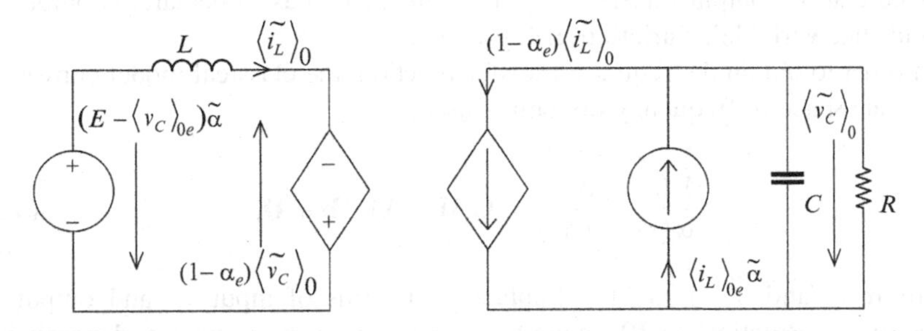

Fig. 4.20 Small-signal equivalent diagram of buck-boost circuit

Equations (4.36) give the nonlinear steady-state behavior of the ideal buck-boost circuit, as illustrated in Fig. 4.19. For example, concerning voltage, circuit output is smaller than input voltage E for $\alpha_e < 0.5$ and larger otherwise (Fig. 4.19b).

(f) Obtaining the small-signal model

Model (4.35) is bilinear and therefore nonlinear. It will be differentiated around the equilibrium point (4.36). As previously, for reasons of simplicity, let the state variables be denoted as $x_1 = \langle i_L \rangle_0$ and $x_2 = \langle v_C \rangle_0$. The index e denotes the equilibrium point and $\tilde{}$ denotes the small variations around it. Proceeding as usual with the differentiation gives

$$\begin{cases} \dot{\tilde{x_1}} = \dfrac{1}{L}(1-\alpha_e)\tilde{x_2} + \dfrac{1}{L}(-x_{2e} + E)\tilde{\alpha} \\ \dot{\tilde{x_2}} = -\dfrac{1}{C}(1-\alpha_e)\tilde{x_1} - \dfrac{1}{RC}\tilde{x_2} + \dfrac{1}{C}x_{1e}\tilde{\alpha}, \end{cases} \tag{4.37}$$

which allows one to obtain the small-signal equivalent diagram in Fig. 4.20.

By replacing the equilibrium points given by Eq. (4.36) into Eq. (4.37) and by introducing $M = \frac{\alpha_e}{1-\alpha_e}$, the small-signal system is described by the following state-space representation:

$$\begin{cases} \dot{\widetilde{\mathbf{x}}} = \mathbf{A}\widetilde{\mathbf{x}} + \mathbf{B}\widetilde{\alpha} \\ \widetilde{y} = \mathbf{C}\widetilde{\mathbf{x}}, \end{cases}$$

where

$$\widetilde{\mathbf{x}} = \begin{bmatrix} \widetilde{x_1} \\ \widetilde{x_2} \end{bmatrix}, \mathbf{A} = \begin{bmatrix} 0 & \dfrac{\alpha_e}{ML} \\ -\dfrac{\alpha_e}{MC} & -\dfrac{1}{RC} \end{bmatrix}, \mathbf{B} = \begin{bmatrix} \dfrac{E}{L} \cdot (1+M) \\ \dfrac{E}{RC} \cdot \dfrac{M^2}{\alpha_e} \end{bmatrix}, \mathbf{C} = [0 \quad 1]. \quad (4.38)$$

Note that the output matrix **C** has been defined so as to declare as output the second state variable's variation, i.e., $\widetilde{y} \equiv \widetilde{v_C}$.

In order to obtain the required transfer function, the classical model conversion from state space to frequency domain is used:

$$\frac{\widetilde{Y}(s)}{\widetilde{\alpha}(s)} = \frac{N(s)}{D(s)} = \mathbf{C}(s\mathbf{I} - \mathbf{A})^{-1}\mathbf{B} + \mathbf{D}, \quad (4.39)$$

where $\widetilde{\alpha}(s)$ and $\widetilde{Y}(s)$ are the Laplace transforms of input $\widetilde{\alpha}$ and output $\widetilde{y}$, respectively. Relation (4.39) denotes a transfer function whose denominator, denoted here by $D(s)$, is the determinant of matrix $(s\mathbf{I} - \mathbf{A})$ and whose numerator, denoted by $N(s)$, is given by

$$N(s) = \mathbf{C}(s\mathbf{I} - \mathbf{A})^{*}\mathbf{B},$$

where $\cdot^{\,*}$ denotes the adjoint matrix. Further, one obtains

$$D(s) = \det[s\mathbf{I} - \mathbf{A}] = \det \begin{bmatrix} s & -\dfrac{\alpha_e}{ML} \\ \dfrac{\alpha_e}{MC} & s + \dfrac{1}{RC} \end{bmatrix} = s^2 + \frac{s}{RC} + \frac{\alpha_e^2}{M^2LC},$$

$$N(s) = [0 \quad 1] \cdot \begin{bmatrix} s + \dfrac{1}{RC} & \dfrac{\alpha_e}{ML} \\ -\dfrac{\alpha_e}{MC} & s \end{bmatrix} \cdot \begin{bmatrix} \dfrac{E}{L}(1+M) \\ \dfrac{E}{RC} \cdot \dfrac{M^2}{\alpha_e} \end{bmatrix} = \frac{EM^2}{RC\alpha_e} s - \frac{\alpha_e}{MC} \cdot \frac{E}{L}(1+M).$$

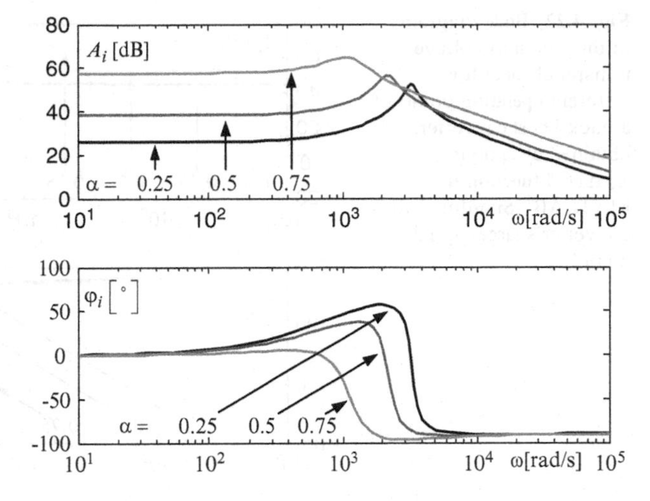

Fig. 4.21 Bode diagrams of duty-ratio-to-current transfer channel for different operating points of a buck-boost converter, obtained by applying `linmod` function in MATLAB®-Simulink® to converter's large-signal model

After the necessary algebra, one finally obtains the required transfer function:

$$\frac{\widetilde{Y}(s)}{\widetilde{U}(s)} = -\frac{M^2 E}{\alpha_e^2} \cdot \frac{1 - \tau s}{s^2 \frac{1}{\omega_n^2} + s \frac{2\xi}{\omega_n} + 1} \tag{4.40}$$

with

$$\tau = \frac{M^2}{\alpha_e} \cdot \frac{L}{R}, \omega_n = \frac{\alpha_e}{M} \cdot \sqrt{\frac{1}{LC}}, \xi = \frac{M}{2\alpha_e} \cdot \sqrt{\frac{L}{R^2 C}}.$$

If one wants to analyze the dynamical transfer between the duty ratio α and the current inductor i_L the above development must be reiterated by defining the output matrix $\mathbf{C} = [1 \quad 0]$ in Eqs. (4.38) and (4.39).

Figures 4.21 and 4.22 show the Bode diagrams of duty-ratio-to-current and duty-ratio-to-voltage transfer channel, respectively, plotted as family curves for different operating points characterized by different values of the duty ratio α.

Notations within these figures denote the following:

$$\begin{cases} A_{i/v} = 20 \cdot \log\left|H_{i/v}(j\omega)\right| \text{ [dB]} \\ \varphi_{i/v} = \arg\left(H_{i/v}(j\omega)\right) \text{ [°]} \end{cases},$$

where $H_{i/v}(j\omega)$ denotes frequency responses of duty-ratio-to-current and duty-ratio-to-voltage influence channel, respectively, in the small-signal model.

(g) Analysis of the nonminimum phase behavior

Analyzing Eq. (4.40) reveals that the transfer function numerator, $N(s)$, has a zero with positive real part (+ τ). This results in the open-loop step response taking a particular shape, namely, the system first evolves in the opposite sense in relation to

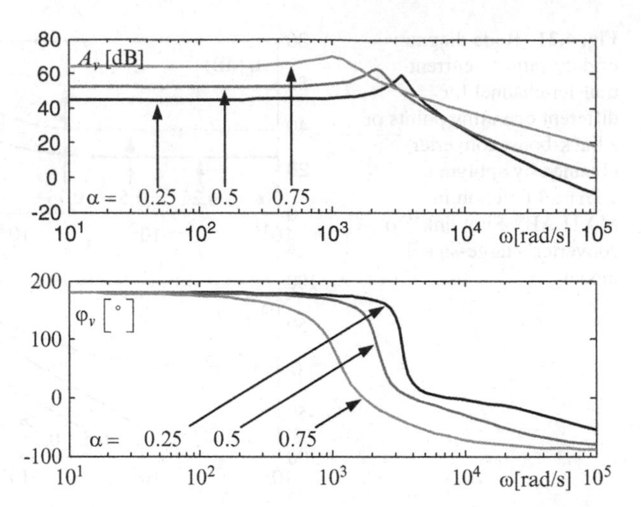

Fig. 4.22 Bode diagrams of duty-ratio-to-voltage transfer channel for different operating points of a buck-boost converter, obtained by applying `linmod` function in MATLAB®-Simulink® to converter's large-signal model

the input variation, then comes back to the positive sense. This is called a *nonminimum-phase response* and it can be observed both in the voltage time evolution (Fig. 4.16b) and in its frequency response (Fig. 4.22).

Suppose that the system operates in a closed loop with a proportional controller K. It will not be possible to compensate exactly this zero in closed loop, because the controller itself will be unstable. If a proportional controller is to be employed, a too small gain leads to a non-negligible steady-state error; the other way round, if the gain is too high, the closed-loop system can become unstable due to plant nonminimum-phase behavior.

(h) Case of discontinuous conduction

Reducing too drastically the switching frequency can result in discontinuous conduction. Figure 4.23 presents such a case for the considered buck-boost converter, where the switching frequency 2 kHz corresponds to the boundary between the continuous- and the discontinuous-conduction mode (ccm and dcm, respectively). One can see that the inductor current becomes zero during certain time intervals (Fig. 4.23a) and the capacitor voltage has a first-order dynamic behavior (Fig. 4.23b). This is due to the fact that the discontinuous-conduction case leads to reduction in the dynamical system's order. Furthermore, the steady-state values of state variables are not correctly obtained through the averaged model. The conclusion is that the averaged model is not suitable for describing this case, which is one of its limitations.

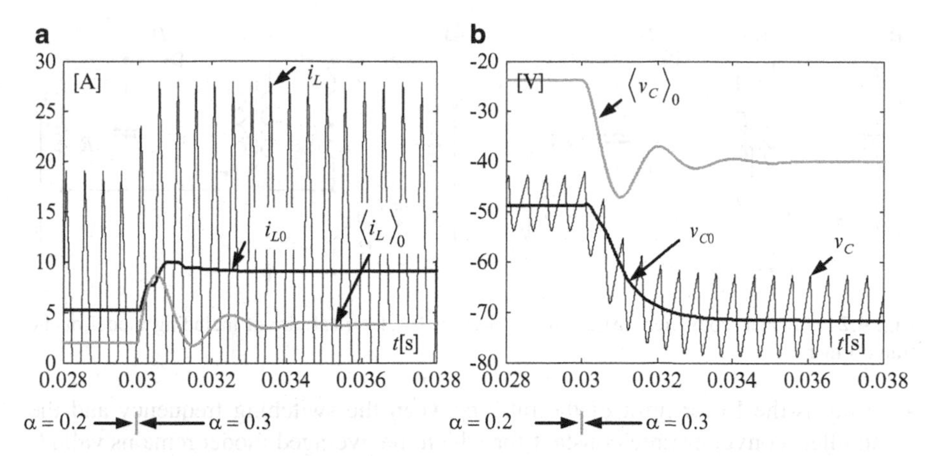

Fig. 4.23 Evolutions of state variables in discontinuous-conduction case, obtained at the boundary between ccm and dcm (2 kHz switching frequency): (**a**) interrupted current; (**b**) first-order behavior of the capacitor voltage

4.7 Advantages and Limitations of the Averaged Model. Conclusion

Averaged models can form the basis for obtaining normalized models by making suitable variable changes in both state variables and time (Sira-Ramírez and Silva-Ortigoza 2006). Discrete-time averaged models, useful for digital control design, can also be obtained based upon sampled-data models (Maksimović and Zane 2007).

Among the advantages of the averaged model in particular is the facility of building and implementing one, as well as its good approximation precision when the state variables are lightly rippled. The other way round, this model begins to lose its precision when good filtering conditions are no longer satisfied. Moreover, the model is not useful for converters having one or multiple AC stages (zero average value of certain variables). It also exhibits difficulties in the discontinuous-conduction case, as illustrated by the presented case study. Thus, as the concerned variable becomes zero at each switching cycle, on a nonzero time interval it cannot be expressed as a large-signal recurrent form, and therefore it cannot be approximated by the averaged model.

Finally, there are some questions that the scientific community has not yet answered (Krein and Bass 1990). The main ones are:

- What conditions guarantee a satisfactory approximation?
- Is the averaged model trajectory a sliding average along the original circuit?
- Are averaged model and original system dynamic characteristics identical?
- Can the original solution be recovered starting from the approximated solution?
- Is the approximation valid for large signals?

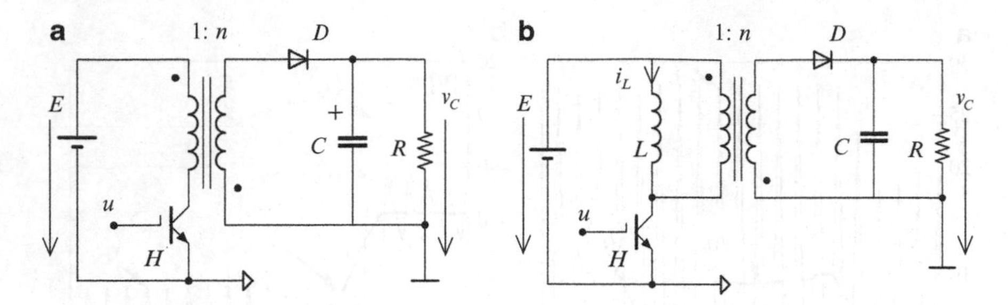

Fig. 4.24 Flyback converter: (**a**) electrical circuit; (**b**) electrical circuit including model of the transformer

- What is the lower limit of the ratio between the switching frequency and the smallest converter time constant for which the averaged model remains valid?
- Can the model be used in closed loop?

Some of these questions have partial answers; the reader is referred to the relevant literature. Taking into account the above issues, alternative solutions to the averaged model have been developed, such as the steady-state *first-order-harmonic model* (Sanders et al. 1990; Kazimierczuk and Wang 1992), the *reduced-order averaged model* (ROAM) (Chetty 1982; Sun and Grotstollen 1992), or generalizations of the averaged model.

Some of these solutions, such as developed by Krein et al. (1990), bring greater precision, but at the price of implementing and building a complexity difficult to accept. The first-order harmonic dynamic approach solves the problem of AC variables presence, but it cannot solve the discontinuous conduction problem. In addition, this latter technique does not add precision in the DC-DC conversion case, although it provides a more complex solution. Nevertheless, modeling in the first-order harmonic dynamic sense finds a natural application in the field of power converters having both DC and AC stages (rectifiers, inverters, resonant power sources, etc.). The discontinuous conduction problem has no solution (in the modeling sense), unless the model of the concerned variable is eliminated. This can be achieved by using the ROAM technique. These techniques will be approached in the following chapters.

Problems

Problems 4.1, 4.2, and 4.3 are given with solutions. Problems 4.4, 4.5, and 4.6 are left as exercises.

Problem 4.1. Flyback converter

Let us consider the flyback converter in Fig. 4.24a, allowing isolated non-inverting boost topology via a transformer with ratio n.

It is supposed that the transformer has negligible primary and secondary winding resistances and leakage inductances, whereas it has nonzero core reluctance. These assumptions lead to transformer modeling as a “two-winding inductor”, i.e., represented by a magnetizing inductance referred to the primary winding coupled

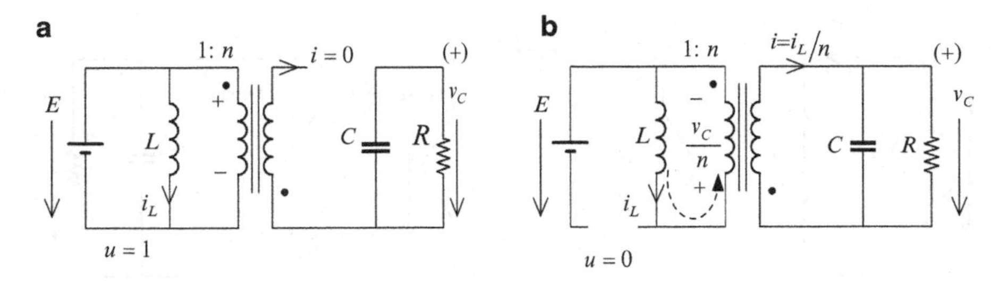

Fig. 4.25 Two configurations of flyback converter, corresponding respectively to the two states of the switch network composed of H and D: (**a**) H turned on and D blocked ($u = 1$); (**b**) H turned off and D in conduction ($u = 0$)

with an ideal transformer, as in Fig. 4.24b (Erikson and Maksimović 2001). It is required to address the following points.

(a) Obtain the averaged model and the corresponding equivalent diagram by using Algorithm 4.1.
(b) Compute the steady-state model.
(c) Considering that input voltage E varies (representing a disturbance input), deduce the small-signal state-space model and draw the associated equivalent diagram.
(d) Using the previously obtained small-signal model, get expressions for the transfer functions representing the influence from the duty ratio (control input) and the input voltage E (disturbance input) to the state variables.

Solution. (a) The averaged model is deduced from the switched model. This latter results from an analysis of the circuit operation; it has two configurations: switch H turned on and diode D blocked (switching function u takes value 1), and switch H turned off and diode D conducting (switching function u takes value 0) (see Fig. 4.25).

When the diode is blocked ($u = 1$), the transformer's primary and secondary currents are zero. When switch H is turned off ($u = 0$) the magnetizing inductor is decoupled from the input source, the primary voltage is v_C/n and the secondary current is i_L/n. Hence, the system has two state variables, current i_L and voltage v_C, whose dynamics are described by the following relations:

$$u = 1 : \begin{cases} L\dfrac{di_L}{dt} = E \\ C\dfrac{dv_C}{dt} = -\dfrac{v_C}{R} \end{cases} \qquad u = 0 : \begin{cases} L\dfrac{di_L}{dt} = -\dfrac{v_C}{n} \\ C\dfrac{dv_C}{dt} = \dfrac{i_L}{n} - \dfrac{v_C}{R}. \end{cases} \tag{4.41}$$

Equations (4.41) can be expressed by a single set, which represents the switched model of the flyback converter:

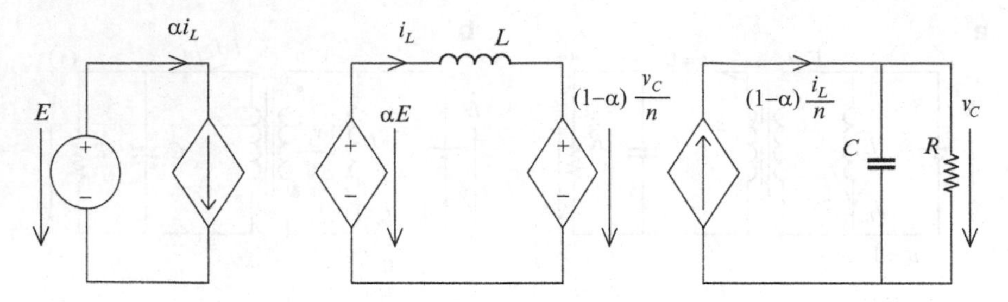

Fig. 4.26 Equivalent diagram of flyback converter averaged model

$$\begin{cases} L \cdot \dot{i_L} = -(1-u) \cdot \dfrac{v_C}{n} + u \cdot E \\ C \cdot \dot{v_C} = (1-u) \cdot \dfrac{i_L}{n} - \dfrac{v_C}{R}. \end{cases} \tag{4.42}$$

The averaged model results directly from model (4.42) by replacing switching function u by its average, i.e., the duty ratio denoted by α. The equivalent diagram of the averaged model can be drawn as shown in Fig. 4.26. The three subcircuits are linked by two couplings, each of which acts as an AC + DC transformer, with ratio α (input voltage side) and $(1 - \alpha)/n$ (load side).

(b) The steady-state model results by zeroing the derivatives in Eq. (4.42). Equilibrium values of the state variables are denoted by subscript e:

$$v_{Ce} = n \cdot \frac{E_e \cdot \alpha_e}{1 - \alpha_e}, \quad i_{Le} = n^2 \cdot \frac{\alpha_e}{(1-\alpha_e)^2} \cdot \frac{E_e}{R}. \tag{4.43}$$

The first equation of (4.43) shows that the output voltage has expression similar to that of the buck-boost case, except it is positive and contains a supplementary multiplying factor n.

(c) The small-signal model may result from perturbation and linearization. Let us replace the perturbed input and state variables into Eq. (4.42): $\alpha = \alpha_e + \widetilde{\alpha}$, $E = E_e + \widetilde{E}$, $i_L = i_{Le} + \widetilde{i_L}$ and $v_C = v_{Ce} + \widetilde{v_C}$. By using (4.43) and neglecting products of small variations, one obtains, after some simple algebra:

$$\begin{cases} L \cdot \dot{\widetilde{i_L}} = -\dfrac{1-\alpha_e}{n} \cdot \widetilde{v_C} + \left(\dfrac{v_{Ce}}{n} + E \right) \cdot \widetilde{\alpha} + \alpha_e \cdot \widetilde{E} \\ C \cdot \dot{\widetilde{v_C}} = \dfrac{1-\alpha_e}{n} \cdot \widetilde{i_L} \quad - \dfrac{1}{RC} \cdot \widetilde{v_C} \quad - \dfrac{i_{Le}}{n} \cdot \widetilde{\alpha}. \end{cases} \tag{4.44}$$

These lead to the small-signal AC equivalent diagram in Fig. 4.27.

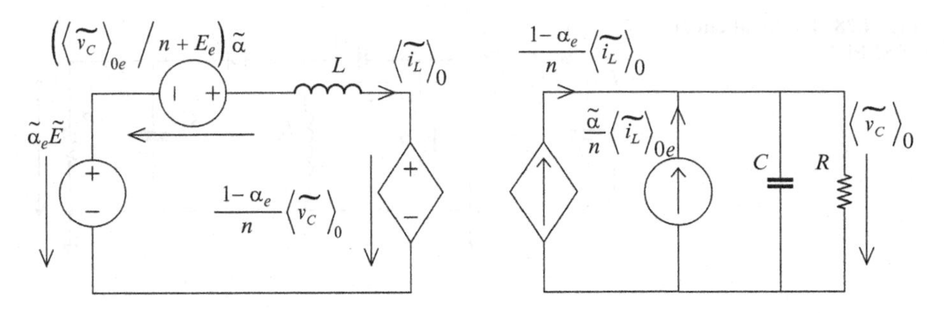

Fig. 4.27 Small-signal AC equivalent diagram of flyback converter

Equations (4.44) allow state-space matrix representation as in Eq. (4.45), emphasizing the small-signal state vector $\mathbf{x} = \begin{bmatrix} \widetilde{i_L} & \widetilde{v_C} \end{bmatrix}^T$, the small-signal input vector $\mathbf{u} = \begin{bmatrix} \widetilde{\alpha} & \widetilde{E} \end{bmatrix}^T$, the state matrix $\mathbf{A}$ and the input matrix $\mathbf{B}$:

$$\underbrace{\begin{bmatrix} \dot{\widetilde{i_L}} \\ \dot{\widetilde{v_C}} \end{bmatrix}}_{\dot{\mathbf{x}}} = \underbrace{\begin{bmatrix} 0 & -\dfrac{1-\alpha_e}{nL} \\ \dfrac{1-\alpha_e}{nC} & -\dfrac{1}{RC} \end{bmatrix}}_{\mathbf{A}} \cdot \underbrace{\begin{bmatrix} \widetilde{i_L} \\ \widetilde{v_C} \end{bmatrix}}_{\mathbf{x}} + \underbrace{\begin{bmatrix} \dfrac{1}{L} \cdot \left(\dfrac{v_{Ce}}{n} + E_e \right) & \dfrac{\alpha_e}{L} \\ -\dfrac{i_{Le}}{nC} & 0 \end{bmatrix}}_{\mathbf{B}} \cdot \underbrace{\begin{bmatrix} \widetilde{\alpha} \\ \widetilde{E} \end{bmatrix}}_{\mathbf{u}}. \tag{4.45}$$

(d) Setting the output vector to be identical to the state vector, i.e., $\mathbf{y} \equiv \mathbf{x}$, results in output matrix $\mathbf{C}$ being the 2×2 identity matrix.

Based upon the matrix state representation in (4.45) and on the definition of the output matrix $\mathbf{C}$, one can compute the transfer matrix $\mathbf{H}(s)$, containing the transfer functions of the four input-to-output channels. To this end, the following well-known formula can be used:

$$\mathbf{H}(s) = \mathbf{C} \cdot (s\mathbf{I} - \mathbf{A})^{-1} \cdot \mathbf{B}.$$

After performing the computation and knowing that element $H_{ij}(s)$ of matrix $\mathbf{H}(s)$ represents the Laplace image of the transfer from input j to output i, the expression of the transfer matrix can be written as

$$\mathbf{H}(s) = \begin{bmatrix} H_{\alpha \to i_L}(s) & H_{E \to i_L}(s) \\ H_{\alpha \to v_C}(s) & H_{E \to v_C}(s) \end{bmatrix}, \tag{4.46}$$

where the transfer functions are

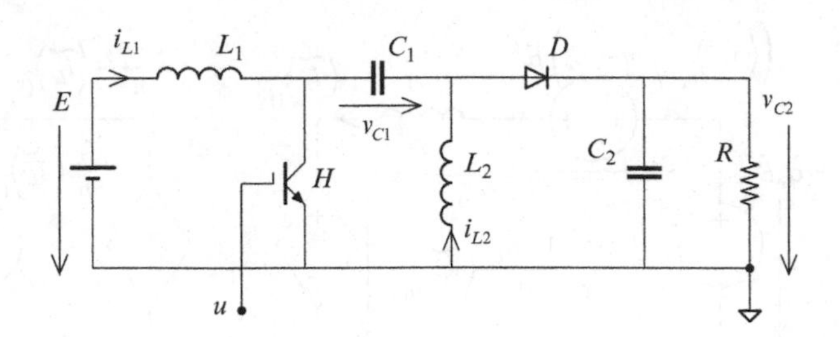

Fig. 4.28 Electrical circuit of SEPIC

$$\begin{cases} H_{\alpha\to i_L}(s) = k_{\alpha i_L}\cdot\dfrac{T_{\alpha i_L}s+1}{s(T_0 s+1)} & H_{E\to i_L}(s)=\dfrac{k_{Ei_L}}{s} \\ H_{\alpha\to v_C}(s) = k_{\alpha v_C}\cdot\dfrac{1-T_{\alpha v_C}s}{s(T_0 s+1)} & H_{Ev_C}(s)=\dfrac{k_{Ev_C}}{s(T_0 s+1)}, \end{cases} \tag{4.47}$$

with the different gain and time constant notations standing for

$$\begin{cases} T_0 = RC \\ k_{\alpha i_L} = \dfrac{v_{Ce}+nE_e+(1-\alpha_e)\cdot i_{Le}/n}{nL} & T_{\alpha i_L} = \dfrac{RC(v_{Ce}+nE_e)}{v_{Ce}+nE_e+(1-\alpha_e)\cdot i_{Le}/n} \\ k_{\alpha v_C} = \dfrac{(1-\alpha_e)\cdot(v_{Ce}/n+E_e)}{n\cdot(L/R)} & T_{\alpha v_C} = \dfrac{i_{Le}L}{(1-\alpha_e)\cdot(v_{Ce}/n+E_e)} \\ k_{Ei_L} = \dfrac{\alpha_e}{L} & k_{Ev_C} = \dfrac{\alpha_e(1-\alpha_e)}{n\cdot(L/R)}. \end{cases} \tag{4.48}$$

Equations (4.47) and (4.48) indicate that the small-signal model is a linear-parameter-varying one because its parameters depend on the steady-state operating point, as expected. Note also that all the four transfer functions have a pole at the origin. As in the case of the buck-boost converter, the capacitor voltage exhibits nonminimum-phase behavior in response to duty ratio variation, as shown by the associated transfer function $H_{\alpha\to v_C}(s)$ having right-half-plane zero (see Eq. (4.47)). Based upon expressions (4.47) of the transfer functions, one can draw the corresponding frequency responses in the form of Bode diagrams.

Problem 4.2. Single-ended primary-inductor converter (SEPIC)
Let us consider the single-ended primary-inductor converter (SEPIC) in Fig. 4.28, allowing noninverting up/down voltage conversion using two uncoupled inductors. The circuit is driven by a single binary switching function and hence has two configurations (Sira-Ramírez and Silva-Ortigoza 2006).

The following points must be addressed.

(a) Obtain the averaged model and the corresponding equivalent diagram.
(b) Compute the steady-state model.

Fig. 4.29 The two configurations of the SEPIC power stage, corresponding respectively to the two states of the switch H: (**a**) H turned on ($u = 1$); (**b**) H turned off ($u = 0$)

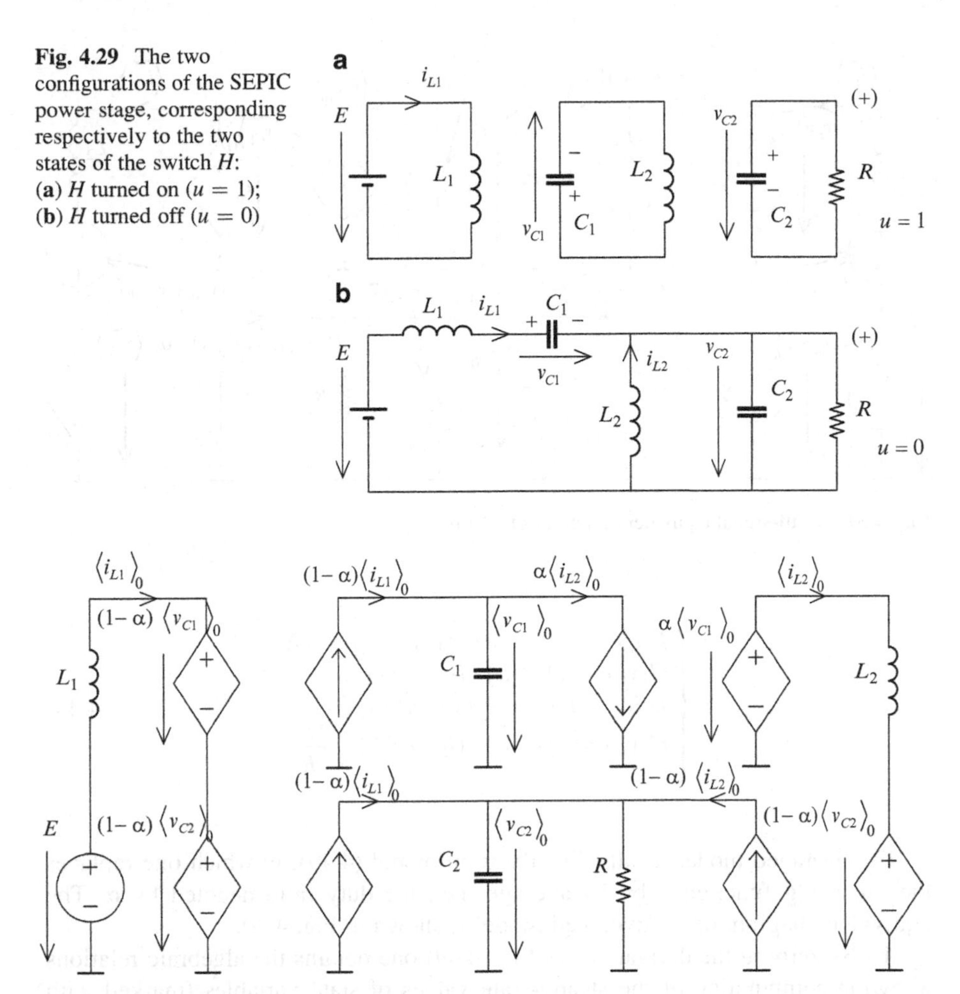

Fig. 4.30 Equivalent diagram of SEPIC averaged model

(c) Assuming that input voltage E varies (being a disturbance input), deduce the small-signal state-space model and draw the associated equivalent diagram.

(d) Using the previously obtained small-signal model, get the expression of the transfer function representing the influence from the duty ratio (control input) to the voltage v_{C2} as output variable.

Solution. (a) The averaged model is deduced from the switched model, which results from analyzing the two configurations of the circuit: for switch H turned on and diode D blocked (switching function u takes value 1 – Fig. 4.29a), and for switch H turned off and diode D conducting (switching function u takes value 0 – Fig. 4.29b).

The circuit is described by four state variables, the two inductor currents i_{L1} and i_{L2}, and the two capacitor voltages v_{C1} and v_{C2}. Operation of the two configurations can be merged to yield the switched model given in Eq. (4.49).

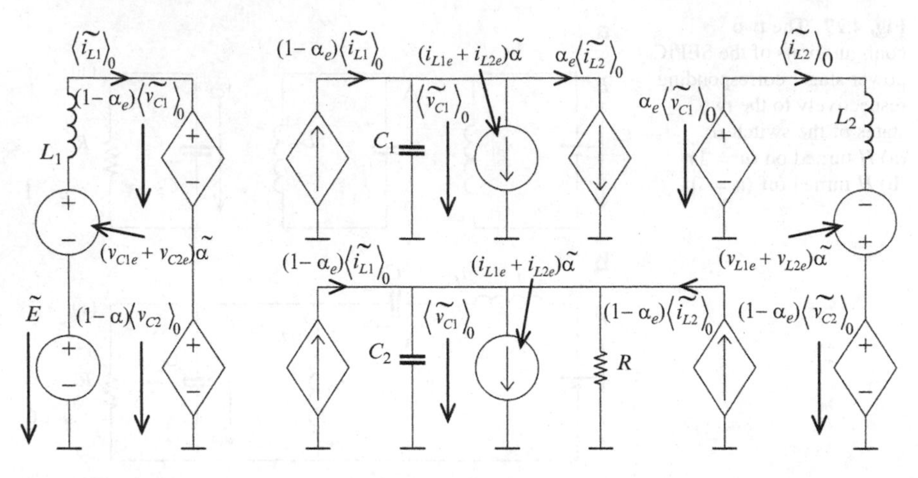

Fig. 4.31 Small-signal equivalent diagram of SEPIC

$$\begin{cases} L_1\dot{i_{L1}} = -(1-u)\cdot(v_{C1}+v_{C2})+E \\ C_1\dot{v_{C1}} = (1-u)\cdot i_{L1} - u\cdot i_{L2} \\ L_2\dot{i_{L2}} = u\cdot v_{C1} - (1-u)\cdot v_{C2} \\ C_2\dot{v_{C2}} = (1-u)\cdot(i_{L1}+i_{L2}) - \dfrac{v_{C2}}{R} \end{cases} \tag{4.49}$$

The averaged model results directly from model (4.49), in which one replaces the switching function u by its average, i.e., the duty ratio denoted by α. The equivalent diagram of the averaged model is shown in Fig. 4.30.

(b) By zeroing the derivatives in Eq. (4.49) one obtains the algebraic relations allowing computation of the steady-state values of state variables (marked with subscript e) as follows:

$$i_{L1e} = \left(\frac{\alpha_e}{1-\alpha_e}\right)^2 \cdot \frac{E_e}{R}, \quad v_{C1e} = E_e, \quad i_{L2e} = \frac{\alpha_e}{1-\alpha_e}\cdot\frac{E_e}{R}, \quad v_{C2e} = \frac{\alpha_e}{1-\alpha_e}\cdot E_e\,. \tag{4.50}$$

Equations (4.50) show that the steady-state value of v_{C1} equals the input voltage E, with the steady-state output voltage v_{C2e} corresponding to the noninverting buck-boost topology.

(c) The perturb-and-linearize method may be used to get the small-signal model. Thus, the perturbed input and state variables $\alpha = \alpha_e + \widetilde{\alpha}$, $E = E_e + \widetilde{E}$, $i_{L1} = i_{L1e} + \widetilde{i_{L1}}$, $v_{C1} = v_{C1e} + \widetilde{v_{C1}}$, $i_{L2} = i_{L2e} + \widetilde{i_{L2}}$ and $v_{C2} = v_{C2e} + \widetilde{v_{C2}}$ are replaced in

Eq. (4.49). After computations, by using Eq. (4.50) and neglecting products of small variations, one obtains Eq. (4.51), whose associated equivalent diagram is given in Fig. 4.31.

$$\begin{cases} L_1\dot{\widetilde{i_{L1}}} = & -(1-\alpha_e)\widetilde{v_{C1}} & -(1-\alpha_e)\widetilde{v_{C2}} & +(v_{C1e}+v_{C2e})\widetilde{\alpha} & +\widetilde{E} \\ C_1\dot{\widetilde{v_{C1}}} = (1-\alpha_e)\widetilde{i_{L1}} & -\alpha_e\widetilde{i_{L2}} & & +(i_{L1e}+i_{L2e})\widetilde{\alpha} & \\ L_2\dot{\widetilde{i_{L2}}} = & \alpha_e\widetilde{v_{C1}} & -(1-\alpha_e)\widetilde{v_{C2}} & +(v_{C1e}+v_{C2e})\widetilde{\alpha} & \\ C_2\dot{\widetilde{v_{C2}}} = (1-\alpha_e)\widetilde{i_{L1}} & +(1-\alpha_e)\widetilde{i_{L2}} & -(1/R)\widetilde{v_{C2}} & +(i_{L1e}+i_{L2e})\widetilde{\alpha} & \end{cases} \tag{4.51}$$

Equations (4.51) can further be put into the matrix form:

$$\dot{\mathbf{x}} = \mathbf{A}\cdot\mathbf{x} + \mathbf{B}\cdot\mathbf{u}, \tag{4.52}$$

where $\mathbf{x} = \begin{bmatrix} \widetilde{i_{L1}} & \widetilde{i_{L2}} & \widetilde{v_{C1}} & \widetilde{v_{C2}} \end{bmatrix}^T$ is the small-signal state vector, $\mathbf{u} = \begin{bmatrix} \widetilde{\alpha} & \widetilde{E} \end{bmatrix}^T$ is the small-signal input vector and state matrix $\mathbf{A}$ and input matrix $\mathbf{B}$ are

$$\begin{cases} \mathbf{A} = \begin{bmatrix} 0 & -(1-\alpha_e)/L_1 & 0 & -(1-\alpha_e)/L_1 \\ (1-\alpha_e)/C_1 & 0 & -\alpha_e/C_1 & 0 \\ 0 & \alpha_e/L_2 & 0 & -(1-\alpha_e)/L_2 \\ (1-\alpha_e)/C_2 & 0 & (1-\alpha_e)/C_2 & -1/(RC_2) \end{bmatrix} \\ \mathbf{B} = \begin{bmatrix} (v_{C1e}+v_{C2e})/L_1 & 1/L_1 \\ -(i_{L1e}+i_{L2e})/C_1 & 0 \\ (v_{C1e}+v_{C2e})/L_2 & 0 \\ -(i_{L1e}+i_{L2e})/C_2 & 0 \end{bmatrix}. \end{cases} \tag{4.53}$$

Relations (4.52) and (4.53) define the SEPIC small-signal state-space model.

(d) Using the matrix state representation in (4.52), the matrix definitions in (4.53) and the definition of the output matrix $\mathbf{C}$ as the four-by-four identity matrix, one can compute the transfer matrix $\mathbf{H}(s)$ as

$$\mathbf{H}(s) = \mathbf{C}\cdot(s\mathbf{I}-\mathbf{A})^{-1}\cdot\mathbf{B}, \tag{4.54}$$

which contains the transfer functions of the eight input-to-output channels. Applying relation (4.54) requires in this case the computation of the inverse of a four-by-four matrix in analytical form, which is quite difficult. In order to get the expression of the transfer function of the channel from the duty ratio $\widetilde{\alpha}$ (first input) to the output voltage $\widetilde{v_{C2}}$ (fourth state variable), the output vector must be set as $\mathbf{y} \equiv v_{C2}$, hence the output matrix must be set as $\mathbf{C} = \begin{bmatrix} 0 & 0 & 0 & 1 \end{bmatrix}^T$. Therefore, computing only the fourth row of matrix $\mathbf{H}(s)$ is sufficient to get the transfer function sought for. A quite laborious but simple computation finally produces the duty-ratio-to-output-voltage transfer function,

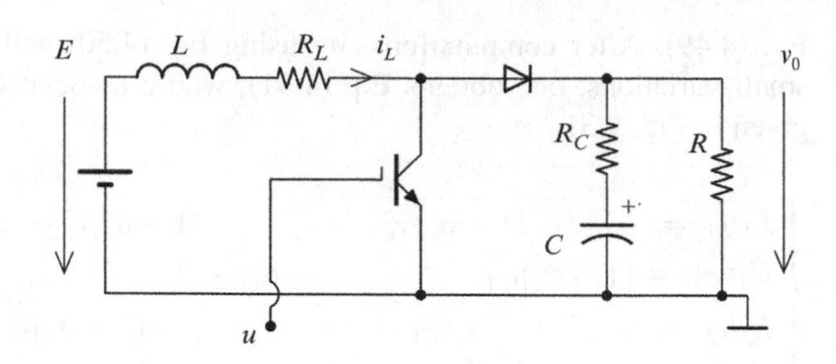

Fig. 4.32 Nonideal boost power stage

$$H_{\alpha \to v_{C2}}(s) = \frac{b_3 s^3 + b_2 s^2 + b_1 s + b_0}{s^4 + a_3 s^3 + a_2 s^2 + a_1 s + a_0}, \tag{4.55}$$

where the different parameters of numerator and denominator are, respectively

$$\begin{cases} b_3 = -\dfrac{i_{L1e} + i_{L2e}}{C_2} & b_2 = (1-\alpha_e)\dfrac{v_{C1e} + v_{C2e}}{L_1 L_2 C_1 C_2}(L_1 + L_2)C_1 \\ b_1 = -\alpha_e L_1 \dfrac{i_{L1e} + i_{L2e}}{L_1 L_2 C_1 C_2} & b_0 = (1-\alpha_e)\dfrac{v_{C1e} + v_{C2e}}{L_1 L_2 C_1 C_2}, \end{cases}$$

$$\begin{cases} a_3 = \dfrac{1}{RC_2} & a_2 = \dfrac{(1-\alpha_e)^2 (L_1 C_1 + L_2 C_2 + L_2 C_1) + \alpha_e^2 L_1 C_2}{L_1 L_2 C_1 C_2} \\ a_1 = \dfrac{1}{R} \cdot \dfrac{\alpha_e^2 L_1 + (1-\alpha_e)^2 L_2}{L_1 L_2 C_1 C_2} & a_0 = \dfrac{(1-\alpha_e)^2}{L_1 L_2 C_1 C_2}. \end{cases}$$

Analyzing the numerator of expression (4.55), one can identify the presence of unstable zeros, characterizing the nonminimum-phase behavior of the output voltage in response to duty ratio variations, similar to the buck-boost and flyback converter cases.

Problem 4.3. Nonideal Boost Converter

Let us consider the boost converter in Fig. 4.32, where the inductor is modeled as a pure inductance $L = 2$ mH and a series resistance $R_L = 0.5\ \Omega$ due to copper losses. The output capacitor has $C = 100\ \mu$F and an equivalent series resistance, $R_C = 0.05\ \Omega$. Both input voltage and load resistor are variant around their respective rated values $E = 5$ V and $R = 10\ \Omega$. It is required to solve the following points.

(a) Deduce the small-signal state-space model and draw the associated equivalent diagram taking into account that both input voltage E and load resistance R vary (they will be represented as disturbance inputs).
(b) Get the expressions of the three following transfer functions: from duty ratio to output voltage, from input voltage to output voltage and from load resistance to output voltage, respectively, by using the previously obtained diagram.
(c) Compute the steady-state model and draw the steady-state characteristics and input–output efficiency curve with respect to the duty ratio.

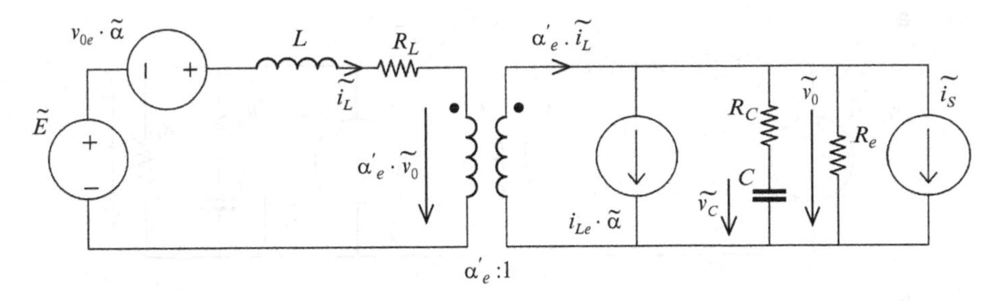

Fig. 4.33 Small-signal model of the nonideal boost circuit: equivalent diagram

(d) Draw the zero-pole diagrams for each of the three above mentioned influence channels using the MATLAB®-Simulink® software.
(e) Simulate the averaged nonlinear (large signal) model and assess the results for small variations around the steady-state operating point corresponding to the full load; make comparisons with the zero-pole diagram.

Solution. (a) For the sake of simplicity, in this example the brackets $\langle \cdot \rangle_0$ will be dropped; therefore, any variable encountered will denote in fact the corresponding average. Following the developments in Sect. 4.5.3 of Chap. 4, the large-signal averaged state-space model may be written as

$$\begin{cases} L\dot{i_L} = E - v_0(1-\alpha) - R_L i_L \\ C\dot{v_C} = i_L(1-\alpha) - \dfrac{v_0}{R} \\ v_0 = CR_C\dot{v_C} + v_C. \end{cases} \tag{4.56}$$

Next, variables describing the steady-state (equilibrium) operating point and rated values bear the subscript e. With notation $\alpha_e' = 1 - \alpha_e$, in the equilibrium point it holds that

$$\begin{cases} E_e - v_{0e}\cdot\alpha_e' - R_L i_{Le} = 0 \\ i_{Le}\cdot\alpha_e' - \dfrac{v_{0e}}{R_e} = 0 \\ v_{0e} = v_{Ce}. \end{cases} \tag{4.57}$$

In order to obtain the small-signal model around the considered operating point, one must differentiate the model (4.56):

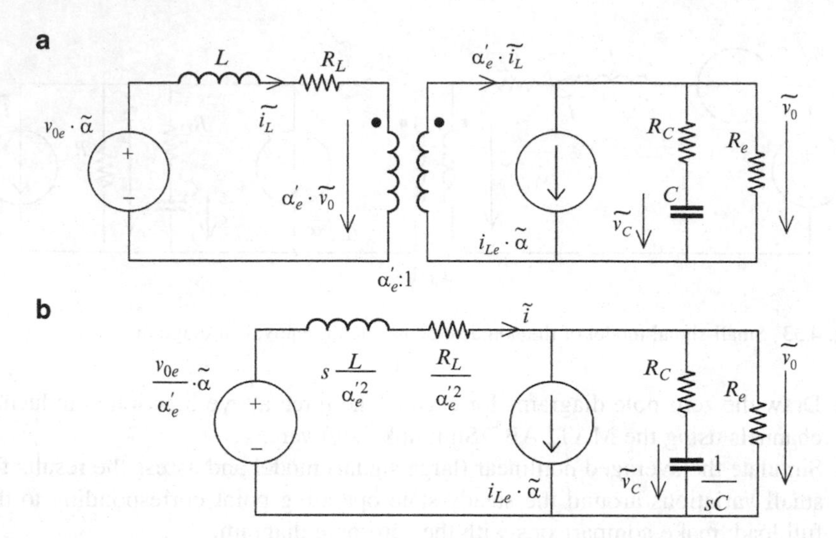

Fig. 4.34 (**a**) Small-signal model of nonideal boost circuit: duty-ratio-to-output-voltage influence; (**b**) push of inductor and voltage source through transformer

$$\begin{cases} L\dot{\widetilde{i_L}} = E_e + \widetilde{E} - (v_{0e} + \widetilde{v_0}) \cdot (1 - \alpha_e - \widetilde{\alpha}) - R_L \cdot \left(i_{Le} + \widetilde{i_L}\right) \\ C\dot{\widetilde{v_C}} = \left(i_{Le} + \widetilde{i_L}\right) \cdot (1 - \alpha_e - \widetilde{\alpha}) - \dfrac{v_{0e}}{R_e} - \dfrac{\widetilde{v_0}}{R_e} + \dfrac{v_{0e}}{R_e^2} \cdot \widetilde{R} \\ v_{0e} + \widetilde{v_0} = CR_C\dot{\widetilde{v_C}} + v_{Ce} + \widetilde{v_C} \, . \end{cases} \tag{4.58}$$

Using relations (4.57) in the system of (4.58) and neglecting small variations, one obtains the small-signal model of the considered boost power stage:

$$\begin{cases} L\dot{\widetilde{i_L}} = \widetilde{E} + v_{0e} \cdot \widetilde{\alpha} - \alpha_e' \cdot \widetilde{v_0} - R_L \cdot \widetilde{i_L} \\ C\dot{\widetilde{v_C}} = -i_{Le} \cdot \widetilde{\alpha} + \alpha_e' \cdot \widetilde{i_L} - \dfrac{\widetilde{v_0}}{R_e} - \widetilde{i_S} \\ \widetilde{v_0} = CR_C\dot{\widetilde{v_C}} + \widetilde{v_C} \, , \end{cases} \tag{4.59}$$

where the variation of the load current due to the load variation has been denoted by $\widetilde{i_S} = -v_{0e}/R_e^2 \cdot \widetilde{R}$. The associated equivalent diagram is given in Fig. 4.33.

This diagram shows the influence in variations of all the exogenous variables over system output $\widetilde{v_0}$: output results from superposition of all input variables. In order to extract a certain transfer function corresponding to one of these influence channels, one must nullify all other input variations.

(b) The duty-ratio-to-output-voltage transfer function is obtained by putting $\widetilde{E} = 0$ and $\widetilde{i_S} = 0$ in Eq. (4.59) or in Fig. 4.33. The result is presented in Fig. 4.34a. Further, the inductor and the voltage source in Fig. 4.34a may be pushed through the transformer; the circuit in Fig. 4.34b results, where the new inductor current has

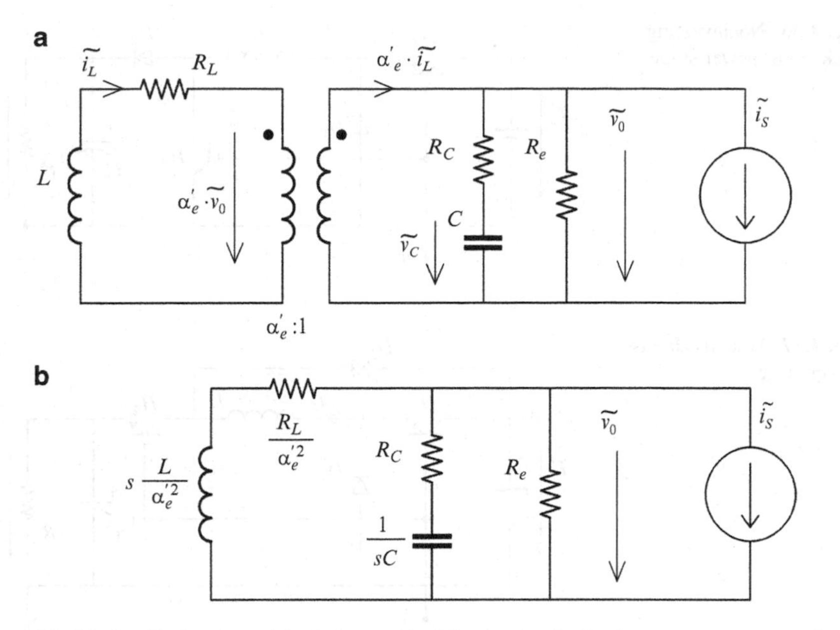

Fig. 4.35 (**a**) Small-signal model of the nonideal boost circuit: load-current-to-output-voltage influence; (**b**) push of inductor through transformer

been denoted by $\widetilde{i}$. Using Kirchhoff's laws, one solves this circuit by expressing output voltage variations $\widetilde{v_0}$ as a function of the duty ratio variations $\widetilde{\alpha}$:

$$\begin{cases} \frac{1}{\alpha'^2_e}(sL+R_L)\cdot\widetilde{i_L} = \frac{v_{0e}}{\alpha'_e}\cdot\widetilde{\alpha} - \widetilde{v_0} \\ \widetilde{i_L} = i_{Le}\cdot\widetilde{\alpha} + \dfrac{\widetilde{v_0}}{R_e\|\left(R_C+\frac{1}{sC}\right)}. \end{cases} \tag{4.60}$$

Combining Eq. (4.60) one finds that

$$\widetilde{v_0}\cdot\left[\frac{1}{\alpha'^2_e}\cdot\frac{sL+R_L}{R_e\|(R_C+\frac{1}{sC})}+1\right] = \widetilde{\alpha}\cdot\left(\frac{v_{0e}}{\alpha'_e} - \frac{sL+R_L}{\alpha'_e}\cdot i_{Le}\right). \tag{4.61}$$

Simple algebra gives the required transfer function, $H_{v_0\alpha}(s) = \frac{\widetilde{V_0}(s)}{\widetilde{\alpha}(s)}$:

$$H_{v_0\alpha}(s) = \frac{R_e(CR_Cs+1)\left(v_{0e}\alpha'_e - i_{Le}R_L - i_{Le}Ls\right)}{CL(R_e+R_C)s^2 + \left[\alpha'^2_eR_eR_CC + CR_L(R_e+R_C)+L\right]s + \alpha'^2_eR_e + R_L}. \tag{4.62}$$

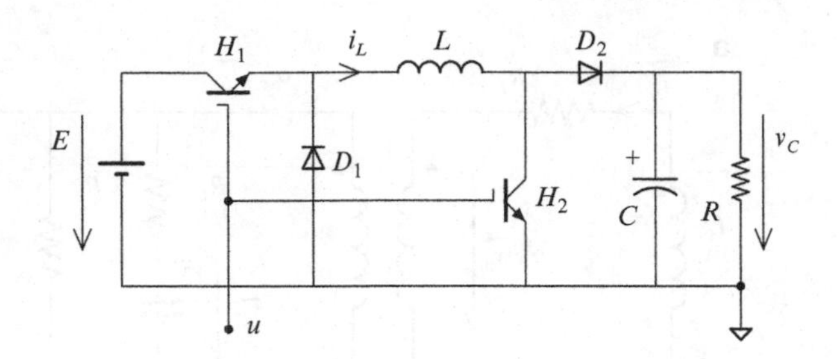

Fig. 4.36 Noninverting buck-boost power stage

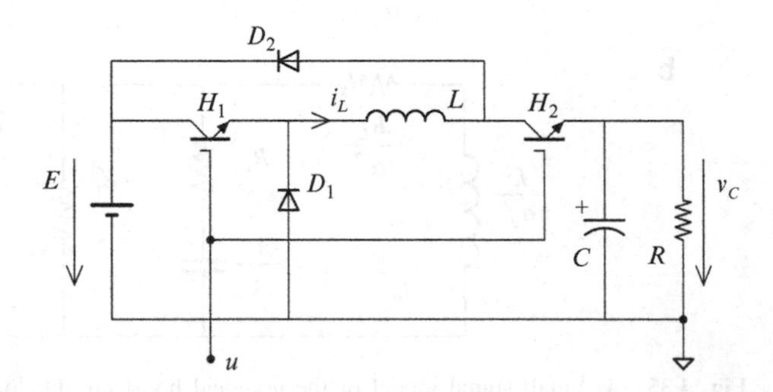

Fig. 4.37 Watkins–Johnson power stage

Similarly, the output-current-to-output-voltage transfer function may be obtained by putting $\widetilde{E} = 0$ and $\widetilde{\alpha} = 0$ in Eq. (4.59) or in Fig. 4.33. The result is shown in Fig. 4.35a. Further, the inductor in Fig. 4.35a may be pushed through the transformer, resulting in the circuit in Fig. 4.35b. One aims at expressing the output voltage variations $\widetilde{v_0}$ as a function of output current variations $\widetilde{i_S}$ and output impedance:

$$\widetilde{v_0} = \widetilde{i_S} \cdot \left[R_e || \left(R_C + \frac{1}{sC}\right) || \left(s\frac{L}{\alpha_e'^2} + \frac{R_L}{\alpha_e'^2}\right)\right]. \tag{4.63}$$

Simple algebra gives the required transfer function, $H_{v_0 i_S}(s) = \frac{\widetilde{V_0}(s)}{\widetilde{I_S}(s)}$:

$$H_{v_0 i_S}(s) = \frac{R_e(CR_C s + 1)(R_L + Ls)}{CL(R_e + R_C)s^2 + \left[\alpha_e'^2 R_e R_C C + CR_L(R_e + R_C) + L\right]s + \alpha_e'^2 R_e + R_L}. \tag{4.64}$$

Computation of the input-voltage-to-output-voltage transfer function results similarly and it is left to the reader. Solutions to the questions proposed in (c), (d) and (e) are also left to the reader, as are the solutions to the following problems.

Problem 4.4. Noninverting Buck-boost Converter
The circuit in Fig. 4.36 has a switching network composed of four switches (two transistors and two diodes). As the transistors are operated synchronously

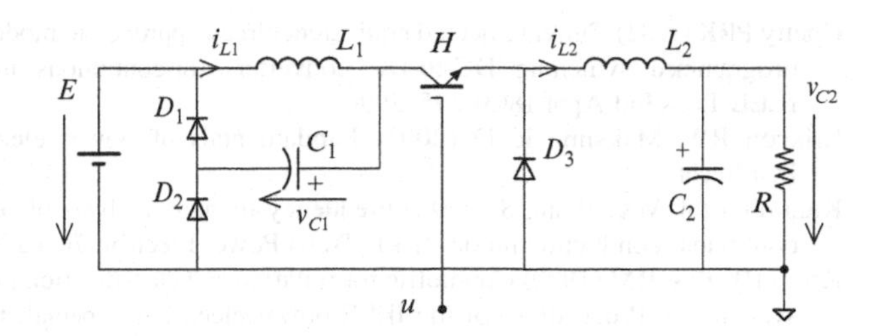

Fig. 4.38 Quadratic buck power stage

with the same binary switching function, $u \in \{0,1\}$, this switching network leads to two circuit configurations.

By taking i_L and v_C (see Fig. 4.30) as state variables

(a) obtain the switched model (bilinear form);
(b) obtain the averaged model and the corresponding equivalent diagram;
(c) compute the steady-state model; draw the static input/output characteristic with respect to the duty ratio;
(d) using the perturb-and-linearize method, deduce the small-signal state-space model and draw the associated equivalent diagram (E is considered constant);
(e) using the previously obtained small-signal model, get the expression of the transfer function representing the influence from the duty ratio (control input) and from the input voltage E (disturbance input) to the output voltage;
(f) analyze the system at point (d) for nonminimum-phase behavior and how its poles and zeroes migrate as the load resistor varies.

Problem 4.5. Watkins–Johnson Converter

Given the circuit in Fig. 4.37, answer the same requirements as in Problem 4.4.

In addition, it is required to simulate numerically (for example, using MATLAB®) the switched and the averaged models in the following case: input voltage $E = 5$ V, inductance $L = 5$ mH, output capacitor $C = 100$ μF and output resistor $R = 10\ \Omega$. Compare the system behavior for duty ratios larger than 0.5 with the behavior for duty ratios smaller than 0.5. Draw the Bode diagram of the transfer from the duty ratio to the output voltage for $\alpha_e = 0.6$ and $\alpha_e = 0.4$.

Problem 4.6. Quadratic Buck Converter

Given the circuit in Fig. 4.38, answer the same requirements as in Problem 4.4.

References

Ben-Yakoov S (1993) Average simulation of PWM converters by direct implementation of behavioral relationships. In: Proceedings of the eighth annual Applied Power Electronics Conference and Exposition – APEC 1993. San Diego, California, USA, pp 510–516

Brown AR, Middlebrook RD (1981) Sampled-data modeling of switching regulators. In: Proceedings of the IEEE Power Electronics Specialists Conference – PESC 8. Boulder, Colorado, USA, pp 349–369

Chetty PRK (1982) Current injected equivalent circuit approach to modelling and analysis of current programmed switching DC-to-DC converters (discontinuous inductor conduction mode). IEEE Trans Ind Appl 18(3):295–299

Erikson RW, Maksimović D (2001) Fundamentals of power electronics, 2nd edn. Kluwer, Dordrecht

Kazimierczuk MK, Wang S (1992) Frequency-domain analysis of series resonant converter for continuous conduction mode. IEEE Trans Power Electron 7(2):270–279

Krein PT, Bass RM (1990) Geometric formulation and classification methods for power electronic systems. In: Proceedings of the IEEE power electronics specialists conference. San Antonio, Texas, USA, pp 499–405

Krein PT, Bentsman J, Bass RM, Lesieutre B (1990) On the use of averaging for the analysis of power electronic systems. IEEE Trans Power Electron 5(2):182–190

Lehman B, Bass RM (1996) Extension of averaging theory for power electronic systems. IEEE Trans Power Electron 11(4):542–553

Maksimović D, Zane R (2007) Small-signal discrete-time modelling of digitally controlled PWM converters. IEEE Trans Power Electron 22(6):2552–2556

Maksimović D, Stanković AM, Thottuvelil VJ, Verghese GC (2001) Modeling and simulation of power electronic converters. Proc IEEE 89(6):898–912

Middlebrook RD (1988) Small-signal modelling of pulse-width modulated switched-mode power converters. Proc IEEE 76(4):343–354

Middlebrook RD, Ćuk S (1976) A general unified approach to modelling switching converter power stages. In: Proceedings of the IEEE power electronic specialists conference. Cleveland, Ohio, USA, pp 18–34

Perard J, Toutain E, Nougaret M (1979) Modelling of energy converters by an equivalent circuit (in French: Modélisation des convertisseurs d'énergie par un schéma equivalent). L'Onde Électrique 59(12)

Rim CT, Joung GB, Cho GH (1988) A state-space modelling of non-ideal DC-DC converters. In: Proceedings of the IEEE Power Electronics Specialists Conference Averaged Model PESC 1988. Kyoto, Japan, pp 943–950

Rodriguez FD, Chen JE (1991) A refined nonlinear averaged model for constant frequency current mode controlled PWM converters. IEEE Trans Power Electron 6(4):656–664

Sanders SR, Verghese GC (1991) Synthesis of averaged circuit models for switched power converters. IEEE Trans Circuit Syst 38(8):905–915

Sanders SR, Noworolski JM, Liu XZ, Verghese GC (1990) Generalized averaging method for power conversion circuits. In: Proceedings of the IEEE power electronics specialists conference. San Antonio, Texas, USA, pp 333–340

Sira-Ramirez H, Silva-Ortigoza R (2006) Control design techniques in power electronics devices. Springer, London

Sun J, Grotstollen H (1992) Averaged modelling of switching power converters: reformulation and theoretical basis. In: Proceedings of the IEEE/PESC 1992 Power Electronics Specialists Conference. Toledo, Spain, pp 1165–1172

Tymerski R, Li D (1993) State-space models for current programmed pulsewidth-modulated converters. IEEE Trans Power Electron 8(3):271–278

Verghese GC, Bruzos CA, Mahabir KN (1989) Averaged and sampled-data models for current mode control: a reexamination. In: Proceedings of the 20th annual IEEE Power Electronics Specialists Conference – PESC 1989. Milwaukee, Wisconsin, USA, vol. 1, pp 484–491

Vuthchhay E, Bunlaksananusorn C (2008) Dynamic modelling of a Zeta converter with state-space averaging technique. In: Proceedings of the 5th international conference on Electrical Engineering/Electronics, Computer, Telecommunications and Information Technology – ECTI-CON 2008. Rhodes Island, Greece, pp 969–972

Wester GV, Middlebrook RD (1973) Low-frequency characterization of switched DC-DC converters. IEEE Trans Aerosp Electron Syst 9(3):376–385

Chapter 5
Generalized Averaged Model

This chapter approaches methodologies of deriving averaged models able to also represent behavior of converters containing AC stages. This time, modeling is not restricted to DC variables and the resultant models – called *generalized averaged models* (GAM) – can handle averages of higher-order harmonics.

The chapter first provides basic ideas of generalized average modeling and the relation of GAM to the previously introduced classical averaged model, followed by examples of obtaining GAM for some simple cases. The GAM application range is identified and its building and implementation simplicity is shown. To this purpose two clear algorithmic approaches will be defined, namely the analytical approach and the graphical one, respectively. The relation between GAM and the real waveforms is shown, as well as the expression of active and reactive AC variable components using GAM. Relations with the classical averaged model (detailed in Chap. 4) and with the first-order-harmonic steady-state model will also be emphasized. Some case studies will serve at illustrating the various approaches. This chapter ends with some problems with solutions and some to be solved.

5.1 Introduction

Each of the various models presented until now applies to somehow dedicated area and therefore has different properties. They have, however, a common aspect: the precision of plant replication grows with model complexity. So, as the modeling effort is sometimes unworthy, especially if the required performances are not high, one should always envisage a trade-off between the two above listed requirements. One should also note that the reciprocal statement is not true: accuracy of the model is not necessarily due to complexity. The model presented by Sanders et al. (1990) – which was resumed later by Noworolsky and Sanders (1991) and by Sanders (1993) – shows this. This model, based on early nineteenth century works of Van der Pol, provides a good trade-off between accuracy and implementation simplicity; this statement is only valuable for converters having both DC stages and AC stages

S. Bacha et al., *Power Electronic Converters Modeling and Control: with Case Studies*, Advanced Textbooks in Control and Signal Processing, DOI 10.1007/978-1-4471-5478-5_5,

where the first-order harmonic prevails. In this context, the chosen application in the cited article is an ideal resonance power supply whose behavior has been illustrated by what the authors have called a generalized averaged model, or GAM. In the same article, the authors have modeled in the same way a DC-DC buck-boost power stage, obtaining satisfactory results only for values of duty ratio of around 50 %. A similar modeling approach was developed by Caliskan et al. (1999) for the DC-DC boost power stage, with similar results.

In the latter literature this kind of modeling has been called *generalized state-space average* (GSSA) modeling (Rimmalapudi et al. 2007), and it is useful in any power electronics application containing AC power stages: in resonant converters (Rim and Cho 1990; Xu and Lee 1998), in active power filters (Nasiri and Emadi 2003; Wong et al. 2006), in electronic ballast for discharge lamps (Yin et al. 2002), and others.

5.2 Principles

5.2.1 Fundamentals

The GAM is based on the waveform representation using the complex Fourier series. Thus, every periodic variable $x(t)$ can be expressed as

$$x(t) = \sum_{k=-\infty}^{+\infty} x_k(t) \cdot e^{jk\omega t}, \tag{5.1}$$

where ω is the fundamental pulsation and x_k is the coefficient of the kth harmonic, whose mathematical definition

$$x_k(t) = \frac{1}{T} \int_{t-T}^{t} x(\tau) \cdot e^{-jk\omega\tau} d\tau, \tag{5.2}$$

with $T = 2\pi/\omega$, corresponds to the definition of a sliding average – in the sense given in Chap. 4 – which will be further called *kth-order sliding harmonic*. The kth-order harmonic – or k-phasor (Maksimović et al. 2001) – coefficient results from the sliding averaging operation. For proof purposes the following notation will be used:

$$x_k(t) = \langle x \rangle_k(t). \tag{5.3}$$

Equations (5.1) and (5.2) lead to two fundamental properties. The first is concerning the derivative of the sliding average, expressed as

$$\frac{d}{dt}\langle x \rangle_k(t) = \left\langle \frac{d}{dt} x \right\rangle_k (t) - jk\omega \langle x \rangle_k(t), \tag{5.4}$$

and the second property concerns the variables product:

$$\langle x \cdot y \rangle_k(t) = \sum_i \langle x \rangle_{k-i}(t) \cdot \langle y \rangle_i(t). \tag{5.5}$$

Interested readers can find a justification of Eq. (5.5) in the Appendix, whereas proof of Eq. (5.4) is presented here. This latter is based on integrating by parts Eq. (5.2), which gives successively

$$x_k(t) = \frac{1}{T} \int_{t-T}^{t} x(\tau) \cdot e^{-jk\omega\tau} d\tau = -\frac{1}{jk\omega T} \int_{t-T}^{t} x(\tau) \cdot \frac{d\left(e^{-jk\omega\tau}\right)}{d\tau} d\tau$$

$$= -\frac{1}{jk\omega T} \left(x(\tau) \cdot e^{-jk\omega\tau} \Big|_{t-T}^{t} \right) + \frac{1}{jk\omega T} \int_{t-T}^{t} \frac{dx(\tau)}{d\tau} \cdot e^{-jk\omega\tau} d\tau. \tag{5.6}$$

Equation (5.6) can further be developed by noting that the second term contains the k-order sliding harmonic of the signal x's time derivative; thus one can write

$$x_k(t) = -\frac{1}{jk\omega T} \left(x(t) \cdot e^{-jk\omega t} - x(t-T) \cdot e^{-jk\omega(t-T)} \right)$$

$$+ \frac{1}{jk\omega} \cdot \underbrace{\frac{1}{T} \int_{t-T}^{t} \frac{dx(\tau)}{d\tau} \cdot e^{-jk\omega\tau} d\tau}_{\left\langle \frac{d}{dt} x \right\rangle_k}$$

$$= -\frac{1}{jk\omega T} \left(x(t) \cdot e^{-jk\omega t} - x(t-T) \cdot e^{-jk\omega(t-T)} \right) + \frac{1}{jk\omega} \left\langle \frac{d}{dt} x \right\rangle_k (t). \tag{5.7}$$

On the other hand, suppose that function $x(t) \cdot e^{-jk\omega t}$ admits the primitive denoted by $F(t)$, i.e., it holds that

$$\frac{dF(t)}{dt} = x(t) \cdot e^{-jk\omega t}. \tag{5.8}$$

Thus, for the same integral in Eq. (5.2) one obtains

$$\langle x \rangle_k(t) = \frac{1}{T} (F(t) - F(t-T)),$$

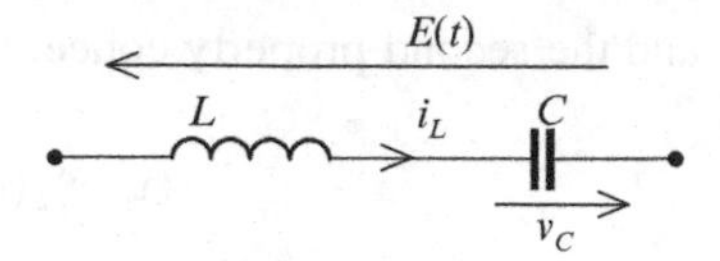

Fig. 5.1 Elementary LC circuit

and the time derivative of the k-order sliding harmonic can further be obtained as depending on the primitive function F:

$$\frac{d}{dt}\langle x\rangle_k(t) = \frac{1}{T}\left(\frac{d}{dt}F(t) - \frac{d}{dt}F(t-T)\right),$$

or equivalently, using Eq. (5.7),

$$\frac{d}{dt}\langle x\rangle_k(t) = \frac{1}{T}\left(x(t)\cdot e^{-jk\omega t} - x(t-T)\cdot e^{-jk\omega(t-T)}\right). \tag{5.9}$$

Equation (5.9) can now be used to replace the first term in Eq. (5.7):

$$\langle x\rangle_k(t) = -\frac{1}{jk\omega}\cdot\frac{d}{dt}\langle x\rangle_k(t) + \frac{1}{jk\omega}\cdot\left\langle\frac{d}{dt}x\right\rangle_k(t)$$

and to finally derive Eq. (5.4).

The modeling in the GAM sense is based on Eqs. (5.4) and (5.5). One can note that Eq. (5.4) is valid only if the pulsation ω is slowly variable. Otherwise, the expression of the derivative $\frac{d}{dt}\langle x\rangle_k(t)$ becomes unusable (Sanders et al. 1990):

$$\frac{d}{dt}\langle x\rangle_k = x\cdot(t-T)\cdot e^{-jk\theta(t-T)}\cdot\omega(t-T)\frac{d}{dt}T + \left\langle\frac{d}{dt}x\right\rangle_k + \left\langle\left(\frac{1}{\omega}\frac{d}{dt}\omega - jk\omega\right)x\right\rangle_k,$$

where ω is the derivative of θ.

5.2.2 Relation with the First-Order-Harmonic Model

By the first-order-harmonic model one understands the steady-state sinusoidal regime of a system characterized by AC variables. If the series expansion in Eq. (5.1) is truncated at the first harmonic ($k = 1$) of pulsation ω and by considering the steady-state regime ($\langle x\rangle_1$ is constant), the conditions of the first-order-harmonic model can be found. Indeed, this can be seen on the elementary circuit from Fig. 5.1.

The circuit equations are:

$$\begin{cases} \frac{d}{dt} i_L = \frac{E}{L} - \frac{v_C}{L} \\ \frac{d}{dt} v_C = \frac{i_L}{C}. \end{cases}$$

By applying the result (5.4), one obtains:

$$\begin{cases} \frac{d}{dt} \langle i_L \rangle_1 = -j\omega \langle i_L \rangle_1 + \frac{\langle E \rangle_1}{L} - \frac{\langle v_C \rangle_1}{L} \\ \frac{d}{dt} \langle v_C \rangle_1 = -j\omega \langle v_C \rangle_1 + \frac{\langle i_L \rangle_1}{C}. \end{cases}$$

By considering the steady-state regime, that is, by putting

$$\frac{d}{dt} \langle i_L \rangle_1 = \frac{d}{dt} \langle v_C \rangle_1 = 0,$$

one finds the complex expression of the impedance of the circuit from Fig. 5.1:

$$\langle E \rangle_1 = \left(j\omega L + \frac{1}{j\omega C} \right) \langle i_L \rangle_1 .$$

This equation shows that the first harmonic model is a particular case of the GAM limited at the fundamental harmonic. This is why it is also called the *harmonic dynamic model.*

5.2.3 Relation with Classical Averaged Model

For showing that the classical averaged model is a particular case of the GAM, one may use Eqs. (5.2), (5.4) and (5.5) and take $k = 0$, that is, one takes into account only the average value of the signal x. Equation (5.2) becomes (T is the averaging time window, e.g., the switching period):

$$\langle x \rangle_0 = \frac{1}{T} \int_{t-T}^{t} x(\tau)\, dt, \tag{5.10}$$

and Eq. (5.4) becomes:

$$\frac{d}{dt} \langle x \rangle_0 (t) = \left\langle \frac{d}{dt} x \right\rangle_0 (t). \tag{5.11}$$

Equation (5.10) is the sliding average in the averaged model sense, whereas Eq. (5.11) represents a fundamental property of this latter, regarding the time derivative.

The fundamental approximation of the classical averaged model (see developments in Chap. 4) finds a justification in the analysis of the product variables of Eq. (5.5). Indeed, if one refers to the sliding average of a product,

$$\begin{aligned}\langle x \cdot y\rangle_0 = \sum_{i+m=0} \langle x\rangle_{m-i}\langle y\rangle_m &= \langle x\rangle_0\langle y\rangle_0 + \langle x\rangle_1\langle y\rangle_{-1} + \langle x\rangle_{-1}\langle y\rangle_1 \\ &\quad + \langle x\rangle_2\langle y\rangle_{-2} + \langle x\rangle_{-2}\langle y\rangle_2 + \cdots,\end{aligned} \tag{5.12}$$

and neglects the variations of variables x and y and therefore their harmonics, then one obtains

$$\langle x \cdot y\rangle_0 \approx \langle x\rangle_0 \cdot \langle y\rangle_0.$$

Averaged model limitations when harmonics are important, or when the impossibility of its use when the state variables vary around zero, can be well understood.

5.3 Examples

Some elementary examples are here approached in the same manner as in the case of the averaged model. First, the kth-order sliding average of a variable is taken into account. Then, the operation is repeated for a passive circuit and for a switch.

5.3.1 *Case of a State Variable*

Let us consider a system described by $dx(t)/dt = F(x(t), u(t))$ with $u(t)$ being a T-periodical input signal. By using Eq. (5.4) this system becomes

$$\frac{d}{dt}\langle x\rangle_k = -jk\omega\langle x\rangle_k + \langle F(x,u)\rangle_k. \tag{5.13}$$

The expression of $\langle x\rangle_k$ is complex, having a real part and an imaginary part. This variable can be developed using the Fourier series (see Eq. (5.1)). As any variable represented by its mean value, first harmonic and superior harmonics, variable $x(t)$ can be written as

$$x(t) \approx \langle x\rangle_0 + 2\left[\mathrm{Re}\left(\langle x\rangle_1\right)\cos(\omega t) - \mathrm{Im}\left(\langle x\rangle_1\right)\sin(\omega t)\right],$$

where the development has been truncated at the first-order harmonic.

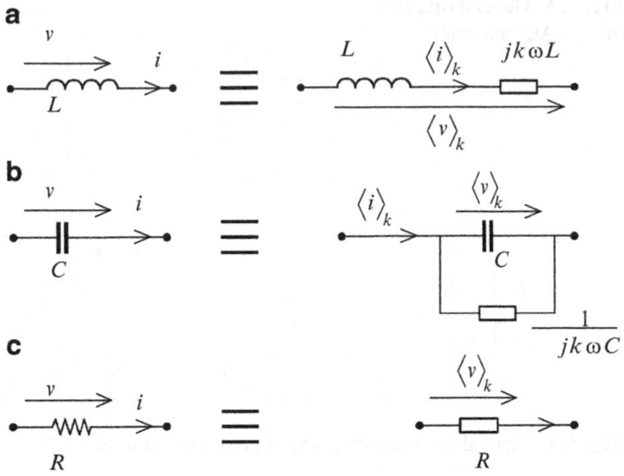

Fig. 5.2 Illustration of applying GAM for passive circuits: (**a**) case of an inductance; (**b**) case of a capacitor; (**c**) case of a resistance

5.3.2 Case of a Passive Circuit

Applying Eq. (5.13) to R, L, or C passive circuit elements, one obtains successively (the time variable being omitted)

- for an inductance:

$$v = L\frac{di}{dt} \Rightarrow \langle v\rangle_k = \left\langle L\frac{di}{dt}\right\rangle_k = jk\omega L\langle i\rangle_k + L\frac{d\langle i\rangle_k}{dt}; \tag{5.14}$$

- for a capacitor:

$$i = C\frac{dv}{dt} \Rightarrow \langle i\rangle_k = \left\langle C\frac{dv}{dt}\right\rangle_k = jk\omega C\langle v\rangle_k + C\frac{d\langle v\rangle_k}{dt}; \tag{5.15}$$

- in the case of a resistance:

$$\langle v\rangle_k = R\langle i\rangle_k. \tag{5.16}$$

Figure 5.2 gives a topological image of Eqs. (5.14), (5.15) and (5.16) respectively.

Remarks. Except for the case of the resistance, the GAM applied to a circuit element changes its configuration.

By putting $k = 0$ one obtains the averaged model diagram of a passive circuit element.

By considering the steady-state regime (i.e., by zeroing the derivatives) one obtains the steady-state harmonic model written by using impedances.

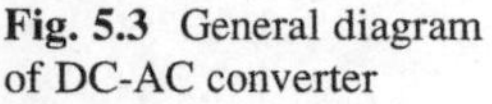

Fig. 5.3 General diagram of DC-AC converter

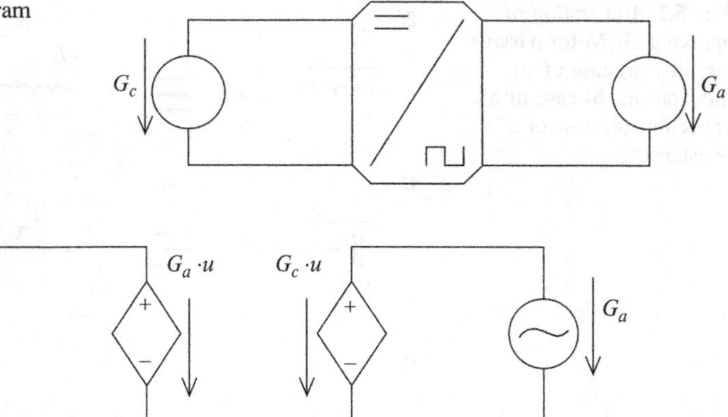

Fig. 5.4 Equivalent exact diagram of converter from Fig. 5.3

5.3.3 *Case of a Coupled Circuit*

The GAM is particularly appropriate for coupled circuit analysis having both DC-type variables – described mainly by their averages – and AC-type variables – described by their first-order harmonic, as well as variables having both DC and AC components.

In the present context, a coupled circuit is a symmetrical-switching converter having a switching function with zero DC component (see Fig. 5.3). This figure shows a generic converter between DC and AC variables without being interested in the power flow direction, either in type of DC source, G_c, or in the AC source, G_a. The represented structure can be a voltage/current inverter or rectifier.

In order to emphasize the couplings, the equivalent switched diagram of the studied structure is built. As usual, one denotes by u the switching function (Fig. 5.4).

Let us consider the first-order harmonic of the AC variable and the average value of the DC variable. One must therefore find the expressions of the following variables:

- in the AC part:

$$\langle G_a \rangle_1, \langle G_c \cdot u \rangle_1;$$

- and in the DC part:

$$\langle G_c \rangle_0, \langle G_a \cdot u \rangle_0.$$

If restraining the model to first-order harmonics and average values, by applying Eq. (5.5) to the coupled expressions one respectively obtains

$$\langle G_c \cdot u \rangle_1 = \langle G_c \rangle_0 \cdot \langle u \rangle_1 + \langle G_c \rangle_1 \cdot \langle u \rangle_0, \tag{5.17}$$

$$\langle G_a \cdot u \rangle_0 = \langle G_a \rangle_0 \cdot \langle u \rangle_0 + \langle G_a \rangle_1 \cdot \langle u \rangle_{-1} + \langle G_a \rangle_{-1} \cdot \langle u \rangle_1. \tag{5.18}$$

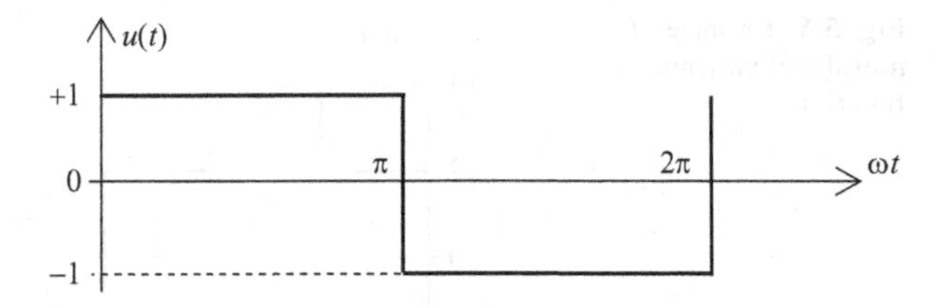

Fig. 5.5 Square-wave switching function

If, moreover, the switching function u has zero average value, these above expressions can be respectively rewritten as:

$$\langle G_c \cdot u \rangle_1 = \langle G_c \rangle_0 \cdot \langle u \rangle_1; \tag{5.19}$$

$$\langle G_a \cdot u \rangle_0 = \langle G_a \rangle_1 \cdot \langle u \rangle_{-1} + \langle G_a \rangle_{-1} \cdot \langle u \rangle_1. \tag{5.20}$$

Equation (5.20) can be rewritten and put into a more suitable form, knowing that terms $\langle x \rangle_k$ and $\langle x \rangle_{-k}$ being conjugated, which represents a real value, namely:

$$\langle G_a \cdot u \rangle_0 = 2\left[\mathrm{Re}\left(\langle G_a \rangle_1\right) \cdot \mathrm{Re}\left(\langle u \rangle_1\right) + \mathrm{Im}\left(\langle G_a \rangle_1\right) \cdot \mathrm{Im}\left(\langle u \rangle_1\right)\right]. \tag{5.21}$$

5.3.4 Switching Functions

The switching functions commonly used exhibit a square-waveform time evolution in the case of symmetrical-switching converters; these functions may depend on state variables or on external inputs. The expressions of their first-order sliding harmonics are useful in developing Eqs. (5.17), (5.18), (5.19), (5.20), and (5.21).

5.3.4.1 Case of Switching Functions Depending on Time

To this class belong the switching functions of forced-commutated inverters, for example.

Let us consider the function $u(t)$ represented in Fig. 5.5, that can be described analytically by $u(t) = \mathrm{sgn}(\sin(\omega t))$.

Using Definition 4.2 particularized to the first-order harmonic gives

$$\langle u \rangle_1 = \frac{2}{\pi j}. \tag{5.22}$$

If this function is phase-shifted in relation to an arbitrarily set origin, its analytical expression is $u(t) = \mathrm{sgn}(\sin(\omega t + \delta))$ and its expression in the sense of the first-order harmonic becomes

$$\langle u \rangle_1 = \frac{2}{\pi j} e^{j\delta}. \tag{5.23}$$

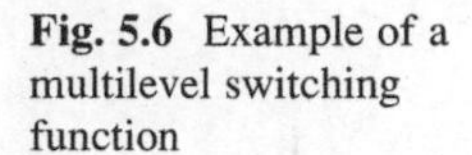

Fig. 5.6 Example of a multilevel switching function

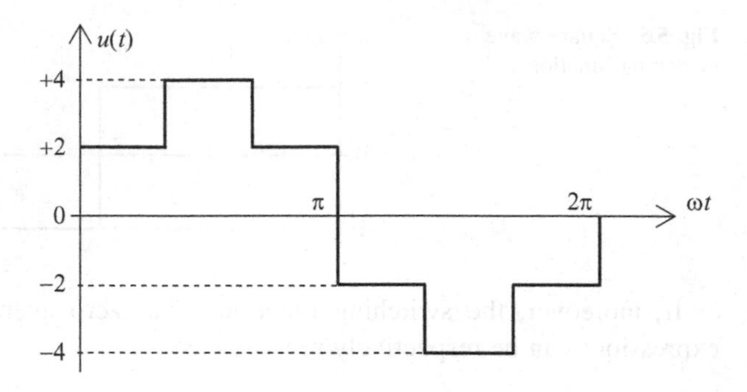

5.3.4.2 Case of Switching Functions Depending on a State Variable

An example of switching function belonging to this class is a diode rectifier. In this case the sign of the AC current shows which branch of the rectifier is in conduction at a given moment. This function is expressed analytically as $u(t) = \text{sgn}(x(t))$, where x is an AC state variable. If variable x is phase-shifted by an angle φ in relation to the phase origin, then the following relation holds:

$$\langle u \rangle_1 = \frac{2}{\pi j} e^{j\varphi}. \tag{5.24}$$

5.3.4.3 Case of Switching Functions Depending on State Variables and Time

A thyristor rectifier is an example belonging to this class because the branch in conduction depends in the meantime on the sign of the AC variable in question and on the moment when the firing order is applied. This dependence may be expressed analytically as $u(t) = \text{sgn}(x(t)) \cdot \nu(\delta)$, where x is an AC state variable and $\nu(\cdot)$ is a delay function.

Under the same assumption of a φ phase shift of variable x, one obtains

$$\langle u \rangle_1 = \frac{2}{\pi j} e^{j(\varphi - \delta)}. \tag{5.25}$$

5.3.4.4 Case of Multilevel Switching Functions

This may be the case of a three-phase inverter.

For this type of switching function one can always write expressions as a finite sum of elementary square-waveform functions. Thus, for the example, from Fig. 5.6 function $u(t)$ can be put into the form of a three-term sum containing elementary switching functions, which are phase-shifted by $2\pi/3$ each in relation to each other. Therefore, one can write

$$u(t) = 2u_1(t) - u_2(t) - u_3(t).$$

Then computation of $\langle u \rangle_1$ leads to

$$\langle u \rangle_1 = \frac{2}{\pi j}\left(2 - e^{-2\pi j/3} - e^{2\pi j/3}\right) = \frac{6}{\pi j}.$$

5.4 Methodology of Averaging

As in the case of the classical averaged model, building of the generalized averaged model can be performed either graphically, by means of the topological diagram, or analytically, based upon the topological exact model.

5.4.1 *Analytical Approach*

Without loss of generality, a given power electronic converter can be described by Eq. (3.2) presented in Chap. 3:

$$\dot{\mathbf{x}} = \sum_{i=1}^{N} (\mathbf{A}_i\mathbf{x} + \mathbf{B}_i\mathbf{e})h_i.$$

A more condensed form describes the converter under a general bilinear form, where matrices **A** and **B** are constant:

$$\dot{\mathbf{x}} = \mathbf{A}\mathbf{x} + \sum_{n=1}^{p} u_n(\mathbf{B}\mathbf{x} + \mathbf{b}) + \mathbf{d},$$

where the p switching functions are emphasized.

If the converter has in the meantime AC stages as well as DC stages, one can apply the Fourier series expansion, comprising the fundamental and the DC-component values of the concerned state variables.

Thus, for the DC components one obtains by averaging

$$\frac{d}{dt}\langle \mathbf{x} \rangle_0 = \mathbf{A}\langle \mathbf{x} \rangle_0 + \sum_{n=1}^{p} \mathbf{B}\langle u_n\mathbf{x} \rangle_0 + \langle u_n\mathbf{b} \rangle_0 + \langle \mathbf{d} \rangle_0,$$

whereas for the fundamental component the following holds (first-order sliding averaging):

$$\frac{d}{dt}\langle \mathbf{x} \rangle_1 = -j\omega\langle \mathbf{x} \rangle_1 + \mathbf{A}\langle \mathbf{x} \rangle_1 + \sum_{n=1}^{p} B\langle u_n\mathbf{x} \rangle_1 + \langle u_n\mathbf{b} \rangle_1 + \langle \mathbf{d} \rangle_1.$$

The mathematical development is further pursued with computation of terms $\langle \mathbf{d} \rangle_0$ and $\langle \mathbf{d} \rangle_1$ and with obtaining the expressions of the coupled terms $\langle u_n \mathbf{x} \rangle_1$, $\langle u_n \mathbf{b} \rangle_1$, $\langle u_n \mathbf{x} \rangle_0$ and $\langle u_n \mathbf{b} \rangle_0$.

By collecting the above expressions, model (5.26) expressed as below represents the generalized averaged model of a given power electronic converter:

$$\begin{cases} \frac{d}{dt} \langle \mathbf{x} \rangle_0 = \mathbf{A} \langle \mathbf{x} \rangle_0 + \sum_{n=1}^{p} \mathbf{B} \langle u_n \mathbf{x} \rangle_0 + \langle u_n \mathbf{b} \rangle_0 + \langle \mathbf{d} \rangle_0 \\ \frac{d}{dt} \langle \mathbf{x} \rangle_1 = -j\omega \langle \mathbf{x} \rangle_1 + \mathbf{A} \langle \mathbf{x} \rangle_1 + \sum_{n=1}^{p} B \langle u_n \mathbf{x} \rangle_1 + \langle u_n \mathbf{b} \rangle_1 + \langle \mathbf{d} \rangle_1 . \end{cases} \tag{5.26}$$

The above computation steps, which may appear complicated at a first sight, can be quickly simplified given that the AC variables have null average values and that the DC variables are sufficiently filtered.

A supplementary step is, however, necessary; in this case it is the separation of complex terms into their respective real and imaginary parts.

5.4.2 *Graphical Approach*

When one uses the graphical approach to deduce the GAM, the starting point is the equivalent exact diagram. The algorithm to follow is resumed below.

Correct use of steps of the Algorithm 5.1 will lead to the same result as that obtained by employing the analytical approach, that is, to model (5.26).

Algorithm 5.1 Building the GAM by using the graphical approach

#1 Build the equivalent exact diagram.
#2 Identify the DC stage and the AC stage of the converter
#3 Within the DC stage:

- replace state variables and coupled products by their sliding averages;
- leave passive elements as they are.

#4 Within the AC stage:

- replace state variables and coupled products by their respective first-order sliding harmonics;
- compute equivalent impedance of inductances and add it in series with them;

(continued)

Algorithm 5.1 (continued)

- compute equivalent impedance of capacitors and add it in parallel with them.
- leave resistances as they are.

#5 Get the expressions of coupled terms.
#6 Based upon the generalized averaged diagram thus obtained, write the characterizing equations.

General Remarks. Irrespective of the approach used, the Fourier series expansion can go beyond the fundamental. In this way, several equivalent generalized averaged diagrams are obtained, which are coupled. Each of these diagrams is valid for any given order harmonic.

5.5 Relation Between Generalized Averaged Model and Real Waveforms

It is sometimes difficult to understand the physical aspect of sliding harmonics. For example, if high-order harmonics appear during design of a control law, one must find a way of reconstruct them based upon different waveforms measured by various transducers. To this end, the relation between the sliding harmonic and the real waveform must be clarified.

Next we take the example of an AC variable not necessarily sinusoidal but whose information of interest is contained in its fundamental term. Let a generic waveform $y(t)$ be considered. One can make the approximation of assimilating it to its first-order sliding harmonic, which is a function of time (see Eq. (5.1)):

$$y(t) \approx \langle y \rangle_1 e^{j\omega t} + \langle y \rangle_{-1} e^{-j\omega t}, \tag{5.27}$$

giving after some algebra,

$$y(t) \approx 2\left[\mathrm{Re}\left(\langle y \rangle_1\right) \cos \omega t - \mathrm{Im}\left(\langle y \rangle_1\right) \sin \omega t\right]. \tag{5.28}$$

For simplicity let us put $x_1 = \mathrm{Re}(\langle y \rangle_1)$ and $x_2 = \mathrm{Im}(\langle y \rangle_1)$. Equation (5.28) will be rewritten under the form

$$y(t) \approx 2(x_1 \cos \omega t - x_2 \sin \omega t). \tag{5.29}$$

Equation (5.29) shows a way in which the real waveform $y(t)$ can be obtained based on the real and the imaginary parts of its first-order sliding harmonic, $\langle y \rangle_1(t)$.

5.5.1 Extracting Real-Time-Varying Signal from GAM

Suppose that the real and the imaginary parts of the first-order harmonic, x_1 and x_2, have already been obtained by solving the circuit model. In this section the aim is to obtain the magnitude $\hat{y}$ and the phase lag φ of the original signal $y(t)$ by using x_1 and x_2. The sought for expression of $y(t)$ is therefore

$$y(t) = \hat{y} \cdot \sin(\omega t + \varphi). \tag{5.30}$$

Supposing that the imaginary part of the first-order sliding harmonic x_2 is nonzero, then from Eq. (5.29) one obtains

$$y(t) \approx 2(x_1 \cos \omega t - x_2 \sin \omega t) = -2x_2\left(-\frac{x_1}{x_2} \cdot \cos \omega t + \sin \omega t\right). \tag{5.31}$$

Because the quantity $-x_1/x_2$ represents a real number, then a unique argument $\psi \in (-\pi/2, \pi/2)$ exists such that:

$$\tan \psi = \frac{\sin \psi}{\cos \psi} = -\frac{x_1}{x_2}. \tag{5.32}$$

Replacing Eq. (5.32) into Eq. (5.31) one obtains successively:

$$y(t) \approx -2x_2 \cdot \left(\frac{\sin \psi}{\cos \psi} \cdot \cos \omega t + \sin \omega t\right)$$

$$= -\frac{2x_2}{\cos \psi} \cdot (\sin \psi \cdot \cos \omega t + \cos \psi \cdot \sin \omega t),$$

and finally

$$y(t) \approx -\frac{2x_2}{\cos \psi} \cdot \sin(\omega t + \psi). \tag{5.33}$$

From Eq. (5.32) it is easy to verify that $\cos^2\psi = x_2^2/(x_1^2 + x_2^2)$ and, taking into account that $\psi \in (-\pi/2, \pi/2)$, it results that $\cos \psi = |x_2|/\sqrt{x_1^2 + x_2^2}$ which is replaced in Eq. (5.33):

$$y(t) \approx \frac{-x_2}{|x_2|} \cdot 2\sqrt{x_1^2 + x_2^2} \cdot \sin(\omega t + \psi). \tag{5.34}$$

A comparison of Eqs. (5.30) and (5.34) shows that the magnitude of signal $y(t)$ is

$$\hat{y} = 2\sqrt{x_1^2 + x_2^2} \tag{5.35}$$

and that the expression of the phase lag depends on the sign of x_2, namely:

$$\varphi = \begin{cases} \arctan(-x_1/x_2) & \text{if } x_2 < 0 \\ \pi + \arctan(-x_1/x_2) & \text{if } x_2 > 0. \end{cases} \tag{5.36}$$

If the nonzero assumption of x_2 does not hold, then from Eq. (5.29) signal $y(t)$ can obviously be expressed as

$$y(t) \approx 2x_1 \cos \omega t = 2x_1 \sin\left(\omega t - \frac{3\pi}{2}\right). \tag{5.37}$$

5.5.2 *Extracting GAM from Real-Time-Varying Signal*

Starting from Eq. (5.29) one aims to extract the real and imaginary parts of the first-order harmonic of signal $y(t)$. Note that the method presented next is valid for harmonics of any order.

5.5.2.1 Extraction of Real Part, $x_1 = \mathrm{Re}(\langle y \rangle_1)$

Equation (5.29) is multiplied by cos ωt, which is in fact multiplication of a real waveform. After performing an elementary trigonometric transform, the following expression is reached:

$$y(t) \cdot \cos(\omega t) = \underbrace{x_1}_{\text{DC term}} \underbrace{+ x_1 \cos 2\omega t - 2x_2 \sin \omega t \cos \omega t}_{\text{AC term}}, \tag{5.38}$$

where two distinct components have been identified: a DC one and a 2ω-pulsation AC component.

In order to extract x_1 it is sufficient to low-pass filtering the signal $y(t) \cdot \cos \omega t$ so as to eliminate the AC component without significantly affecting the dynamic of the continuous component (e.g., by using conveniently sized Butterworth filters). One must however pay attention to the delays inherently introduced by filtering and compensate for them if they are too significant.

5.5.2.2 Extraction of Imaginary Part, $x_2 = \mathrm{Im}(\langle y \rangle_1)$

This time the waveform (5.29) will be multiplied by sin ωt.

Figure 5.7 shows an example of what is called *amplitude demodulation* (Oppenheim et al. 1997), used for extracting the real and imaginary parts of the kth-order sliding harmonic. Obviously, information about the phase origin of the AC variable is necessary; usually this is obtained using phase-locked loops (PLL).

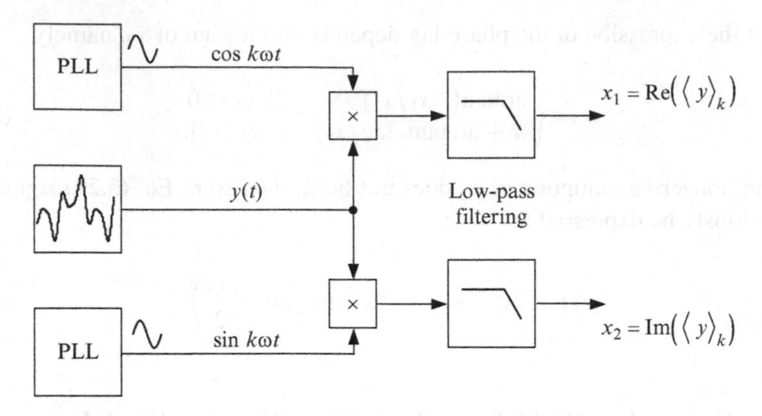

Fig. 5.7 Example of extracting real and imaginary parts of a generic sliding harmonic by amplitude demodulation

Figure 5.8 shows an example of applying the principle described in Fig. 5.7 to extracting the real and imaginary parts of harmonic components of a periodical signal (Fig. 5.8a). Second-order Butterworth low-pass filters have been used. Figures 5.8b–d show the result of demodulation in the case of the first-order, third-order and fifth-order harmonic, respectively. The original signal $y(t)$ variation occurring at about 0.5 s is due to modification of the real part of its first-order harmonic. One can also note slight dynamic variations occurring in the real and imaginary parts of all the other harmonic components, as well as the filter effect on the harmonic variations and on their ripples.

5.6 Using GAM for Expressing Active and Reactive Components of AC Variables

The circuit in Fig. 5.9 contains a coupling between the AC and DC parts, which is controlled via a switching function u taking values in the discrete set $\{-1; +1\}$. Two differential equations describing energy accumulation in inductance L and capacitor C compose the circuit switched model:

$$\begin{cases} L \cdot \dfrac{di_L}{dt} = e - v_0 \cdot u \\ C \cdot \dfrac{dv_0}{dt} = i_L \cdot u - i_S. \end{cases} \tag{5.39}$$

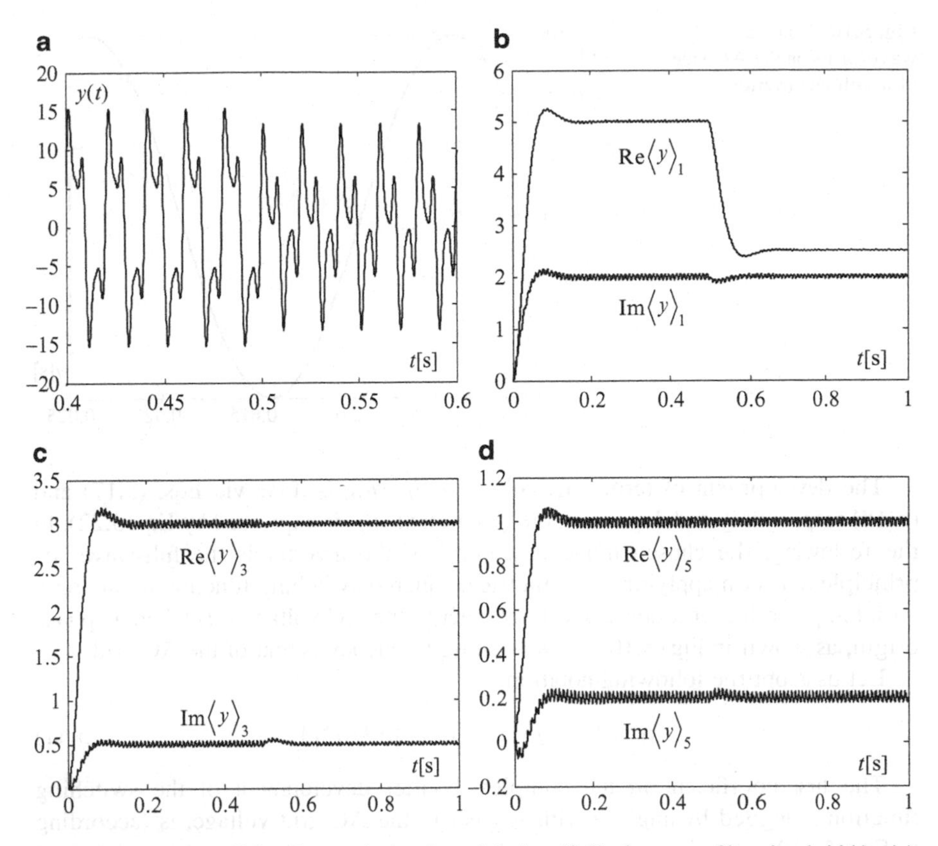

Fig. 5.8 Extracting sliding harmonic by demodulation (8 Hz low-pass filter bandwidth): (**a**) original signal; (**b**) first-order harmonic averaged components; (**c**) third-order harmonic averaged components; (**d**) fifth-order harmonic averaged components

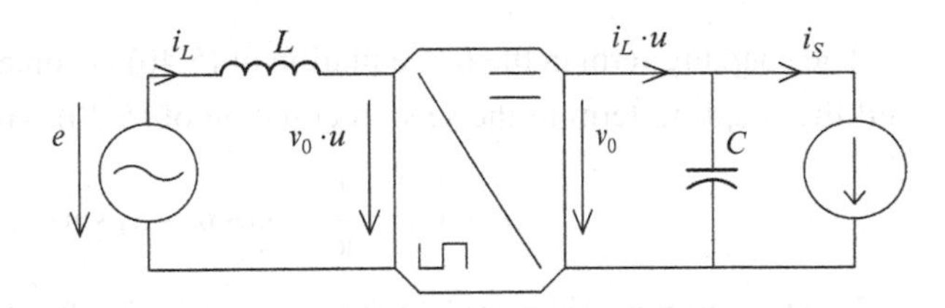

Fig. 5.9 Voltage-source inverter block diagram

If only the first harmonic of the AC variables and the average of the DC variables are of interest, the GAM applied to the voltage inverter is obtained by averaging the above equations:

$$\begin{cases} \dfrac{d\langle i_L\rangle_1}{dt} = -j\omega\cdot\langle i_L\rangle_1 + \dfrac{1}{L}\left(\langle e\rangle_1 - \langle v_0\cdot u\rangle_1\right) \\ \dfrac{d\langle v_0\rangle_0}{dt} = \dfrac{1}{C}\left(\langle i_L\cdot u\rangle_0 - \langle i_S\rangle_0\right). \end{cases} \tag{5.40}$$

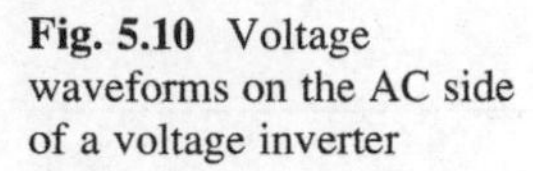

Fig. 5.10 Voltage waveforms on the AC side of a voltage inverter

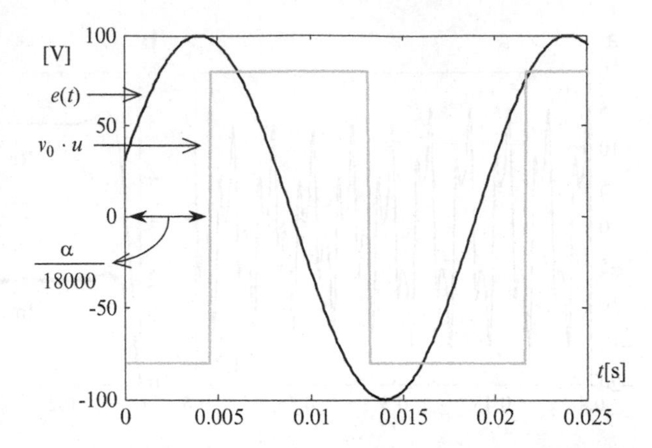

The development of terms $\langle v_0 \cdot u \rangle_1$ and $\langle i_L \cdot u \rangle_0$ is done via Eqs. (5.17) and (5.18) respectively and depend on the function u; $\langle u \rangle_1$ is expressed by Eq. (5.22). In the following, the chosen mode of control of the inverter is the full-wave. Its principle relies on applying a symmetrical squared switching function u, having a variable phase lag of α degrees with respect to the AC voltage $e(t)$ taken as phase origin, as shown in Fig. 5.10. The switching frequency is that of the AC grid.

Let us adopt the following notation:

$$x_1 = \mathrm{Re}\left(\langle i_L \rangle_1\right), \quad x_2 = \mathrm{Im}\left(\langle i_L \rangle_1\right). \tag{5.41}$$

The first coefficient in the complex Fourier development of the switching function u, lagged by angle α with respect to the AC grid voltage, is (according to Eq. (5.24)):

$$\langle u \rangle_1 = \frac{2}{j\pi} e^{j\alpha}.$$

The coupling term in the first equation of (5.40) becomes $\langle v_0 \cdot u \rangle_1 = \frac{2}{j\pi} e^{j\alpha} \cdot \langle v_0 \rangle_0$ and the coupling term in the second equation of (5.40) – using Eq. (5.29) – gives

$$\langle i_L \cdot u \rangle_0 = \frac{4}{\pi} (x_2 \cos\alpha - x_1 \sin\alpha).$$

If $e(t)$ is supposed sinusoidal with magnitude $\hat{E}$, after having identified the real and imaginary parts, Eq. (5.40) become

$$\begin{cases} \dfrac{dx_1}{dt} = \omega \cdot x_2 - \langle v_0 \rangle_0 \cdot \dfrac{2}{\pi L} \cdot \sin\alpha \\ \dfrac{dx_2}{dt} = -\omega \cdot x_1 - \dfrac{\hat{E}}{2L} + \langle v_0 \rangle_0 \cdot \dfrac{2}{\pi L} \cdot \cos\alpha \\ \dfrac{d\langle v_0 \rangle_0}{dt} = \dfrac{4}{\pi C} \cdot (x_2 \cos\alpha - x_1 \sin\alpha) - \dfrac{\langle i_s \rangle_0}{C}. \end{cases} \tag{5.42}$$

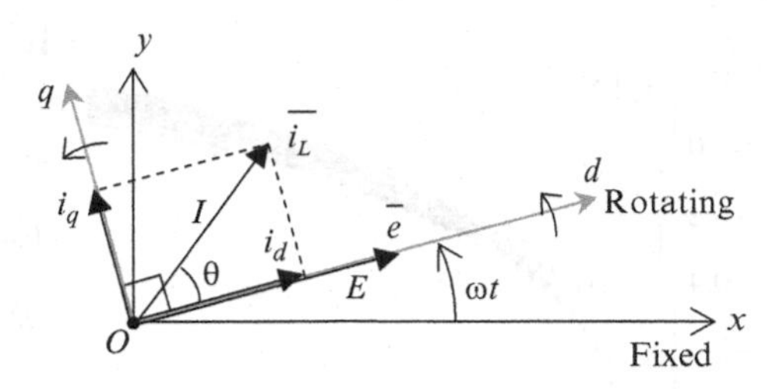

Fig. 5.11 Phasor diagram introducing the *dq* frame

This model can be written in a more "conventional" form by identifying the active and reactive inductor current components, i_d and i_q, respectively, instead of using the real and imaginary parts of the first Fourier coefficient. Thus, taking into account that i_L has sinusoidal variation of pulsation ω, the relation $i_L = \langle i_L \rangle_{-1} e^{-j\omega t} + \langle i_L \rangle_1 e^{j\omega t}$ holds (Eq. (5.28)). Knowing that $\langle i_L \rangle_1 = x_1 + jx_2$ and that harmonic coefficients of opposite sign are complex conjugated, reveals that $\langle i_L \rangle_{-1} = x_1 - jx_2$; then, simple algebra leads to (see also Eq. (5.29)):

$$i_L = 2(x_1 \cos \omega t - x_2 \sin \omega t). \tag{5.43}$$

A sinusoidal AC variable is a rotating vector (phasor) with pulsation ω; its active and reactive components can be obtained by projecting this vector on axes of a frame rotating synchronously with ω. The active component (on abscissa axis) is usually denoted by subscript *d*, whereas reactive component (on ordinate axis) is denoted by subscript *q*. These notations give the name of *dq* frame (Bose 2001).

In order to write current i_L in the *dq* frame, the phasor diagram in Fig. 5.11 is introduced, where voltage phasor $\overline{e}$ and current phasor $\overline{i_L}$ rotate with the grid pulsation with respect to an absolute (stationary) frame *xOy*, the phase lag between them being denoted by θ. This means that $e(t) = E \sin \omega t$ and $i_L(t) = I_L \sin(\omega t + \theta)$.

Further, a *dq* frame synchronous with phasor $\overline{e}$ and having the same instantaneous phase is considered. The two phasors are projected on the axis of this new frame; the *d*- and *q*-components of i_L are denoted by i_d and i_q, respectively, and serve at expressing the active and reactive power – e.g., $P = E \cdot I_L \cdot \cos \theta = E \cdot i_d$.

Now, let us again refer to the original stationary frame *xOy* in which both components, i_d and i_q, rotate with the grid pulsation, their sum equaling i_L:

$$i_L = i_d \sin \omega t + i_q \sin(\omega t + \pi/2) = i_d \sin \omega t + i_q \cos \omega t. \tag{5.44}$$

Therefore, on one hand, the inductor current is given by Eq. (5.43), on the other hand it is expressed as depending on its active and reactive components, i_d and i_q, according to Eq. (5.44).

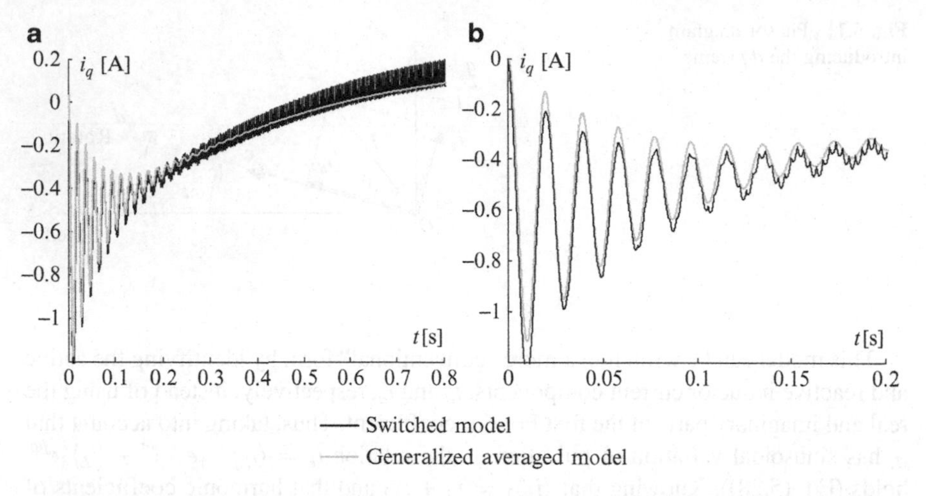

Fig. 5.12 (**a**) Reactive current waveform at start-up (full-wave control): GAM model vs. switched model; (**b**) zoom of reactive current waveform (Bacha and Gombert 2006)

By identifying analogous terms of Eqs. (5.43) and (5.44), one obtains the following relations:

$$\begin{cases} i_d = -2x_2 \\ i_q = 2x_1, \end{cases} \quad \text{or equivalently} \begin{cases} x_1 = \dfrac{i_q}{2} \\ x_2 = -\dfrac{i_d}{2}, \end{cases}$$

which can be viewed as a variable change in the state model (5.42). In this way, a new state model results, having as states the active and reactive components of current i_L as well as the DC-link voltage average $\langle v_0 \rangle_0$:

$$\begin{cases} \dfrac{di_d}{dt} = \omega i_q + \dfrac{\hat{E}}{L} - \langle v_0 \rangle_0 \cdot \dfrac{4}{\pi L} \cdot \cos\alpha \\ \dfrac{di_q}{dt} = -\omega i_d - \langle v_0 \rangle_0 \cdot \dfrac{4}{\pi L} \cdot \sin\alpha \\ \dfrac{d\langle v_0 \rangle_0}{dt} = -\dfrac{2}{\pi C} \cdot \left(i_d \cos\alpha + i_q \sin\alpha \right) - \dfrac{\langle i_s \rangle_0}{C}. \end{cases} \tag{5.45}$$

This large signal model is nonlinear; it can be useful for performing fast numerical simulations, for control law design, for supporting the circuit design or establishing the state-space or frequency small-signal model.

Figures 5.12 and 5.13 show the concordance between the switched and the generalized averaged models. The differences are mainly due to the fact that the GAM takes into account only the first harmonic in the AC variables and the average

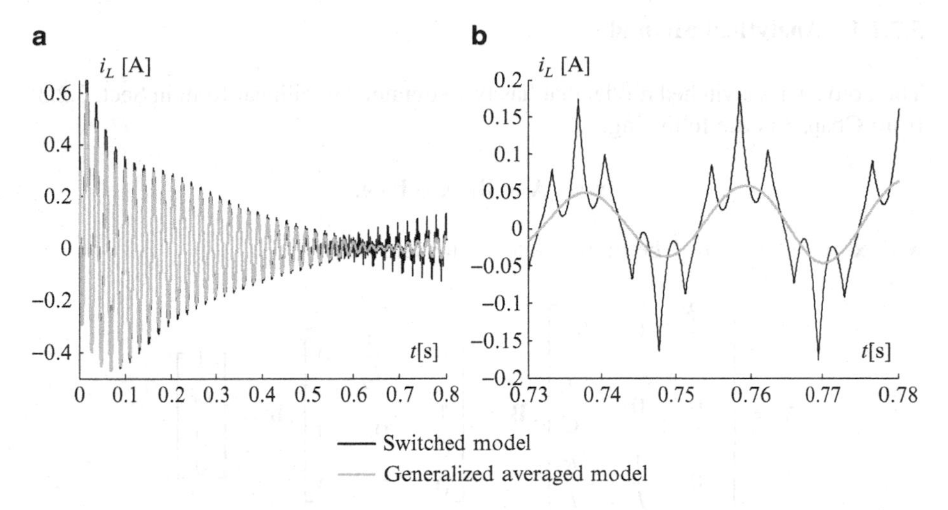

Fig. 5.13 (**a**) AC current waveform at start-up (full-wave control): GAM model vs. switched model; (**b**) zoom of the current waveform (Bacha and Gombert 2006)

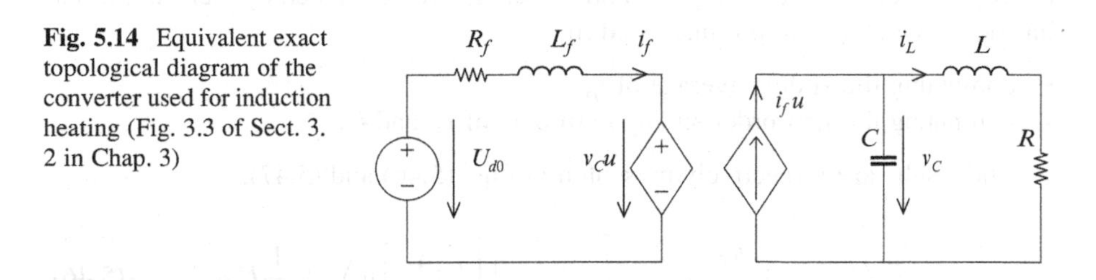

Fig. 5.14 Equivalent exact topological diagram of the converter used for induction heating (Fig. 3.3 of Sect. 3.2 in Chap. 3)

of the DC variables (higher-order harmonics are neglected). The precision of GAM increases with the filtering quality in the AC part.

5.7 Case Studies

The following case studies will illustrate the performance of the generalized averaged model against that of the switched model.

5.7.1 Current-Source Inverter for Induction Heating

One can start directly with the switched model of the converter used for induction heating, studied in Chap. 3, Sect. 3.2, and repeated in Fig. 5.14.

We will illustrate the application of both analytical and graphical approach for obtaining the GAM of this converter (Bacha et al. 1995; Bendaas et al. 1995). The equivalent exact topological diagram of this converter can be seen in Fig. 5.8.

5.7.1.1 Analytical Method

The converter's switched model that has been obtained in bilinear form in Sect. 3.2.3 from Chap. 3 is the following:

$$\dot{\mathbf{x}} = (\mathbf{A} + \mathbf{B}u)\mathbf{x} + \mathbf{b} \cdot \mathbf{e},$$

with $\mathbf{x} = \begin{bmatrix} i_f & v_C & i_L \end{bmatrix}^T$ being the state vector and

$$\mathbf{A} = \begin{bmatrix} -\frac{R_f}{L_f} & 0 & 0 \\ 0 & 0 & -\frac{1}{C} \\ 0 & \frac{1}{L} & -\frac{R}{L} \end{bmatrix}; \mathbf{B} = \begin{bmatrix} 0 & -\frac{1}{L_f} & 0 \\ \frac{1}{C} & 0 & 0 \\ 0 & 0 & 0 \end{bmatrix}; \mathbf{b} = \begin{bmatrix} \frac{1}{L_f} \\ 0 \\ 0 \end{bmatrix}.$$

The continuous current i_f through the filter is essentially characterized by its average value, whereas voltage v_C and current i_L are dominated by their first-order harmonic. One is therefore interested in

- computing the sliding average of i_f;
- computing the first-order sliding harmonic of v_C and i_L.

The results are respectively presented in Eqs. (5.46) and (5.47).

$$\frac{d}{dt}\langle i_f \rangle_0 = -\frac{R_f}{L_f}\langle i_f \rangle_0 + \begin{bmatrix} -\frac{1}{L_f} & 0 \end{bmatrix} \left\langle \begin{bmatrix} v_C \\ i_L \end{bmatrix} u \right\rangle_0 + \frac{1}{L_f} U_{d0}, \tag{5.46}$$

$$\frac{d}{dt}\left\langle \begin{bmatrix} v_C \\ i_L \end{bmatrix} \right\rangle_1 = -j\omega \left\langle \begin{bmatrix} v_C \\ i_L \end{bmatrix} \right\rangle_1 + \begin{bmatrix} \frac{1}{C} \\ 0 \end{bmatrix} \langle i_f u \rangle_1 + \begin{bmatrix} 0 & -\frac{1}{C} \\ \frac{1}{L} & -\frac{R}{L} \end{bmatrix} \left\langle \begin{bmatrix} v_C \\ i_L \end{bmatrix} \right\rangle_1. \tag{5.47}$$

By developing the expression of the first-order sliding average of vector $\begin{bmatrix} v_C & i_L \end{bmatrix}^T$ and by distinguishing between the real parts and the imaginary parts of complex variables, one can put

$$\begin{cases} \langle v_C \rangle_1 = x_2 + jx_3 \\ \langle i_L \rangle_1 = x_4 + jx_5. \end{cases}$$

For a more homogenous writing, the following notation is adopted:

$$\langle i_f \rangle_0 = x_1.$$

Knowing that the switching function has a square waveform and by taking it as the phase origin, one deduces that $\langle u\rangle_1 = -\ 2j/\pi$. Expressing the coupled terms leads to

$$\begin{aligned}\langle v_C u\rangle_0 &= \langle v_C\rangle_1\langle u\rangle_{-1} + \langle v_C\rangle_{-1}\langle u\rangle_1 \\ &= 2\left[\mathrm{Re}(\langle v_C\rangle_1)\mathrm{Re}(\langle u\rangle_1) + \mathrm{Im}(\langle v_C\rangle_1)\mathrm{Im}(\langle u\rangle_1)\right] = -\frac{4}{\pi}\mathrm{Im}(\langle v_C\rangle_1) = -\frac{4}{\pi}x_3\end{aligned}$$

and to

$$\langle i_f u\rangle_1 = \langle i_f\rangle_1\langle u\rangle_0 + \langle i_f\rangle_0\langle u\rangle_1 = \langle i_f\rangle_0\langle u\rangle_1 = -\frac{2}{\pi}j\langle i_f\rangle_0 = -\frac{2}{\pi}jx_1.$$

After some computation, systems (5.46) and (5.47) can be merged so as to obtain the final form of the converter's GAM:

$$\begin{cases}\dot{x_1} = -\dfrac{R_f}{L_f}x_1 + \dfrac{4}{\pi L_f}x_3 + \dfrac{U_{d0}}{L_f} \\ \dot{x_2} = \omega x_3 - \dfrac{1}{C}x_4 \\ \dot{x_3} = -\dfrac{2}{\pi C}x_1 - \omega x_2 - \dfrac{1}{C}x_5 \\ \dot{x_4} = \dfrac{1}{L}x_2 - \dfrac{R}{L}x_4 + \omega x_5. \\ \dot{x_5} = \dfrac{1}{L}x_3 - \omega x_4 - \dfrac{R}{L}x_5\end{cases} \tag{5.48}$$

The output models result according to the following set of equations (based upon Eq. (5.29)):

$$\begin{cases} i_f = x_1 \\ v_C = 2(x_2\cos\omega t - x_3\sin\omega t) \\ i_L = 2(x_4\cos\omega t - x_5\sin\omega t).\end{cases} \tag{5.49}$$

5.7.1.2 Graphical Method

The same result can be obtained by the graphical approach. By applying the previously detailed Algorithm 5.1 on the equivalent diagram in Fig. 5.14, gives the diagram in Fig. 5.15. The changes brought by adding AC impedances are emphasized by gray-filled areas. Using the equivalent diagram from Fig. 5.15, one need only write down Kirchhoff's equations to obtain the analytical model in forms (5.46) and (5.47) and to finalize the computation in forms (5.48) and (5.49).

Remark. If considering the steady-state regime –derivatives of state variables are null – of the diagram in Fig. 5.15, all the dynamical elements, i.e., inductances and capacitor, no longer appear. From here the first-order-harmonic model results, as shown in Fig. 5.16.

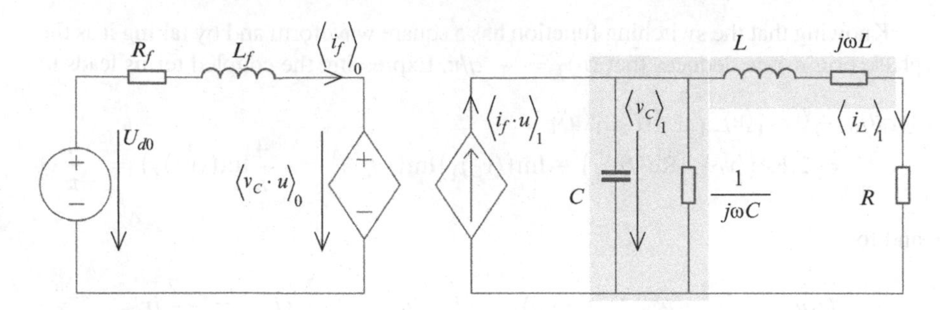

Fig. 5.15 Equivalent diagram corresponding to the GAM of the converter

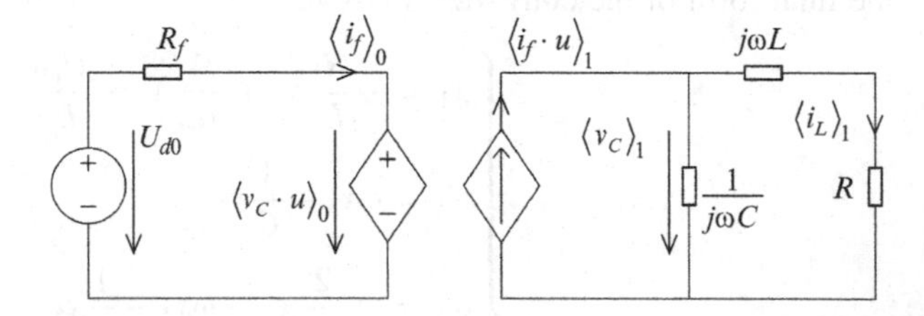

Fig. 5.16 Equivalent diagram of first-order-harmonic model for converter used for induction-based heating

The results obtained so far provide an opportunity for some remarks.

First, note that the final GAM shown in Eq. (5.48) represents a continuous invariant system, which is also bilinear because it has as control input the pulsation. One can linearize this system around an equilibrium point for purpose of modal analysis or for designing a linear control law.

The GAM (5.48) is characterized by having taken into account all the dynamics of the original system. Its order is, however, higher than that of the switched model because it captures not only the dynamic regime but also the steady-state regime of those AC variables which depend on time.

Figures 5.17, 5.18, and 5.19 serve to make a comparison between the "exact" behavior of the induction-based heating converter and its behavior predicted by the GAM.

As suggested by this comparison, the two models have quite different accuracy properties when good filtering conditions are not met. The initial conditions used in simulation have been: $U_{d0} = 297\,\text{V}, R = 1\Omega, C = 6\,\mu\text{F}, R_f = 0.01\,\Omega, L = 100\,\mu\text{H}$. Two values of the filtering inductance have been considered, namely, $L_f = 600\,\mu\text{H}$ (weak filtering) and $L_f = 6\,\text{mH}$ (strong filtering). A step of control input (switched function u) frequency has been applied in the middle of the time range.

Figure 5.17 shows the filter current i_f (to the left) and capacitor voltage v_C (right) evolutions as the control input frequency significantly decreases. Both the i_f average and the ripple diminish. The v_C envelope changes with a significant nonminimum-phase behavior. A zoom of these latter waveforms can be seen in Fig. 5.18, left side. One can note that the switched model is richer in harmonics than the variable issued from GAM. The right plot in Fig. 5.18 contains variations of the output inductor current i_L, as given by the switched model and by the GAM (one can note they are practically identical).

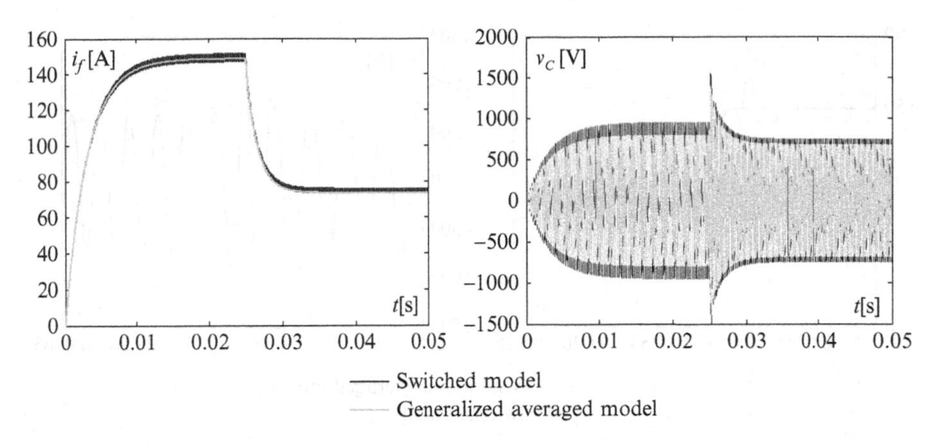

Fig. 5.17 GAM vs. switched model of an induction-based heating converter: comparison by simulation in the case of strong filtering

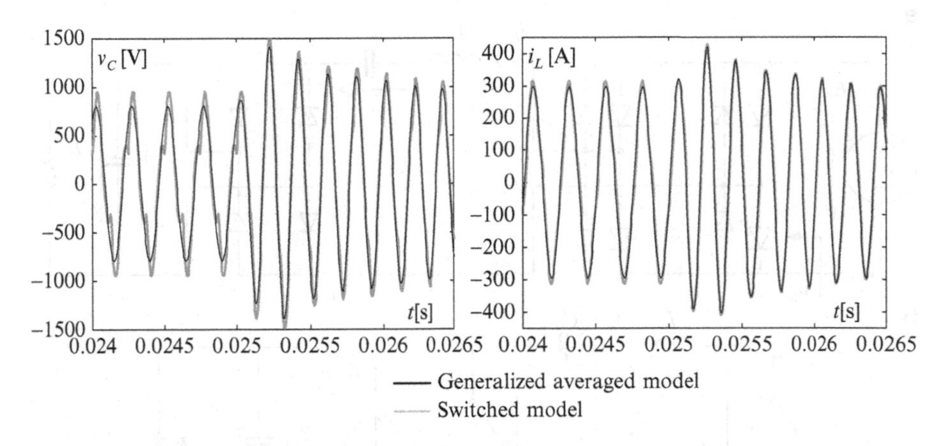

Fig. 5.18 GAM vs. switched model of an induction-based heating converter: comparison by simulation in case of strong filtering (detail)

Figure 5.19 deals with the case of weak filtering. On the left a large i_f ripple together with a fast response can be seen. A zoom of the v_C waveforms gives the difference qualitatively between the switched and GAM outputs.

5.7.2 Series-Resonant Converter

Let the series-resonant power supply with diode rectifier be considered, whose circuit diagram and equivalent diagram are shown in Figs. 5.20a, b, respectively.

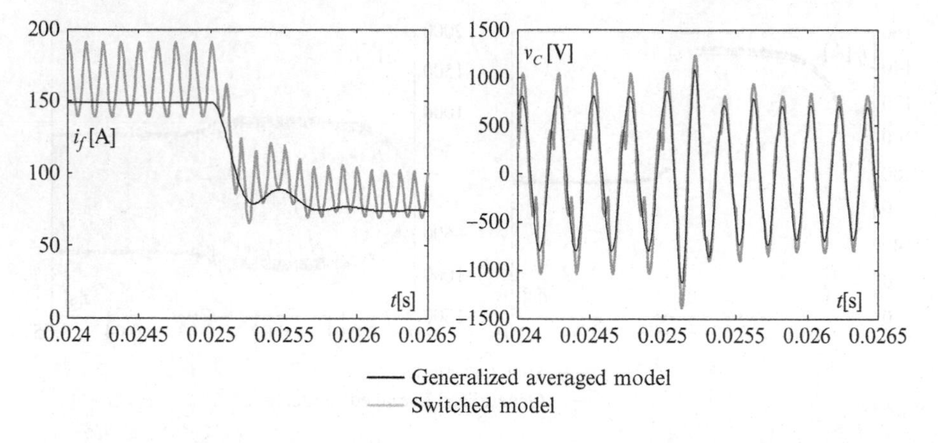

Fig. 5.19 GAM vs. switched model of an induction-based heating converter: comparison by simulation in case of weak filtering (detail)

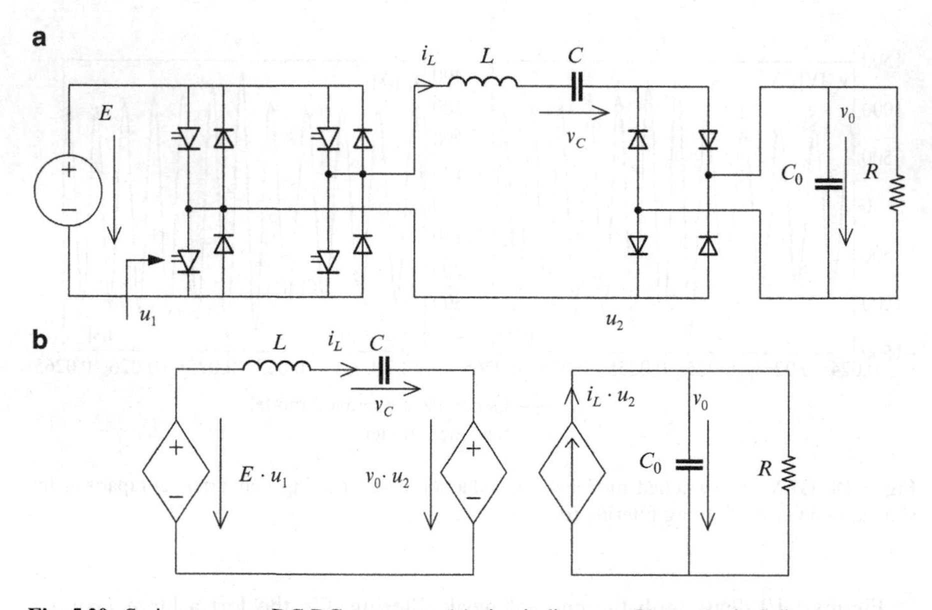

Fig. 5.20 Series-resonant DC-DC converter: (**a**) circuit diagram; (**b**) topological diagram

In order to present general results, one normalizes the switched model by adopting the following notations, representing normalized values (Bacha 1993; Sira-Ramírez and Silva-Ortigoza 2006):

$$\xi = \frac{1}{2} r \sqrt{\frac{C}{R}}, \quad i = \frac{i_L}{\omega_0 C E}, \quad v = \frac{v_C}{E}, \quad v_0 = \frac{v_{C0}}{E},$$

$$\omega = \frac{\omega_S}{\omega_0}, \quad \omega_0 = \frac{1}{\sqrt{LC}}, \quad \tau = \omega_0 t, \quad \theta_0 = \omega_0 R_0 C_0,$$

where r is the inductor resistance, ω_S is the switching frequency in radians, ω_0 is the series resonance frequency, Θ_0 is the circuit quality factor. A new time variable, τ, has thus been obtained, which is expanded with respect to the original time variable t. The normalized model will operate at the normalized frequency ω. The switching functions u_1 and u_2 have the following analytical expressions:

$$u_1 = \text{sgn}(\sin \omega\tau), u_2 = \text{sgn}(i\omega_0 CE).$$

The equations describing the converter operation according to the switched model become by normalization

$$\begin{cases} \dfrac{d}{d\tau}i = -2\xi i - v + u_1 - v_0 u_2 \\ \dfrac{d}{d\tau}v = i \\ \dfrac{d}{d\tau}v_0 = \dfrac{C}{C_0}iu_2 - \dfrac{v_0}{\theta_0}. \end{cases} \tag{5.50}$$

As in the previous case, one can obtain the GAM of model (5.50) (see the analytical approach in Sect. 5.4.1). The continuous state variable is v_0; we are interested in its sliding average. The AC variables are represented by their first sliding harmonic. Employing Eq. (5.3), model (5.50) becomes

$$\begin{cases} \dfrac{d}{d\tau}\langle i\rangle_1 = -j\omega\langle i\rangle_1 - 2\xi\langle i\rangle_1 - \langle v\rangle_1 + \langle u_1\rangle_1 - \langle v_0 \cdot u_2\rangle_1 \\ \dfrac{d}{d\tau}\langle v\rangle_1 = -j\omega\langle v\rangle_1 + \langle i\rangle_1 \\ \dfrac{d}{d\tau}\langle v_0\rangle_0 = \dfrac{C}{C_0}\langle i \cdot u_2\rangle_0 - \dfrac{\langle v_0\rangle_0}{\theta_0}. \end{cases} \tag{5.51}$$

The term $\langle u_1\rangle_1$ is computed as above: $\langle u_1\rangle_1 = -2j/\pi$. If current i_L represented by its first harmonic is phase-shifted by an angle φ in relation to the phase origin, then

$$\langle u_1\rangle_1 = -2je^{j\varphi}/\pi.$$

The following notations are adopted:

$$\langle i\rangle_1 = x_1 + jx_2, \langle v\rangle_1 = x_3 + jx_4, \langle v_0\rangle_0 = x_5.$$

The angle φ is given by Eq. (5.36), namely

$$\varphi = -\text{atan}(x_1/x_2).$$

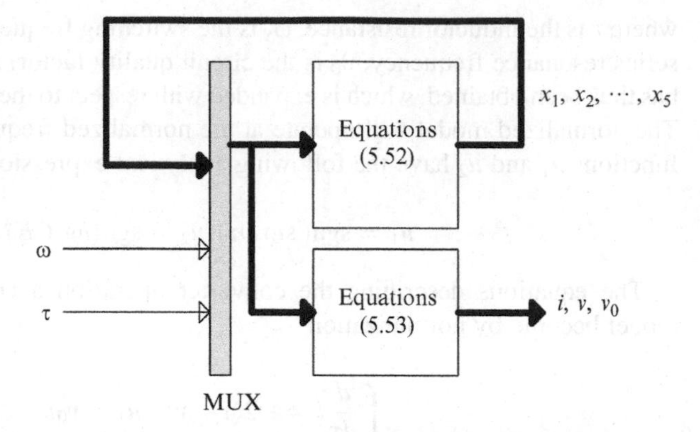

Fig. 5.21 Implementation of model represented by Eqs. (5.52) and (5.53)

The coupled terms may also be computed as in the previous example, with the particular feature that function u_2 depends on the sign of the current through the resonant circuit. After separating the real and the imaginary part of the complex terms, one finally obtains a five-state GAM in the form given by Eq. (5.52). This model is ready to be implemented in a simulation software package, e.g., MATLAB®-Simulink®. The model inputs are the normalized time variable τ and the control input ω (normalized frequency), whereas its outputs are given by Eq. (5.53).

$$\begin{cases} \dot{x_1} = -2\xi x_1 + \omega x_2 - x_3 - \dfrac{2}{\pi}\dfrac{x_5}{\sqrt{x_1^2+x_2^2}}x_1 \\ \dot{x_2} = -\omega x_1 - 2\xi x_2 - x_4 - \dfrac{2}{\pi} - \dfrac{2}{\pi}\dfrac{x_5}{\sqrt{x_1^2+x_2^2}}x_2 \\ \dot{x_3} = x_1 + \omega x_4 \\ \dot{x_4} = x_2 - \omega x_3 \\ \dot{x_5} = \dfrac{4C}{\pi C_0}\cdot\sqrt{x_1^2+x_2^2} - \dfrac{x_5}{\theta_0}, \end{cases} \tag{5.52}$$

$$\begin{cases} i = 2(x_1 \cos\omega\tau - x_2 \sin\omega\tau) \\ v = 2(x_3 \cos\omega\tau - x_4 \sin\omega\tau) \\ v_0 = x_5. \end{cases} \tag{5.53}$$

An example of implementation is presented in Fig. 5.21.

5.7.3 *Limitations of GAM: Example*

The goal of this example is to enrich the averaged model of a DC-DC converter (Fig. 5.22) such as to render it closer to its switched model, i.e., to add information about the state variable ripple.

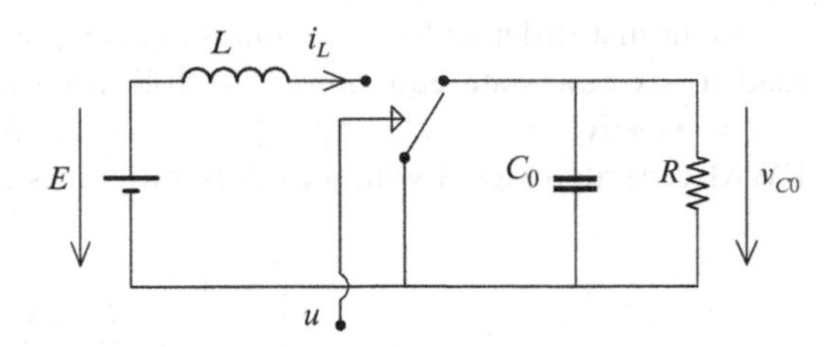

Fig. 5.22 Boost converter schematics

Switching functions have not always square waveforms; thus, in asymmetrical switching converters, these functions have variable duty ratio. Hereafter, one can follow the approach of obtaining a continuous model of a boost converter assumed to operate in continuous conduction.

The switched model of this converter was already been deduced in Sect. 3.2 from Chap. 3, see Eq. (3.11). For the sake of simplicity, let us now suppose that $u = 1$ when the switch is turned off and $u = 0$ otherwise. Therefore, α is defined here as the ratio of the time interval during which the switch is turned off and the switching period.

As in the previous example, a normalization of this model is envisioned. Voltages are normalized to the supply voltage E, and the current is divided by $C\omega_0 E$ (where ω_0 is the resonance frequency of circuit L, C_0). By adopting the notation $\theta_0 = \omega_0 R_0 C_0$ (representing the quality factor of the circuit) the model describing the converter is composed of the following normalized equations:

$$\begin{cases} \dfrac{d}{d\tau} i = 1 - v \cdot u \\ \dfrac{d}{d\tau} v = i \cdot u + \dfrac{v}{\theta_0}, \end{cases} \tag{5.54}$$

where i and v are the normalized values of the inductor current and capacitor voltage. Each state variable has a DC component and an AC component at the switching frequency. The normalized time variable τ is defined as $\tau = \omega_0 t$. The switching frequency also is normalized, i.e., $\omega = \omega_S/\omega_0$, with ω_S being the first-order harmonic pulsation. In order to obtain the GAM, one performs zero-order and the first-order averaging of both terms of each equation from (5.54):

$$\begin{cases} \dfrac{d}{d\tau}\langle i\rangle_0 = 1 - \langle v \cdot u\rangle_0 \\ \dfrac{d}{d\tau}\langle v\rangle_0 = \langle i \cdot u\rangle_0 + \dfrac{1}{\theta_0}\langle v\rangle_0 \\ \dfrac{d}{d\tau}\langle i\rangle_1 = -j\omega\langle x\rangle_1 + \langle v \cdot u\rangle_1 \\ \dfrac{d}{d\tau}\langle v\rangle_1 = -j\omega\langle v\rangle_1 + \langle i \cdot u\rangle_1 + \dfrac{1}{\theta_0}\langle v\rangle_1. \end{cases} \tag{5.55}$$

As the first-order sliding harmonics are complex, the two equations of (5.54) will lead to six new state equations. The following notations are adopted: $x_1 = \langle i\rangle_0$, $\langle i\rangle_1 = x_2 + jx_3$, $x_4 = \langle v\rangle_0$ and $\langle v\rangle_1 = x_5 + jx_6$. As the switching function u is a PWM-generated signal with α as duty ratio, it is easy to show that

$$\begin{cases} \langle u\rangle_0 = \alpha \\ \langle u\rangle_1 = \dfrac{M + jN}{2\pi}, \end{cases}$$

where $M = \sin(2\pi\alpha)$ and $N = \cos(2\pi\alpha) - 1$.

Further, the terms of products $\langle y \cdot u\rangle_0$ and $\langle y \cdot u\rangle_1$ appearing in the averaged state Eq. (5.55) – where y denotes either i or v – are respectively given by

$$\begin{cases} \langle y \cdot u\rangle_0 \approx \langle y\rangle_1 \langle u\rangle_{-1} + \langle y\rangle_{-1} \langle u\rangle_{+1} + \langle y\rangle_0 \langle u\rangle_0 \\ \langle y \cdot u\rangle_1 \approx \langle y\rangle_0 \langle u\rangle_1 + \langle y\rangle_1 \langle u\rangle_0. \end{cases}$$

Computation may be simplified by noting that the complex harmonics of the same order and opposite sign are conjugates. Each state variable is described by its average to which the first-order harmonic is added. This can be written as

$$\begin{cases} i \approx x_1 + 2(x_2 \cos\omega t - x_3 \sin\omega t) \\ v \approx x_4 + 2(x_5 \cos\omega t - x_6 \sin\omega t). \end{cases} \tag{5.56}$$

In this case, the boost converter GAM turns out to be a more accurate model – it has the added precision of the first harmonic –described by six state equations:

$$\begin{cases} \dot{x_1} = 1 - \alpha x_4 - (2Mx_5 + 2Nx_6)/(2\pi) & \dot{x_2} = \omega x_3 - Mx_4/(2\pi) - \alpha x_5 \\ \dot{x_3} = -\omega x_2 - Nx_4/(2\pi) - \alpha x_6 & \dot{x_4} = \alpha x_1 + (2Mx_2 + 2Nx_3)/(2\pi) - x_4/\theta_0 \\ \dot{x_5} = Mx_1/(2\pi) + \alpha x_2 - x_5/\theta_0 + \omega x_6 & \dot{x_6} = Nx_1/(2\pi) + \alpha x_3 - \omega x_5 - x_6/\theta_0. \end{cases} \tag{5.57}$$

Model (5.57) has α as control input and outputs given by (5.56). More simplified versions of model (5.56) can be obtained, for example, by ignoring the effect of output voltage oscillations v. Then derivatives and values of variables x_5 and x_6 are zeroed and the model is reduced to fourth order:

$$\begin{cases} \dot{x_1} = 1 - \alpha x_4 & \dot{x_2} = \omega x_3 - Mx_4/(2\pi) \\ \dot{x_3} = -\omega x_2 - Mx_4/(2\pi) & \dot{x_4} = \alpha x_1 + (2Mx_2 + 2Nx_3)/(2\pi) - x_4/\theta_0. \end{cases} \tag{5.58}$$

Finally, by only taking a first-order approximation, one can ignore the oscillations of current i (x_2 and x_3 are zeroed) and the classical averaged model is obtained:

$$\{\dot{x_1} = 1 - \alpha x_4 \quad \dot{x_4} = \alpha x_1 - x_4/\theta_0. \tag{5.59}$$

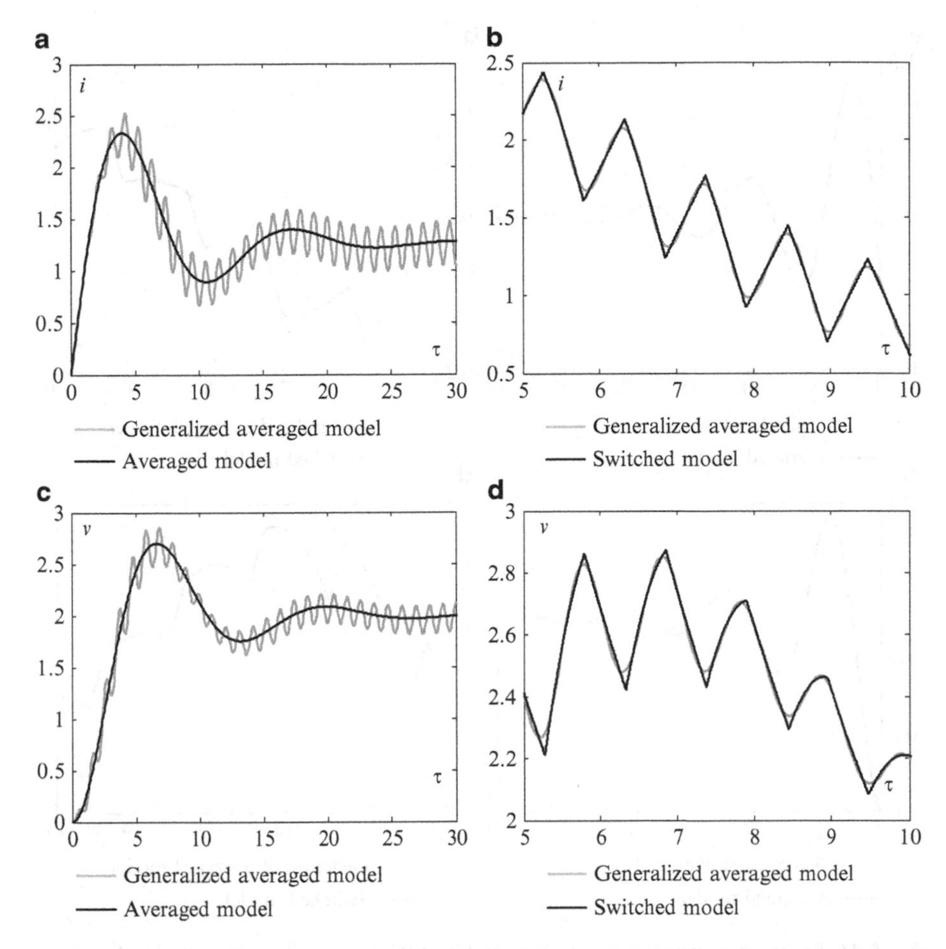

Fig. 5.23 Response at turn on of circuit from Fig. 5.22 for $\alpha = 0.5$: (**a**) normalized current through inductance L, i – GAM vs. averaged model; (**b**) i as given by GAM vs. switched model; (**c**) output normalized voltage, v – GAM vs. averaged model; (**d**) v as given by GAM vs. switched model

In conclusion, three models and the switched model have been obtained by now, all four usable to describe the boost converter behavior. Numerical simulations can help us explore the information added by the GAM – in its two forms (5.57) and (5.58) – against the classical averaged model (5.59).

Figures 5.23 and 5.24 show the boost converter GAM output vs. its switched model. Input data are $E = 5$ V, $L = 1$ mH, $C_0 = 100$ μF, $R = 10$ Ω, $f_S = 3$ kHz ($\omega \approx 6$ and $\theta_o \approx 3$). Simulations have been performed for two different values of duty ratio, namely $\alpha = 0.5$ and $\alpha = 0.8$.

The above results lead to the following remarks. The case of a 0.5 duty ratio is ideal, since the output of model (5.57) is practically identical with the switched model output. If one solves the model (5.58), then the difference noted in the

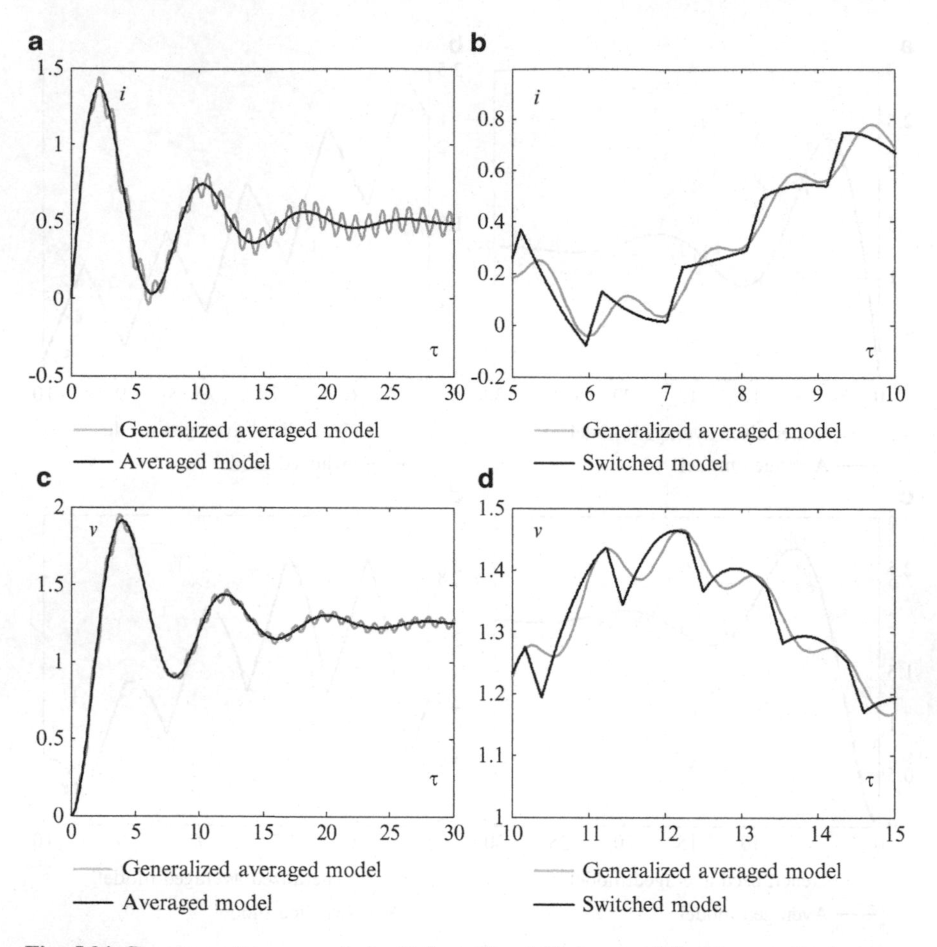

Fig. 5.24 Response at turn on of circuit from Fig. 5.22 for $\alpha = 0.8$: (**a**) normalized current through inductance L, i – GAM vs. averaged model; (**b**) i as given by GAM vs. switched model; (**c**) output normalized voltage, v – GAM vs. averaged model; (**d**) v as given by GAM vs. switched model

current evolution is negligible. However, model (5.57) provides a more realistic representation of the output voltage, since oscillations are not neglected.

The accuracy of generalized averaged models deteriorates when the harmonic content of function u becomes richer, corresponding to duty ratio values other than 0.5 – this can be seen in Fig. 5.24.

Note that information about the pulse-width modulation carrier phase is required for a coherent comparison of GAM and the switched model outputs to be made. Simulation results suggest that the harmonic approach is not totally suitable for modeling asymmetrical converters.

Indeed, in order to obtain an acceptable modeling precision irrespective of the value of the duty ratio, one must extend the investigation to higher-order harmonics,

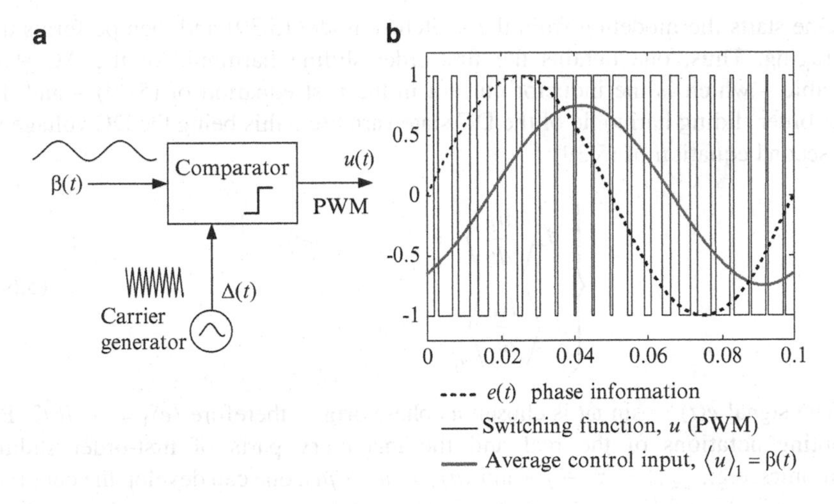

Fig. 5.25 Pulse-width modulation: (**a**) basic diagram; (**b**) main signals

which complicates significantly the computation. For example, considering second-order harmonics would result in a ninth-order model.

5.7.4 PWM-Controlled Converters

The model of a given power electronic converter significantly depends on the nature of the switching function, e.g., the PWM inverters behave differently than the full-wave operated inverter. This section deals with two types of voltage-source converters, corresponding to the single-phase and three-phase case, respectively.

5.7.4.1 Single-Phase Case

The converter circuit has the same topology as depicted in Fig. 5.9 and the same switched model. The difference lies in the control signal, as the converter is not controlled in full wave anymore, but with a pulse-width-modulated signal u.

There are many possibilities for producing the PWM switching function; Fig. 5.25 shows the basic analog solution, where

$$u(t) = \mathrm{sgn}(\beta(t) - \Delta(t)) \text{ with } \mathrm{sgn}(x) = \begin{cases} 1 & \text{if } x \geq 0 \\ -1 & \text{if } x < 0. \end{cases}$$

Converter behavior is controlled by modifying either the amplitude of the input signal $\beta(t)$ in order to increase the AC power, or its phase (measured with respect to the grid voltage phase) in order to vary the balance between active and reactive AC power.

One starts the modeling from the switched model (5.39) and then performs the averaging. Thus, one obtains the first-order sliding harmonic of the AC state variable – which is the inductor current in the first equation of (5.39) – and the zero-order sliding harmonic of the DC state variable – this being the DC voltage in the second equation of (5.39):

$$\begin{cases} L\left\langle \dfrac{di_L}{dt} \right\rangle_1 = \langle e \rangle_1 - \langle v_O u \rangle_1 \\ C\left\langle \dfrac{dv_O}{dt} \right\rangle_0 = \langle i_L u \rangle_0 - \langle i_S \rangle_0. \end{cases} \tag{5.60}$$

The signal $e(t) = \sin \omega t$ is chosen as phase origin, therefore $\langle e \rangle_1 = -jE/2$. By adopting notations of the real and the imaginary parts of first-order sliding harmonics, e.g., $\langle i_L \rangle_1 = x_1 + jx_2$ and $\langle u \rangle_1 = u_1 + ju_2$, one can develop the coupling terms $\langle v_O \cdot u \rangle_1$ and $\langle i_L \cdot u \rangle_0$ using the fundamental property expressed by Eq. (5.5). Further, the application of property (5.4) in Eq. (5.60) leads to

$$\begin{cases} L\dot{x}_1 + jL\dot{x}_2 = -j\omega L(x_1 + jx_2) - j\dfrac{E}{2} - \langle v_O \rangle_0 (u_1 + ju_2) \\ C\dot{\langle v_O \rangle}_0 = 2x_1u_1 + 2x_2u_2 - \langle i_S \rangle_0, \end{cases}$$

and next, by equaling real and imaginary parts of complex quantities, to

$$\begin{cases} L\dot{x}_1 = L\omega x_2 - \langle v_O \rangle_0 u_1 \\ L\dot{x}_2 = -L\omega x_1 - \dfrac{E}{2} - \langle v_O \rangle_0 u_2 \\ C\dot{\langle v_O \rangle}_0 = 2x_1u_1 + 2x_2u_2 - \langle i_S \rangle_0. \end{cases} \tag{5.61}$$

Similarly as in Sect. 5.6, one can express the inductor current by employing its dq components in the form $i_L = i_q \cos \omega t + i_d \sin \omega t$; then, according to the result in Eq. (5.44), $i_d = -2x_2$ and $i_q = 2x_1$. Note that the first-order sliding harmonic of the signal u computed on a switching period is the intelligence signal, $\beta(t)$:

$$\langle u \rangle_1 = \beta(t).$$

Being harmonic, signal $\beta(t)$ can similarly be expressed in the dq frame as $\beta = \beta_q \cos \omega t + \beta_d \sin \omega t$. Using the same approach as in the case of i_L, gives $\beta_d = -2u_2$ and $\beta_q = 2u_1$. The following relation can be written in conclusion:

$$\begin{cases} \langle i_L \rangle_1 = \dfrac{i_q}{2} - j\dfrac{i_d}{2} \\ \langle u \rangle_1 = \beta(t) = \dfrac{\beta_q}{2} - j\dfrac{\beta_d}{2}. \end{cases} \tag{5.62}$$

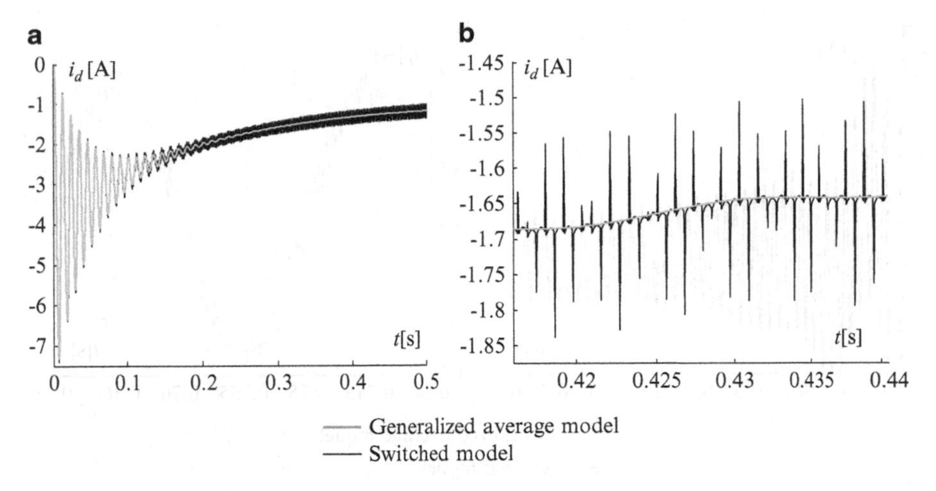

Fig. 5.26 Single-phase PWM-controlled converter: (**a**) reactive current waveforms GAM vs. switched model; (**b**) detail (Bacha and Gombert 2006)

Replacing (5.62) into (5.61) gives the final state-space equations:

$$\begin{cases} \dfrac{di_d}{dt} = \omega i_q + \dfrac{E}{L} - \dfrac{\langle v_0 \rangle_0}{L} \cdot \beta_d \\ \dfrac{di_q}{dt} = -\omega i_d - \dfrac{\langle v_0 \rangle_0}{L} \cdot \beta_q \\ \dfrac{d\langle v_0 \rangle_0}{dt} = \dfrac{1}{2C}\left(i_d\beta_d + i_q\beta_q\right) - \dfrac{1}{C}\langle i_S \rangle_0. \end{cases} \tag{5.63}$$

Equations (5.63) describe the dynamical behavior of the PWM-controlled voltage-source inverter in the dq frame (variables being approximated with the sums of their averages and their first-order harmonics). Note that there are two components of the control input on each axis of the dq frame allowing the possibility of vector control of the power converter.

Figures 5.26 and 5.27 are obtained by solving system (5.63) and show that the GAM has a satisfying precision with respect to the switched model.

This good result is explained by the fact that the high-order harmonics due to the switching frequency are placed quite far in relation to the system bandwidth and contains only little supplementary energy. In this way, the averaged model takes into account the quasi-totality of the energy transferred by the converter.

5.7.4.2 Three-Phase Case

The three-phase voltage-source converter in Fig. 5.28 is controlled by three switching functions, u_1, u_2, u_3, obtained by using pulse-width modulation,

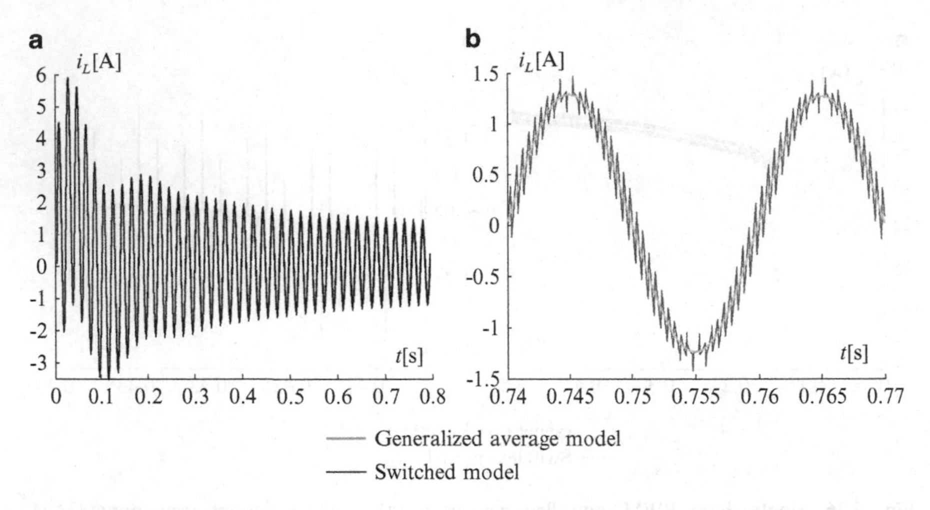

Fig. 5.27 Single-phase PWM-controlled converter: (**a**) inductor current waveforms GAM vs. switched model; (**b**) detail (Bacha and Gombert 2006)

similarly as in the single-phase case. As was true in the previous case, converter behavior is controlled by modifying either the magnitude or the phase of the control input signals, which are the averages of the three switching functions. Here also the modeling ignores the harmonics of order higher than one.

The switched model can be obtained by emphasizing the switched variables (see Sect. 3.3 from Chap. 3). To this end, the vector is defined as $\mathbf{x} = \begin{bmatrix} i_1 & i_2 & i_3 & v_0 \end{bmatrix}^T$, where i_k are the three-phase AC currents and v_0 is the DC capacitor voltage. The switched variables are voltages v_1, v_2 and v_3 of the points a, b and c, respectively (Fig. 5.28), considered with respect to the neutral N and the current feeding the C filter, i_0. By writing Kirchhoff's laws (inductor resistances ignored), the state equations can be written as

$$\begin{cases} L\dot{i}_k = e_k - v_k, \quad k = 1, 2, 3, \\ C\dot{v}_0 = i_0 - i_S. \end{cases} \tag{5.64}$$

The following relations hold for voltages v_1, v_2 and v_3:

$$v_k = \begin{cases} v^+ & \text{if } H_k \text{ is turned on} \\ v^- & \text{if } H_k \text{ is turned off} \end{cases} \quad k = 1, 2, 3,$$

where voltages v^+ and v^- are relative to the grid neutral point N.

If one assumes a symmetrical three-phase voltage system $\{e_k\}_{k=1,2,3}$ – for which it is true that $\sum_{k=1}^{3} e_k = 0$ the currents sum is also zero: $\sum_{k=1}^{3} i_k = 0$. Consequently, according to the first three Eq. (5.64), the sum of voltages v_k is also null: $\sum_{k=1}^{3} v_k = 0$.

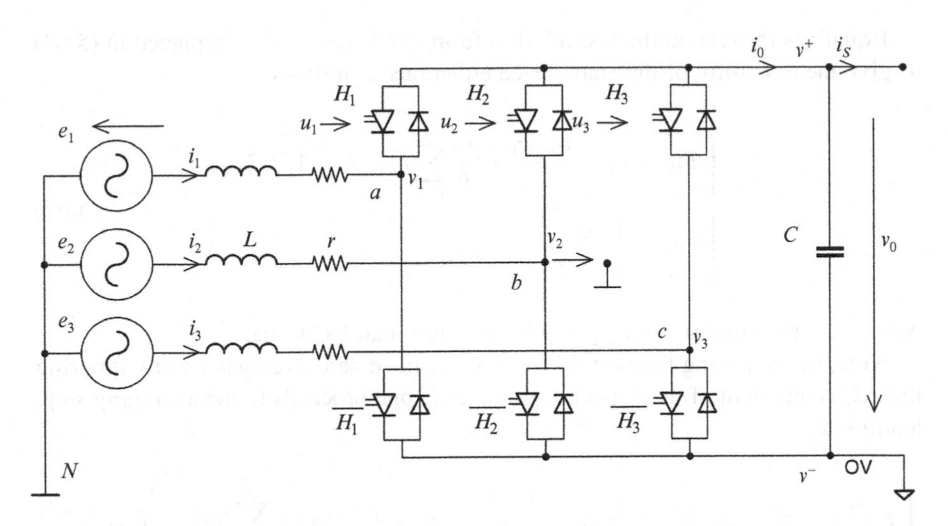

Fig. 5.28 Three-phase voltage-source converter diagram

By using bipolar switching functions defined as $u_k \in \{-1; 1\}$, $k = 1, 2, 3$, the switched variables can be expressed as depending on the state variables. Thus, by knowing that $v^+ - v^- = v_0$ (see Fig. 5.28) and that $\sum_{k=1}^{3} i_k = 0$, one obtains successively:

$$\begin{cases} v_k = v^+ \cdot \dfrac{1+u_k}{2} + v^- \cdot \dfrac{1-u_k}{2} = \dfrac{v^+ + v^-}{2} + \dfrac{v_0 u_k}{2}, \quad k = 1,2,3, \\ i_0 = \sum_{k=1}^{3} i_k \cdot \dfrac{1+u_k}{2} = \dfrac{1}{2}(i_1 \cdot u_1 + i_2 \cdot u_2 + i_3 \cdot u_3). \end{cases} \tag{5.65}$$

Now the sum $v^+ + v^-$ must be expressed appropriately. To this end, the first three equations from (5.65) are summed up:

$$0 = \sum_{k=1}^{3} v_k = \frac{3(v^+ + v^-)}{2} + \frac{v_0}{2} \cdot \sum_{k=1}^{3} u_k,$$

hence $v^+ + v^- = -\dfrac{v_0}{3} \cdot \sum_{k=1}^{3} u_k$, which is a result ready to be replaced in the first three Eq. (5.64) to give

$$v_k = \frac{v_0 \cdot u_k}{2} - \frac{v_0}{6} \sum_{k=1}^{3} u_k, \quad k = 1,2,3. \tag{5.66}$$

Equations (5.66) and the last relation from (5.65) can now be replaced in (5.64) to give the new form of the state-space equations as follows:

$$\begin{cases} L\dot{i_k} = e_k - \dfrac{v_0 \cdot u_k}{2} + \dfrac{v_0}{6}\displaystyle\sum_{k=1}^{3} u_k, \quad k = 1,2,3, \\ C\dot{v_0} = \dfrac{1}{2} \cdot \displaystyle\sum_{k=1}^{3} i_k \cdot u_k - i_S, \end{cases} \tag{5.67}$$

where i_S is the current exchanged with an additional DC stage.

Now, by supposing that the AC currents i_k have zero averages and by ignoring the AC component of the capacitor voltage v_0, one proceeds to the averaging step, leading to

$$\begin{cases} L\dfrac{d}{dt}\langle i_k\rangle_1 = -jL\omega\langle i_k\rangle_1 + \langle e_k\rangle_1 - \dfrac{1}{2}\langle v_0\rangle_0 \cdot \langle u_k\rangle_1 + \dfrac{1}{6}\langle v_0\rangle_0 \cdot \displaystyle\sum_{k=1}^{3}\langle u_k\rangle_1, \quad k = 1,2,3, \\ C\dfrac{d}{dt}\langle v_0\rangle_0 = \dfrac{1}{2} \cdot \displaystyle\sum_{k=1}^{3}\langle i_k \cdot u_k\rangle_0 - \langle i_S\rangle_0. \end{cases} \tag{5.68}$$

Before proceeding with the computation, let the notations

$$\begin{cases} \langle i_1\rangle_1 = x_1 + jx_2 \\ \langle i_2\rangle_1 = y_1 + jy_2 \\ \langle i_3\rangle_1 = z_1 + jz_2 \end{cases} \begin{cases} \langle u_1\rangle_1 = a_1 + ja_2 \\ \langle u_2\rangle_1 = b_1 + jb_2 \\ \langle u_3\rangle_1 = c_1 + jc_2 \end{cases} \tag{5.69}$$

be introduced. Note that the first-order harmonics of signals u_k, $k = 1, 2, 3$, represent the control inputs denoted by β_k, $k = 1, 2, 3$, respectively. Note that system $\{\beta_k\}_{k\,=\,1,2,3}$ is a three-phase system, therefore

$$\sum_{k=1}^{3}\beta_k = \sum_{k=1}^{3}\langle u_k\rangle_1 = 0. \tag{5.70}$$

Note also that the first-order sliding harmonics of the three-phase voltage system $\{e_k\}_{k\,=\,1,2,3}$ are

$$\langle e_1\rangle_1 = -j\frac{E}{2}, \quad \langle e_2\rangle_1 = -\frac{\sqrt{3}E}{4} + j\frac{E}{4}, \quad \langle e_3\rangle_1 = \frac{\sqrt{3}E}{4} + j\frac{E}{4}. \tag{5.71}$$

Let us now analyze the coupled terms in the DC voltage equation from (5.68). By applying the fundamental result given in Eq. (5.5), one writes successively

$$\langle i_k \cdot u_k\rangle_0 = \langle i_k\rangle_1\langle u_k\rangle_{-1} + \langle i_k\rangle_{-1}\langle u_k\rangle_1 = 2\mathrm{Re}\big(\langle i_k\rangle_1\langle u_k\rangle_{-1}\big), \quad k = 1,2,3,$$

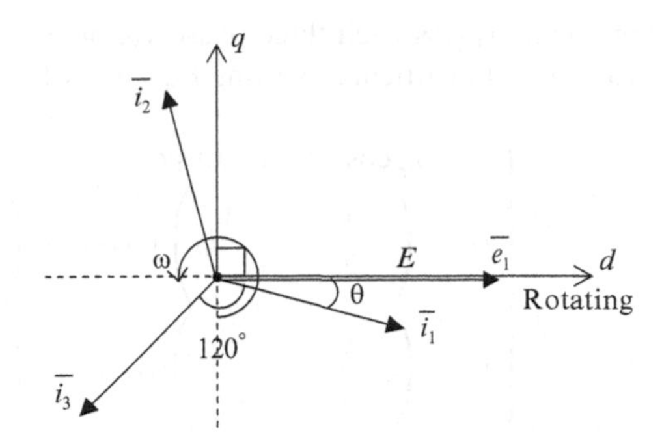

Fig. 5.29 Phasor diagram introducing the dq frame in the three-phase case

which, taking into account notations (5.69), gives further

$$\begin{cases} \langle i_1 \cdot u_1 \rangle_0 = 2(x_1 a_1 + x_2 a_2) \\ \langle i_2 \cdot u_2 \rangle_0 = 2(y_1 b_1 + y_2 b_2) \\ \langle i_3 \cdot u_3 \rangle_0 = 2(z_1 c_1 + z_2 c_2). \end{cases} \tag{5.72}$$

Introducing Eqs. (5.70), (5.71), (5.72) and notations given in (5.69) into Eq. (5.68) and equaling the real and the imaginary parts, one can write

$$\begin{cases} L\dot{x}_1 = \omega L x_2 - \dfrac{1}{2}\langle v_O \rangle_0 a_1, & L\dot{x}_2 = -\omega L x_1 - \dfrac{1}{2}\langle v_O \rangle_0 a_2 - \dfrac{E}{2} \\ L\dot{y}_1 = \omega L y_2 - \dfrac{1}{2}\langle v_O \rangle_0 b_1 - \dfrac{\sqrt{3}E}{4}, & L\dot{y}_2 = -\omega L y_1 - \dfrac{1}{2}\langle v_O \rangle_0 b_2 + \dfrac{E}{4} \\ L\dot{z}_1 = \omega L z_2 - \dfrac{1}{2}\langle v_O \rangle_0 c_1 + \dfrac{\sqrt{3}E}{4}, & L\dot{z}_2 = -\omega L z_1 - \dfrac{1}{2}\langle v_O \rangle_0 c_2 + \dfrac{E}{4} \\ C\langle \dot{v_O} \rangle_0 = (x_1 a_1 + x_2 a_2 + y_1 b_1 + y_2 b_2 + z_1 c_1 + z_2 c_2) - \langle i_S \rangle_0. & \end{cases} \tag{5.73}$$

At this point the connection between the three-phase and the dq variables is needed. An illustration of the relation between the dq frame and the three-phase variables represented as phasors is given in Fig. 5.29.

The corresponding transform matrices are (Bose 2001)

$$\begin{bmatrix} f_q \\ f_d \end{bmatrix} = \frac{2}{3} \cdot \begin{bmatrix} \cos \omega t & \cos(\omega t - 2\pi/3) & \cos(\omega t + 2\pi/3) \\ \sin \omega t & \sin(\omega t - 2\pi/3) & \sin(\omega t + 2\pi/3) \end{bmatrix} \cdot \begin{bmatrix} f_1 \\ f_2 \\ f_3 \end{bmatrix}, \tag{5.74}$$

$$\begin{bmatrix} f_1 \\ f_2 \\ f_3 \end{bmatrix} = \begin{bmatrix} \cos \omega t & \sin \omega t \\ \cos(\omega t - 2\pi/3) & \sin(\omega t - 2\pi/3) \\ \cos(\omega t + 2\pi/3) & \sin(\omega t + 2\pi/3) \end{bmatrix} \cdot \begin{bmatrix} f_q \\ f_d \end{bmatrix}, \tag{5.75}$$

where $\{f_k\}_{k=1,2,3}$ make up a generic three-phase system (currents, voltages or control inputs) with f_d and f_q being the associated dq components. Using Eq. (5.75) one can express each three-phase variable as depending on the same couple of dq variables. In particular, writing Eq. (5.75) for the three-phase currents, gives

$$\begin{cases} i_1 = i_q \cos\omega t + i_d \sin\omega t \\ i_2 = \left(-\frac{1}{2}i_q - \frac{\sqrt{3}}{2}i_d\right)\cos\omega t + \left(\frac{\sqrt{3}}{2}i_q - \frac{1}{2}i_d\right)\sin\omega t \\ i_3 = \left(-\frac{1}{2}i_q + \frac{\sqrt{3}}{2}i_d\right)\cos\omega t + \left(-\frac{\sqrt{3}}{2}i_q - \frac{1}{2}i_d\right)\sin\omega t. \end{cases} \tag{5.76}$$

On the other hand, knowing that each three-phase current is a zero-mean AC variable – therefore approximated by its first-order harmonic – it can be expressed as depending of the real and imaginary parts of its first-order sliding harmonic, as shown in Eq. (5.29). Thus, taking account of notations (5.69), the following relations hold:

$$\begin{cases} i_1 = 2x_1 \cos\omega t - 2x_2 \sin\omega t \\ i_2 = 2y_1 \cos\omega t - 2y_2 \sin\omega t \\ i_3 = 2z_1 \cos\omega t - 2z_2 \sin\omega t. \end{cases} \tag{5.77}$$

By comparing Eqs. (5.76) and (5.77) one can deduce the relations expressing the real and the imaginary parts of the first-order sliding harmonics of each of the three-phase currents as depending on the currents i_d and i_q. Thus,

$$\begin{cases} x_1 = \frac{1}{2}i_q, & x_2 = -\frac{1}{2}i_d \\ y_1 = -\frac{1}{4}i_q - \frac{\sqrt{3}}{4}i_d, & y_2 = -\frac{\sqrt{3}}{4}i_q + \frac{1}{4}i_d \\ z_1 = -\frac{1}{4}i_q + \frac{\sqrt{3}}{4}i_d, & z_2 = \frac{\sqrt{3}}{4}i_q + \frac{1}{4}i_d. \end{cases} \tag{5.78}$$

In an analogous way one obtains the relations expressing the real and the imaginary parts of the first-order sliding harmonics of each of the three control inputs, β_k, $k = 1, 2, 3$, as depending on the associated dq components, β_d and β_q. Taking account of notations (5.69), the following relations are valid:

$$\begin{cases} a_1 = \frac{1}{2}\beta_q, & a_2 = -\frac{1}{2}\beta_d \\ b_1 = -\frac{1}{4}\beta_q - \frac{\sqrt{3}}{4}\beta_d, & b_2 = -\frac{\sqrt{3}}{4}\beta_q + \frac{1}{4}\beta_d \\ c_1 = -\frac{1}{4}\beta_q + \frac{\sqrt{3}}{4}\beta_d, & c_2 = \frac{\sqrt{3}}{4}\beta_q + \frac{1}{4}\beta_d. \end{cases} \tag{5.79}$$

The following step uses the equation sets (5.78) and (5.79) in order to transpose the state-space Eq. (5.73) into dynamic equations that have currents i_d and i_q and the average value of the DC voltage $\langle v_0\rangle_0$ as states, and β_d and β_q as control inputs. It is quite easy to verify that to this end it is sufficient to use only equations related to variables x_1, x_2 and $\langle v_0\rangle_0$. Indeed, the same set of equations is obtained if equations concerning either y_1, y_2 and $\langle v_0\rangle_0$, or z_1, z_2 and $\langle v_0\rangle_0$ are used. The final state-space dq model of the three-phase voltage-source PWM-controlled converter from Fig. 5.28 is

$$\begin{cases} \dot{i_d} = \omega i_q - \dfrac{\langle v_0\rangle_0}{2L}\beta_d + \dfrac{E}{L} \\ \dot{i_q} = -\omega i_d - \dfrac{\langle v_0\rangle_0}{2L}\beta_q \\ \dot{\langle v_0\rangle_0} = \dfrac{3}{4C}\left(i_d\beta_d + i_q\beta_q\right) - \dfrac{\langle i_S\rangle_0}{C}. \end{cases} \tag{5.80}$$

In conclusion, a large-signal model of the PWM three-phase converter depicted in Fig. 5.28 has been obtained. This model refers to a rotating dq frame, synchronized with the grid voltage vector e_1. Note that there are two control inputs β_d and β_q allowing the active and reactive components of the current interchanged with the grid to be influenced.

The first two equations in model (5.80) are coupled; this further complicates the separate control of the two currents (decoupling may be necessary). The evolution of the DC-bus voltage with respect to the components of the current interchanged with the grid – expressed by the third relation – makes one think to employ a regulation loop in order to avoid DC-bus over- or under voltage.

So, control laws are to be developed in the dq frame, and the effective $\{\beta_k\}_{k=1,2,3}$ inputs will be computed by using the inverse Park transform based upon a measure of the dq frame instantaneous angle (the same as e_1), which is obtained by means of a phase-locked loop (PLL).

Remark

Note that a slightly different final form of model (5.80) may be found in the literature, which is due to the fact that the switching functions u_k, $k = 1, 2, 3$, take values in the set $\{0;1\}$ instead of $\{-1;\ 1\}$. In this case current i_0 and voltages v_k have expressions different from those given in (5.65) and (5.66), respectively, which are based on the following relations:

$$\begin{cases} v_k = v^+ \cdot u_k + v^- \cdot (1 - u_k) = v_0 u_k + v^-, \quad k = 1,2,3, \\ i_0 = \sum_{k=1}^{3} i_k \cdot u_k. \end{cases}$$

Following a similar inference, one can deduce that in this case, $v^- = -\frac{v_0}{3}\sum_{k=1}^{3} u_k$ and further obtain the switched model in the form

$$\begin{cases} L\dot{i}_k = e_k - v_0 \cdot u_k + \dfrac{v_0}{3}\displaystyle\sum_{j=1}^{3} u_j, \quad k = 1, 2, 3, \\ C\dot{v}_0 = \displaystyle\sum_{j=1}^{3} i_j \cdot u_j - i_S. \end{cases}$$

Further computation follows the same steps as above and the complete development is left to the reader. The final state-space *dq* model results as (Blasko and Kaura 1997)

$$\begin{cases} \dot{i}_d = \omega i_q - \dfrac{\langle v_0 \rangle_0}{L}\beta_d^{'} + \dfrac{E}{L} \\ \dot{i}_q = -\omega i_d - \dfrac{\langle v_0 \rangle_0}{L}\beta_q^{'} \\ \dot{\langle v_0 \rangle}_0 = \dfrac{3}{2C}\left(i_d\beta_d^{'} + i_q\beta_q^{'}\right) - \dfrac{\langle i_S \rangle_0}{C}, \end{cases} \tag{5.81}$$

where the new control inputs $\beta_d^{'}$ and $\beta_q^{'}$ are related to β_d and β_q given in model (5.80) by the relations $\beta_d^{'} = \beta_d/2$ and $\beta_q^{'} = \beta_q/2$ because the sinusoidal set of functions $\{\beta_k\}_{k=1,2,3}$ has half the amplitude seen in the previous case.

5.8 Conclusion

The generalized averaged model is suitable to symmetrical PWM converter analysis, whereas the averaged model is more adapted to asymmetrical converters (e.g., DC-DC power stages).

The GAM proves to be an accurate tool for obtaining large-signal converter models and also allows replicating their dynamical behavior at small time scales close to the switching period. Being flexible, this type of model can be used even if the frequency is changing slightly. Of course, this change must be slower than the system dynamics in order to ensure acceptable errors.

The approach is valid also for asymmetrical converters. However, the gain in accuracy is not justified with respect to the complexity engendered by the addition of supplementary variables.

There are multiple possibilities of employment of these models, among which one can cite:

- *Steady-state analysis*: This steady-state approach is in fact the classical method of the first-harmonic model (permanent sinusoidal regime). In this context, the model provides results easy to obtain because once the equivalent diagram has been built one must only apply the classical rules of electrical circuits. This advantage, of the simplicity, makes the model more attractive than the

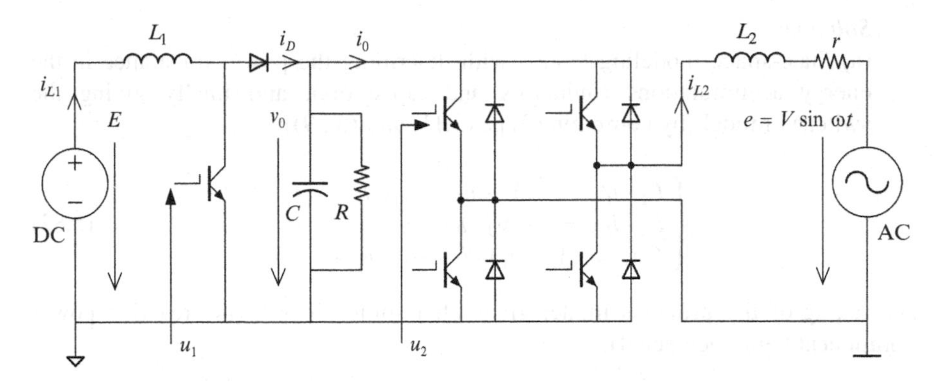

Fig. 5.30 Grid interface circuit used as power conveyor between variable DC source and AC power grid

phase-plane model. Obtaining the analytical expressions, despite taking into account a damping in the resonant circuits, allows one to obtain the gain sensitivity in relation with a given parameter (switching frequency, load, etc.).

- *Dynamical analysis*: Good dynamical precision ensures a good description of the input-output dynamics, as well as of the internal dynamics. This feature recommends the GAM as an analysis tool at least as useful as the large-signal model. Small-signal models can be built based upon continuous models without making use of too heavy and complicated computation methods. To this end it is sufficient to differentiate the large-signal model around an equilibrium operating point.

Problems

Problem 5.1

The diagram in Fig. 5.30 shows a circuit used to convey the energy between a variable DC source (e.g., a PV array) and a single-phase strong AC grid. There are two power stages, a DC-DC boost converter and a voltage-source inverter (VSI) connected by means of a DC-bus. Therefore, the system has two control inputs – u_1 acting on the DC-DC converter and u_2 acting on the VSI – and the disturbance input E, due to changes in the primary resource.

Modeling assumptions limit the variables spectra at the DC component for the boost power stage and at the first harmonic for the voltage-source inverter. The inverter provides only active power to the grid. The purpose of control is to ensure the power flux between the primary DC source and the AC grid, while preserving the rated operating conditions (in terms of voltage and current).

Using the information in Chaps. 3, 4, and 5, perform the following actions:

(a) obtain the large-signal system model using the generalized average modeling technique;
(b) obtain the small-signal model by perturbation and linearization around a typical operating point. Give the transfer function having $\langle u_2 \rangle_1$ as input and $\langle v_0 \rangle_0$ as output.

Solution
(a) State-space modeling begins with describing the power imbalance in the energy accumulations (inductors and capacitors), and finally giving the switched model (by combining Eqs. (3.11) and (5.39)):

$$\begin{cases} L_1 \cdot \dot{i_{L1}} = -(1-u_1)\cdot v_0 + E \\ L_2 \cdot \dot{i_{L2}} = e - v_0 \cdot u_2 - i_{L2}\cdot r \\ C\cdot \dot{v_0} = (1-u_1)\cdot i_{L1} - v_0/R - i_L \cdot u_2. \end{cases} \tag{5.82}$$

Averaging of the previous model gives after some algebra (the reactive power component being neglected):

$$\begin{cases} \dfrac{d\langle i_{L1}\rangle_0}{dt} = \dfrac{1}{L_1}[E - v_{DC}(1-\alpha)] \\ \dfrac{di_d}{dt} = \dfrac{V}{L_2} - \dfrac{v_{DC}}{L_2}\cdot \beta_d - \dfrac{i_d \cdot r}{L_2} \\ \dfrac{dv_{DC}}{dt} = \dfrac{1}{C}\langle i_{L1}\rangle_0 (1-\alpha) - \dfrac{1}{RC} v_{DC} - \dfrac{i_d}{2C}\cdot \beta_d, \end{cases} \tag{5.83}$$

where $v_{DC} = \langle v_0\rangle_0$, $i_d = \langle i_{L2}\rangle_1$ represents the magnitude of the grid inductor current and $\beta_d = \langle u_2\rangle_1$ is the inverter's single control input (the active component on the d-axis). The whole system's state vector is $\mathbf{x} = [\langle i_{L1}\rangle_0 \quad i_d \quad v_{DC}]^T$ and its input vector is $\mathbf{u} = [\alpha \quad \beta_d \quad E]^T$, which is composed of the control input vector $[\alpha \quad \beta_d]^T$ and the disturbance input E.

(b) Let us consider a steady-state operating point determined by the input $[\alpha_e \quad \beta_{de} \quad E_e]^T$ and the state $[i_{L1e} \quad i_{de} \quad v_{DCe}]^T$. By zeroing the derivatives in Eq. (5.83), one can write

$$\begin{cases} E_e - (1-\alpha_e)v_{DCe} = 0 \\ V - v_{DCe}\beta_{de} - i_{de}r = 0 \\ i_{L1}(1-\alpha_e) - v_{DCe}/R - i_{de}\beta_{de}/2 = 0. \end{cases} \tag{5.84}$$

Let us consider small variations around this operating point: $\alpha = \tilde{\alpha} + \alpha_e$, $\beta_d = \tilde{\beta_d} + \beta_{de}$, $E = \tilde{E} + E_e$ and $i_{L1} = \widetilde{i_{L1}} + i_{L1e}$, $i_d = \tilde{i_d} + i_{de}$, $v_{DC} = \widetilde{v_{DC}} + v_{DCe}$. Perturbing the state-space equations from (5.83) around the previously defined steady-state operating point gives the small-signal state-space equations in the form

$$\begin{cases} L_1 \cdot \dot{\widetilde{i_{L1}}} = -(1-\alpha_e)\cdot \widetilde{v_{DC}} + v_{DCe}\cdot \tilde{\alpha} + \tilde{E} \\ L_2 \cdot \dot{\tilde{i_d}} = -r\cdot \tilde{i_d} - \beta_{de}\cdot \widetilde{v_{DC}} - v_{DCe}\cdot \tilde{\beta_d} \\ C\cdot \dot{\widetilde{v_{DC}}} = (1-\alpha_e)\cdot \widetilde{i_{L1}} - \dfrac{\beta_{de}}{2}\cdot \tilde{i_d} - \dfrac{1}{R}\cdot \widetilde{v_{DC}} - i_{L1e}\cdot \tilde{\alpha} - \dfrac{i_{de}}{2}\cdot \tilde{\beta_d}. \end{cases}$$

With $\widetilde{\mathbf{x}} = \begin{bmatrix} \widetilde{i_{L1}} & \widetilde{i_d} & \widetilde{v_{DC}} \end{bmatrix}^T$ as the state vector and by $\widetilde{\mathbf{u}} = \begin{bmatrix} \widetilde{\alpha} & \widetilde{\beta_d} & \widetilde{E} \end{bmatrix}^T$ the input vector of the small-signal model, one can write the small-signal state-space matrix representation, as

$$\dot{\widetilde{\mathbf{x}}} = \underbrace{\begin{bmatrix} 0 & 0 & -\frac{1-\alpha_e}{L_1} \\ 0 & -\frac{r}{L_2} & -\frac{\beta_{de}}{L_2} \\ \frac{1-\alpha_e}{C} & -\frac{\beta_{de}}{2C} & -\frac{1}{2C} \end{bmatrix}}_{\mathbf{A}} \cdot \widetilde{\mathbf{x}} + \underbrace{\begin{bmatrix} \frac{v_{DCe}}{L_1} & 0 & \frac{1}{L_1} \\ 0 & -\frac{v_{DCe}}{L_2} & 0 \\ -\frac{i_{Le}}{C} & -\frac{i_{de}}{2C} & 0 \end{bmatrix}}_{\mathbf{B}} \cdot \widetilde{\mathbf{u}}, \tag{5.85}$$

which makes apparent the state matrix **A** and the output matrix **B**. One is interested in considering $\widetilde{v_{DC}}$ as the output variable; consequently, to the state-space equation in (5.85) the output equation is added, which introduces the output matrix **C**:

$$\mathbf{y} = \widetilde{v_{DC}} = \underbrace{\begin{bmatrix} 0 & 0 & 1 \end{bmatrix}}_{\mathbf{C}} \cdot \widetilde{\mathbf{x}}. \tag{5.86}$$

Note that the 1×3 transfer matrix

$$\mathbf{H}(s) = \mathbf{C} \cdot (s\mathbf{I} - \mathbf{A})^{-1} \cdot \mathbf{B} = \begin{bmatrix} H_{11}(s) & H_{12}(s) & H_{13}(s) \end{bmatrix}$$

contains the transfer functions from all the inputs to the output $\widetilde{v_{DC}}$. Here the interest is focused on the transfer from $\widetilde{\beta_d}$ to $\widetilde{v_{DC}}$, which is the second element of the transfer matrix $\mathbf{H}(s)$: $H_{\widetilde{\beta_d} \to \widetilde{v_{DC}}}(s) = H_{12}(s)$. Taking into account the expressions of matrices **A**, **B** and **C** given in Eqs. (5.85) and (5.86), quite laborious but straightforward algebra leads to the final form of the required transfer function:

$$H_{\widetilde{\beta_d} \to \widetilde{v_{DC}}}(s) = -\frac{s\frac{\beta_{de}}{2C}}{s^3 + s^2\left[\frac{1}{RC} + \frac{r}{L_2}\right] + s\left[\frac{r}{RCL_2} + \frac{(1-\alpha_e)^2}{L_1C} - \frac{\beta_{de}^2}{2L_2C}\right] + \frac{(1-\alpha_e)^2 r}{L_1L_2C}}.$$

The reader is invited to solve the following problems.

Problem 5.2 Modeling a basic structure in the GAM sense

Let us consider the diagram from Fig. 5.31 and assume that the switching function acting on the switches takes value 1 if H_1 is turned on and -1 if H_2 is turned on. One should note that the two switches cannot be simultaneously turned on.

(a) By taking current i_a sinusoidal and voltage v_0 continuous, draw on one plot v_a, i_a and i_0 such that u is delayed by angle φ in relation to i_a.

Fig. 5.31 Basic converter

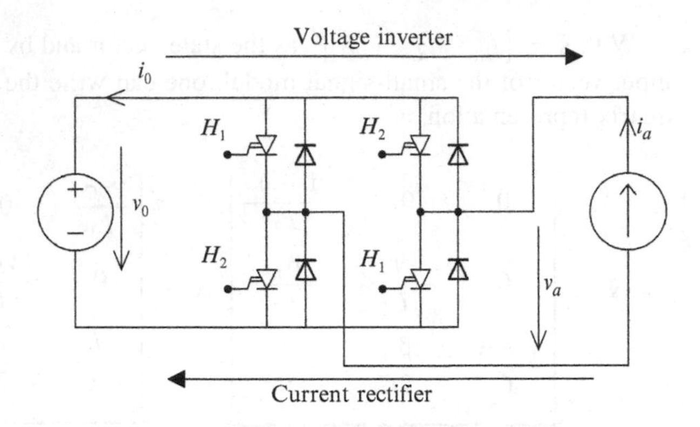

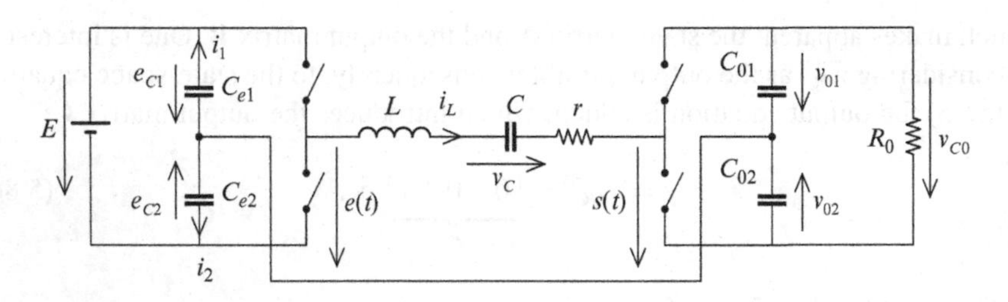

Fig. 5.32 Series-resonant switching power supply

(b) Establish the generalized averaged diagram by employing the necessary topological transform but without computing the terms $\langle v_a \rangle_1$ and $\langle i_0 \rangle_0$.

(c) In the case where the rectifier current (voltage inverter) is not controlled (i.e., it is a diode rectifier), compute the terms $\langle v_a \rangle_1$ and $\langle i_0 \rangle_0$, then give the expressions of these terms in the general case when the control input is the AC current-voltage phase lag φ.

Problem 5.3 Capacitive half-bridge series-resonant power supply
Given the circuit in Fig. 5.32,

(a) by using the results provided in Chap. 3, establish the equivalent exact topological diagram of the converter by defining the switching functions;

(b) deduce its generalized averaged model equivalent diagram without developing the coupling terms;

(c) develop the terms $\langle e(t) \rangle_1$ and $\langle s(t) \rangle_1$ and compute the coupling terms if the phase reference is the angle δ between the inverter commutations and those of the rectifier;

(d) write the GAM equations of the converter and compute its equilibrium points. Using these results, build the small-signal model (its state-space representation) by taking as control input the variations of δ and as output the variations of v_{Co}.

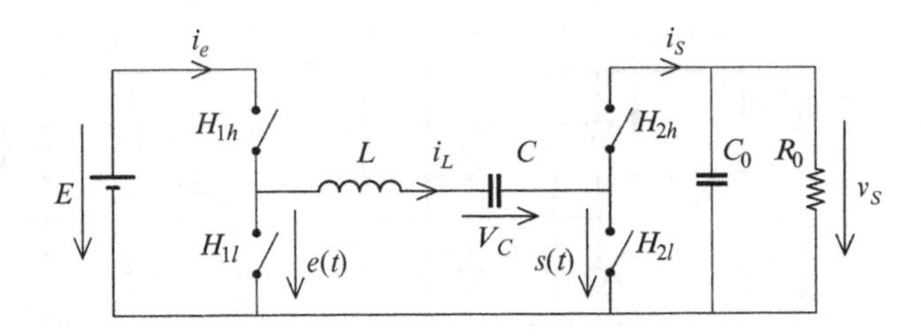

Fig. 5.33 Resonant power supply

Problem 5.4

A dynamical model of the resonant power supply depicted in Fig. 5.33 is sought. In this topology, both the inverter and the rectifier are based only on switches, without employing capacitive half-bridges.

The following assumptions are adopted:

- operating frequency is quite close to resonance; this allows i_L and v_C fundamental harmonics prevalence in relation to their higher-order harmonics;
- operating frequency and output capacitor values are sufficiently high in order to ignore the output voltage v_S ripple (it is superposed on its average value);
- switches H_{1h} and H_{1l} are complementary and driven by the switching function u_1 such that $u_1 = 1$ if H_{1h} is turned on and $u_1 = 0$ if H_{1h} is turned off; the switching function u_2 plays the same role for the complementary switches H_{2h} and H_{2l}: $u_2 = 1$ yields H_{2h} being turned on and $u_2 = 0$ yields H_{2h} being turned off. Both switching functions have pulsation ω and duty ratio 0.5 such that u_1 is the phase origin and u_2 is delayed by an angle φ with respect to u_1.

(a) Write the differential equations that govern the evolutions of the capacitor C voltage v_C and of the inductance L current i_L in the resonant circuit. These equations must be written as depending on the inverter output voltage $e(t)$ and on rectifier input voltage $s(t)$. Express the dynamic of the output voltage v_S as a function of the output current $i_s(t)$.

(b) Write the voltages $e(t)$ and $s(t)$ as functions of E, u_1, u_2 and v_S. Express the currents $i_e(t)$ and $i_s(t)$ as functions of i_L, u_1 and u_2.

(c) Deduce the switching model of the converter in Fig. 5.19 having the switching functions u_1 and u_2. Deduce also the equivalent topological diagram that makes apparent the coupling terms.

(d) With the variable changes

$$\begin{cases} u_1^{'} = 2u_1 - 1 \\ u_2^{'} = 2u_2 - 1 \end{cases}$$

(i) compute the terms $\langle u_1^{'} \rangle_0$, $\langle u_1^{'} \rangle_1$, $\langle u_2^{'} \rangle_0$ and $\langle u_2^{'} \rangle_1$;

(ii) express the coupling terms and the converter switched model having these new switching functions;

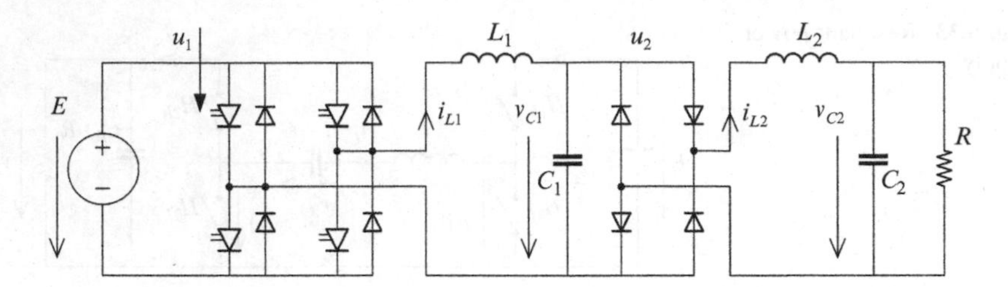

Fig. 5.34 Parallel-resonant switching power supply

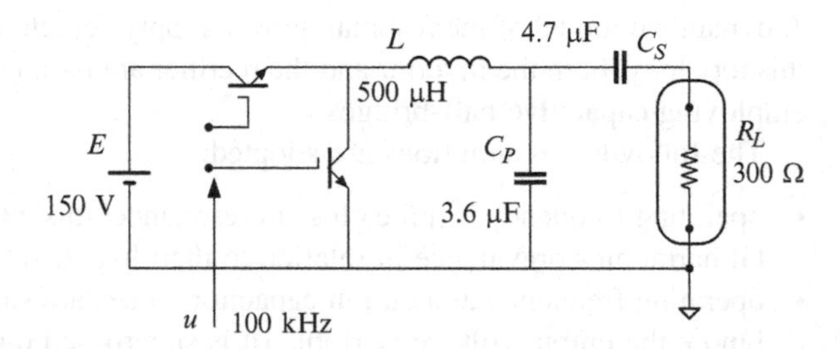

Fig. 5.35 Electronic ballast feeding a HID lamp

(iii) give the dynamical expressions of $\langle i_L \rangle_0$, $\langle i_L \rangle_1$, $\langle v_C \rangle_0$ and $\langle v_C \rangle_1$ as functions of the previous coupling terms and make the associated computations. Knowing that the current i_L cannot have a DC component, show that $\langle v_C \rangle_0$ has no dynamic and deduce its steady-state value;

(iv) give the expression of the output voltage dynamic v_S (identical with its average value $\langle v_S \rangle_0$). Show that the term whose average is $\langle v_C \rangle_0$ has no influence on v_S;

(v) give the large-signal generalized averaged model of the entire converter by making all the associated computations.

(e) Build the small-signal model resulted from the large-signal model obtained at point (v).

(f) Compare by simulation the time behavior of the various models obtained.

Problem 5.5

The DC-DC parallel-resonant converter in Fig. 5.34 is controlled via a squared switching function u_1. Develop the large-signal GAM of this structure.

Problem 5.6

Figure 5.35 shows an electronic circuit that supplies a high-intensity discharge (HID) lamp. The switching function u is a high-frequency squared signal of 0.5 duty ratio. When one transistor is turned on the other one is turned off, and inversely,

(a) develop the large-signal generalized averaged model;
(b) deduce the small-signal averaged model;
(c) compute the transfer function between the control input frequency and the lamp current.

Appendix

Justification of equation (5.5)

$$\langle x \cdot y \rangle_k(t) = \sum_i \langle x \rangle_{k-i}(t) \cdot \langle y \rangle_i(t)$$

This relation is the Fourier-series correspondent of the convolution theorem holding for the Fourier transforms of nonperiodical signals. Note that it is valid for signals having the same fundamental frequency.

Let us first consider that x and y are both DC signals, i.e., they are equal to their zero-order sliding harmonics. Then, the maximal-degree harmonic of their product will be zero as well. In this case it is obvious that

$$\langle x \cdot y \rangle_0(t) = \langle x \rangle_0(t) \cdot \langle y \rangle_0(t)$$

If now one supposes that x is a biased sinusoidal signal of pulsation ω (therefore containing harmonics of maximal degree equal to one) and y contains harmonics until second order (time variable is skipped), then

$$\begin{aligned} x &= \langle x \rangle_0 + \langle x \rangle_{-1} e^{-j\omega t} + \langle x \rangle_1 e^{j\omega t}, \\ y &= \langle y \rangle_0 + \langle y \rangle_{-1} e^{-j\omega t} + \langle y \rangle_1 e^{j\omega t} + \langle y \rangle_{-2} e^{-2j\omega t} + \langle y \rangle_2 e^{2j\omega t}. \end{aligned}$$

Their product will contain harmonics of maximal degree equal to three. One obtains successively

$$\begin{aligned} x \cdot y = {} & \langle x \rangle_0 \langle y \rangle_0 + \langle x \rangle_{-1} \langle y \rangle_1 + \langle x \rangle_1 \langle y \rangle_{-1} \\ & + \big(\langle x \rangle_{-1} \langle y \rangle_0 + \langle x \rangle_0 \langle y \rangle_{-1} + \langle x \rangle_1 \langle y \rangle_{-2} \big) e^{-j\omega t} \\ & + \big(\langle x \rangle_{-1} \langle y \rangle_2 + \langle x \rangle_0 \langle y \rangle_1 + \langle x \rangle_1 \langle y \rangle_0 \big) e^{j\omega t} \\ & + \big(\langle x \rangle_{-1} \langle y \rangle_{-1} + \langle x \rangle_0 \langle y \rangle_{-2} \big) e^{-2j\omega t} + \big(\langle x \rangle_1 \langle y \rangle_1 + \langle x \rangle_0 \langle y \rangle_2 \big) e^{2j\omega t} \\ & + \langle x \rangle_{-1} \langle y \rangle_{-2} e^{-3j\omega t} + \langle x \rangle_1 \langle y \rangle_2 e^{3j\omega t}, \end{aligned}$$

which can be posed under the form

$$\begin{aligned} x \cdot y = {} & \langle xy \rangle_{-3} e^{-3j\omega t} + \langle xy \rangle_{-2} e^{-2j\omega t} + \langle xy \rangle_{-1} e^{-j\omega t} + \langle xy \rangle_0 \\ & + \langle xy \rangle_1 e^{j\omega t} + \langle xy \rangle_2 e^{2j\omega t} + \langle xy \rangle_3 e^{3j\omega t}, \end{aligned}$$

where

$$\begin{cases} \langle xy \rangle_0 = \langle x \rangle_{-1} \langle y \rangle_1 + \langle x \rangle_0 \langle y \rangle_0 + \langle x \rangle_1 \langle y \rangle_{-1} \\ \langle xy \rangle_1 = \langle x \rangle_{-1} \langle y \rangle_2 + \langle x \rangle_0 \langle y \rangle_1 + \langle x \rangle_1 \langle y \rangle_0 \\ \langle xy \rangle_2 = \langle x \rangle_1 \langle y \rangle_1 + \langle x \rangle_0 \langle y \rangle_2 \\ \langle xy \rangle_3 = \langle x \rangle_1 \langle y \rangle_2. \end{cases} \tag{5.90}$$

The result given by (5.90) can be generalized to the product of two periodical signals having the same fundamental frequency x and y containing harmonics up to degree n, respectively m. Thus, one can formulate the following proposition about the expression of the kth-order harmonic of the product xy:

$$P(n,m): \langle x \cdot y \rangle_k = \sum_{\substack{-n \le k-i \le n \\ -m \le i \le m}} \langle x \rangle_{k-i} \cdot \langle y \rangle_i$$

and must prove by mathematical induction that it is true for any n and m. To this end, one supposes that $\mathcal{P}(n, m)$ is true and must prove that either $\mathcal{P}(n+1, m)$ or $\mathcal{P}(n, m+1)$ is true. One can note that adopting the variable change $z = e^{j\omega t}$, signal x can be expressed as a sum of two n-degree polynomials in z and $1/z$, whereas signal y is a sum of two m-degree polynomials in z and $1/z$. Results concerning the general form of coefficients of products of polynomials can further be applied in order to get the expression of the harmonics of the two signals expressed as polynomials (Osborne 2000).

References

Bacha S (1993) On the modelling and control of symmetrical switching power supplies (in French: "Sur la modélisation et la commande des alimentations à découpage symétrique"). Ph.D. thesis, Grenoble National Institute of Technology, France

Bacha S, Gombert C (2006) Modelling of basic elements (in French: Modélisation des briques de base). In: Crappe M (ed) Exploitation of electrical grids by means of power electronics (in French: L'exploitation des réseaux électriques avec l'électronique de puissance). Hermès Lavoisier, Paris

Bacha S, Rognon J-P, Ferrieux J-P, Bendaas ML (1995) Dynamical approach of first-order harmonic for modelling of AC-AC converters with DC-link. Application to induction heating (in French: Approche dynamique du premier harmonique pour la modélisation de convertisseurs AC-AC à étage intérmédiaire continu. Application au chauffage à induction). Journal de Physique III 5:145–160

Bendaas ML, Bacha S, Ferrieux J-P, Rognon J-P (1995) Safe and time-optimized power transfer between two induction loads supplied by a single generator. IEEE Trans Ind Electron 42(5):539–544

Blasko V, Kaura V (1997) A new mathematical model and control of a three-phase AC-DC voltage source converter. IEEE Trans Power Electron 12(1):116–123

Bose BK (2001) Modern power electronics and AC drives. Prentice-Hall, Upper Saddle River

Caliskan VA, Verghese GC, Stanković AM (1999) Multifrequency averaging of DC/DC converters. IEEE Trans Power Electron 14(1):124–133

Maksimović D, Stanković AM, Thottuvelil VJ, Verghese GC (2001) Modeling and simulation of power electronic converters. Proc IEEE 89(6):898–912

Nasiri A, Emadi A (2003) Modeling, simulation, and analysis of active filter systems using generalized state space averaging method. In: Proceedings of the 29th annual Conference of the Industrial Electronics Society – IECON 2003. Roanoke, Virginia, USA, vol. 3, pp 1999–2004

Noworolsky JM, Sanders SR (1991) Generalized in-place circuit averaging. In: Proceedings of the 6th annual Applied Power Electronics Conference and Exposition – APEC 1991. Dallas, Texas, USA, pp 445–451

Oppenheim AV, Willsky AS, Hamid S (1997) Signals and systems, 2nd edn. Prentice-Hall, Upper Saddle River

Osborne MS (2000) Basic homological algebra. Graduate texts in mathematics. Springer, Berlin/ New York

Rim CT, Cho GH (1990) Phasor transformation and its application to the DC/AC analyses of frequency phase-controlled series resonant converters (SRC). IEEE Trans Power Electron 5 (7):201–211

Rimmalapudi SR, Williamson SS, Nasiri A, Emadi A (2007) Validation of generalized state space averaging method for modeling and simulation of power electronic converters for renewable energy systems. J Electr Eng Technol 2(2):231–240

Sanders SR (1993) On limit cycles and the describing function method in periodically switched circuits. IEEE Trans Circuit Syst I Fundam Theory Appl 40(9):564–572

Sanders SR, Noworolski JM, Liu XZ, Verghese GC (1990) Generalized averaging method for power conversion circuits. In: Proceedings of the IEEE power electronics specialists conference. San Antonio, Texas, USA, pp 333–340

Sira-Ramírez H, Silva-Ortigoza R (2006) Control design techniques in power electronics devices. Springer, London

Wong S-C, Tse CK, Orabi M, Ninomiya T (2006) The method of double averaging: an approach for modeling power-factor-correction switching converters. IEEE Trans Circuit Syst I Regul Pap 53(2):454–462

Xu J, Lee CQ (1998) A unified averaging technique for the modeling of quasi-resonant converters. IEEE Trans Power Electron 13(3):556–563

Yin Y, Zane R, Glaser J, Erickson RW (2002) Small-signal analysis of frequency-controlled electronic ballasts. IEEE Trans Circuit Syst I Fundam Theory Appl 50(8):1103–1110

Chapter 6
Reduced-Order Averaged Model

This chapter deals with modeling methodologies used for obtaining simplified – in the sense of reduced order – power electronic converter models, which are able to represent their low-frequency average behavior and are more easily employed in simulation or control law design.

Reduced-order averaged modeling relies on splitting the converter dynamics in the frequency domain and preserving the main, low-frequency dynamic.

This chapter attempts to bring these modeling approaches together under a single general methodology. First, the principles are introduced and the general methodology is derived. Some examples and case studies illustrate the application of this methodology for both AC and DC power stages. Finally, some problems are proposed to the reader.

6.1 Introduction

The *reduced-order averaged model* (denoted here by ROAM) presented by Chetty (1982) gave the solution to the problem that arose from modeling DC-DC power stages operating in discontinuous-conduction mode (where the classical averaging method failed). The principle of ROAM is to eliminate the incriminated variable and replace it by a function of other state variables; hence, a reduced-order model is obtained. This modeling framework has been further extended (see, for example, Maksimović and Ćuk 1991; Sun et al. 2001). The effect of algebraically linking two variables has been encountered for DC-DC converters controlled in (peak) current-programmed mode (Middlebrook 1985, 1989).

Applications for the ROAM are larger and include:

- the application domain of the classical averaged model (see Chap. 4);
- power electronic converters having both DC and AC stages (Sun and Grotstollen 1992);

S. Bacha et al., *Power Electronic Converters Modeling and Control: with Case Studies*, Advanced Textbooks in Control and Signal Processing, DOI 10.1007/978-1-4471-5478-5_6,

- converters whose switching is state-controlled and is also controlled by an independent input (e.g., thyristor-based converters);
- converters having both DC and AC stages that operate in discontinuous-conduction mode, which is a particular case of the previous class.

Therefore, in general, this averaged modeling applies in cases where the power stage presents – either intrinsically or as the effect of a control loop – a dominant low-frequency dynamic and a high-frequency behavior that can be neglected. Reduced-order averaged modeling thus relies on splitting converter dynamics in the frequency domain and preserving the main, low-frequency dynamic.

The ROAM is simple in both its construction and use. On the other hand, this model loses in precision what it gains in simplicity. If in certain cases this loss is acceptable, it could happen that in other cases dynamics that may be important by their effects (e.g., closed-loop instability) to be neglected.

6.2 Principle

The principle of ROAM relies upon separating the switched model into two dynamics:

- one that must be kept, representing the main dynamic,
- another that must be eliminated because it does not significantly contribute to the equivalent low-frequency behavior of the converter.

This separation depends on each particular case, as illustrated below.

(a) It may be the result of a modal analysis of the converter, emphasizing the two types of dynamics belonging to clearly separable frequency ranges. One must therefore identify a very slow dynamic prevailing over another much faster one. The separation may not result from the initial circuit design, but it can later be ensured by control action (presence of an inner control loop).
(b) It may be applied in the case of variables being zeroed periodically as a consequence of discontinuous-conduction mode operation. It can be shown that these variables do not significantly influence the converter main dynamics (see, for example, the case study in Sect. 4.6 of Chap. 4).

Now, let us consider a power electronic converter taking N configurations described by the general form of the switched model given by Eq. (3.2) from Chap. 3:

$$\dot{\mathbf{x}}(t) = \sum_{i=1}^{N} (\mathbf{A}_i \mathbf{x}(t) + \mathbf{B}_i \mathbf{e}(t)) \cdot h_i, \tag{6.1}$$

where $\mathbf{A}_i$ and $\mathbf{B}_i$ are the $n \times n$ state matrix and $n \times p$ input matrix respectively, corresponding to configuration i (with i ranging from 1 to N), $\mathbf{x}(t)$ is the n-length state vector and h_i are the T-periodical validation functions. We separate system

(6.1) into two systems according to one of the two methods stated above. To this end, each subsystem will be indexed differently, namely:

- the *slower* or *low-frequency* (LF) subsystem will be indexed by "c"; this subsystem has dimension nc;
- the *faster* or *high-frequency* (HF) subsystem will be indexed by "a"; this subsystem has dimension na and will be called *disturbing*.

The corresponding state vectors of the two subsystems can be identified within the total n-length state vector $\mathbf{x}$:

$$\mathbf{x} = \begin{bmatrix} \mathbf{x}_a & \mathbf{x}_c \end{bmatrix}^T.$$

Equation system (6.1) can be arranged, consequently, as

$$\begin{cases} \dot{\mathbf{x}_a} = \sum_{i=1}^{N} \left(\mathbf{A}_{aa}^{(i)} \mathbf{x}_a + \mathbf{A}_{ac}^{(i)} \mathbf{x}_c + \mathbf{B}_a^{(i)} \mathbf{e} \right) \cdot h_i \\ \dot{\mathbf{x}_c} = \sum_{i=1}^{N} \left(\mathbf{A}_{ca}^{(i)} \mathbf{x}_a + \mathbf{A}_{cc}^{(i)} \mathbf{x}_c + \mathbf{B}_c^{(i)} \mathbf{e} \right) \cdot h_i, \end{cases} \tag{6.2}$$

where matrices $\mathbf{A}_{aa}{}^{(i)}$, $\mathbf{A}_{ac}{}^{(i)}$, $\mathbf{A}_{ca}{}^{(i)}$, $\mathbf{A}_{cc}{}^{(i)}$, $\mathbf{B}_a{}^{(i)}$ and $\mathbf{B}_c{}^{(i)}$ result from appropriately arranging matrices $\mathbf{A}_i$ and $\mathbf{B}_i$ of configuration i:

$$\mathbf{A}_i = \begin{bmatrix} \mathbf{A}_{aa}^{(i)} & \mathbf{A}_{ac}^{(i)} \\ \mathbf{A}_{ca}^{(i)} & \mathbf{A}_{cc}^{(i)} \end{bmatrix}, \mathbf{B}_i = \begin{bmatrix} \mathbf{B}_a^{(i)} \\ \mathbf{B}_c^{(i)} \end{bmatrix}.$$

From the first equation in (6.2) one can determine the response of fast variables, $\mathbf{x}_{a\_st}(\mathbf{x}_c,t)$ by supposing that variables $\mathbf{x}_c$ remain unchanged, which holds if the two dynamics can indeed be sharply enough separated. This value, $\mathbf{x}_{a\_st}$, is then substituted into the dynamic equation of the slow subsystem as follows:

$$\dot{\mathbf{x}_c} = \sum_{i=1}^{n} \left(\mathbf{A}_{cc}^{(i)} \mathbf{x}_c + \mathbf{A}_{ca}^{(i)} \cdot \mathbf{x}_{a\_st}(t, \mathbf{x}_c) + \mathbf{B}_c^{(i)} \mathbf{e} \right) \cdot h_i. \tag{6.3}$$

To Eq. (6.3) one can now apply the classical averaged model by averaging on time slot T. Thus, the averaged behavior is obtained as

$$\dot{\langle \mathbf{x}_c \rangle_0} = \frac{1}{T} \int_{t_{i-1}}^{t_i} \left(\mathbf{A}_{cc}^{(i)} \mathbf{x}_c + \mathbf{A}_{ca}^{(i)} \cdot \mathbf{x}_{a\_st}(t, \mathbf{x}_c) + \mathbf{B}_c^{(i)} \mathbf{e} \right) \cdot h_i \cdot dt, \tag{6.4}$$

which, by introducing switching function u as depending on validation functions h_i, can further be presented under the form

$$\dot{\langle \mathbf{x}_c \rangle_0} = f\left(\langle \mathbf{x}_c \rangle_0, \mathbf{x}_{a\_st}(t, \mathbf{x}_c), u \right). \tag{6.5}$$

Model (6.5) ignores the faster dynamics and preserves only the low-frequency dynamic of the slower subsystem. Note that in power converters having only AC variables, the prevalent low-frequency dynamic is obviously of AC type and supposes magnitude and phase variations. Hence, the reduced-order model must be obtained by applying generalized averaging to Eq. (6.3).

In conclusion, the ROAM is applicable if one can identify the presence of two dynamics sufficiently clearly separable. An advantage of this model is that it has reduced order, i.e., $n_c = n - n_a$.

6.3 General Methodology

A general procedure of obtaining reduced-order averaged models is given below. It represents a generalization of the procedure of Sun and Grotstollen (1992) – originally proposed for DC-DC converters – which was next resumed by Sun and Grotstollen (1997) and then by Sun et al. (2001).

In order to illustrate the application of Algorithm 6.1, two examples respectively corresponding to the above mentioned cases will be discussed next:

(a) the case of a current inverter for induction heating, where the AC stage exhibits a fast dynamic and the DC stage has a slow dynamic;
(b) the case of a buck-boost converter operating in discontinuous inductor conduction: on one hand the output voltage and on the other hand the current responsible for the discontinuous conduction.

These two examples, for which the classical averaged model does not apply, are different from the point of view of how the principle of ROAM applies. Indeed, in the first case it is about the AC-type variables that are disturbing, whereas in the second case it is the discontinuous current that must be eliminated.

Algorithm 6.1.

Computation of the reduced-order averaged model (ROAM) of a given power electronic converter

#1 Write the switched model of the studied power electronic converter.
#2 Split the model into two parts, respectively corresponding to two subsystems, one that exhibits slow dynamics (LF) and the other having fast dynamics (HF).
#3 Solve the fast-dynamic subsystem by considering constant the slow variables (at their averaged values) and compute its averaged response depending on the values of the slow variables.
#4 Apply either the classical or the generalized averaged model to this latter resulting subsystem, depending on the power stage type.
#5 Replace the fast variables by their averaged values, previously computed at #3, into the slow-dynamic subsystem.

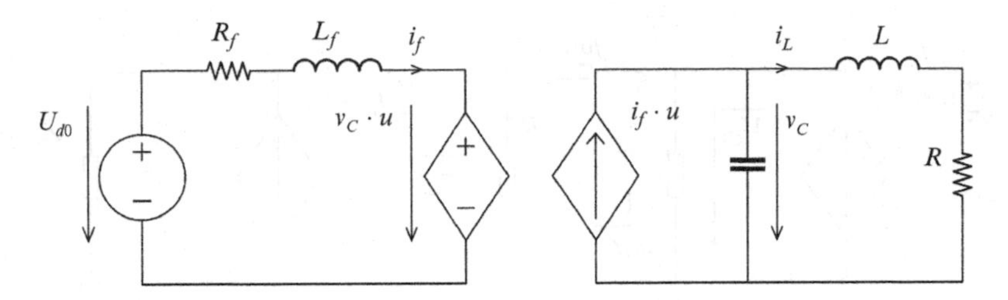

Fig. 6.1 Switched model equivalent diagram of current-source inverter

6.3.1 *Example with Alternating Variables: Current-Source Inverter for Induction Heating*

Let us consider the case of the current-source inverter for induction heating analyzed in Sect. 3.2.3 of Chap. 3, whose schematics are presented in Fig. 3.3. Its equivalent diagram corresponding to the switched model is given in Fig. 6.1.

The classical averaged model cannot apply to this converter because of the presence of AC state variables. One can build a reduced-order averaged model by neglecting the dynamics of AC state variables with respect to the DC filter dynamic.

Remember that this converter is described by the system of equations:

$$\dot{\mathbf{x}} = (\mathbf{A} + u\mathbf{B}) \cdot \mathbf{x} + \mathbf{b} \cdot \mathbf{e}, \tag{6.6}$$

where the state vector is $\mathbf{x}_c = [\, i_f \quad v_C \quad i_L \,]^T$ with

$$\mathbf{A} = \begin{bmatrix} -\dfrac{R_f}{L_f} & 0 & 0 \\ 0 & 0 & -\dfrac{1}{C} \\ 0 & \dfrac{1}{L} & -\dfrac{R}{L} \end{bmatrix}, \mathbf{B} = \begin{bmatrix} 0 & -\dfrac{1}{L_f} & 0 \\ \dfrac{1}{C} & 0 & 0 \\ 0 & 0 & 0 \end{bmatrix}, \mathbf{b} = \begin{bmatrix} \dfrac{1}{L_f} \\ 0 \\ 0 \end{bmatrix}, \mathbf{e} = U_{d0}.$$

As already detailed in Sect. 5.7.1 of Chap. 5, the DC current i_f through the filter has slow dynamic and the AC state variables – the voltage v_C and the current i_L – are characterized by fast variations because of their high-frequency sinusoidal nature. This is the reason the separation principle is clearly applicable: the filter current i_f exhibiting the prevalent LF dynamic will be kept, the other two variables being excluded from the dynamic model (Bacha et al. 1995).

By using notations previously introduced and by taking $\mathbf{x}_c = i_f$ and $\mathbf{x}_a = [\, v_C \quad i_L \,]^T$, one can separate the HF variables v_C and i_L from the continuous variable i_f.

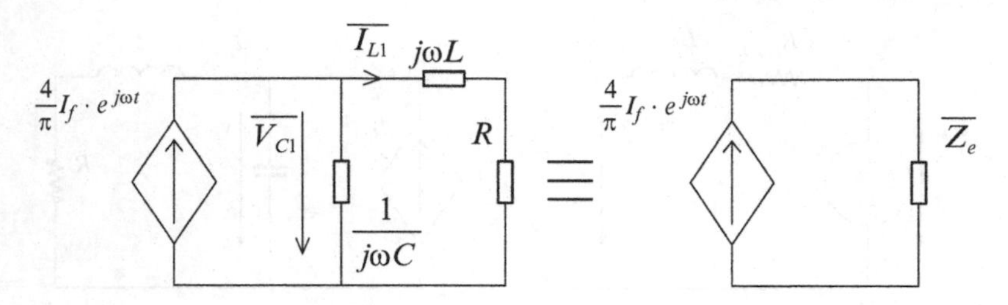

Fig. 6.2 Current-source inverter: equivalent diagram of AC part in the sense of the first-order harmonic

Two subsystems are consequently obtained:

$$\dot{\mathbf{x}}_c = -\frac{R_f}{L_f}\mathbf{x}_c + \begin{bmatrix} -\frac{1}{L_f} & 0 \end{bmatrix} \mathbf{x}_a \cdot u + \frac{1}{L_f}\mathbf{e} \tag{6.7}$$

and

$$\dot{\mathbf{x}}_a = \begin{bmatrix} \frac{1}{C} \\ 0 \end{bmatrix} \mathbf{x}_c \cdot u + \begin{bmatrix} 0 & -\frac{1}{C} \\ \frac{1}{L} & -\frac{R}{L} \end{bmatrix} \mathbf{x}_a, \tag{6.8}$$

where u is the symmetrical switching function having pulsation $\omega = 2\pi/T$.

Let us note that from point of view of the AC variables (at small time range), current i_f can be considered constant and therefore denoted by I_f. In order to compute the response of the HF variables, denoted by $\mathbf{x}_{a\_st}(I_f, t)$, one can employ the first-order harmonic method. The justification consists in the fact that Eq. (6.8), describing the AC part, represents a differential equation system with constant coefficients and periodical input $\mathbf{x}_c \cdot u \equiv I_f \cdot u$, whose state variables are dominated by their fundamental terms. Moreover, the power is transmitted to the load essentially through the first-order harmonic. By taking as origin of phases the moment when u takes the value 1, the equivalent diagram – in terms of impedances – of the AC part is given in Fig. 6.2.

By solving the equations of the above diagram, one obtains the response of the AC part, $\mathbf{x}_{a\_st}(I_f, t)$ defined by i_{L1} and v_{C1} which are ω-pulsation sinusoidal variables representing fundamentals of current i_L and voltage v_C, respectively, in the circuit presented in Fig. 6.2.

By replacing $\mathbf{x}_a$ by $\mathbf{x}_{a\_st}(I_f, t)$ in Eq. (6.7) one obtains

$$\dot{\mathbf{x}}_c = -\frac{R_f}{L_f}\mathbf{x}_c + \begin{bmatrix} -\frac{1}{L_f} & 0 \end{bmatrix} \cdot \mathbf{x}_{a\_st}(I_f, t) \cdot u + \frac{1}{L_f}\mathbf{e}. \tag{6.9}$$

Now one applies the sliding average to Eq. (6.9). The second term of the right side becomes

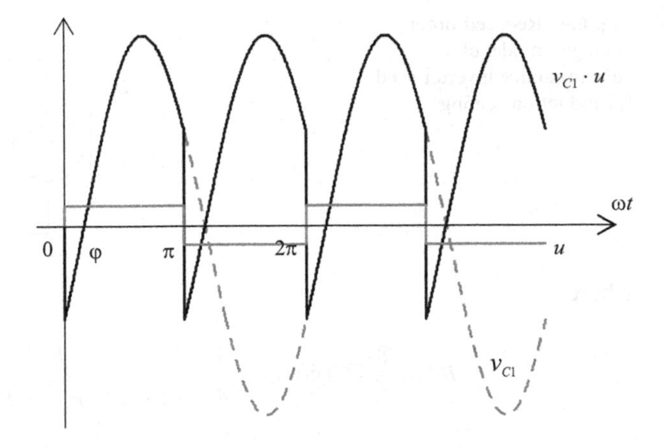

Fig. 6.3 Waveforms of switching function and capacitor voltage fundamental

$$
\begin{aligned}
\left\langle \begin{bmatrix} -\frac{1}{L_f} & 0 \end{bmatrix} \cdot \mathbf{x}_{a\_st}(I_f, t) \cdot u \right\rangle_0 &= \frac{1}{T} \int_0^T \left(\begin{bmatrix} -\frac{1}{L_f} & 0 \end{bmatrix} \cdot \begin{bmatrix} v_{C1} \\ i_{L1} \end{bmatrix} \cdot u \right) dt \\
&= -\frac{1}{TL_f} \cdot \int_0^T (v_{C1} \cdot u) dt.
\end{aligned}
\tag{6.10}
$$

Note that, according to Fig. 6.2, the phasor expression of the capacitor voltage fundamental is

$$
\overline{V_{C1}} = \frac{4}{\pi} I_f e^{j\omega t} \cdot |Z_e| e^{j\varphi} = \frac{4}{\pi} I_f |Z_e| e^{j(\omega t + \varphi)}, \tag{6.11}
$$

where $|Z_e|$ and φ are the modulus and the argument of impedance Z_e, respectively, depending of the circuit element values, R, L and C.

Taking into account Eq. (6.11) and Fig. 6.3, Eq. (6.10) can further be developed:

$$
\begin{aligned}
-\frac{1}{TL_f} \cdot \int_0^T (v_{C1} \cdot u) \cdot dt &= -\frac{1}{TL_f} \cdot \int_0^T \left(\frac{4}{\pi} I_f |Z_e| \sin(\omega t + \varphi) \cdot u \right) \cdot dt \\
&= -\frac{4 I_f |Z_e|}{\pi T L_f} \left[\int_0^{T/2} \sin(\omega t + \varphi) dt - \int_{T/2}^{T} \sin(\omega t + \varphi) dt \right] \\
&= -\frac{8 I_f |Z_e|}{\pi^2 L_f} \cos \varphi,
\end{aligned}
$$

leading to the final form:

$$
\left\langle \begin{bmatrix} -\frac{1}{L_f} & 0 \end{bmatrix} \cdot \mathbf{x}_{a\_st}(I_f, t) \cdot u \right\rangle_0 = -\frac{8|Z_e|}{\pi^2 L_f} \cos \varphi \cdot I_f = -\frac{R_e}{L_f} I_f, \tag{6.12}
$$

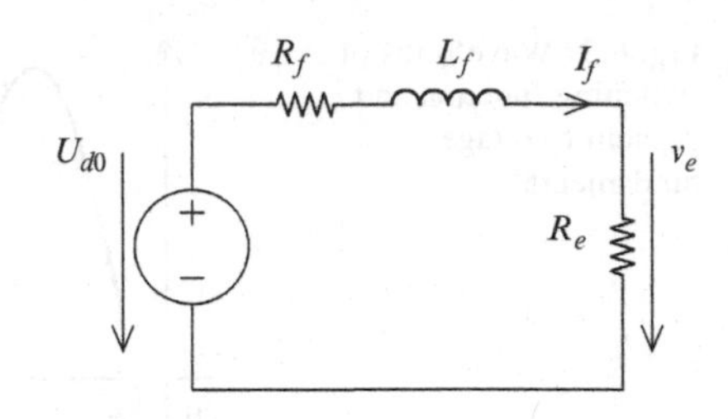

Fig. 6.4 Reduced-order averaged model of a current-source inverter used for induction heating

where

$$R_e = \frac{8}{\pi^2}|Z_e|\cos\varphi = \frac{8}{\pi^2}\cdot\frac{R}{(1-LC\omega^2)^2+R^2C^2\omega^2}. \tag{6.13}$$

Now, changing the time scale to a large time range, variations of i_f are rendered visible; in the meantime Eq. (6.12) remains valid. Therefore, by replacing the result from (6.12) in Eq. (6.9) averaged, one obtains

$$\langle \dot{i_f}\rangle_0 = -\frac{(R_f+R_e)}{L_f}\cdot\langle i_f\rangle_0 + \frac{1}{L_f}\cdot U_{d0}, \tag{6.14}$$

which leads to the equivalent circuit presented in Fig. 6.4.

Remarks. The order of the resulting model has been reduced by two. The final model is first-order, thus easier to use. Since the control input of the converter is the switching frequency ω of the current-source inverter, model (6.14) is highly nonlinear: it depends on R_e, so finally on a nonlinear function of ω (see (6.13)).

The dynamics of the AC variables are totally ignored, and sometimes this could have non-negligible consequences.

6.3.2 *Example with Discontinuous-Conduction Mode: Buck-Boost Converter*

This example deals with a buck-boost converter (Fig. 6.5) operating in discontinuous-conduction mode (dcm). In the literature it has been shown that the dcm significantly affects converter behavior (Ćuk and Middlebrook 1977; Vorpérian 1990). Circuit-oriented analysis on DC-DC converters has proved that in general the poles of its second-order dynamic are not complex-conjugated anymore (as in continuous conduction), but placed on the real axis (Maksimović and Ćuk 1991). The one corresponding to inductor energy accumulation is located in high frequency close to the switching frequency (Erikson and Maksimović 2001). Therefore, the output voltage response will consequently become

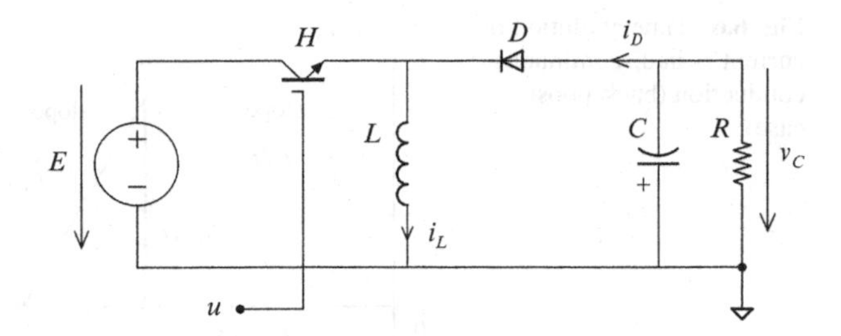

Fig. 6.5 Electrical circuit of a buck-boost converter

aperiodical – see for reference Fig. 4.23b in Chap. 4. When focusing on controlling the output voltage, such a system can easily be assumed as first-order (because of neglecting the HF pole). System-oriented analysis of these aspects can be found in Sun et al. (2001).

As inductor current becomes zero periodically in dcm operation, the classical average modeling no longer applies. So, in this example the average (LF) behavior is expressed mathematically by considering that the output capacitor is fed by an equivalent current generator (Chetty 1981, 1982).

To solve the modeling problem, one follows the step procedure synthesized in Algorithm 6.1, namely:

- compute the average value of current i_D through diode D as a function of capacitor voltage, which is equivalent to solving the high-frequency subsystem;
- average the dynamic equation of output voltage v_C;
- replace the previously computed value of current i_D in the dynamic equation of the averaged output voltage;
- obtain the classical averaged model of the resulting system (which will be in this case of first order).

The switched model of the circuit from Fig. 6.5 operating in continuous-conduction mode was already given in Eq. (4.34) of Sect. 4.6 (Chap. 4). The operation in dcm can be described by either adding a supplementary equation – expressing the constraint of current zeroing – or by modifying the dynamic equation of the inductor current, as shown below:

$$\begin{cases} \dot{i_L} = \dfrac{1}{L}(Eu + v_C(1-u) - ri_L) \cdot \dfrac{1+\mathrm{sgn}(i_L)}{2} \\ \dot{v_C} = \dfrac{1}{C}\left(-i_L(1-u) - \dfrac{v_C}{R}\right), \end{cases} \tag{6.15}$$

where

$$\mathrm{sgn}(i_L) = \begin{cases} 1 & \text{if } i_L > 0 \\ -1 & \text{if } i_L \le 0 \end{cases}$$

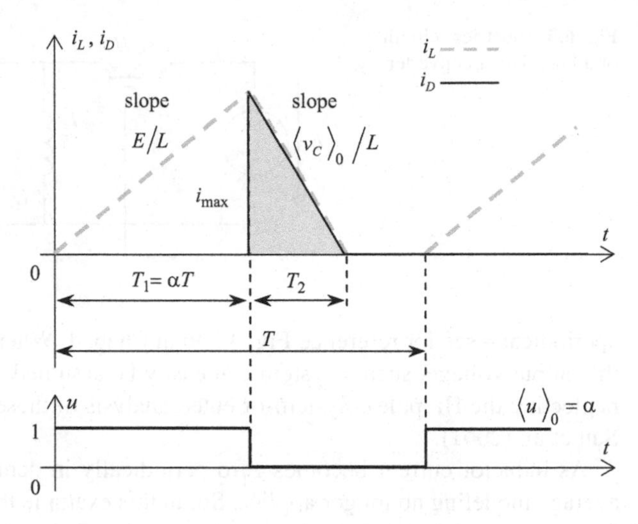

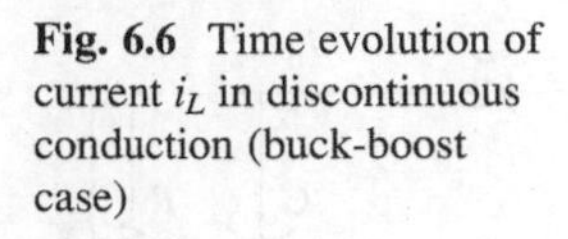
Fig. 6.6 Time evolution of current i_L in discontinuous conduction (buck-boost case)

and u is the T-periodical switching function taking values in the set {0;1} and having α as duty ratio. The other notations have the usual meaning assigned to circuit element characteristics. Note that the model given in (6.15) is general – it does not adopt any simplifying assumption, except for what is stated in Sect. 2.2 of Chap. 2.

Now, the search for an equivalent low-frequency model justifies the adoption of an approximation: as the switching frequency is large enough with respect to the output voltage evolution, one can assume that the current i_L decreases linearly in time when the switching function u is equal to zero. Omitting inductor resistance r, the discontinuous-conduction time evolution of current i_L can be viewed in Fig. 6.6. Three phases can be identified in this figure:

- from 0 to $T_1 = \alpha T$, the switch H being turned on, the current increases linearly and slope $E/L > 0$, according to the first equation from (6.15) with $u = 1$, which corresponds to the energy being accumulated in the inductor L;
- during the time interval denoted by T_2, after turning off the current switch H, the current decreases linearly and slope $\langle v_C \rangle_0 / L < 0$ (note that $\langle v_C \rangle_0 < 0$) according to the first equation from (6.15) with $u = 0$, which corresponds to using the previously accumulated inductor energy to charge the capacitor C;
- finally, during the last time subinterval the current remains zero until the switch H is again turned on.

One can see in Fig. 6.5 that in dcm the effective conversion ratio is no longer equal to α and depends on the circuit operating point (subinterval T_2 is a variable of v_C).

The goal is to transform the initial model into a model of an equivalent average current generator supplying a capacitor C and replacing the averaged current through diode D, as shown in Fig. 6.7 (Chetty 1982).

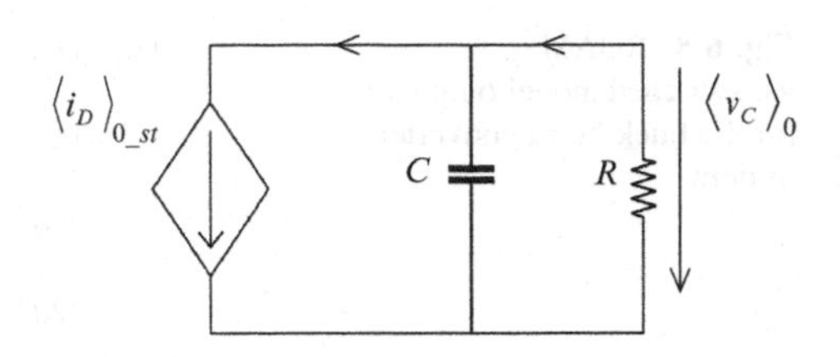

Fig. 6.7 Reduced-order averaged model of a buck-boost converter

The averaged output current – computed by assuming $\langle v_C \rangle_0$ constant and denoted by $\langle i_D \rangle_{0\_st}$ – corresponds to area of the gray filled surface in Fig. 6.6 divided by the time interval T. Thus, noting that the height of the filled triangle is $i_{\max} = T_1 \cdot E/L$, one obtains

$$\langle i_D \rangle_{0\_st} = \frac{E}{2LT} T_1 T_2. \tag{6.16}$$

The value of T_2 is found from Fig. 6.6 by noting that in the filled triangle it is true that $i_{\max}/T_2 = - \langle v_C \rangle_0 / L$; by further replacing $i_{\max} = T_1 \cdot E/L$ one obtains

$$T_2 = -\frac{E}{\langle v_C \rangle_0} T_1. \tag{6.17}$$

By replacing the value of T_2 as obtained from (6.17) and $T_1 = \alpha \cdot T$ into (6.16), it can be seen that the averaged values of i_D and v_C are related *algebraically*:

$$\langle i_D \rangle_{0\_st} = -\frac{E^2 T \alpha^2}{2L \langle v_C \rangle_0}, \tag{6.18}$$

thus indicating as a matter of fact (if $\langle v_C \rangle_0$ is considered constant on a switching period), that *the two state variables exhibit the same dynamic*. This remark justifies saying that the equivalent low-frequency converter dynamics operating in dcm are of first order instead of second order, as explained next.

Let us consider the second equation from (6.15), emphasizing the diode current i_D:

$$C\dot{v_C} = -i_D - v_C/R.$$

The latter relation is averaged, the average value of i_D being replaced by Eq. (6.18):

$$C\langle \dot{v_C} \rangle_0 = \frac{E^2 T}{2L \langle v_C \rangle_0} \alpha^2 - \frac{\langle v_C \rangle_0}{R}. \tag{6.19}$$

Equation (6.19) is nonlinear and represents the ROAM of the buck-boost converter operating in discontinuous-conduction mode. It also shows a first-order linear dynamic between the squared duty ratio α^2 and the output voltage squared $\langle v_C \rangle_0^2$.

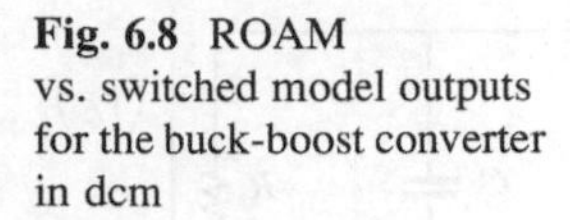

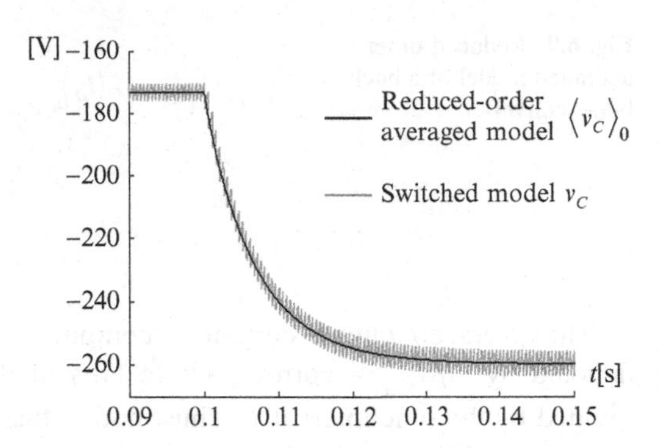

Fig. 6.8 ROAM vs. switched model outputs for the buck-boost converter in dcm

Remarks. In Fig. 6.8 is given the converter output voltage evolution in response to a step of the duty ratio during operation in dcm (for reference see Fig. 4.23b in Sect. 4.6 of Chap. 4). This figure allows a comparison of the ROAM with the switched model output.

6.4 Case Studies

6.4.1 Thyristor-Controlled Reactor Modeling

This case study is dedicated to the situation when the inductor current becomes zero periodically, this time within an AC power stage whose circuit diagram is presented in Fig. 6.9 (*thyristor-controlled reactor* – TCR).

Three circuits like the one in Fig. 6.9 are connected in delta to a three-phase grid. Therefore line voltage $v(t)$ is applied to the circuit. By varying the thyristor firing angles within ($\pi/2$, π) in relation to the line voltage angle, one can vary the inductor current and hence the reactive power interchanged with the grid (Mohan et al. 2002). The switching function u depends on both the firing angle, denoted by β_0, and the moment of current zeroing.

This operation is described briefly in Fig. 6.10. By analyzing this figure one can anticipate that the first harmonic of inductor current i lags the grid voltage v by $\pi/2$. Hence, this component influences the reactive power. For this reason, the next analysis will be focused on obtaining analytically the TCR first-order harmonic behavior.

Note that the inductor current resets to zero in every switching cycle; this means that energy flow in the inductor is independent from cycle to cycle. Therefore the current inductor carries no information from cycle to cycle, which means it can no longer represent a state variable (Sun et al. 2001). This is equivalent to stating that the low-frequency current components – and therefore their first harmonics – have no dynamic, i.e., they evolve instantaneously in response to control input (β_0) changes. Within a switching cycle, the inductor current is a state variable that

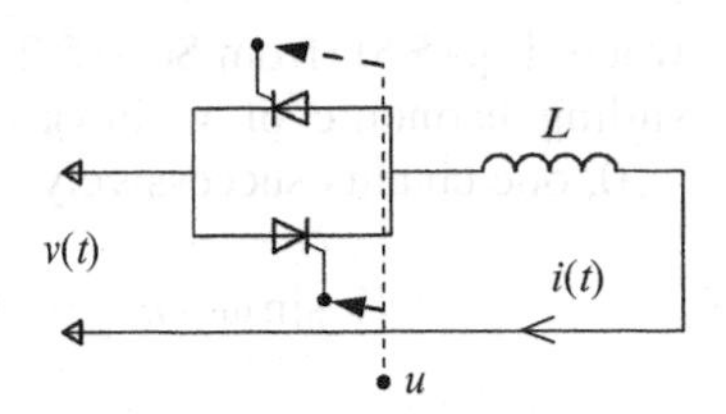

Fig. 6.9 TCR circuit diagram

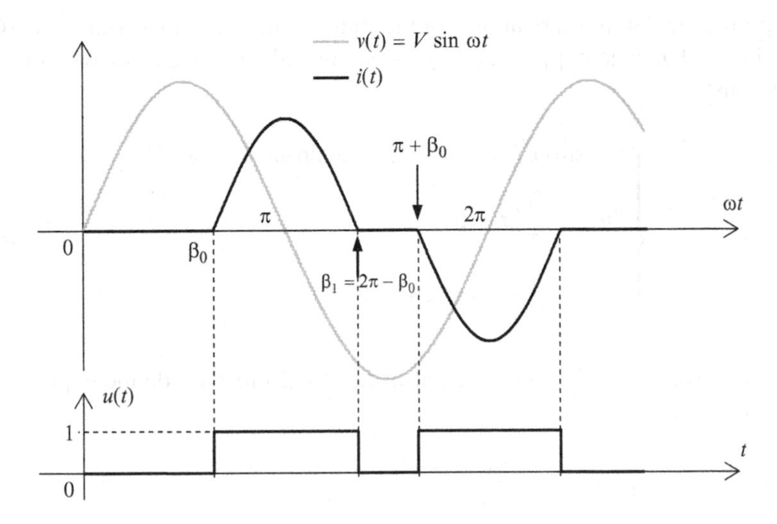

Fig. 6.10 TCR operation – main waveforms

corresponds to very fast dynamics, which can be neglected from the perspective of designing control laws dedicated to managing the equivalent LF behavior.

By taking the input sinusoidal voltage $v(t) = V \sin \omega t$ as the phase origin and having chosen to delay the firing orders of thyristors by angle β_0, the dynamic equation (switched model) of the circuit is

$$L\frac{di(t)}{dt} = V \sin \omega t \cdot u(t) - r \cdot i(t), \tag{6.20}$$

where $u(t) = \frac{1+\mathrm{sgn}[V\sin(\omega t - \beta_0)]\cdot\mathrm{sgn}[i(t)]}{2}$ (see Fig. 6.10). For the sake of simplicity, suppose that inductor resistance is negligible, i.e., $r = 0$. Note that in this case the moment of current zeroing corresponds to the angle $\beta_1 = 2\pi - \beta_0$. If this assumption does not hold, further computation is rendered difficult because the moment of current zeroing, being part of a transcendent equation, cannot be determined analytically (Bacha et al. 1995).

By applying the first-order sliding average to Eq. (6.20) with $r = 0$, one obtains (time variable skipped for simplicity):

$$L\frac{d\langle i\rangle_1}{dt} = -j\omega L\langle i\rangle_1 + \langle V \sin \omega t \cdot u\rangle_1, \tag{6.21}$$

where Eq. (5.5) (from Sect. 5.2.1 in Chap. 5) can be used to compute the first-order sliding harmonic of $V \sin \omega t \cdot u$. Thus, knowing that $\langle V \sin \omega t\rangle_0 = \langle V \sin \omega t\rangle_2 = 0$, one obtains successively:

$$\begin{aligned}\langle V \sin \omega t \cdot u\rangle_1 &= \langle V \sin \omega t\rangle_0 \langle u\rangle_1 + \langle V \sin \omega t\rangle_1 \langle u\rangle_0 \\ &\quad + \langle V \sin \omega t\rangle_{-1} \langle u\rangle_2 + \langle V \sin \omega t\rangle_2 \langle u\rangle_{-1} \\ &= \langle V \sin \omega t\rangle_1 \langle u\rangle_0 + \langle V \sin \omega t\rangle_{-1} \langle u\rangle_2.\end{aligned} \tag{6.22}$$

Applying the definition relations for the sliding harmonics involved in Eq. (6.22) – see Fig. 6.10, where $\beta_1 = 2\pi - \beta_0$ – simple algebra leads to the following expressions:

$$\begin{cases} \langle V \sin \omega t\rangle_1 = -jV/2, \quad \langle V \sin \omega t\rangle_{-1} = jV/2 \\ \langle u\rangle_0 = \dfrac{2}{\pi}(\pi - \beta_0) \\ \langle u\rangle_2 = -\dfrac{1}{2\pi} \sin(2\beta_0) \cdot e^{-2j\omega t}. \end{cases} \tag{6.23}$$

Equation (6.23) are further replaced in Eq. (6.22) to provide the expression of $\langle V \sin \omega t \cdot u\rangle_1$:

$$\begin{aligned}\langle V \sin \omega t \cdot u\rangle_1 &= j\frac{V}{2} \cdot (\langle u\rangle_2 - \langle u\rangle_0) \\ &= -j\frac{V}{\pi} \cdot \left(\frac{\sin(2\beta_0)}{2} e^{-2j\omega t} + 2\pi - 2\beta_0\right).\end{aligned} \tag{6.24}$$

Equation (6.24) is at its turn used in Eq. (6.21) to yield the description of the first-order sliding harmonic of current i:

$$L\frac{d\langle i\rangle_1}{dt} = -j\omega L\langle i\rangle_1 - j\frac{V}{\pi} \cdot \left(\frac{\sin(2\beta_0)}{2} e^{-2j\omega t} + 2\pi - 2\beta_0\right). \tag{6.25}$$

By denoting $\langle i\rangle_1 = x_1 + jx_2$, one can deduce from Eq. (6.25) the dynamics of the real and imaginary parts respectively:

$$\begin{cases} L\dfrac{dx_1}{dt} = \omega L x_2 - \dfrac{V}{2\pi} \cdot \sin 2\omega t \cdot \sin 2\beta_0 \\ L\dfrac{dx_2}{dt} = -\omega L x_1 - \dfrac{V}{\pi}(\pi - \beta_0) - \dfrac{V}{2\pi} \cdot \cos 2\omega t \cdot \sin 2\beta_0. \end{cases} \tag{6.26}$$

As stated before, the inductor current has no dynamic, therefore one proceeds to zero the derivatives in Eq. (6.26). The steady-state regimes of the real and imaginary parts of the inductor current's first-order sliding harmonic are therefore

$$\begin{cases} \omega L x_{2\_st} - \dfrac{V}{2\pi} \cdot \sin 2\omega t \cdot \sin 2\beta_0 = 0 \\ -\omega L x_{1\_st} - \dfrac{V}{\pi}(\pi - \beta_0) - \dfrac{V}{2\pi} \cdot \cos 2\omega t \cdot \sin 2\beta_0 = 0, \end{cases}$$

which further gives:

$$\begin{cases} x_{1\_st} = -\dfrac{V(\pi - \beta_0)}{\pi\omega L} - \dfrac{V \sin 2\beta_0}{2\pi\omega L} \cdot \cos 2\omega t \\ x_{2\_st} = \dfrac{V \sin 2\beta_0}{2\pi\omega L} \cdot \sin 2\omega t. \end{cases} \tag{6.27}$$

Using Eq. (5.29) from Chap. 5, which gives the connection between a signal and its first-order sliding harmonic, and knowing that signal $i(t)$ is sinusoidal, one can write that $i(t) = 2x_1 \cos \omega t - 2x_2 \sin \omega t$. Finally, taking into account (6.27), the expression of the steady-state current regime is obtained as

$$i_{st}(t) = -\frac{V(2\pi - 2\beta_0 + \sin 2\beta_0)}{\pi\omega L} \cdot \cos \omega t. \tag{6.28}$$

Equation (6.28) shows that the fundamentals of current $i(t)$ and of input voltage $v(t)$ are quadrature signals, as expected, and that the current fundamental amplitude depends on initial angle lag β_0 and on frequency ω. This equation represents the ROAM of the TCR because the system order is reduced by one. The model given by (6.28) can be directly used to control larger power electronic structures as static VAR compensators (SVC).

Next, simulations of TCR behavior have been carried out using MATLAB®-Simulink®. Input voltage is perfectly sinusoidal, having 400 V RMS and 50 Hz, and the inductor is of 25 mH. The ROAM from (6.28) is hence compared with the circuit switched model giving the inductor real current waveform.

Figure 6.11 shows the TCR behavior at constant control input, $\beta_0 = 135°$. As expected, the current first-order harmonic lags the voltage phase by $\pi/2$, see Fig. 6.11a. Current magnitude $I_1 \approx 13.1$ A, which is confirmed by the spectral analysis in Fig. 6.11b.

Figure 6.12 presents the TCR behavior when β_0 decreases in step by 10°, from 135° to 125°. The effective voltage applied to the inductor, $v \cdot u$, is shown in Fig. 6.12a. Figure 6.12b shows that the current magnitude increases as β_0 decreases and the current fundamental remains in quadrature with respect to the voltage. Also, the current first-order harmonic amplitude increases *instantaneously* from 13.1 to 22.5 A in response to β_0 variation, thus justifying the ROAM validity.

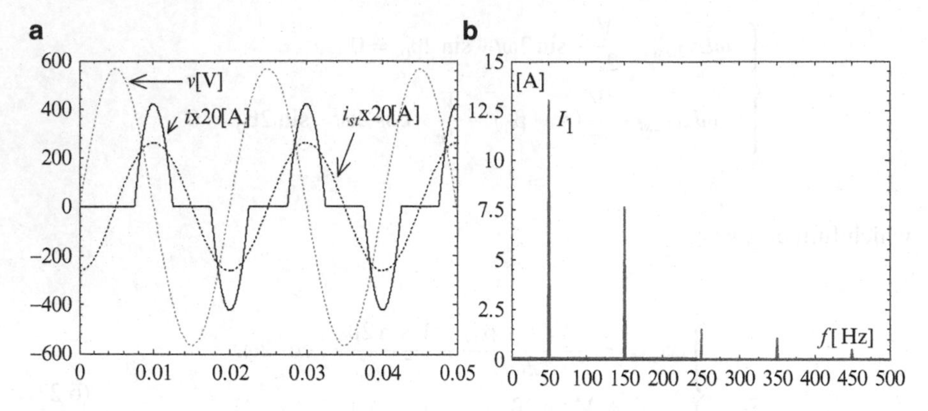

Fig. 6.11 TCR behavior at $\beta_0 = 135°$: (**a**) output of ROAM ($i_{st}(t)$) vs. output of switched model ($i(t)$); (**b**) inductor current spectrum

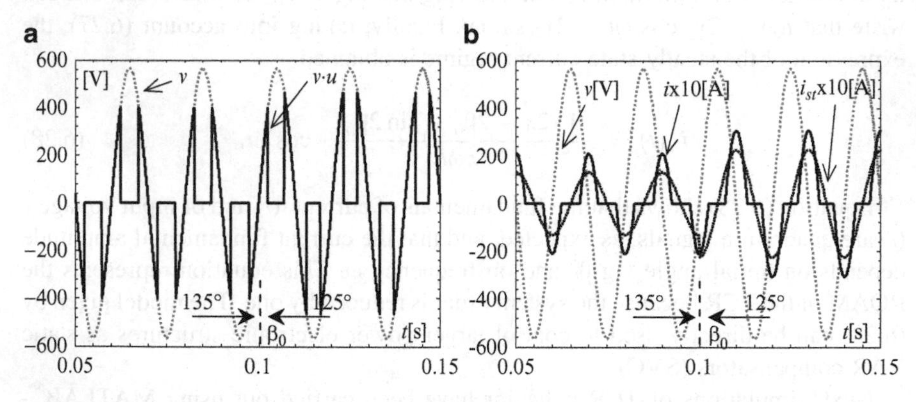

Fig. 6.12 TCR behavior at β_0 step change: (**a**) the effective voltage, $v \cdot u$, applied to the inductor; (**b**) output of ROAM ($i_{st}(t)$) vs. output of switched model ($i(t)$)

6.4.2 DC-DC Boost Converter Operating in Discontinuous-Conduction Mode

Given a DC-DC boost power stage operating in dcm, one must deduce its averaged dynamic model that represents output voltage variation as a function of averaged control input (duty ratio). The circuit diagram is presented in Fig. 6.13, where inductor resistance and capacitor equivalent series resistance are neglected.

This case study is placed in the same context as the example from Sect. 6.3.2. Therefore, in order to deduce its ROAM one must consider the equivalent averaged current that feeds the output capacitor and express it as a function of output voltage.

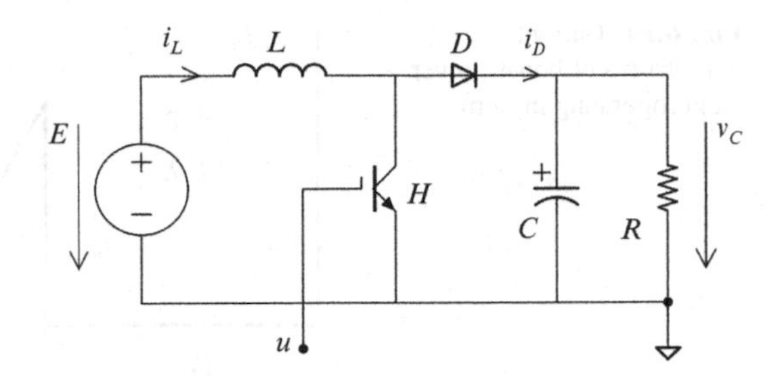

Fig. 6.13 Boost power stage diagram

The switched model is given by (see for reference Eq. (3.11) of Sect. 3.2.3, Chap. 3):

$$\begin{cases} \dot{i_L} = \dfrac{1}{L}(E - v_C(1-u) - ri_L) \cdot \dfrac{1+\mathrm{sgn}(i_L)}{2} \\ \dot{v_C} = \dfrac{1}{C}\left(i_L(1-u) - \dfrac{v_C}{R}\right), \end{cases} \tag{6.29}$$

with

$$\mathrm{sgn}(i_L) = \begin{cases} 1 & \text{if } i_L > 0 \\ -1 & \text{if } i_L \le 0 \end{cases}$$

and u being the T-periodical switching function taking values in the set $\{0;1\}$ and having α as duty ratio, i.e., $\langle u \rangle_0 = \alpha$.

Discontinuous-conduction operation is characterized by the existence of three time subintervals within a switching period T, which describe the time evolution of the inductor current (see Fig. 6.14), similar to the ones of the buck-boost converter (depicted in Fig. 6.6). The assumption of output voltage v_C being constant at the evolution time scale of current i_L is adopted. The expression of $i_{\max}$ can be written in two ways corresponding to the two triangles; thus

$$i_{\max} = \frac{E}{L} \cdot T_1 = \frac{v_C - E}{L} \cdot T_2,$$

hence

$$T_2 = \frac{E}{v_C - E} \cdot T_1. \tag{6.30}$$

The averaged value of the diode current over a T-length time window can be expressed using the gray filled area of Fig. 6.14, namely,

$$\langle i_D \rangle_0 = \frac{i_{max} \cdot T_2}{2T},$$

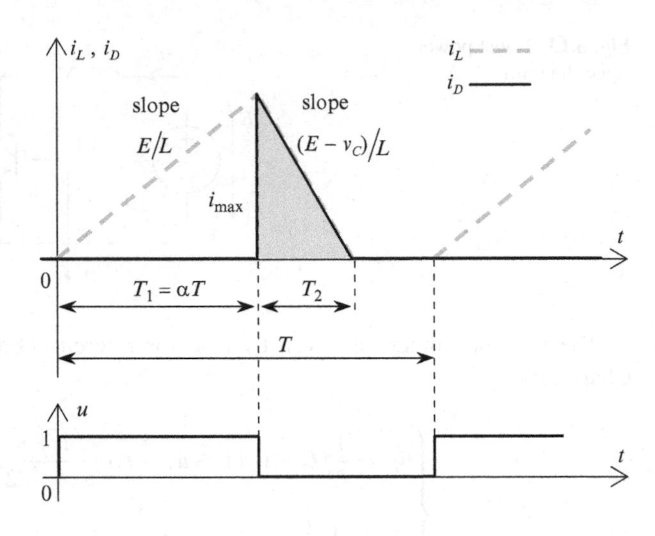

Fig. 6.14 Current waveforms of boost power stage operating in dcm

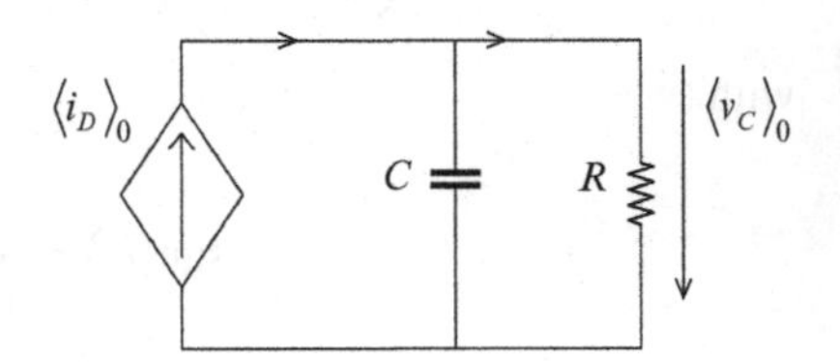

Fig. 6.15 Boost power stage – equivalent circuit corresponding to ROAM

in which Eq. (6.30) is replaced; therefore,

$$\langle i_D \rangle_0 = \frac{E^2 T \alpha^2}{2L(v_C - E)}. \tag{6.31}$$

Now let us return to the second equation from (6.29), where the inductor current is expressed depending on the diode current, i.e., $i_L(1 - u) = i_D$; this equation is further averaged and relation (6.31) is used for replacing the average value of i_D:

$$C\dot{\langle v_C \rangle}_0 = \frac{E^2 T \alpha^2}{2L(\langle v_C \rangle_0 - E)} - \frac{\langle v_C \rangle_0}{R}. \tag{6.32}$$

Relation (6.32) represents the ROAM of the boost converter operating in discontinuous-conduction mode. One can note that this relation describes the nonlinear dynamic dependence of the averaged output voltage $\langle v_C \rangle_0$ on the duty ratio α. This model corresponds to the diagram depicted in Fig. 6.15.

Relation (6.32) can be linearized around an equilibrium point in order to simplify further analysis and control law design. The equilibrium operating point is computed by zeroing the derivative in Eq. (6.32) for a fixed value of the duty ratio α_e.

In this way, by supposing that voltage E is constant, the steady-state value of $\langle v_C \rangle_0$ results from solving a second-order polynomial equation, namely,

$$2L\langle v_C \rangle_{0e}^2 - 2EL\langle v_C \rangle_{0e} - E^2RT\alpha_e^2 = 0.$$

The above equation has two solutions; the positive one is chosen:

$$\langle v_C \rangle_{0e} = \frac{E}{2} + \frac{E}{2}\sqrt{1 + \frac{2RT\alpha_e^2}{L}}. \tag{6.33}$$

In order to proceed with the linearization around the operating point given by (6.33), the following change of variable is used:

$$v = \langle v_C \rangle_0 - E. \tag{6.34}$$

Note that $\dot{v} = \dot{\langle v_C \rangle}_0$. Using this and introducing notation (6.34) into (6.32), one obtains

$$C\dot{v} = \frac{E^2T\alpha^2}{2Lv} - \frac{v+E}{R}$$

or, equivalently,

$$C\dot{v^2} = \frac{E^2T}{L}\alpha^2 - \frac{2}{R}v^2 - \frac{2E}{R}v. \tag{6.35}$$

Equation (6.35) is further linearized around the equilibrium point given by (6.33). Notation $\tilde{}$ denoting small variation around the chosen equilibrium point is adopted. Note that $v_e = \langle v_C \rangle_{0e} - E$ and $\tilde{v} = \widetilde{\langle v_C \rangle}_0$. Then straightforward computation leads to

$$\dot{\tilde{v}} = \frac{E^2T\alpha_e}{LCv_e}\tilde{\alpha} - \frac{2}{RC}\tilde{v} - \frac{E}{RCv_e}\tilde{v},$$

which, according to notation (6.34), is equivalent to

$$\dot{\widetilde{\langle v_C \rangle}}_0 = \frac{E^2T\alpha_e}{LC(\langle v_C \rangle_{0e} - E)}\tilde{\alpha} - \frac{2\langle v_C \rangle_{0e} - E}{RC(\langle v_C \rangle_{0e} - E)}\widetilde{\langle v_C \rangle}_0. \tag{6.36}$$

Relation (6.36) represents the linearized ROAM of the boost converter operating in discontinuous conduction. One can see that the output voltage dynamic is of first order in relation to the duty ratio, having $\frac{E^2RT\alpha_e}{L(2\langle v_C \rangle_{0e} - E)}$ as gain and $\frac{RC(\langle v_C \rangle_{0e} - E)}{2\langle v_C \rangle_{0e} - E}$ as time constant.

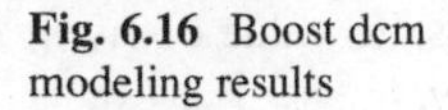

Fig. 6.16 Boost dcm modeling results

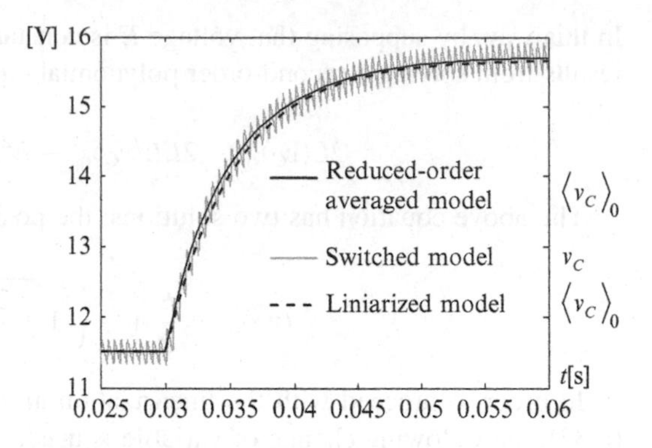

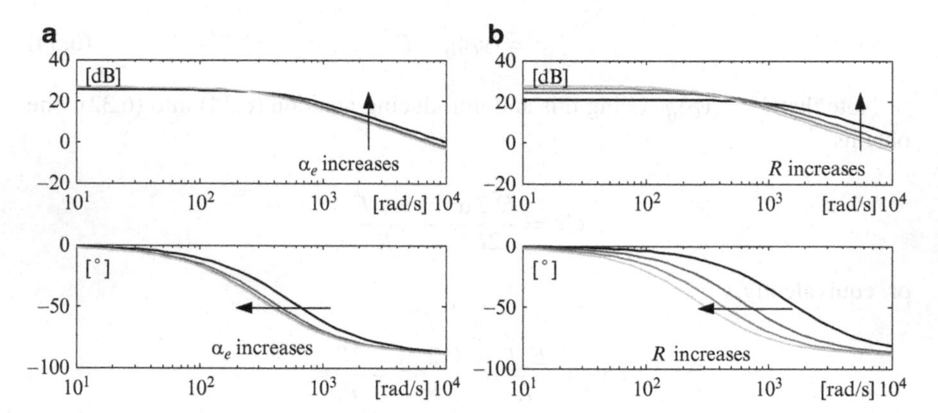

Fig. 6.17 Bode diagrams for various equilibrium points of the boost converter operating in dcm: **(a)** curve family for various α_e; **(b)** curve family for various load resistances R

Next, simulations of the boost converter operating in dcm have been carried out using MATLAB®-Simulink®. The input voltage is constant, $E = 5$ V, the switching frequency is 2000 kHz. The switches are perfect and the circuit parameters are $L = 0.5$ mH, $C = 100$ μF and $R = 150$ Ω. The ROAM from (6.32) is assessed against its linearized version from (6.36) and the switched model of the circuit – see Fig. 6.16.

Figure 6.16 shows the capacitor voltage evolution as the duty ratio is changed from 0.2 to 0.3 in step. Consequently, the average voltage evolves between 11.51 and 15.73 V – see expression (6.33). As the linearized model (6.36) is expressed in variations, one must add the steady-state value given by expression (6.33) in order to compare model outputs.

Bode diagrams of the linearized model (6.36) have also been plotted for various equilibrium points. Figure 6.17a shows how Bode diagrams change as the operating point changes for duty ratios between 0.1 and 0.4 at constant load. Figure 6.17b shows how Bode diagrams change as operating point changes for load resistances between 50 and 200 Ω. Note that system bandwidth decreases and steady-state gain increases slightly in both cases: when equilibrium duty ratio increases and when load resistance increases.

6.5 Conclusion

At this point let us formulate some summary remarks about the necessity and utility of the reduced-order averaged model. There are at least three cases where the use of ROAM is recommended:

- if the converter operates in discontinuous conduction;
- when the converter has both alternating and DC variables, and one wishes to neglect the former;
- if the converter dynamic is altered by means of a control structure, i.e., some of the variables are rendered significantly faster than others and hence can be neglected.

The mechanism of effectively reducing the system order consists in emphasizing an algebraic relationship between some fast variables and some slow variables, which makes the differential equations of the fast variables drop out. The result is a reduced-order model.

Reduced-order averaged modeling has the strength of being easy to compute and easy to use for control purposes – due to elimination of undesirable variables; hence it is widely applicable. Its drawbacks are related to ignoring the dynamics that can be qualified as "internal", exposing the control law designer to instability problems that are not predictable by this kind of modeling.

A final remark is that the ROAM provides the most suitable mechanism for modeling converters operating in discontinuous conduction. This model can be used in other cases, especially when a simpler model is needed, but it is recommended that one first analyze the dynamics likely to be neglected.

Problems

Problem 6.1. Linearized ROAM of buck-boost converter in dcm
Let us consider the buck-boost circuit in Fig. 6.5. It is required to obtain the small-signal ROAM.

Solution. In order to derive the small-signal ROAM, Eq. (6.19) – representing the nonlinear ROAM of the buck-boost converter in discontinuous conduction – should be linearized around an equilibrium point:

$$C\langle \dot{v_C} \rangle_0 = \frac{E^2 T}{2L\langle v_C \rangle_0}\alpha^2 - \frac{\langle v_C \rangle_0}{R}, \tag{6.37}$$

where notations have the meanings as introduced in the example detailed in Sect. 6.3.2. Equation (6.37) describes the first-order averaged dynamic of the capacitor voltage in relation to the input represented by duty ratio α. Supposing that voltage E is constant, the equilibrium point results from zeroing the time derivative in Eq. (6.37) for a fixed value α_e; hence

$$2L\langle v_C\rangle_{0e}^2 = E^2RT\alpha_e^2,$$

which gives

$$\langle v_C\rangle_{0e} = E\alpha_e\sqrt{(RT)/(2L)}. \tag{6.38}$$

Equation (6.37) can be written as the dynamic equation of the variable $\langle v_C\rangle_0{}^2$ depending linearly on α^2:

$$C\left(\dot{\langle v_C\rangle_0^2}\right) = \frac{E^2T}{L}\alpha^2 - \frac{2\langle v_C\rangle_0^2}{R}. \tag{6.39}$$

Let $\tilde{}$ be the notation dedicated to denoting small variations around a given equilibrium point. Notations $\langle v_C\rangle_0 = \langle v_C\rangle_{0e} + \widetilde{\langle v_C\rangle_0}$ and $\alpha = \alpha_e + \tilde{\alpha}$ are adopted; hence, $\dot{\langle v_C\rangle_0} = \dot{\widetilde{\langle v_C\rangle_0}}$ and $\dot{\alpha} = \dot{\tilde{\alpha}}$. Linearization uses the first-order Taylor series approximation of function x^2, namely $x^2 \approx x_0{}^2 + 2x_0(x - x_0)$. Thus, Eq. (6.39) gives by linearization around the equilibrium operating point (6.38)

$$\dot{\widetilde{\langle v_C\rangle_0}} = \frac{E^2T\alpha_e^2}{2LC\langle v_C\rangle_{0e}} + \frac{E^2T\alpha_e}{LC\langle v_C\rangle_{0e}}\cdot\tilde{\alpha} - \frac{\langle v_C\rangle_{0e}}{RC} - \frac{2}{RC}\cdot\widetilde{\langle v_C\rangle_0},$$

where $\frac{E^2T\alpha_e^2}{2LC\langle v_C\rangle_{0e}} - \frac{\langle v_C\rangle_{0e}}{RC} = 0$ according to Eq. (6.38). Therefore:

$$\dot{\widetilde{\langle v_C\rangle_0}} = \frac{E^2T\alpha_e}{LC\langle v_C\rangle_{0e}}\cdot\tilde{\alpha} - \frac{2}{RC}\cdot\widetilde{\langle v_C\rangle_0}. \tag{6.40}$$

Equation (6.40) corresponds to a first-order linear dynamic system having as input small variations of duty ratio $\tilde{\alpha}$ and as output small variations of averaged capacitor voltage $\widetilde{\langle v_C\rangle_0}$. That is, the ROAM dynamic of a buck-boost converter around a given equilibrium point is of first order, linear, having $\frac{E^2RT\alpha_e}{2L\langle v_C\rangle_{0e}}$ as gain and $\frac{RC}{2}$ as time constant.

Problem 6.2
Consider the system in Fig. 5.30 from Problem 5.1 in Chap. 5. It is about a circuit used to convey the energy between a variable DC source and a single-phase strong AC grid via a DC link, the reactive power being zero.

Given the system's large-signal model developed in the solution to Problem 5.1, obtain a small-signal reduced-order averaged model used for control purpose by taking into account the following considerations.

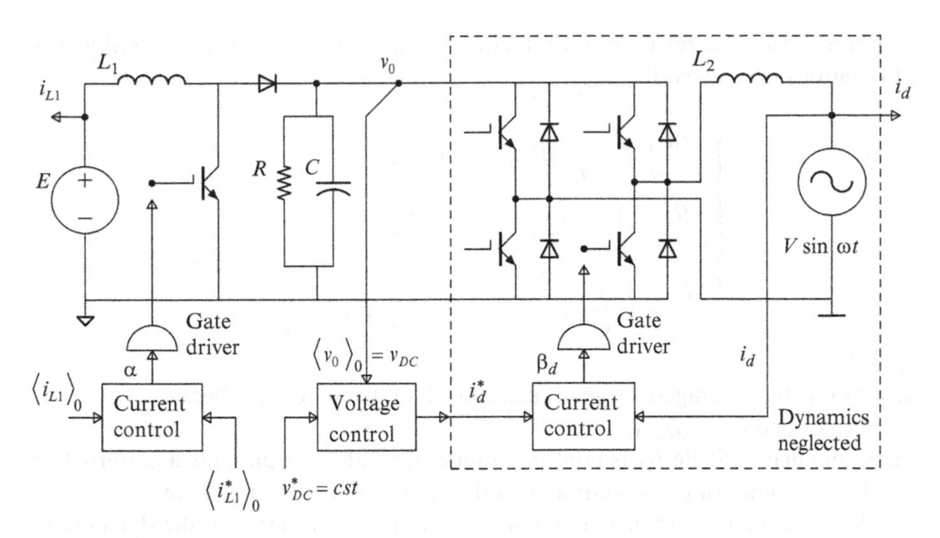

Fig. 6.18 Simplified diagram of cascaded control structure for the DC-AC converter given in Fig. 5.30 (Chap. 5). Auxiliary elements – filters, i_d computation using PLL, etc. – have been omitted

The system must have the possibility of varying the input power; this is done by controlling the DC-source current by means of control input u_1. The primary control purpose is to ensure the power flux between the primary DC source and the AC grid while preserving the rated operating conditions in terms of voltage and current; this is done by means of control input u_2. To this end, a widely employed cascaded control structure can be used (Aström and Hägglund 1995), in which the DC-link voltage controller imposes the set-point of the grid inductor current. Therefore, the control structure is composed of two loops: the inner loop used for controlling the grid-side inductor current and the outer loop for controlling the DC-link voltage.

Solution. As shown in the solution of (b) in Problem 5.1, computation of the small-signal model and of its associated transfer functions may be a difficult way to obtain necessary information for designing control laws. Depending on each particular control objective, some remarks may be useful in order to introduce simplifying assumptions before starting control design procedures. In our case, let us focus on the regulation of the average value of the DC-link voltage, v_{DC}. As AC voltage is stiff, the (variable) output power can be controlled by means of the inductor L_2 current (the output current loop). This supposes a variable DC current drain from the DC-link. In order to maintain the value of v_{DC} within operating limits, a voltage loop should also be used. In fact, the invariance of v_{DC} guarantees the input–output power balance. Figure 6.18 presents briefly this control approach.

Let us review the result from Problem 5.1 (Eq. 5.83), preserving the significance of notations introduced there:

$$\begin{cases} \dfrac{d\langle i_{L1}\rangle_0}{dt} = \dfrac{1}{L_1}[E - v_{DC}(1-\alpha)] \\ \dfrac{di_d}{dt} = \dfrac{V}{L_2} - \dfrac{v_{DC}}{L_2}\cdot\beta_d - \dfrac{i_d\cdot r}{L_2} \\ \dfrac{dv_{DC}}{dt} = \dfrac{1}{C}\langle i_{L1}\rangle_0(1-\alpha) - \dfrac{1}{RC}v_{DC} - \dfrac{i_d}{2C}\cdot\beta_d, \end{cases} \tag{6.41}$$

describing the averaged large-signal model of the circuit, where $v_{DC} = \langle v_0\rangle_0$, $\beta_d = \langle u_2\rangle_1$ and $i_d = \langle i_{L2}\rangle_1$.
First note that variable i_{L1} is controlled independently; it represents a perturbation for the remainder of the system and its dynamic is not of interest here.

The dynamic of current i_d is rendered significantly faster than the dynamic of v_{DC} by means of the inner control loop, i.e., i_d equals i_d^* practically instantaneously in relation to the dynamic of v_{DC}. Therefore, one can assume that variable i_d influences v_{DC} by means of its steady-state regime, characterized by zeroing the derivative of the corresponding dynamic equation in relations (6.41) and taking into account neglecting the resistance of inductor L_2, $r = 0$. Thus, the steady-state value of β_d is reached, which depends on the value of v_{DC}, namely $\beta_{d\_st} = V/v_{DC}$. At its turn, the latter relation is used in the dynamic equation of v_{DC} to obtain

$$\dot{v_{DC}} = \frac{1}{C}\langle i_{L1}\rangle_0(1-\alpha) - \frac{1}{RC}v_{DC} - \frac{i_d V}{2Cv_{DC}}$$

or, equivalently,

$$\dot{v_{DC}^2} = \frac{2P_{in\_DC}}{C} - \frac{2}{RC}v_{DC}^2 - \frac{V}{C}i_d, \tag{6.42}$$

where $P_{in\_DC} = \langle i_{L1}\rangle_0 \cdot v_{DC}(1-\alpha)$ signifies the power supplying the DC link.

Relation (6.42) represents the *linear* dynamical equation of the DC-link voltage squared. This relation also represents the *nonlinear* ROAM of the circuit because the resulting system order has been decreased by one.

One can use relation (6.42) to design linear control laws to regulate v_{DC}^2 (instead of v_{DC}) by using i_d as control input, knowing that P_{in_DC} is a perturbation. The plant transfer function used to this end is

$$H_{i_d\to v_{DC}^2}(s) = \frac{v_{DC}^2}{i_d} = -\frac{V\cdot R}{s\frac{RC}{2}+1}. \tag{6.43}$$

If the transfer from i_d to v_{DC} is sought, then linearization of Eq. (6.42) around an equilibrium operating point v_{DCe} is necessary. This operating point obeys the following relation:

$$\frac{2P_{in\_DCe}}{C} - \frac{2}{RC}v_{DCe}^2 - \frac{V}{C}i_{de} = 0. \tag{6.44}$$

As in the previous solved problem, the linearization makes use of the first-order Taylor series approximation of function x^2, namely $x^2 \approx {x_0}^2 + 2x_0(x - x_0)$. Variables of interest can respectively be written as the sum of their equilibrium values and their small variations:

$$\begin{cases} v_{DC} = v_{DCe} + \widetilde{v_{DC}} \\ i_d = i_{de} + \widetilde{i_d} \,. \end{cases}$$

Therefore, linearization of Eq. (6.42) yields

$$2v_{DCe} \cdot \dot{\widetilde{v_{DC}}} = \frac{2P_{in\_DCe}}{C} - \frac{4v_{DCe}}{RC}\widetilde{v_{DC}} - \frac{2v_{DCe}^2}{RC} - \frac{V}{C}i_{de} - \frac{V}{C}\widetilde{i_d}\,, \tag{6.45}$$

where the power supplying the DC link has been supposed a sufficiently slowly variable, i.e., constant: $\widetilde{P_{in\_DCe}} = 0$. By substituting relation (6.44) into (6.45) one obtains

$$2v_{DCe} \cdot \dot{\widetilde{v_{DC}}} = -\frac{4v_{DCe}}{RC}\widetilde{v_{DC}} - \frac{V}{C}\widetilde{i_d}\,,$$

from which the transfer function results as

$$H_{\widetilde{i_d} \to \widetilde{v_{DC}}}(s) = \frac{\widetilde{v_{DC}}}{\widetilde{i_d}} = -\frac{V \cdot R}{4v_{DCe}} \cdot \frac{1}{s\frac{RC}{2} + 1}. \tag{6.46}$$

Equation (6.46) gives the *small-signal* ROAM that expresses the *linear first-order* dynamic of the DC voltage as depending on current i_d. Its utility consists in providing the basis for the design of a DC-link voltage regulator that imposes the i_d value in response to v_{DC} evolution (see Fig. 6.18).

Note that transfer functions given by relations (6.43) and (6.46) have the same time constant – meaning that they correspond to the same dynamic – but different gains (in the latter case this depends on the chosen equilibrium point).

The following problems are proposed to the reader to solve.

Problem 6.3. Modeling a buck power stage operating in dcm
Let us consider the buck converter with capacitive output filter having the circuit diagram presented in Fig. 2.10 of Chap. 2. The converter operates in discontinuous-inductor-current mode. The following points should be addressed.

(a) Deduce the converter ROAM by taking the duty ratio as control input and the output voltage as controlled variable.

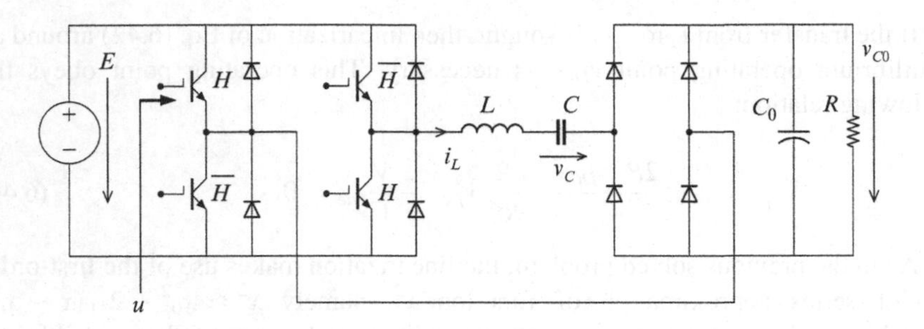

Fig. 6.19 Series-resonance supply of a diode rectifier

(b) Determine the conversion ratio between the input and the output voltages.
(c) Deduce the small-signal ROAM.

Problem 6.4. Modeling a flyback converter operating in dcm
Answer the same questions as in Problem 6.3 for the case of a flyback converter having the circuit diagram given in Fig. 4.24a of Chap. 4.

Problem 6.5. ROAM of a series-resonance supply of a diode rectifier
Let us consider the circuit composed of a voltage inverter, a resonant tank and a diode rectifier given in Fig. 6.19.

The voltage inverter is assumed to operate in full wave at a frequency close to the resonance frequency of the alternative circuit. Address the following points.

(a) Deduce the switched model.
(b) Explain why the classical first-order harmonic approach is suitable for studying the resonant circuit.
(c) Write the ROAM of this converter by taking the output voltage v_{C0} as a controlled variable and the inverter frequency as a control input variable.
(d) Deduce the small-signal model of the previously obtained ROAM.
(e) Give the equivalent diagram corresponding to the ROAM of the converter.

References

Aström KJ, Hägglund T (1995) PID controllers: theory, design and tuning, 2nd edn. Instrument Society of America, Research Triangle Park

Bacha S, Rognon JP, Hadj Said N (1995) Averaged modelling for AC converters working under discontinuous conduction mode – application to thyristors controlled series compensators. In: Proceedings of European Power Electronics Conference – EPE 1995. Sevilla, Spain, pp 1826–1830

Chetty PRK (1981) Current injected equivalent circuit approach to modeling switching DC-DC converters. IEEE Trans Aerosp Electron Syst 17(6):802–808

Chetty PRK (1982) Current injected equivalent circuit approach to modelling and analysis of current programmed switching DC-to-DC converters (discontinuous inductor conduction mode). IEEE Trans Ind Appl 18(3):295–299

Ćuk S, Middlebrook RD (1977) A general unified approach to modelling switching dc-to-dc converters in discontinuous conduction mode. In: Proceedings of IEEE Power Electronics Specialists Conference – PESC 1977. Palo Alto, California, USA, pp 36–57

Erikson RW, Maksimović D (2001) Fundamentals of power electronics, 2nd edn. Kluwer, Dordrecht

Maksimović D, Ćuk S (1991) A unified analysis of PWM converters in discontinuous modes. IEEE Trans Power Electron 6(3):476–490

Middlebrook RD (1985) Topics in multiple-loop regulators and current mode programming. In: Proceedings of IEEE Power Electronics Specialists Conference – PESC 1985. Toulouse, France, pp 716–732

Middlebrook RD (1989) Modelling current-programmed buck and boost regulators. IEEE Trans Power Electron 4(1):36–52

Mohan N, Undeland TM, Robbins WP (2002) Power electronics: converters, applications and design, 3rd edn. Wiley, Hoboken

Sun J, Grotstollen H (1992) Averaged modeling of switching power converters: reformulation and theoretical basis. In: Proceedings of IEEE Power Electronics Specialists Conference – PESC 1992. Toledo, Spain, pp 1166–1172

Sun J, Grotstollen H (1997) Symbolic analysis methods for averaged modeling of switching power converters. IEEE Trans Power Electron 12(3):537–546

Sun J, Mitchell DM, Greuel MF, Krein PT, Bass RM (2001) Averaged modeling of PWM converters operating in discontinuous conduction mode. IEEE Trans Power Electron 16 (4):482–492

Vorpérian V (1990) Simplified analysis of PWM converters using model of PWM switch. Part II: discontinuous conduction mode. IEEE Trans Aerosp Electron Syst 26(3):497–505

Part II
Control of Power Electronic Converters

Chapter 7
General Control Principles of Power Electronic Converters

This chapter aims at offering a synthetic perspective over control goal formulation and control design methods for power electronic converters in the context of their specific features and constraints. It concerns intrinsically variable-structure systems with fast nonlinear dynamics that are potentially subject to significant noise disturbance because of modulated control signals. Such plants are challenging from the control perspective and so a large palette of control methods will be explored.

7.1 Control Goals in Power Electronic Converter Operation

Generally speaking, power electronic converters are key elements in power systems. Besides conveying electrical power with high efficiency, they offer the possibility of controlling internal variables in order to ensure both safe operation and output regulation. In the quasi-totality of their applications, power electronic converter operation requires a form of control intended not only for attaining the operating objective but also for safety. Control specifications are quite diversified, depending on the particular converter role.

As an example, a DC-DC converter operating in a switch-mode power supply feeding a certain variable load may need duty ratio adjustments in order to ensure a constant output voltage for the entire operating range (voltage regulation). Alternately, a grid-tie inverter fed by a renewable energy system must output a desired AC current to the grid in order to satisfy certain power transfer requirements, thus behaving as a controlled current source. Active power filters must provide the necessary current/voltage spectrum in order to cancel undesired harmonic content produced by a polluting load, all by maintaining the grid-load power balance.

Figure 7.1 shows a generic application of power electronics control. In most cases the control algorithm outputs the duty ratio in response to the circuit state/output evolution. The duty ratio needs modulation (conversion to the ON/OFF

S. Bacha et al., *Power Electronic Converters Modeling and Control: with Case Studies*, Advanced Textbooks in Control and Signal Processing, DOI 10.1007/978-1-4471-5478-5_7,

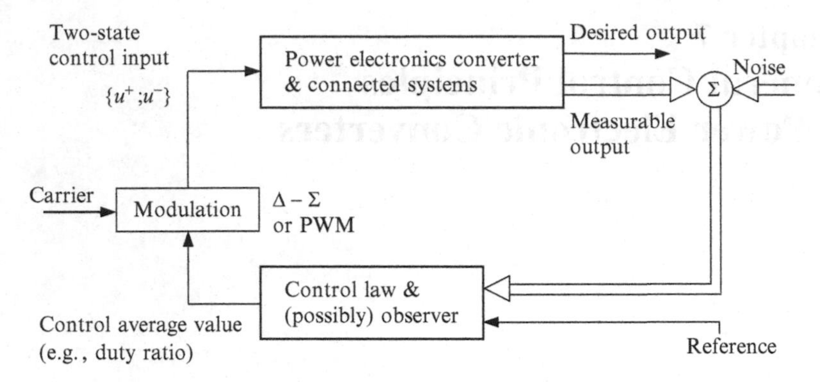

Fig. 7.1 Generic control application of a system including power electronic converter

signal) to be applied to the power switch base driver. To this end, various types of modulation may be used, of which the most frequently employed are pulse-width and sigma-delta modulations (in their analog or digital version) and space-vector modulation (SVM) (Leon et al. 2010). Hysteretic modulators or constant turn-on modulators can also be used (Corradini et al. 2011). However, the diagram in Fig. 7.1 is not general, as there are control laws that directly output the two-state control signal (e.g., the sliding-mode control).

In either case, the controller must respond to significant system perturbation, that is the load variation, according to certain a priori specified control objectives. As usual in control systems technology, these objectives include output regulation/tracking, internal variable dynamics compensation, limitation of variables to admissible values, etc. For purposes of zero steady-state error, integral control action is mandatory in most cases, at least in the outer control loops.

Generally speaking, converter control design is focused on imposing desired low-frequency (macroscopic or otherwise equivalent) behavior with respect to the specified requirements (Kassakian et al. 1991). In this context, the converter averaged model is an important instrument. The frequency domain of interest is placed well below the switching frequency; it concerns the compensation of inertia introduced by passive circuit elements.

In power electronic converters the use of passive elements – capacitors and/or inductors – is mandatory, not only for filtering high (switching) frequencies but also for control purposes (inserting larger inertia leads to more controllable plants). The presence of such inertial elements leads to variations of electrical variables in the case of power imbalance (e.g., capacitors DC voltage increasing to dangerous levels). Therefore, in order to preserve the proper operation of converters one must ensure the power balance between the various sections of the circuit. This is usually done by regulating some internal variables that are sensitive to the electrical power imbalance, to certain typical operating values.

Figure 7.2 illustrates a power supply based on a boost DC-DC power stage. This plant must be controlled in order to maintain a constant output voltage irrespective of load value (within acceptable limits). The two-loop cascaded control structure is charged not only with output regulation at the desired value but also with inductor current tracking and limitation.

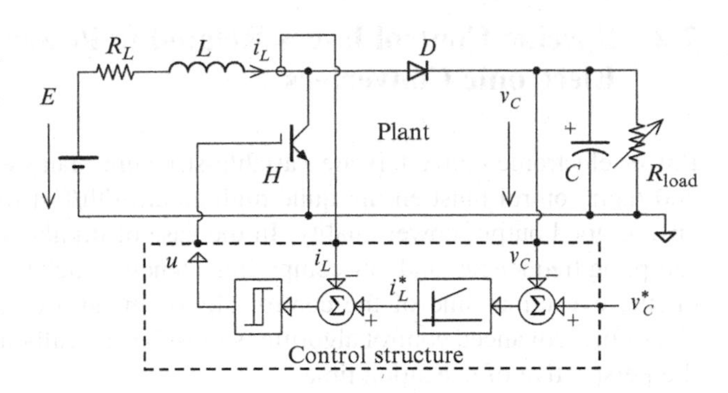

Fig. 7.2 Example of regulating output voltage of a DC-DC boost power stage

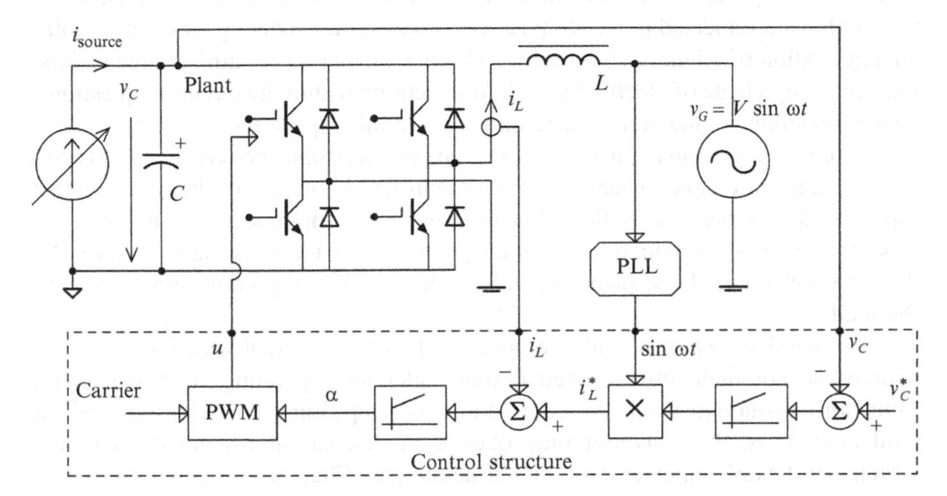

Fig. 7.3 Example of application requiring power electronic converter control (gate driver omitted): renewable energy source feeding a single-phase AC grid

Figure 7.3 contains an example of a grid-connected inverter used for conveying power from a renewable energy source (let us suppose a PV array). This role is translated into regulating the DC-link voltage v_C to a constant value, as this latter is sensitive at power imbalance (Hur et al. 2001). Therefore, the outermost control loop deals with DC-link voltage regulation, which can be achieved by injecting the corresponding amount of current into the power grid. In this case, the controlled variable is the grid current i_G that is "naturally" filtered by the inductor, the current extracted from the DC link being indirectly controlled.

To conclude, the outer voltage loop should impose the grid current reference that must be followed (in the inner control loop) in order to maintain the power input-output balance. Its magnitude follows DC current input random variations. Concerning circuit safety, as the inductor current is controlled, its value (and its DC image) cannot exceed the safety effective values. Also, the DC voltage regulation avoids capacitor rated voltage excess, thus protecting the DC link. Moreover, this control structure adds flexibility to the application as it can feed reactive power into the grid by varying the phase lag between the injected current and the AC grid voltage.

7.2 Specific Control Issues Related to Power Electronic Converters

Power electronic converters are variable-structure plants exhibiting fast dynamics and their control must ensure quite high bandwidth/fast time response in order to ensure good output power quality. In the case of digital control this requires high sampling frequencies and low computing latency, which further induces restrictions on the execution time of the control algorithm on a certain hardware platform. Therefore, advanced control algorithms must be carefully used and optimized from the perspective of execution time.

Power electronic converter behavior is relevant only in the context of its interaction with interconnected power conversion systems; the modeling may thus require an aggregation of all these components. Often, control of a certain variable is done indirectly (by virtue of electric variable interactions within the circuit in question), so the modeling approach must take into account this aspect, too.

As previously shown, in most cases power electronic converter models are expressed using coupled sources that are controlled by means of the same control input, which is the duty ratio. This induces severe nonlinearities in converter models (e.g., due to products between the control input and some states) that render dynamic behavior of the plant dependent on the operating point variation (e.g., the load).

Also, passive elements could be variant when their current varies and the real sources do not maintain the rated output under any operating conditions (e.g., inductance reduction when the core saturates). Depending on its design and on load conditions, the converter may operate in discontinuous-conduction mode, which significantly changes its model structure (see Chap. 6).

As the closed-loop high bandwidth is a basic requirement for converter control, parameter/model variations may reduce the stability phase margin to dangerously low levels (Morroni et al. 2009b). Therefore, a “strongly” tuned controller may easily lead to instabilities in certain operating regimes. For these reasons, the designer may sometimes sacrifice performance for the sake of robustness.

All these aspects suggest that a unique linear controller is not likely to maintain control performance (dynamic response, stability, etc.) for the entire operating range. Solutions to this problem suppose the design of the controller to ensure optimal closed-loop behavior at a certain operating point and the use of an adaptive algorithm for other operating points or ranges (e.g., gain scheduling).

To conclude, closed-loop system robustness and sensitivity to parameter variations is an open issue in power electronic converter control.

There is another specificity arising from the modulated control input and the converter’s specific operation: the controller must deal with larger spectra (with respect to baseband control applications) in the measurable variables that may significantly affect its performance. Supplementary filtering and synchronous data acquisition may be employed in order to reduce modulation effects. The noise additive to the output may also contain spectral components (current/voltage ripple), placed within the system's closed-loop bandwidth, whose rejection requires the controller complexity be increased.

Depending on converter type, the current/voltage may be physically limited (e.g., a one-quadrant buck power stage cannot have negative voltage). Also, for safety requirements, some of the variables must be limited to certain rated values. By virtue of power converter physical implementation, the duty ratio takes values in the interval (0,1). In the case of certain converters this interval may be even smaller, for efficiency reasons (e.g., the boost power stage). All these limitations induce nonlinearities that must be dealt with within the control structure. In this context, effective control structures employ antiwindup schemes used in conjunction with controllers that use integral components, especially in the inner/faster control loops.

Converters with a "boost" effect display nonminimum-phase behavior on the duty-ratio-to-output-voltage channel. Their control requires specific solutions that preserve control quality, meanwhile avoiding closed-loop instabilities, e.g., optimum modulus criterion (Ceangă et al. 2001).

Converter operation often induces nonessential disturbances at frequencies within the closed-loop system bandwidth that may adversely affect controller operation. For example, the inverter in Fig. 7.3 induces quite a low-frequency (two times the grid frequency) ripple in the DC-link voltage due to the modulation effect (a product between the duty ratio and the grid current, both being sinusoidal variables). This ripple has a spectrum within the system bandwidth, and, if taken into account by the control law, it may induce undesired grid current amplitude variations (Bratcu et al. 2008).

In the case of converters that have more than two energy accumulations (states) some economic issues prevent control of the entire state vector: not all the states are sensed because transducers may be expensive. In this case partial (incomplete) state feedback may be used for control purposes, which employs observers for state reconstruction. The use of integral reconstruction may prove to be a useful solution to this problem (Sira-Ramirez and Silva-Ortigoza 2006). See, for example, the application in Fig. 7.4, where the full-state feedback requires an unaffordable number of transducers; in this case partial-state observers may be a cost-effective solution.

7.3 Different Control Families

In the relevant literature, one can find a wide plethora of control structures that have been employed for power electronic converter control. Without being exhaustive, the next chapters of this book visit some of them, linear as well as nonlinear.

The so-called *standard control* employs simple and robust classical invariant PID controllers that are tuned based upon linearized single-input–single-output averaged models of converters (small-signal models). These output a continuous control signal (the duty ratio), which needs modulation so it may be applied to the power switching gates (PWM, sigma-delta, SVM, etc.). Once the worst operating

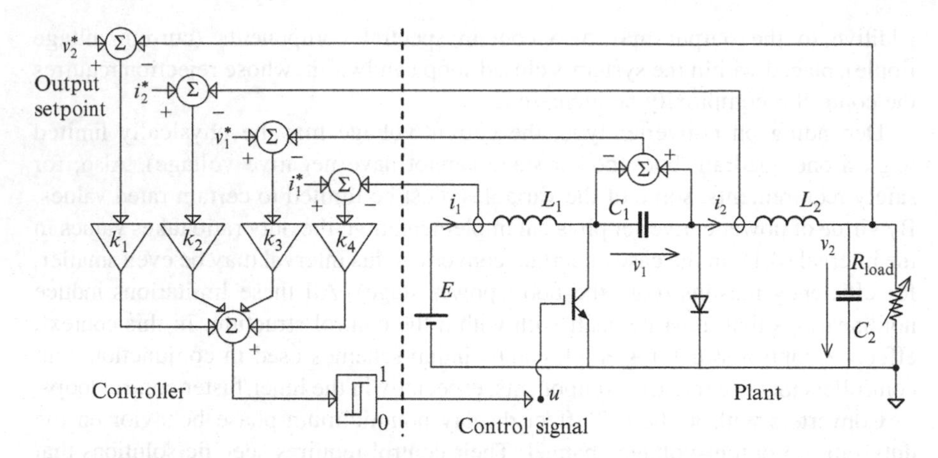

Fig. 7.4 Example of control application requiring a significant number of transducers: output voltage regulation of Cúk DC-DC converter (Malesani et al. 1995)

conditions have been identified, controller tuning is done with classical procedures such as loop shaping or pole-zero allocation (d'Azzo et al. 2003). In the following step the control solution is checked to see if it preserves acceptable performance for the remainder of the operating regime, not only in continuous-conduction but also in discontinuous-conduction mode.

A widely used and effective control structure is one that employs *nested* (or *cascaded*) *loops* (Aström and Hägglund 1995). This allows the output circuit regulation/tracking, meanwhile preserving the internal variables within specified safety limits. For example, in a converter having two states – the current inductor and the capacitor voltage – the outer control loop deals with voltage regulation imposing low-frequency dynamics and the inner loop concerns the faster current control. The voltage controller provides the setpoint of the current variable, and this latter acts as the control input of the outer voltage loop.

Another linear control structure is the *full-state feedback* (d'Azzo et al. 2003). The duty ratio is obtained as a linear combination of the converter's states, and the controller gains result from imposing the poles of the closed-loop system. In some cases full-state measurement is not possible and state observers must be employed for reconstructing nonmeasured state variables (Friedland 2011). However, an outer integral loop is needed in order to drive the output towards a desired value. As the internal variables are no longer explicitly controlled, supplementary actions must be taken in order to maintain them within safe limits.

Also, in the case of three-phase AC/DC or DC/AC converters, control may be accomplished by using linearized averaged modeling in the rotating *dq* frame (see Chap. 5). This allows separate control of the active and reactive power flows, respectively, by using simple proportional-integral (PI) controllers on each of the channels (Bose 2001). A decoupling structure may be needed to cancel the reciprocal influence between channels *d* and *q*.

Linearized averaged models may also be used for designing *resonant controllers* for converters having AC power stages. These controllers employ generalized integrators having infinite gain at a certain frequency (Etxeberria-Otadui et al. 2006). This allows harmonic reference tracking (in this case the grid frequency), while rejecting all other signals that influence the control goal (e.g., higher-order harmonics that pollute the grid). The implementation of these controllers may either be continuous (in the simpler cases) or discrete (i.e., digital). In this latter case, the linearized model must be discretized using the sampling time of the computing system (Maksimović and Zane 2007; Morroni et al. 2009a).

Nonlinear averaged models (large-signal models) are useful for deriving advanced nonlinear controllers (as also stated in Chap. 2). By using the converter model's intrinsic properties, these control structures strive to maintain closed-loop performance for the entire operating range. This book will look at the so-called *stabilizing control* and *passivity-based control* that result from applying the more general Lyapunov stability theory (Sanders and Verghese 1992; Ortega et al. 1998). In addition to these, the *feedback-linearization control* approach may be used for yielding a closed-loop system consisting of a series of integrators (Isidori 1989; Sira-Ramirez and Silva-Ortigoza 2006). In some of the cases these approaches may lead to quite complicated mathematical expressions of the duty ratio and may pose software implementation problems.

Last, but not least, the *sliding-mode control* for power electronic converters will be detailed. This is a natural approach since power electronic converters are essentially variable-structure systems (see switched models in Chap. 3). The resulting controller is simple and robust in both analog and discrete implementations and does not require duty cycle modulation (Tan et al. 2011). However, supplementary operations are needed for confining the (otherwise free) switching frequency in a typical spectral domain suitable for a certain power switching device.

7.4 Conclusion

Far from being exhaustive, the control methods mentioned here have a common control goal: to impose the desired low-frequency power electronic converter behavior. This goal is rather general and frequently relies upon a control structure with more than one variable to control. Other control methods suitable to power electronic converter control – e.g., the straightforward method of peak current control (Middlebrook 1985) – will not be considered because the scope of this book is narrower.

References

Aström KJ, Hägglund T (1995) PID controllers: theory, design and tuning, 2nd edn. Instrument Society of America, Research Triangle Park

Bose BK (2001) Modern power electronics and AC drives. Prentice-Hall, Upper Saddle River

Bratcu AI, Munteanu I, Bacha S, Raison B (2008) Maximum power point tracking of grid-connected photovoltaic arrays by using extremum seeking control. Control Eng Appl Inform 10(4):3–12
Ceangă E, Protin L, Nichita C, Cutululis NA (2001) Theory of control systems (in French: Théorie de la Commande des Systèmes). Technical Publishing House, Bucharest
Corradini L, Bjeletic A, Zane R, Maksimović D (2011) Fully digital hysteretic modulator for DC-DC switching converters. IEEE Trans Power Electron 26(10):2969–2979
d'Azzo JJ, Houpis CH, Sheldon SN (2003) Linear control system analysis and design with MATLAB, 5th edn. Marcel-Dekker, New York
Etxeberria-Otadui I, Lopez de Heredia A, Gaztanaga H, Bacha S, Reyero R (2006) A single synchronous frame hybrid (SSFH) multi-frequency controller for power active filters. IEEE Trans Ind Electron 53(5):1640–1648
Friedland B (2011) Observers. In: Levine WS (ed) The control handbook–control system advanced methods. CRC Press/Taylor & Francis Group, Boca Raton, pp 15-1–15-23
Hur N, Jung J, Nam K (2001) A fast dynamic DC-link power-balancing scheme for a PWM converter-inverter system. IEEE Trans Ind Electron 48(4):794–803
Isidori A (1989) Nonlinear control systems, 2nd edn. Springer, Berlin
Kassakian JG, Schlecht MF, Verghese GC (1991) Principles of power electronics. Addison-Wesley, Reading
Leon JI, Vazquez S, Sanchez JA, Portillo R, Franquelo LG, Carrasco JM, Dominguez E (2010) Conventional space-vector modulation techniques versus the single-phase modulator for multilevel converters. IEEE Trans Ind Electron 57(7):2473–2482
Maksimović D, Zane R (2007) Small-signal discrete-time modelling of digitally controlled PWM converters. IEEE Trans Power Electron 22(6):2552–2556
Malesani L, Rossetto L, Spiazzi G, Tenti P (1995) Performance optimization of Ćuk converters by sliding-mode control. IEEE Trans Power Electron 10(3):302–309
Middlebrook RD (1985) Topics in multiple-loop regulators and current mode programming. In: Proceedings of IEEE Power Electronics Specialists Conference – PESC 1985. Toulouse, France, pp 716–732
Morroni J, Zane R, Maksimović D (2009a) Design and implementation of an adaptive tuning system based on desired phase margin for digitally controlled DC-DC converters. IEEE Trans Power Electron 24(2):559–564
Morroni J, Zane R, Maksimović D (2009b) An online stability margin monitor for digitally controlled switched-mode power supplies. IEEE Trans Power Electron 24(11):2639–2648
Ortega R, Loria A, Nicklasson PJ, Sira-Ramirez H (1998) Passivity-based control of Euler-Lagrange systems: mechanical, electrical and electromechanical applications. Springer, London
Sanders SR, Verghese GC (1992) Lyapunov-based control for switched power converters. IEEE Trans Power Electron 7(1):17–24
Sira-Ramirez H, Silva-Ortigoza R (2006) Control design techniques in power electronics devices. Springer, London
Tan S-C, Lai Y-M, Tse C-K (2011) Sliding mode control of switching power converters: techniques and implementation. CRC Press/Taylor & Francis Group, Boca Raton

Chapter 8
Linear Control Approaches for DC-DC Power Converters

This chapter covers the topic of DC-DC power electronic converter control in continuous-conduction mode. It aims at presenting how converter averaged (low-frequency) behavior can be tailored by means of feedback control structures. There is no unique control paradigm that solves this problem. To this end, various control structures with different design methods that have become classical in control systems theory can be employed with a similar degree of efficiency. It is generally assumed that the obtained control input is applied to converter switches by using pulse-width modulation. This chapter aims at giving a comprehensive view of this topic without exhausting the entire range of control approaches dedicated to the subject. The chapter presents the main principles, the design procedures and provides some pertinent examples, ending with two case studies and a set of problems.

8.1 Linearized Averaged Models. Control Goals and Associated Design Methods

As seen before (Chap. 4), it is beneficial to describe the average behavior of a converter by means of a linearized model; it provides good insight on plant performance in terms of steady-state gains, bandwidth, damping and other dynamic effects (e.g., nonminimum-phase behavior). This allows a full assessment of converter (plant) properties leading to control loop design that ensures the achievement of a certain goal (Stefani 1996). The drawback of this modeling approach is the fact that these plants are nonlinear (Verghese et al. 1986). As it has already been discussed, their linearized models have coefficients whose values depend on one (or more) time-varying exogenous signals, i.e., they change with the actual operating point and so does the linear-controller-based closed-loop system (Philips and François 1981; Mitchell 1988; Kislovsky et al. 1991). Therefore, linear controllers suffer from lack of robustness with respect to load and input voltage mean values, which must be

S. Bacha et al., *Power Electronic Converters Modeling and Control: with Case Studies*, Advanced Textbooks in Control and Signal Processing, DOI 10.1007/978-1-4471-5478-5_8,

taken into account when one approaches worst-case design, gain scheduling linear-parameter-varying or other adaptive techniques (Lee et al. 1980; Shamma and Athans 1990; Morroni et al. 2009; Algreer et al. 2011).

The main control goal in DC-DC converters may vary with the converter role, but, in general, the aim is to regulate/track either the output or the input converter voltage (with respect to the power flow), while meeting a set of imposed performance requirements. The control loop may be more complicated as the primary goal may generate subordinate objectives produced when other issues are revealed during modeling (e.g., stability aspects). Also, the chosen control structure may be hierarchically organized for tracking some internal variables in order to achieve better dynamic response or/and to keep them within safety limits (Åström and Hägglund 1995).

Therefore, the control paradigm and associated design method chosen for solving the control problem of a DC-DC converter depend on a complex of factors that include the converter role, its original dynamics, desired closed-loop dynamics, operating range, safety issues, control input limitations, etc.

In the following sections, some classical design methods of control theory (Levine 2011) will be examined with respect to basic power electronic converters. Time-domain design techniques that use *temporal response specifications* – e.g., settling time, overshoot, integral performance indices – are useful when the closed-loop plant can be described as a dynamical system having second-order dynamics dominant behavior and no critical stability issues. When stability must be ensured by a supplementary degree of freedom in the control law, frequency-domain techniques such as *loop shaping* may be useful. *Pole-placement method* is useful for conveniently place the closed-loop eigenvalues when full-state feedback information is available. This method is usually employed in conjunction with an outer control loop that guarantees the fulfillment of the main operating goal. In the case of high-order plants the *root locus method* provides a powerful tool for deriving controllers that ensure robust closed-loop behavior (d'Azzo et al. 2003). Prediction-based methods (e.g., *internal model control*) are suitable for dealing with converter models having right-half-plane zeros.

Next, for simplicity, the brackets denoting averaged values will be dropped. For example, the averaged inductor current, $\langle i_L \rangle_0(t)$ will be denoted simply by $i_L(t)$, the duty ratio as average of the switching function u by $\alpha(t) = \langle u \rangle_0(t)$, and so on. Notations with subscript e are assumed to denote steady-state values (e.g., the steady-state capacitor voltage will be denoted as v_{Ce}).

8.2 Direct Output Control

In direct output control only one loop is used to control the output variable, either the voltage or the current. In most of DC-DC power electronic converter applications the output variable is voltage. Hence, this approach is also named *voltage-mode control*. It is a classical one in DC-DC converter control and with it

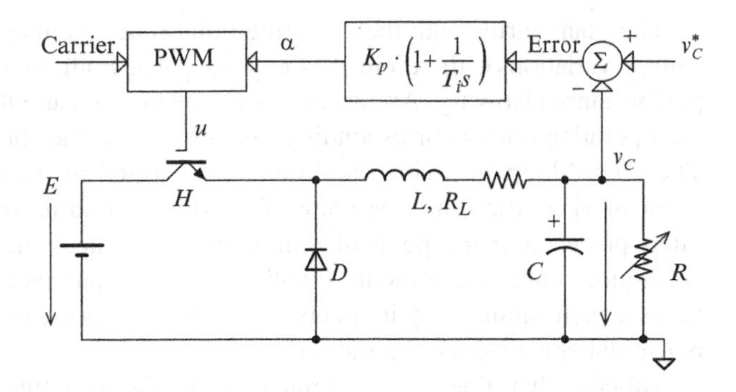

Fig. 8.1 Voltage direct control – buck converter case

no current need to be measured in the computation of the required duty ratio. It aims at regulating output voltage to an imposed value irrespective of load or input voltage variations. Output voltage variations are allowed within a specified band (e.g., within ±2 %) and depend on the closed-loop bandwidth.

8.2.1 Assumptions and Design Algorithm

The basic requirement of direct output control is that output voltage can be measured. Also, it is assumed that the small ripple condition is fulfilled and the converter operates in ccm. Output voltage control is achieved by means of the simple but very powerful idea of negative feedback and is based on the converter averaged linear model. This allows closing the system loop and presents a way to conveniently change its static and dynamic proprieties.

As the complete converter dynamics may prove quite complex, the system may be "difficult" to handle by such a direct approach. Therefore, the simplicity of the control structure has to determine the complexity of the controller in order to obtain suitable closed-loop performance. For example, in Sect. 4.6 from Chap. 4 it has been shown that the buck-boost transfer function from the duty ratio to the output voltage presents a right-half-plane zero. This requires appropriate compensation in order to obtain a stable closed loop with sufficiently large bandwidth. Therefore, the controller structure may contain a phase compensator (usually a lead-lag or a derivative component) or another kind of prediction based upon the plant model. Also, an integral component is required if the aim is voltage steady-state error zeroing.

Following comments from Chap. 7, Fig. 8.1 shows the control structure in the case of a buck converter. The controller outputs the duty ratio α, which is further pulse-width modulated so it may be applied to the power switch gate.

Once the controller structure has been chosen, its parameters can be obtained by equating the closed-loop transfer function parameters with the ones imposed by the dynamic performance specifications, either in the time- or frequency-domain.

The plant varies with the operating point (due to either load or primary voltage source variations); therefore it is expected that both its stability and closed-loop performance also vary. As consequence, one must ensure that stability is verified at the operating points corresponding to extremum values of the varying parameters. The closed-loop behavior should also be verified at these points. From the load point of view, these are the point of maximum load (corresponding to the circuit rated power) and the point of minimum load, where the circuit still operates in continuous-conduction mode. Finally, controller parameters result from ensuring that the performance requirements are met in the worst case, which depends on each particular converter configuration.

Suppose that linearization around a certain operating point leads to transfer function $H_{v\alpha}(s)$ of the duty-ratio-to-output-voltage channel. The design of a controller starts from the observation that the plant $H_{v\alpha}(s)$ is very weakly damped, therefore very close to instability (see the results presented in Chap. 4). In this context, the requirement of ensuring zero steady-state error is quite constraining because it requires an integral component in the controller that worsens the stability margin. One can however succeed in finding a trade-off in choosing the integral time constant so as to ensure zero steady-state error with not very large, but satisfactory, convergence speed. Some closed-loop dynamic performance is in this way sacrificed for the sake of ensuring good stability margin (more precisely, phase margin). Consequently, controller design is based upon loop shaping, i.e., shaping of the open-loop Bode diagrams in such a way as to ensure the largest bandwidth possible while still keeping sufficient phase margin. An extended lecture on this topic may be found in Erikson and Maksimović (2001).

The main steps of the design algorithm are listed next.

Algorithm 8.1. Design of direct control structure

#1 Describe the averaged model of the circuit. Choose the quiescent operating point where the control design will be made. Linearize the circuit model around the selected operating point.

#2 Deduce the transfer functions from the duty ratio to the output voltage and from the output current to the output voltage (in variations). If necessary, also deduce the transfer function from the input voltage to the output voltage.

#3 Decide if the controller will include an integral component or not.

#4 Plot the Bode diagrams for the open-loop controller-and-plant system. Deduce the number of necessary folding frequencies and their convenient ordering so as to ensure stability in the first place (Nyquist criterion fulfilled).

#5 Tune the folding frequencies in order to obtain sufficient stability margin. In particular, reduce the phase lag around the open-loop cut-off frequency (e.g., by introducing a lead component into the controller).

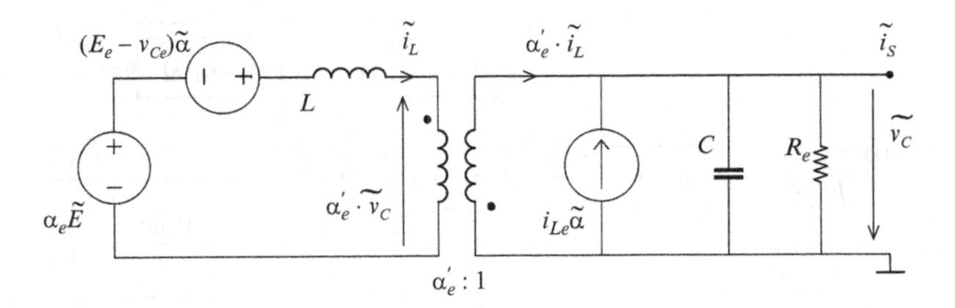

Fig. 8.2 Buck-boost small-signal equivalent diagram

Remark. The design operating point may be chosen either at the worst or at a typical (most probable) operating point that usually involves both the load condition and the supply voltage value at reduced loads DC-DC converters are highly undamped, whereas at full loads they may have a significant – i.e., low-frequency – right-half-plane zero (case of converters having boost capability).

8.2.2 *Example of a Buck-Boost Converter*

Let us consider the buck-boost power stage supplying a variable resistive load. Suppose the control goal is to maintain the output voltage at a given setpoint with zero steady-state error despite load resistance R_e (within rated limits) and source voltage E variations (at about 30 % around the rated) and irrespective of operating mode, as either buck or boost converter.

One follows steps of Algorithm 8.1. The converter is modeled according to details provided in Sect. 4.6 from Chap. 4. Consider a quiescent operating point determined by steady-state duty ratio α_e, inductor current i_{Le}, voltage capacitor v_{Ce}, voltage source E_e and load resistance R_e. The linearized averaged model around this operating point is

$$\begin{cases} L \cdot \dot{\widetilde{i_L}} &= \alpha_e \widetilde{E} + (E_e - v_{Ce})\widetilde{\alpha} + (1 - \alpha_e)\widetilde{v_C} \\ C \cdot \dot{\widetilde{v_C}} &= -(1 - \alpha_e)\widetilde{i_L} + i_{Le}\widetilde{\alpha} - \dfrac{1}{R_e}\widetilde{v_C} - \widetilde{i_S}, \end{cases} \tag{8.1}$$

where the variable $\widetilde{i_S} = -\frac{v_{Ce}}{R_e^2} \cdot \widetilde{R}$ represents output current variation due to load variation $\widetilde{R}$. The corresponding equivalent electrical diagram is given in Fig. 8.2. The notation α_e' stands for $1 - \alpha_e$.

By zeroing $\widetilde{E}$ and $\widetilde{i_S}$ one can compute the transfer function describing the channel from the control input (duty ratio) α to the capacitor (output) voltage v_C as

$$H_{v\alpha}(s) = \frac{\widetilde{v_C}(s)}{\widetilde{\alpha}(s)} = -\frac{E_e - v_{Ce}}{\alpha_e'} \cdot \frac{-\dfrac{i_{Le}L}{(E_e - v_{Ce})\alpha_e'}s + 1}{\dfrac{LC}{\alpha_e'^2}s^2 + \dfrac{L}{R_e \alpha_e'^2}s + 1}. \tag{8.2}$$

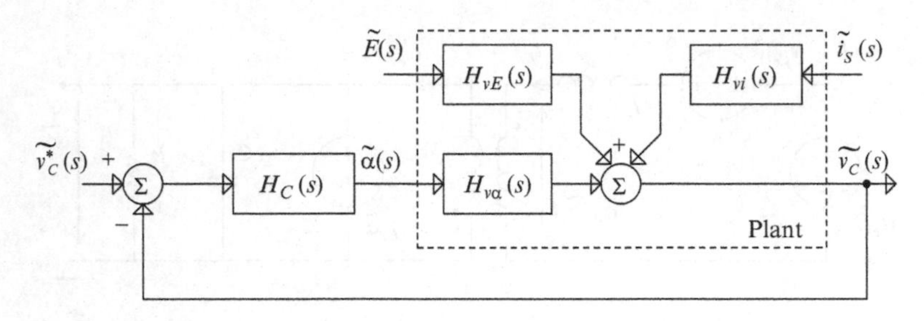

Fig. 8.3 The closed-loop system, emphasizing influence channels from reference and from disturbances respectively

The channel from the input supply E to the output voltage is described by the transfer ($\widetilde{\alpha}$ and $\widetilde{i_S}$ being zeroed):

$$H_{vE}(s) = \frac{\widetilde{v_C}}{\widetilde{E}} = -\frac{\alpha_e}{\alpha_e'} \cdot \frac{1}{\frac{LC}{\alpha_e'^2}s^2 + \frac{L}{R_e\alpha_e'^2}s + 1}. \tag{8.3}$$

The influence of output current i_S over output voltage is determined by Eq. (8.4) and actually represents circuit output impedance ($\widetilde{\alpha}$ and $\widetilde{E}$ being zeroed).

$$H_{vi}(s) = \frac{\widetilde{v_C}}{\widetilde{i_S}} = -\frac{1}{\alpha_e'^2} \cdot \frac{1}{\frac{LC}{\alpha_e'^2}s^2 + \frac{L}{R_e\alpha_e'^2}s + 1}. \tag{8.4}$$

These transfer functions can be computed either from relation (8.1) or by conveniently transforming the diagram in Fig. 8.2 (as detailed in Sect. 4.6 from Chap. 4). Using the principle of linear superposition one may add all these three effects in order to obtain the complete plant (see Fig. 8.3). Note that in both of these transfer functions the value v_{Ce} is negative.

In order to make the idea clear, let us consider that the plant $H_{v\alpha}(s)$ from relation (8.2) is written as

$$H_{v\alpha}(s) = K_1 \cdot \frac{1 - T_1 s}{T^2 s^2 + 2\zeta T s + 1}, \tag{8.5}$$

where all its parameters depend on the operating point: gain $K_1 = -\frac{E_e - v_{Ce}}{\alpha_e'} < 0$, right-half-plane zero time constant $T_1 = \frac{i_{Le}L}{(E_e - v_{Ce})\alpha_e'}$, second-order time constant $T = \frac{\sqrt{LC}}{\alpha_e'}$ and damping coefficient $\zeta = \frac{1}{2R_e\alpha_e'}\sqrt{\frac{L}{C}}$. One can note that plant (8.5) becomes more undamped as load R_e gets larger. Plant (8.5) is coupled with a PI controller with lead-lag, having transfer function

$$H_c(s) = K_p\left(1 + \frac{1}{T_i s}\right) \cdot \frac{T_b s + 1}{0.05 T_b s + 1}, \tag{8.6}$$

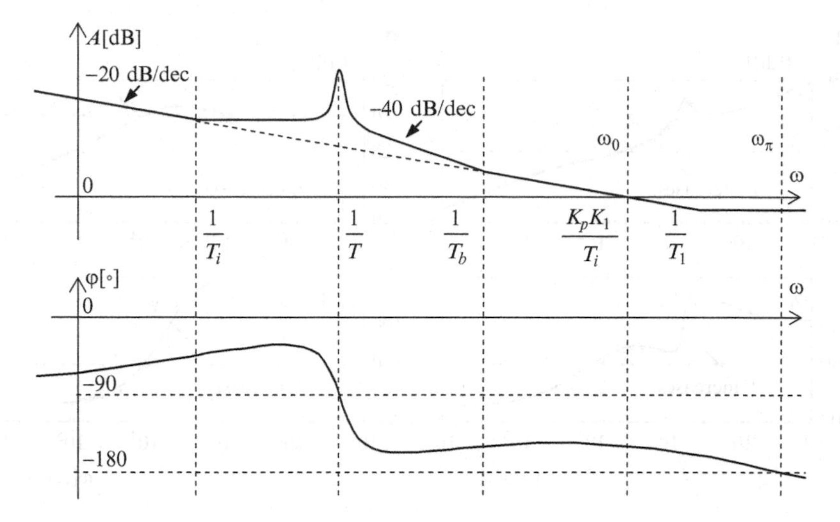

Fig. 8.4 Convenient shaping of open-loop Bode plot corresponding to plant $H_{v\alpha}(s)$ coupled with a PI controller with lead-lag

where the lag time constant is much smaller (i.e., twenty times smaller) than the lead one, T_b (it mainly helps to render causal the controller). The open-loop Bode plots corresponding to $H_c(s) \cdot H_{v\alpha}(s)$ are shown in Fig. 8.4. The influence of the introduced lag is not shown on the plot, as it is placed in the high-frequency range. Note that the integrator severely affects the phase characteristic and further induces reduction of the phase margin.

An important remark is that the place of the first folding frequency, $1/T_i$, determines the open-loop bandpass gain, which must be large in order to ensure large closed-loop bandwidth. At the same time, one must ensure that the cross-over frequency is smaller than the folding frequency introduced by the right-half-plane zero, $1/T_1$, in order to limit the nonminimum phase influence. The time constant T_1 is usually small in relation to the main time constant T; in this way, a sufficiently large bandwidth can be ensured. Figure 8.4 suggests a convenient way of placing folding frequencies in relation to the plant's lowest folding frequency (the one giving the dominant dynamic, $1/T$), so as to ensure requirements of both satisfactory bandwidth and phase margin:

$$\frac{1}{T} = \frac{2}{T_i}, \quad \frac{1}{T_b} = \frac{2}{T}, \quad \frac{K_p K_1}{T_i} = \frac{5}{T}, \tag{8.7}$$

thus allowing one to compute values of the controller parameters:

$$T_i = 2T, \quad T_b = \frac{T}{2}, \quad K_p = \frac{10}{K_1}. \tag{8.8}$$

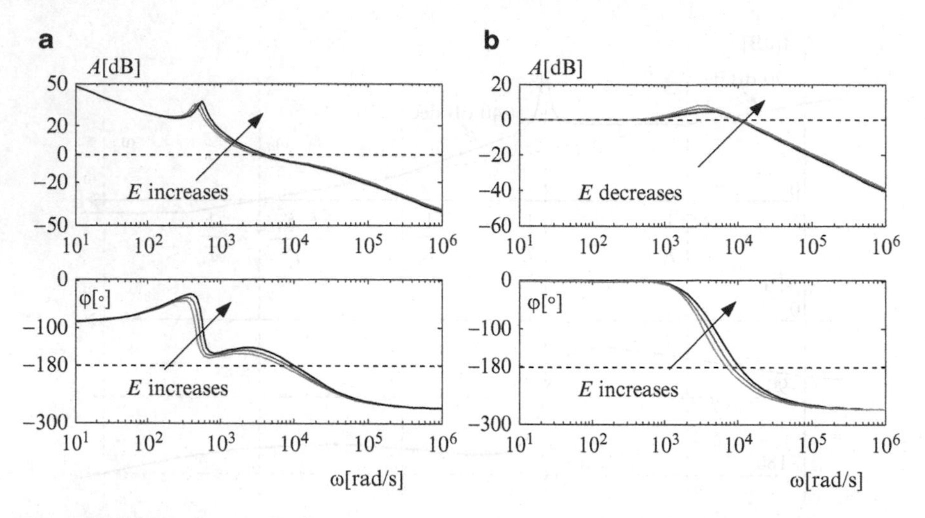

Fig. 8.5 (**a**) Open-loop Bode plots for different supply voltages E_e; (**b**) closed-loop Bode plots for different supply voltages E_e

Remarks Note that controller choice according to (86) does not allow for the nonminimum-phase behavior – induced by the right-half-plane zero of plant (8.5) – being alleviated in the closed loop. Choosing controller parameters according to (8.7) and (8.8) normally leads to increased damping in the closed loop. Note also that choice (8.7) is not unique and in any case must be completed with numerical simulation, which allows assessing the closed-loop dynamic performance. If needed, this latter can further be improved by conveniently adjusting parameters obtained with (8.8).

Figure 8.5 shows the typical variation of the open-loop (a) and the closed-loop (b) Bode plots in the full-load case, as the input supply voltage E varies. For low supply voltages the circuit operates in boost mode and the phase plot shows that it represents a most unfavorable case from the stability viewpoint. Note that the closed-loop bandwidth does not vary significantly.

Figure 8.6a shows the closed-loop response at step load variations for different values of input supply (steady-state). The steady-state load R_e is at the rated value. Figure 8.6b shows the closed-loop response at step input voltage variations for different steady-state values of input supply E_e. The best response is obtained for the buck mode.

8.3 Indirect Output Control: Two-Loop Cascaded Control Structure

Cascaded control is built by nesting several – generally two – control loops and is used when there is more than one measured variable – usually the system states – and a single output. It provides tighter control and improves overall dynamics as it

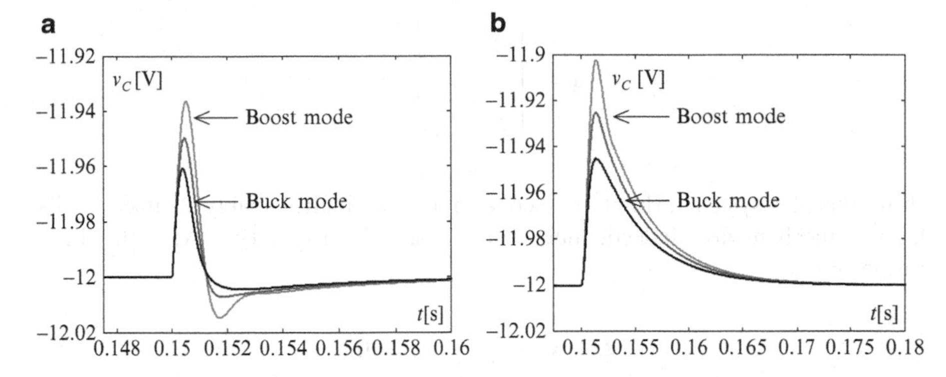

Fig. 8.6 (**a**) Closed-loop output voltage evolutions at step load variations for different values of E_e; (**b**) closed-loop output voltage evolution at step input voltage variations for different values of E_e (steady-state load R_e is at its rated value)

uses an intermediate measured signal that responds faster to the control demand. However, this structure requires separation of the dynamics in the sense that the inner loop must act much faster than the outer loop (Åström and Hägglund 1995).

8.3.1 Assumptions and Design Algorithm

In most DC-DC power electronic converter applications the output variable is the voltage and is involved in the outer loop; therefore, the variable within the inner loop is the current. This is the reason this technique is also called *averaged current-mode control* (Dixon 1991). It applies easily to converters having two states (independent energy accumulations), but it can be extended to higher-order converters. In order to solidify the idea, let us consider the example of a buck converter. Its averaged model, obtained by using the procedure from Chap. 4, and its linearized version around a steady-state operating point are given by expressions (8.9) and (8.10) respectively, with source E being constant, inductor resistance R_L considered negligible and $\widetilde{i_S} = -v_{Ce}^2/R_e \cdot \widetilde{R}$ being the load current small variation due to the load resistance variation $\widetilde{R}$:

$$\begin{cases} L \cdot \dot{i_L} = \alpha \cdot E - v_C - R_L i_L \\ C \cdot \dot{v_C} = i_L - v_C/R, \end{cases} \tag{8.9}$$

$$\begin{cases} L \cdot \dot{\widetilde{i_L}} = E \cdot \widetilde{\alpha} - \widetilde{v_C} \\ C \cdot \dot{\widetilde{v_C}} = \widetilde{i_L} - \widetilde{v_C}/R_e - \widetilde{i_S}\,. \end{cases} \tag{8.10}$$

The eigenvalues of the plant (8.10) state matrix

$$\mathbf{A} = \begin{bmatrix} 0 & -\dfrac{1}{L} \\ \dfrac{1}{C} & -\dfrac{1}{CR_e} \end{bmatrix}$$

show that the system exhibits a second-order, habitually weakly damped oscillatory time behavior. Indeed, their values result from $\det(s\mathbf{I}_2 - \mathbf{A}) = 0$, which further gives

$$LC \cdot s^2 + \frac{L}{R_e} \cdot s + 1 = 0.$$

This equation corresponds to a second-order dynamic having $T = \sqrt{LC}$ as time constant and $\xi = \frac{1}{2R_e}\sqrt{\frac{L}{C}}$ as damping coefficient. A typical numerical application for a buck circuit having $L = 0.5$ mH and $C = 470\ \mu F$ in the steady-state operating point with $R_e = 5\ \Omega$ gives $T = 1.5$ ms and $\xi = 0.33$, which indeed describes a weakly damped time behavior. Note that for larger values of R_e the system becomes even more undamped. Therefore, the dynamics of the two state variables (the output voltage and the inductor current) are in fact strongly coupled, meaning that the assumption of clear separation of the dynamics is not a priori met.

For purposes of comparison, let us consider the well-known case of a DC motor two-loop speed control, where the two states – the armature current and the rotational speed – correspond respectively to electrical and mechanical energy accumulations. In this case each of the state variables introduces a real pole, with the dominant one corresponding to shaft rotational speed. The result is that the dynamics of the two states are (in general) naturally clearly separable. This is not the case for power electronic converters; here, dynamics separation is forced by control design.

This separation is effectively achieved by splitting the original plant (averaged circuit) into two plants: the faster inner plant, which is driven by the slower outer plant. In general, in the case of the two-state power electronic converter, the inner plant captures the inductor current dynamics, while the outer plant embeds the equivalent capacitor voltage dynamics. The main control goal applies to the latter plant.

In the general case, the duty ratio may influence not only the inner plant but also the outer plant, inducing a nonminimum-phase effect and rendering their dynamics not clearly separable. But, for converters for which this behavior is not significant, a simplified algorithm can be stated, as detailed further in this section.

Now, suppose that one of the circuit variables (capacitor voltage v_C) must be maintained by the control action to a certain constant value or must track a slowly variable signal. Supposing this goal achieved, this slow variable is practically constant (at the equilibrium value) in relation to the inner plant dynamic; so, this plant may be described by a simpler equation. Taking the buck converter as an example, the first equation of (8.9) becomes

$$L \cdot \frac{di_L}{dt} = \alpha \cdot E - v_{Ce} - R_L i_L,$$

and, by further linearization, one obtains the transfer function describing the inner plant having as input the duty ratio and as output the averaged inductor current (v_{Ce} is constant):

$$H_{i\alpha}(s) = \frac{\widetilde{i_L}}{\widetilde{\alpha}} = \frac{E}{\frac{L}{R_L}s + 1}. \tag{8.11}$$

Inductor current i_L is driven by its reference i_L^* within the inner loop such that its dynamic becomes much faster than that of v_C. From the outer plant point of view, if the closed-loop dynamic of the inner variable is very fast, it may be neglected and $i_L = k \cdot i_L^*$, where $k < 1$ but close to 1. By zeroing the left side of the first equation of (8.9) one obtains the steady-state algebraic equation of the converter linking the outer output variable (v_C) with the inner control variable (α): $v_C = \alpha \cdot E$ (inductor losses being neglected). Using these two conditions, the equivalent dynamic of the converter results from the second equation of (8.9):

$$C \cdot \frac{dv_C}{dt} = k \cdot i_L^* - \frac{v_C}{R},$$

which gives by linearization around the operating point described by load R_e:

$$C \cdot \dot{\widetilde{v_C}} = k \cdot \widetilde{i_L^*} - \frac{\widetilde{v_C}}{R_e} - \widetilde{i_S},$$

where $\widetilde{i_S} = -v_{Ce}/R_e^2 \cdot \widetilde{R}$ is the load current variation due to load variation $\widetilde{R}$. The outer plant, which has as control input the current reference i_L^* and as output v_C, is described by the transfer function

$$H_{vi_L}(s) = \frac{\widetilde{v_C}(s)}{\widetilde{i_L^*}(s)} = \frac{kR_e}{CR_e s + 1}. \tag{8.12}$$

This outer plant describes the equivalent dynamic of the capacitor voltage in response to the inductor current reference. It is employed in the primary loop design, which must perform a tracking of a certain reference v_C^* with a dynamic much slower (at least five times) than the one obtained for the inner loop.

The cascaded control structure is given in Fig. 8.7. The plant describes here the averaged behavior of the converter. $H_{CV}(s)$ and $H_{CI}(s)$ represents the voltage and current controller transfer functions, respectively, which must be designed in order to solve the control problem. They are computed in conjunction with the transfer functions of the outer and inner plant, respectively (relations (8.11) and (8.12) in the previous example), and with some imposed time- or frequency-domain closed-loop requirements. Note that the output of the outer loop controller represents the reference to track by the inner loop. The prefilter is intended to cancel the zero of the inner closed loop.

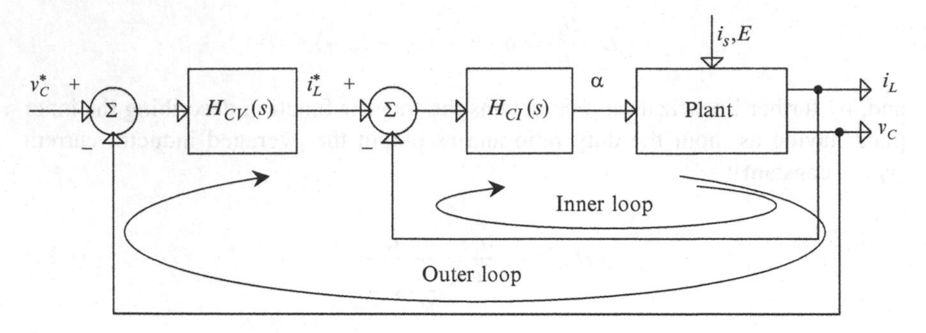

Fig. 8.7 Cascaded control structure for voltage control purpose

Note that in Eq. (8.12) the ratio between the gain and the time constant does not depend on the operating point variation. A PI controller having the transfer function $H_{CV}(s) = K_p(1 + 1/(T_i s))$ in the outer loop will lead to a closed-loop transfer function as

$$H_0(s) = \frac{T_i s + 1}{\frac{T_i T_V}{K_p K_V} s^2 + T_i\left(1 + \frac{1}{K_p K_V}\right)s + 1},$$

where $K_V = kR_e$ and $T_V = CR_e$ (see Eq. (8.12)). Indeed, the coefficient of s^2, which determines the bandwidth of the closed-loop system, does not depend on R_e. On the contrary, the damping coefficient depends on the operating point, but its variations may be alleviated by choosing large values of K_p. These remarks are quite general, i.e., they apply to any two-state power electronic converter.

To conclude the above discussion: a quasi-general algorithm used for designing the controllers within the inner and outer loops, respectively, can be formulated for converters having two states. In order to ease the presentation, the inner-loop controlled variable is denoted by v_1 and the outer variable by v_2.

Algorithm 8.2 Design of cascaded-loop control structure when dynamics separation is possible

#1 Write the averaged large-signal model of the converter and emphasize the variable to be controlled (main control goal). This will be the outer-loop controlled variable, v_1, whereas the other will be the inner-loop controlled variable, v_2.

#2 In the equation describing the dynamic of variable v_1 consider that variable v_2 is constant; rewrite this equation.

#3 Linearize this equation and obtain the open-loop transfer function having the duty ratio as input and v_1 as output.

(continued)

Algorithm 8.2 (continued)

#4 Analyze the transfer function parameters, emphasizing the inner plant gain and the dominant time constant. Impose the inner closed-loop static and dynamic performance – usually its bandwidth must be significantly enlarged in relation to the open-loop one.

#5 Design the inner control loop by choosing a controller appropriate to the previously imposed requirements.

#6 Inside the outer loop consider that v_1 has infinite bandwidth; this is mathematically expressed by zeroing the v_1 derivative in the inner equation. Write the algebraic equation expressing the duty ratio as a function of the variable v_2.

#7 In the outer equation expressing the dynamics of v_2 replace the duty ratio by the previously obtained algebraic relation and obtain the equivalent dynamics of the outer variable v_2.

#8 Linearize the outer variable equation; extract the transfer function from the inner variable reference to v_2.

#9 Analyze the obtained dynamic by emphasizing the outer plant gain and the dominant time constant. Impose the outer closed-loop static and dynamic performance so as to simultaneously satisfy two requirements: first, the original main control goal, in terms of steady-state error and dynamic response, must be fulfilled, and second, the outer closed loop should be at least five times slower than the inner closed loop.

Remarks In the case of a two-state converter, the inner and the outer plants end up as first-order filters. The controllers are proportional, or proportional-integral if zero steady-state error is required. Their parameters results from identifying the closed-loop transfer function parameters with the ones corresponding to the imposed dynamic performance specifications. Note also that one could aim at ensuring a slower dynamic for the outer plant from the very beginning, namely by choosing a large value for the capacitor in the initial circuit sizing stage.

This algorithm also works for boost-type converters having the right-half-plane zero at high frequency.

8.3.2 Example of a Bidirectional-Current DC-DC Converter

Let us consider the synchronous converter in Fig. 8.8 that feeds a motor drive by means of a DC-link. The current can be either drawn from or injected into the DC-link by the motor drive operation. It is required to ensure the power balance between the primary power source E and the motor drive (load) by regulating the DC voltage to a fixed value.

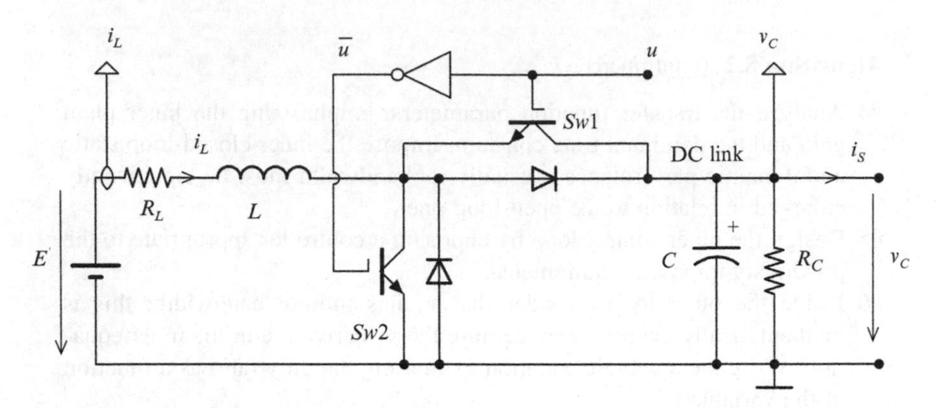

Fig. 8.8 Two-quadrant DC-DC converter, $E < v_C$

Note that the power switches are controlled complementarily – by means of u and $\overline{u}$, respectively – and the converter always operates in current continuous-conduction mode (ccm), allowing the circulation of a bidirectional current. The averaged model is given by

$$\begin{cases} L\dfrac{di_L}{dt} = E - \alpha \cdot v_C - R_L \cdot i_L \\ C\dfrac{dv_C}{dt} = \alpha \cdot i_L - \dfrac{v_C}{R_C} - i_S, \end{cases} \tag{8.13}$$

where $\alpha = \langle u \rangle_0$ and $\langle \overline{u} \rangle_0 = 1 - \alpha$.

Now, suppose that the DC-link voltage v_C has significantly slower variations than the inductor current i_L; moreover, if the overall control goal is fulfilled, the DC-link voltage v_C should remain almost constant. So the outer controlled variable is v_C and the inner variable is i_L. Consequently, by linearizing the first equation of (8.13) around a quiescent operating point (E is constant) one obtains

$$L\dot{\tilde{i}}_L = -v_{Ce} \cdot \tilde{\alpha} - R_L \cdot \tilde{i}_L .$$

The transfer function relating the inductor current with the duty ratio variations (inner plant) is computed straightforwardly

$$H_{i\alpha}(s) = \frac{\tilde{i}_L(s)}{\tilde{\alpha}(s)} = \frac{K_C}{T_C s + 1}, \tag{8.14}$$

where $K_C = -\, v_C^*/R_L$ and $T_C = L/R_L$.

If one considers that $v_{Ce} \equiv v_C^*$ (the main control goal is fulfilled), then the inner plant is invariant with the operating point.

Further, consider a PI control of the inner plant such that the inductor current follows its reference i_L^*. Also, within the outer loop consider that the inner loop

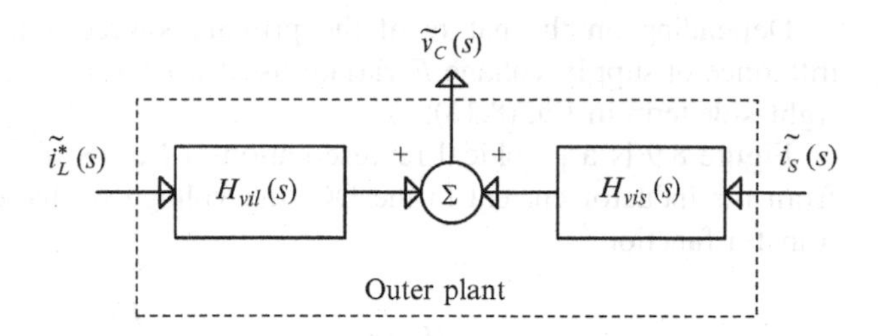

Fig. 8.9 Outer plant structure; $H_{vis}(s)$ is the transfer function multiplying the load current variations i_S in Eq. (8.16)

bandwidth is sufficiently enlarged such that the inductor current varies much faster than the DC-link voltage ($i_L \equiv i_L^*$). By zeroing its derivative and ignoring the losses, the first equation of (8.13) becomes $L \cdot di_L/dt = E - \alpha \cdot v_C = 0$, leading to

$$\alpha = \frac{E}{v_C},$$

with v_C being nonzero for typical operation. By replacing the latter expression of α in the second equation of (8.13) one may emphasize the DC-link voltage v_C variation with respect to the inductor current (reference):

$$C \cdot \frac{dv_C}{dt} = \frac{E}{v_C} \cdot i_L - \frac{v_C}{R_C} - i_S. \tag{8.15}$$

Multiplying this equation with v_C one obtains

$$\frac{C}{2} \cdot \dot{v_c^2} = E \cdot i_L - \frac{v_C^2}{R_C} - i_S \cdot v_C.$$

Linearizing this expression around an equilibrium point leads to

$$C \cdot v_{Ce} \cdot \dot{\widetilde{v_C}} = E \cdot \widetilde{i_L} - \frac{2v_{Ce}\widetilde{v_C}}{R_C} - i_{Se} \cdot \widetilde{v_C} - v_{Ce} \cdot \widetilde{i_S},$$

that further gives the outer plant transfer in variations from both inductor current (control input) and load current i_S (perturbation) to the output voltage:

$$\widetilde{v_C} = \frac{E}{Cv_{Ce}s + \frac{2v_{Ce}}{R_C} + i_{Se}}\widetilde{i_L} - \frac{v_{Ce}}{Cv_{Ce}s + \frac{2v_{Ce}}{R_C} + i_{Se}}\widetilde{i_S}.$$

Note that for efficiency reasons the resistor R_C is much larger (it is usually several tens of kiloohms) than v_C^*/i_{Se} for normal operating points, and therefore term $2v_{Ce}/R_C$ may easily be neglected, hence:

$$\widetilde{v_C} = \frac{E}{i_{Se}} \cdot \frac{1}{\frac{Cv_{Ce}}{i_{Se}}s + 1} \cdot \widetilde{i_L} - \frac{v_{Ce}}{i_{Se}} \cdot \frac{1}{\frac{Cv_{Ce}}{i_{Se}}s + 1} \cdot \widetilde{i_S}. \tag{8.16}$$

Depending on the nature of the primary source, one may also consider the influence of supply voltage E variations, if necessary, leading to a supplementary right-side term in Eq. (8.16).

Figure 8.9 is a graphical representation of Eq. (8.16). The outer plant channel from the inductor current to the DC-link voltage is characterized by a first-order transfer function

$$H_{vil}(s) = \frac{\widetilde{v_C}(s)}{\widetilde{i_L}(s)} = \frac{K_V}{T_V s + 1}, \tag{8.17}$$

where $K_V = E/i_{Se}$ and $T_V = Cv_C^*/i_{Se}$. Note that the outer plant depends on the steady-state operating point (i_{Se}).

Because inner and outer plants are of first order, PI controllers may be effectively used to ensure both zero steady-state error and controlled bandwidth.

By choosing the current controller as

$$H_{RC}(s) = K_{pC} \cdot \left(1 + \frac{1}{T_{iC}s}\right),$$

the closed-loop transfer function of the inner control loop (see Fig. 8.7) turns out to be

$$H_{0C}(s) = \frac{T_{iC}s + 1}{\frac{T_{iC}T_C}{K_{pC}K_C}s^2 + T_{iC}\left(1 + \frac{1}{K_{pC}K_C}\right)s + 1}. \tag{8.18}$$

By studying relations (8.14) and (8.16) one compares the two open-loop time constants of the inner and outer plants. Then – depending also on these plants' gains – the closed-loop time constants must be chosen so that the inner loop is at least five times faster than the inner loop. Suppose that the inner closed-loop dynamical performance is described by the bandwidth $1/T_{0C}$ and a convenient damping coefficient ζ_{0C} (e.g., 0.7). Expression (8.18) allows identification of controller parameters that satisfy the desired performance:

$$T_{iC} = 2\zeta_{0C}T_{0C} - \frac{T_{0C}^2}{T_C}, K_{pC} = \frac{T_C}{K_C T_{0C}^2}\left(2\zeta_{0C}T_{0C} - \frac{T_{0C}^2}{T_C}\right). \tag{8.19}$$

The first-order prefilter in Fig. 8.7 has unit gain and time constant is equal to T_{iC}. It is used in order to cancel the inner closed-loop zero (see relation (8.18)).

Remark Depending on the application, one may consider a proportional controller that drives the inner loop. In this case, the closed inner loop transfer function is

$$H_{0C}(s) = \frac{1}{\frac{T_C}{1 + K_{pC}K_C}s + 1}$$

and the controller gain results is then

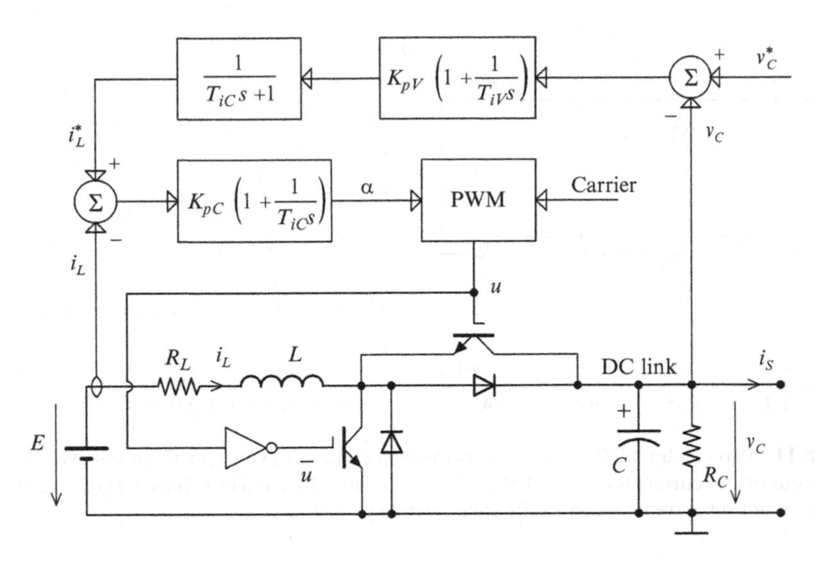

Fig. 8.10 Two-loop cascaded control structure of bidirectional-current DC-DC converter

$$K_{pC} = \frac{T_C - T_{0C}}{T_{0C}K_C}.$$

The outer loop is designed by using a PI-type controller that allows one to achieve the main control goal and to obtain a conveniently slow outer dynamics with respect to the inner loop. The transfer function of this controller is

$$H_{RV}(s) = K_{pV} \cdot \left(1 + \frac{1}{T_{iV}s}\right).$$

Similarly as in the case of the inner loop, the controller parameters turn out to be:

$$T_{iV} = 2\zeta_{0V}T_{0V} - \frac{T_{0V}^2}{T_V}, K_{pV} = \frac{T_V}{K_V T_{0V}^2}\left(2\zeta_{0V}T_{0V} - \frac{T_{0V}^2}{T_V}\right), \tag{8.20}$$

where $1/T_{0V}$ is the bandwidth and ζ_{0V} is the damping imposed for the outer closed loop.

Figure 8.10 synthesizes the numerical simulation results obtained through the above detailed control design procedure. Figures 8.11 and 8.12 show some results concerning the above example for a typical application having a millisecond-sized outer loop. The ratio between the inner loop and the outer loop bandwidths has been chosen as 20.

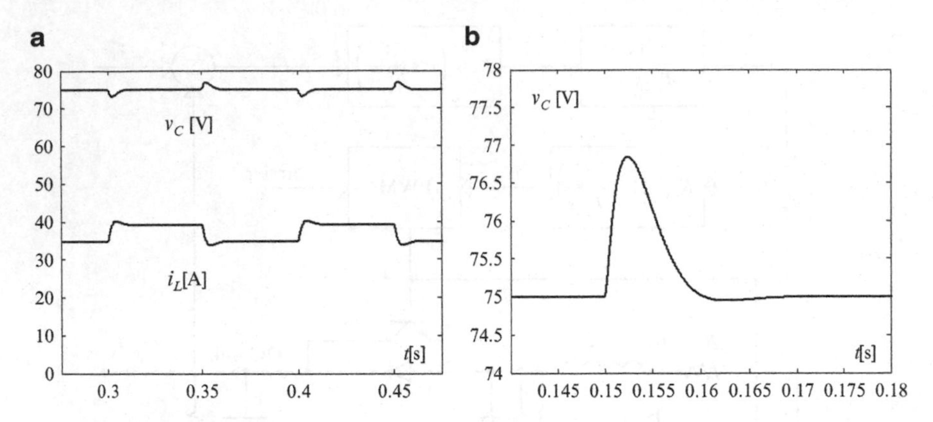

Fig. 8.11 Two-quadrant DC-DC converter two-loop control: (**a**) voltage and current evolutions to unit steps of load current for $\zeta_{0V} = 0.85$ and $T_{0V} = 2.5$ ms; (**b**) detail of voltage response (negative unit step of load current around equilibrium operating point)

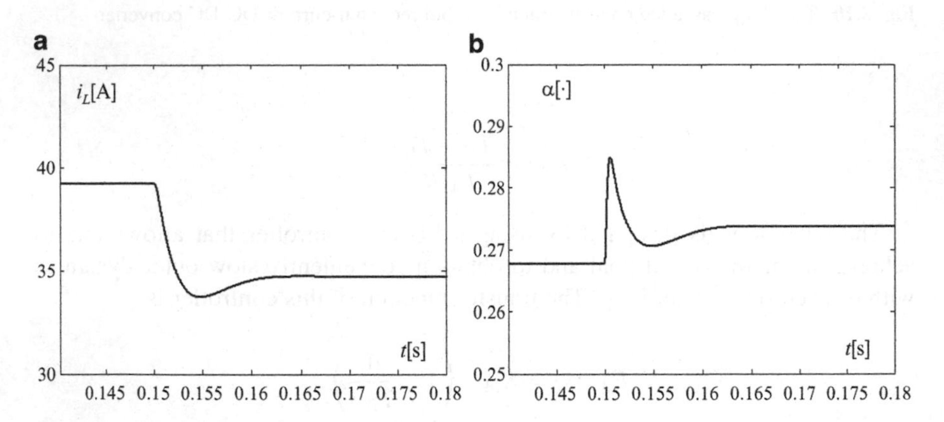

Fig. 8.12 Two-quadrant DC-DC converter two-loop control: (**a**) detail of current response; (**b**) detail concerning the duty ratio evolution (negative unit step of load current around equilibrium operating point)

8.3.3 Two-Loop Cascaded Control Structure for DC-DC Converters with Nonminimum-Phase Behavior

DC-DC converters having boost capability exhibit nonminimum-phase behavior, which can be recognized through a right-half-plane zero in the transfer function from the duty ratio to the output voltage around a certain operating point, depending on that point. In this case, classical control design procedures employed in the two-loop structure – usually not aiming at alleviating the influence of this zero – may encounter some difficulties, especially when it happens that the dynamics within the two loops are not rendered as clearly separable. This is the case, for

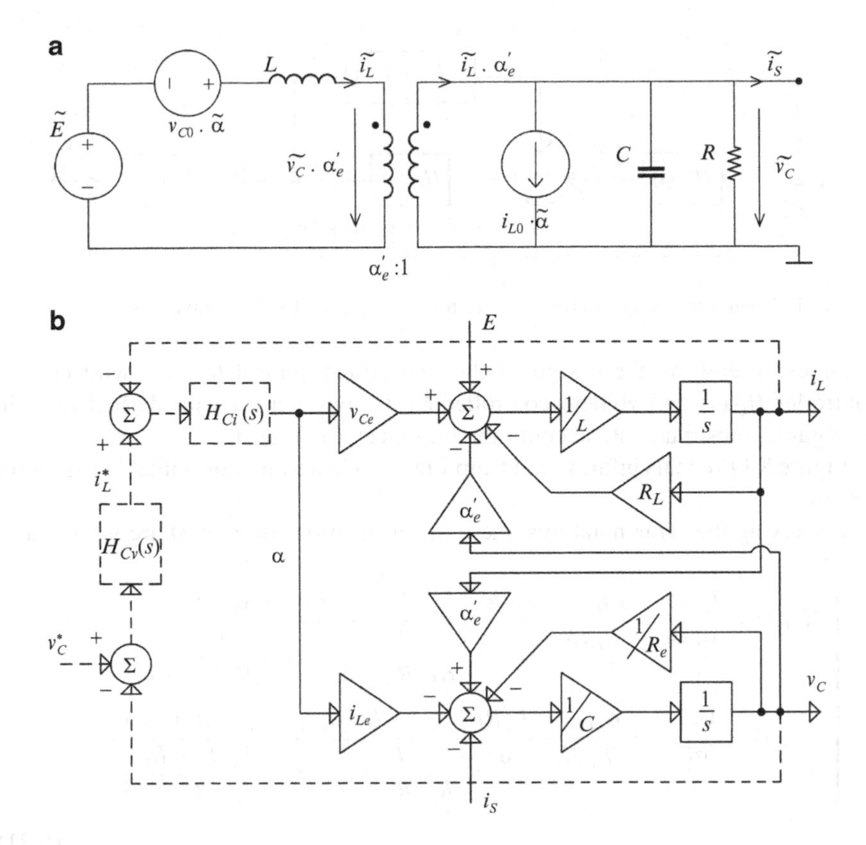

Fig. 8.13 Linearized model of the DC-DC boost converter: (**a**) small-signal equivalent diagram; (**b**) corresponding block diagram

example, when the right-half-plane zero present in the outer voltage loop corresponds to a time constant near to that of the inner closed loop. For such converters, Algorithm 8.2, providing the steps of the averaged current-mode control based on a two-loop structure, must suffer some changes.

Figure 8.13 presents the case of a boost DC-DC converter, where the influence channels from the duty ratio α as input to the two states, the inductor current i_L and the capacitor voltage v_C, around a given operating point have been emphasized; L and R_L characterize the inductor and C is the capacitor's capacity. Figure 8.13a shows the small-signal equivalent diagram of the circuit (repeated from Fig. 4.9 in Chap. 4), where $\alpha' = 1 - \alpha$, and Fig. 8.13b represents the block diagram representation of relation (4.33) from Chap. 4. Subscript e denotes values at the considered operating point, $\widetilde{E}$ is supply voltage variation and $\widetilde{i_S}$ is load current variation around this point.

Diagram in Fig. 8.13 shows that the duty ratio's variations are transmitted to the outer loop before the variations of current i_L are, which come from the inner loop. This means that the more important the contribution of the right-half-plane zero in high frequency is, the less clear the dynamic separation within the two nested loops

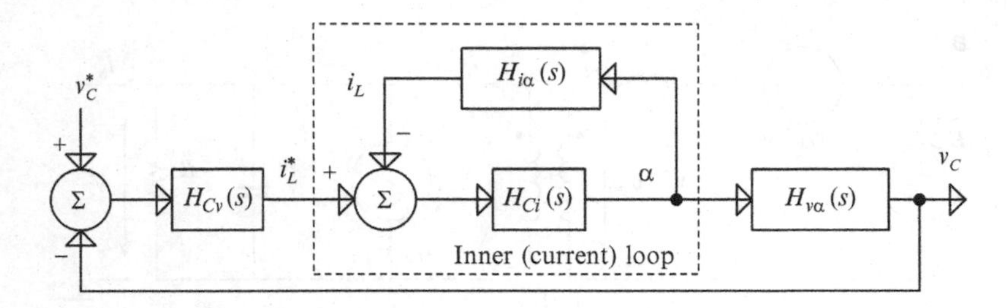

Fig. 8.14 General two-loop control structure for second-order DC-DC converters

becomes. Therefore, the design of the two nested control loops – with current controller $H_{Ci}(s)$ and voltage controller $H_{Cv}(s)$ represented with dashed lines in the figure – must take into account the presence of this zero.

Figure 8.14 repeats information from Fig. 8.13 in a form more suitable to control design.

Preserving the same notations, the transfer functions of interest are written as

$$\begin{cases} H_{i\alpha}(s) = \dfrac{I_L(s)}{\alpha(s)} = \dfrac{\alpha_e' i_{Le} + v_{Ce}/R_e}{R_L/R_e + \alpha_e'^2} \cdot \dfrac{Cv_{Ce}/\left(\alpha_e' i_{Le} + v_{Ce}/R_e\right) \cdot s + 1}{\dfrac{LC}{R_L/R_e + \alpha_e'^2} \cdot s^2 + \dfrac{L/R_e + R_L C}{R_L/R_e + \alpha_e'^2} \cdot s + 1}, \\ H_{v\alpha}(s) = \dfrac{V_C(s)}{\alpha(s)} = \dfrac{\alpha_e' v_{Ce} - i_{Le} R_L}{R_L/R_e + \alpha_e'^2} \cdot \dfrac{-L/\left(\alpha_e' v_{Ce}/i_{Le} - R_L\right) \cdot s + 1}{\dfrac{LC}{R_L/R_e + \alpha_e'^2} \cdot s^2 + \dfrac{L/R_e + R_L C}{R_L/R_e + \alpha_e'^2} \cdot s + 1}. \end{cases} \tag{8.21}$$

Expression of $H_{v\alpha}(s)$ in (8.21) indeed shows the unavoidable presence of a right-half-plane zero in the linearized dynamics of the output voltage, corresponding to a time constant equal to $L/(\alpha_e' v_{Ce}/i_{Le} - R_L)$ depending on the operating point. Its contribution is more important when the current i_{Le} is large (heavy-load case). Moreover, this right-half-plane zero cannot be displaced by means of control action. If this zero is placed near the inner loop's bandwidth, it prevents the clear separation of the dynamics within the two nested loops, which is a basic assumption for the control design. If this assumption does not hold any longer, then the imposed performance cannot be guaranteed either.

Figure 8.15 presents comparatively two cases of output voltage regulation using the two-loop control structure, with the right-half-plane zero's influence being more obvious in the first case – i.e., for reduced values of the load current – than in the second case – large values of the load current – for the same step variation of the voltage reference v_C^*.

The same set of desired performance features (same bandwidth and same damping coefficient) has been imposed on the outer closed loop and controllers have been computed using Algorithm 8.2 from Sect. 8.3.1 (which does not take into account the presence of the right-half-plane zero). The time evolutions in Fig. 8.15 confirm the behavior predicted by the modeling given by expressions (8.21) and Fig. 8.14.

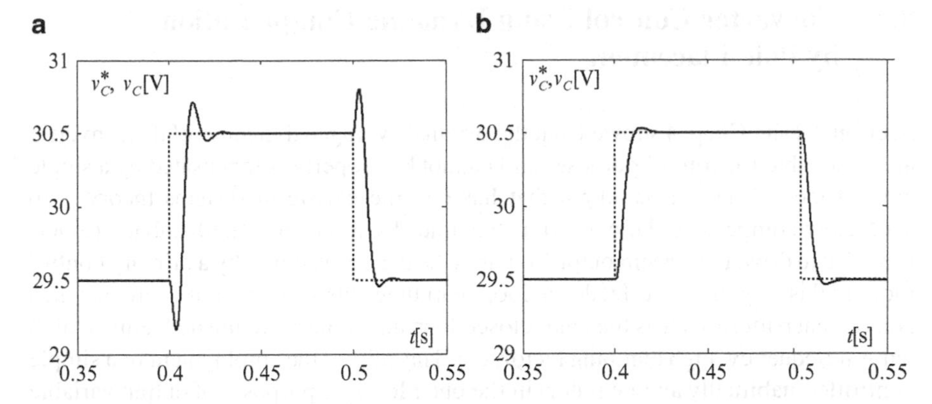

Fig. 8.15 Closed-loop behavior of a boost DC-DC converter: (**a**) heavy-load case; (**b**) light-load case

One can also note that, whereas in the second case the desired performance is met (damping coefficient 0.8), in the first case the performance is altered due to the nonminimum-phase behavior exhibited in high frequency. This latter case corresponds to the assumption of separable dynamics not being met; as a consequence, the dynamic performance imposed on the outer control loop cannot be fulfilled.

A modified version of Algorithm 8.2 can be proposed for second-order DC-DC converters in the general case – including dealing with the nonminimum-phase behavior – as summarized below in the form of Algorithm 8.3.

Algorithm 8.3 Design of the cascaded control structure in the general case

#1 Write the linearized small-signal model of the converter by expressing the transfer functions from the duty ratio to the two state variables.

#2 Design the inner current control loop by choosing an appropriate controller and by placing the inner closed-loop poles. The root locus method is a useful tool to this end.

#3 Express the outer plant transfer function by using the voltage transfer function and the inner loop transfer function – see Fig. 8.14.

#4 Choose an appropriate controller allowing the stated global control goal being fulfilled. Impose the set of dynamic performances (bandwidth and overshoot) in relation with the inner closed-loop dynamics.

#5 Use the root locus method in order to establish the dominant pole's position (the outer loop bandwidth). For example, if a PI controller is chosen, then it is required that the root locus pass through the pole corresponding to the imposed bandwidth and overshoot. In this way, the left-half-plane zero position results, providing the corresponding value of the controller's integral gain, and also the proportional gain of the controller (Ceangă et al. 2001).

8.4 Converter Control Using Dynamic Compensation by Pole Placement

Section 8.2 and Chap. 4 showed that the original averaged dynamics of the converter are unsuitable for control purposes and cannot be properly compensated by a single control loop. A useful paradigm that has become classic in systems theory is to separately compensate the converter dynamics by a second state-feedback (inner) loop and to drive the system output towards the desired setpoint by a primary control loop. In this way, the state-feedback loop rearranges the systems' pole-zero map in a convenient pattern such as the inner closed loop has some prerequisite features: high natural frequency, high damping factor, etc. This allows the employment of a simple controller (habitually an integrator) in the outer loop for purposes of output variable regulation or tracking (d'Azzo et al. 2003; Rytkonen and Tymerski 2012). Extremization of a suitably chosen performance index also leads to an optimally sized full-state-feedback control loop (Leung et al. 1991).

8.4.1 Assumptions and Design Algorithm

The same assumptions as in the previous sections are adopted. This technique applies also to high-order converters if all states are available for measure. State reconstructors (observers) may be employed if one wishes to estimate some of the states and hence to save transducers (reducing overall circuit cost, but adversely affecting its time response).

Use of integral control that ensures zero steady-state error within a reasonable time for a highly undamped plant quickly results in stability loss. This can easily be shown by using the root locus method (d'Azzo et al. 2003). It is thus preferable to impose first some convenient dynamic by using the pole-placement technique (Sobel et al. 2011). In this way, the desired poles can be placed by a full-state-feedback gain, denoted by **K** in Fig. 8.16. To simplify the design further, it is useful to impose a pair of complex-conjugated poles at this stage.

The regulation task can further be effectively accomplished if an integrator – with high integral gain – is coupled with the plant having the desired second-order dominant dynamic, as imposed in the previous stage and shown in Fig. 8.16. The third-order closed-loop dynamic may subsequently be adjusted via the root locus method – which is suitable for plants having complex-conjugated-poles

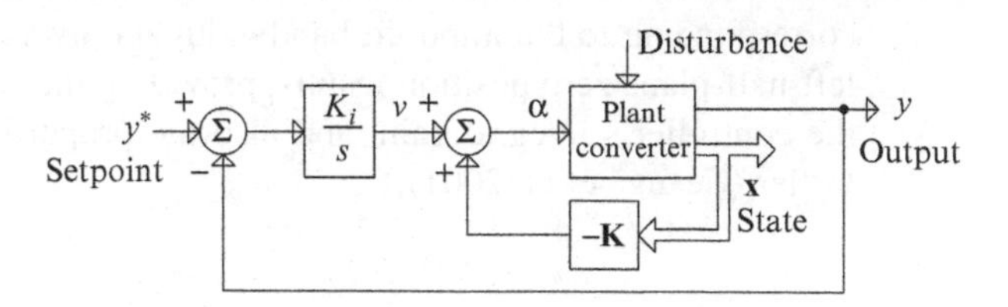

Fig. 8.16 Output regulation structure using state-feedback-based inner loop

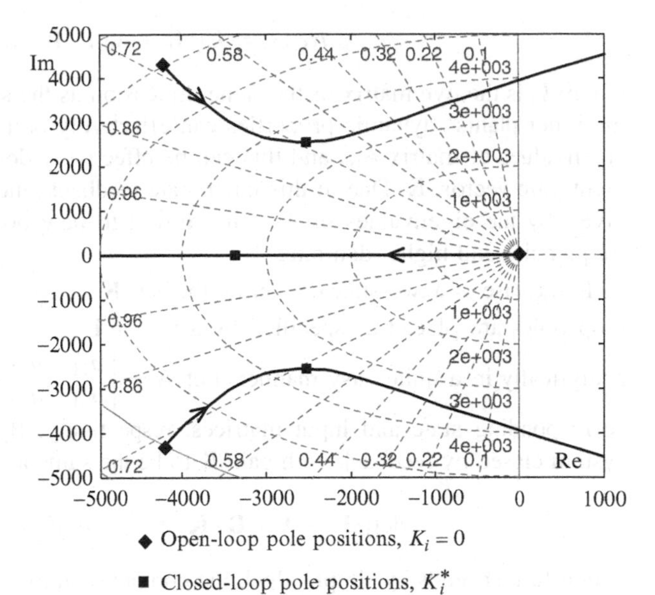

Fig. 8.17 Example of closed-loop pole positions on root locus as controller gain K_i varies for control structure in Fig. 8.16

dominant dynamic – by choosing a convenient closed-loop pole placement as the integral gain K_i varies. One can note that, given the single degree of freedom in the control law, this last design step can entail imposing either a given closed-loop bandwidth or a desired closed-loop overshoot, depending on the shape of the root locus when gain K_i varies.

For example, one can impose the maximum bandwidth for which the overshoot value does not change in relation to the one imposed in the state-feedback design stage, as shown in Fig. 8.17. In this figure, the root locus corresponds to some quasi-constant value of the damping coefficient for a quite large variation domain of K_i, whose limit K_i^* is emphasized. If K_i is increased beyond K_i^* the damping decreases; meanwhile, the natural frequency of the complex-conjugated poles – giving the dominant dynamic or the bandwidth – decreases. In conclusion, the value K_i^* suitably trades off the two requirements of the closed-loop dynamic performance.

Now, suppose that the original plant (converter) is described by an averaged linearized model:

$$\begin{cases} \dot{\mathbf{x}} = \mathbf{A} \cdot \mathbf{x} + \mathbf{B} \cdot \alpha \\ y = \mathbf{C} \cdot \mathbf{x}, \end{cases}$$

where the state includes measurable currents and voltages and α is the duty ratio.

The inner plant having variable v as input, variable y as output and vector $\mathbf{x}$ as state is described by the state matrix $\mathbf{A}_{0i} = \mathbf{A} - \mathbf{B} \cdot \mathbf{K}$, the input matrix $\mathbf{B}$ and the output matrix $\mathbf{C}$. One imposes the desired set of eigenvalues on the closed-loop state matrix $\mathbf{A}_{0i}$ and obtains the necessary feedback gain $\mathbf{K}$. The associated transfer function can be obtained as

$$H_{0i}(s) = \mathbf{C}(s\mathbf{I}_n - \mathbf{A}_{0i})^{-1}\mathbf{B}, \tag{8.22}$$

where $\mathbf{I}_n$ is the eye matrix of the same dimension as the state matrix $\mathbf{A}$. Therefore the inner plants' dynamic properties can effectively be adjusted by imposing the eigenvalues of matrix $\mathbf{A}_{0i}$, and this can be effectively done by choosing a convenient gain vector, $\mathbf{K}$. Due to this inner state feedback, the plant's original poles – given by the eigenvalues of $\mathbf{A}$ – are moved to new positions at higher natural frequencies and higher damping.

In the case of a two-state converter, the gain $\mathbf{K} = [k_1 \quad k_2]^T$ ensuring the closed-loop poles are placed as specified by a vector $\mathbf{P} = [p_1 \quad p_2]^T$ can be computed analytically in a quite easy manner. Let $\mathbf{A} = \begin{bmatrix} a_{11} & a_{12} \\ a_{21} & a_{22} \end{bmatrix}$ and $\mathbf{B} = \begin{bmatrix} b_1 \\ b_2 \end{bmatrix}$ be the corresponding state and input matrices, respectively. By imposing poles of the system closed by feedback with gain $\mathbf{K}$ to be the same as vector $\mathbf{P}$, one obtains

$$\det(s\mathbf{I}_2 - \mathbf{A} + \mathbf{B} \cdot \mathbf{K}) \equiv (s - p_1)(s - p_2),$$

which further leads by simple algebraic computation to

$$\begin{aligned} s^2 + (b_1k_1 + b_2k_2 - a_{11} - a_{22})s + a_{11}a_{22} - a_{12}a_{21} + a_{12}b_2k_1 \\ -a_{22}b_1k_1 + a_{21}b_1k_2 - a_{11}b_2k_2 \equiv s^2 - (p_1 + p_2)s + p_1p_2. \end{aligned}$$

Identification of polynomial coefficients allows for a second-order linear equation system being obtained as

$$\begin{cases} b_1k_1 + b_2k_2 = a_{11} + a_{22} - (p_1 + p_2) \\ (a_{12}b_2 - a_{22}b_1)k_1 + (a_{21}b_1 - a_{11}b_2)k_2 = p_1p_2 - (a_{11}a_{22} - a_{12}a_{21}), \end{cases} \tag{8.23}$$

which must be solved for k_1 and k_2. By adopting notations

$$\mathbf{M} = \begin{bmatrix} b_1 & b_2 \\ a_{12}b_2 - a_{22}b_1 & a_{21}b_1 - a_{11}b_2 \end{bmatrix}, \mathbf{N} = \begin{bmatrix} \mathrm{Tr}(\mathbf{A}) - (p_1 + p_2) \\ p_1p_2 - \det(\mathbf{A}) \end{bmatrix}, \tag{8.24}$$

where $\mathrm{Tr}(\mathbf{A}) = a_{11} + a_{22}$ is the trace of matrix $\mathbf{A}$ and $\det(\mathbf{A}) = a_{11}a_{22} - a_{12}a_{21}$ is its determinant, system (8.23) has a unique solution if and only if matrix $\mathbf{M}$ is invertible, which is equivalent to the controllability of matrix pair $(\mathbf{A},\mathbf{B})$. Then the gain matrix $\mathbf{K}$ can be computed as

$$\mathbf{K}^T = \mathbf{M}^{-1} \cdot \mathbf{N}. \tag{8.25}$$

In the particular case of complex conjugated poles the following relations hold: $-(p_1 + p_2) = 2\zeta_0\omega_{n0}$ and $p_1p_2 = \omega_{n0}^2$ with ω_{n0} being the desired closed-loop bandwidth and ζ_0 being the desired closed-loop damping coefficient.

In the case of higher-order converters the solution to the problem relies on the possibility of effectively inverting the matrix $\mathbf{M}$. Computer-aided design (CAD) procedures that solve this problem are available within dedicated software (e.g., `acker` function in MATLAB®).

Algorithm 8.4 Control design for output voltage regulation using intermediary state feedback and integral control

#1 Choose operating point where control design must be done (e.g., one corresponding to the full load). Obtain converter's averaged linearized model and draw pole-zero map.
#2 Choose new set of poles (conveniently placed at a higher natural frequency and higher damping in relation to those of open-loop system).
#3 Obtain gain vector that ensures new map of poles by closed-loop state feedback. This may be computed either analytically or by using dedicated CAD software (e.g., function `acker` in MATLAB®).
#4 Draw open-loop transfer function root locus. Choose compensator integral gain K_i such that closed-loop bandwidth is desired one and plant has reasonable damping.
#5 If no solution can be found, change the new set of poles (e.g., their associated damping) and reiterate from #3.
#6 If a good solution is found, verify that gain K_i gives good results for any operating point.

Next, having imposed suitable inner dynamics and having computed the closed-loop transfer function $H_{0i}(s)$ according to relation (8.22), one may approach the outer loop, dedicated to outer voltage regulation. The associated open loop transfer function is

$$H_{co}(s) = \frac{K_i}{s} \cdot H_{0i}(s).$$

The goal is to choose a sufficiently high value for the integrator gain K_i so as to ensure the required closed-loop bandwidth with convenient damping. As stated before, this operation is done by using the root locus method. A CAD solution (e.g., `rltool` in MATLAB®) may be of great help in this operation (Mathworks 2012).

Algorithm 8.4 synthesizes all aspects of control loop design using an inner state feedback.

8.4.2 *Example of a Buck Converter*

A control structure diagram like the one given in Fig. 8.16 is proposed. The inductor current and the capacitor (output) voltage are the measured state variables (see also Fig. 2.10).

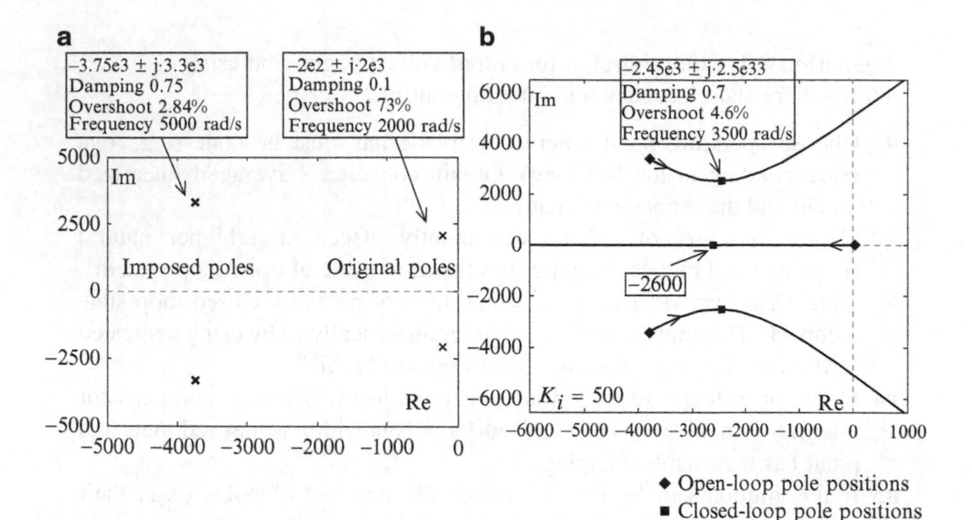

Fig. 8.18 Buck converter controlled with inner state feedback: (**a**) pole-zero maps for the original, open-loop buck converter and inner state-feedback plant, respectively; (**b**) root locus and pole positions for $K_i = 500$

One begins the control design with the averaged model of the buck converter (see Chap. 4 and relation (8.9)), which must be linearized in order to obtain the following state-space model as a version of Eq. (8.10):

$$\begin{bmatrix} \dot{\widetilde{i_L}} \\ \dot{\widetilde{v_C}} \end{bmatrix} = A \underbrace{\begin{bmatrix} -\dfrac{R_L}{L} & -\dfrac{1}{C} \\ \dfrac{1}{C} & -\dfrac{1}{CR_e} \end{bmatrix}}_{\mathbf{A}} \cdot \begin{bmatrix} \widetilde{i_L} \\ \widetilde{v_C} \end{bmatrix} + B \underbrace{\begin{bmatrix} \dfrac{E}{L} & 0 \\ 0 & -\dfrac{1}{C} \end{bmatrix}}_{\mathbf{B}} \cdot \begin{bmatrix} \widetilde{\alpha} \\ \widetilde{i_S} \end{bmatrix}. \tag{8.26}$$

The transfer function from the duty ratio variations to the output voltage variations is

$$H_{v\alpha}(s) = \frac{\widetilde{v_C}(s)}{\widetilde{\alpha}(s)} = \frac{E}{LCs^2 + \dfrac{L}{R_e}s + 1}. \tag{8.27}$$

The pole-zero map of the original plant (see denominator in Eq. (8.27)) for typical values of L and C and a typical operating point (in terms of values of E and R_0) is given in Fig. 8.18a. The imposed poles of inner plant $H_{0i}(s)$ at higher values of frequency and damping are also visible.

Figure 8.18b shows the root locus of the outer plant. The controller gain K_i establishes pole positions on this locus. For $K_i = 0$ the loop is not closed and the pole positions correspond to the plant $H_{0i}(s)/s$. For higher values of K_i the poles

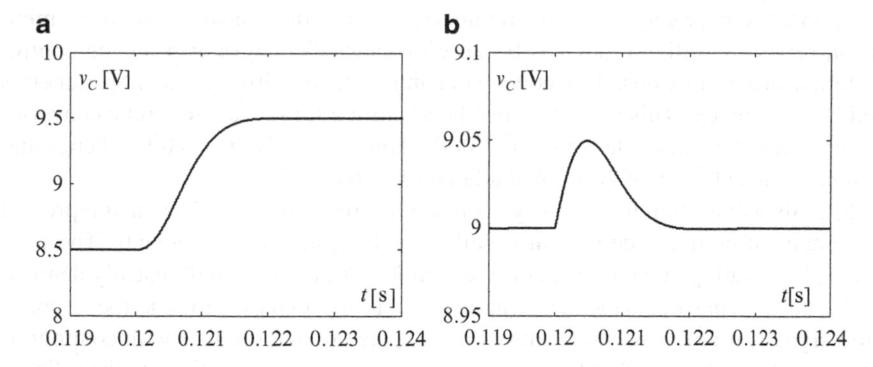

Fig. 8.19 Simulation results for the buck converter controlled with inner state feedback – output voltage response in: (**a**) tracking (unit step of voltage reference), (**b**) regulation (unit step of load resistance)

migrate on the locus, the position of the real pole (which is still the dominant one) corresponds to a higher bandwidth and the original damping of the complex poles begins to deteriorate. For sufficiently high K_i it is possible that the complex poles become the dominant ones and the system may eventually lose its stability (as poles pass to the right-half plane).

Remarks The controller gain K_i is chosen so as to have the largest bandwidth possible with a good damping. Obviously, the choice of inner plant poles $H_{0i}(s)$ depends on each converter and hence on each root locus shape; for example, if the root locus slopes are higher, obtaining a higher closed-loop bandwidth for the outer plant may require choosing a lower damping for the inner plant $H_{0i}(s)$.

The associated closed-loop step responses for the chosen value of K_i corresponding to Fig. 8.18b are depicted in Fig. 8.19 for both tracking and regulation modes. Note that the transient response lasts 2 ms, as predicted by the bandwidth and the damping values deducible from Fig. 8.18b.

Remark In the first pole-placement stage one can intent placing the desired closed-loop poles quite far, so as to correspond to large bandwidth, thus compensating for its reduction in the second design stage. In this latter stage, because the final closed-loop dynamic is of third order, the dominant dynamic might need to be that of the real pole, which further gives the value of the controller gain K_i. The damping value also results from such a choice, with no possibility of its being imposed separately.

8.5 Digital Control Issues

Digital control systems are based upon digital hardware, usually represented by a programed digital computer as the controller core. Digital controllers perform numerical computations and interact with the analog plant by means of analog components composing its peripheral interface. Nowadays most control implementation relies upon digital devices because of their advantages over analog control systems, such as low cost, low power consumption, zero drift of system parameters and high accuracy. This trend can also be identified for power electronic converter control (Prodić and Maksimović 2000; Hung and Nelms 2002; Peng and Maksimović 2005; Wen 2007; Arikatla and Qahouq 2011).

Signals within digital control systems are discrete-time signals, which represent sequences of numbers defined at usually evenly spaced time moments. The time interval separating such moments is the sampling time (or period), usually denoted by T_s. The Z transform is the mathematical tool most often used to describe discrete-time signals; it plays the same role as the Laplace transform does with continuous systems (Santina and Stubberud 2011). In a similar way, discrete-time linear systems can be described by z-transfer functions, defined as the ratios between the Z transforms of the input and output signals, respectively. There are well-known methods for obtaining the Z transform of a discrete-time signal starting from the Laplace transform of the corresponding continuous-time signal; the same holds for continuous-time linear systems and their discrete-time versions in terms of transfer functions.

8.5.1 Approaches in Digital Control Design

There are two main approaches to designing digital controllers for continuous-time plants. A first approach – the traditional one – starts from designing an analog controller for the plant; its digital counterpart is further derived by means of the discretization operation. Even if in general it closely approximates the behavior of the original analog controller, it may happen that the digital one performs less well than its analog counterpart, even for small sampling periods. The second approach to deriving digital controllers for continuous-time plants is first to compute a discrete-time equivalent of the plant and then to design a digital controller directly to control the discretized plant (Landau and Zito 2005).

Some of the most common methods employing the first approach – i.e., discretizing analog controllers – will be outlined next.

By digitizing a given continuous transfer function one obtains a discrete transfer function approximation of it. Obviously, as such approximation is not unique, it can be done in multiple ways, among which one can cite the numerical approximation of differential equations, matching time responses or pole-zero matching (Tan 2007; Santina and Stubberud 2011). Differential equations can be approximated numerically by either numerical integration or by numerical differentiation. From these two latter possibilities, numerical integration has been more satisfactory for obtaining discrete-time equivalents in terms of both simplicity and approximation accuracy.

Usually, numerical integration is performed as follows: the interval of integration is divided into many subintervals; then the contribution of each subinterval is approximated by the integral of an approximating polynomial. The so-called trapezoidal method – called also Tustin's method or else bilinear transform – is an illustrative example which uses two samples to update the approximation within a sampling interval. This method leads to a discrete-time transfer function $H_C(z)$ being computed based on its continuous counterpart $H_C(s)$ and the following variable change:

$$H_C(z) = H_C(s)|_{s=\frac{2}{T_s}\cdot\frac{z-1}{z+1}}, \tag{8.28}$$

where T_s is the sampling period. It is common to express discrete-time transfer functions as functions of the one-sample delay operator z^{-1} because in this way one can deduce easily the finite-differences equation that describes the digital controller in time domain.

Note that the Tustin's transform maps the stability region of the s-plane (its left half) to the interior of the unit circle on the z-plane, whereas the right half of the s-plane is mapped into the exterior of the unit circle and the imaginary axis of the s-plane to the unit circle's boundary. In this way, the necessary and sufficient stability condition in the z-plane requires that the poles of the discrete transfer function be placed inside the unit circle.

A frequency prewarping operation is required in many control and signal processing applications when the frequency response of the digital filter (or controller) ω_d should approximate the frequency response of the analog one, ω_c. They are related through the following nonlinear relation:

$$\omega_c = \frac{2}{T_s}\tan\left(\frac{\omega_d T_s}{2}\right),$$

justified by the fact that the entire imaginary axis of the s-plane is mapped into one complete revolution of the unit circle in the z-plane.

Higher-order approximations result in higher-order digital controllers, which track better the samples of the analog controller output for any input. The digital controller is usually equipped with a sample-and-hold output between samples; therefore, the approximation accuracy is less important than the sample rate choice. In the particular case of power electronic systems, it is good practice to choose the sampling period to be the same order of magnitude as the switching frequency (Peng and Maksimović 2005).

The example of discretizing a PID continuous controller described by the transfer function $H_{C\_PID}(s) = k_p\left(1 + \frac{1}{T_i s} + T_d s\right)$ is considered next. By employing Tustin's method (8.28), simple algebra leads to the following discrete transfer function:

$$H_{C\_PID}\left(z^{-1}\right) = \frac{k_{1z} + k_{2z}z^{-1} + k_{3z}z^{-2}}{1 - z^{-2}}, \tag{8.29}$$

where $k_{1z} = k_p\left(1 + \frac{T_s}{2T_i} + \frac{2T_d}{T_s}\right)$, $k_{2z} = k_p\left(\frac{T_s}{T_i} - \frac{4T_d}{T_s}\right)$, $k_{3z} = -k_p\left(1 - \frac{T_s}{2T_i} - \frac{2T_d}{T_s}\right)$. In the particular case of power converter control, by denoting by $E(z^{-1})$ and $\alpha(z^{-1})$ the Z transforms of the controller input (error) and output (duty ratio α), respectively, one can write

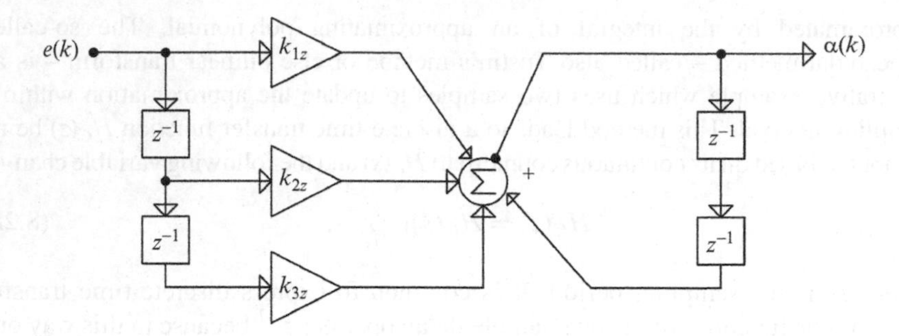

Fig. 8.20 Diagram of a digital PID controller

$$H_{C\_PID}\left(z^{-1}\right) = \frac{\alpha(z^{-1})}{E(z^{-1})} = \frac{k_{1z} + k_{2z}z^{-1} + k_{3z}z^{-2}}{1 - z^{-2}},$$

or equivalently

$$\alpha\left(z^{-1}\right) = k_{1z}E\left(z^{-1}\right) + k_{2z}z^{-1}E\left(z^{-1}\right) + k_{3z}z^{-2}E\left(z^{-1}\right) + z^{-2}\alpha\left(z^{-1}\right),$$

from which the digital controller's input–output time-domain equation can be obtained by taking into account that z^{-1} denotes the one-sample delay operator

$$\alpha(kT_s) = k_{1z}e(kT_s) + k_{2z}e((k-1)T_s) + k_{3z}e((k-2)T_s) + \alpha((k-2)T_s). \quad (8.30)$$

Equation (8.30) corresponds to a second-order digital filter; it shows how the current output sample $\alpha(kT_s)$ depends on the two-step-before output sample $\alpha((k-2)T_s)$, and current and previous input samples, $e(kT_s)$, $e((k-1)T_s)$ and $e((k-2)T_s)$, respectively. To the controller described by (8.29) and equivalently by (8.30) corresponds the block diagram of Fig. 8.20.

8.5.2 *Example of Obtaining Digital Control Laws for Boost DC-DC Converter Used in a Photovoltaic Application*

Figure 8.21 presents the circuit diagram of a photovoltaic (PV) panel supplying power to a DC bus by means of a boost DC-DC power converter. Supposing a grid-connected architecture, the DC-bus voltage is kept constant at v_{DC}^* by another control loop, and the DC-DC converter must be controlled so as to impose the operating point of the PV panel. To this end, the PV voltage must be controlled, and this is achieved by means of the boost converter current control. A two-loop cascaded control structure having the voltage control loop as the outer loop and the current control loop as the inner loop is built, as detailed in Fig. 8.21.

Following the steps of Algorithm 8.2 presented in Sect. 8.3.1 and guidelines given in the example from Sect. 8.3.2, one computes the duty-ratio-to-boost-current

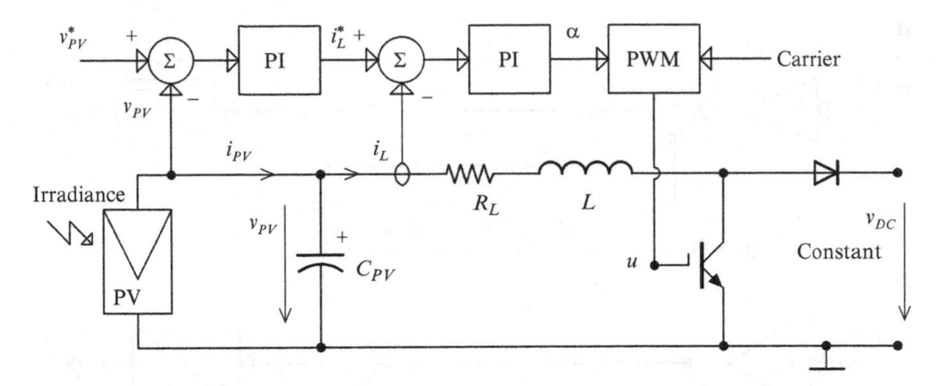

Fig. 8.21 Block diagram of a two-loop control structure of a boost DC-DC converter used in a PV application

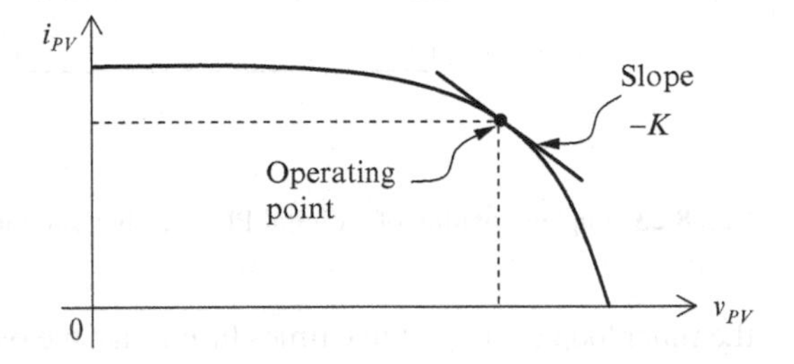

Fig. 8.22 Current–voltage steady-state characteristic of a PV panel for a given irradiance level: emphasizing the slope of the curve around a given operating point

transfer function and the boost-current-to-PV-voltage transfer function around the chosen operating point defined by a pair of PV current–voltage values (e.g., for grid-connected PV applications this is the point corresponding to maximum power). These transfer functions are respectively

$$H_{i_L\alpha}(s) = \frac{I_L(s)}{\alpha(s)} = \frac{v^*_{DC}}{Ls + R_L}, H_{v_{PV}i_L}(s) = \frac{V_{PV}(s)}{I_L(s)} = \frac{-1}{C_{PV}s + K}, \tag{8.31}$$

where the meanings of the different notations are clear from Fig. 8.21 and K is the absolute value of the slope of the PV current–voltage curve, which depends on the irradiance level and on the operating point (see Fig. 8.22). Relations (8.31) show that the current plant is practically invariant and that the voltage plant is variant through parameter K. Supposing that the operating point is known – as resulting from voltage regulation task, that is, from imposing a certain value of desired voltage – then K can be estimated based upon the steady-state PV current–voltage characteristics for different irradiance values.

Two continuous PI controllers can be computed for the two nested loops respectively, starting by imposing a performance on each one in terms of zero steady-state error and suitable transients (trade-off between overshoot and settling time), provided

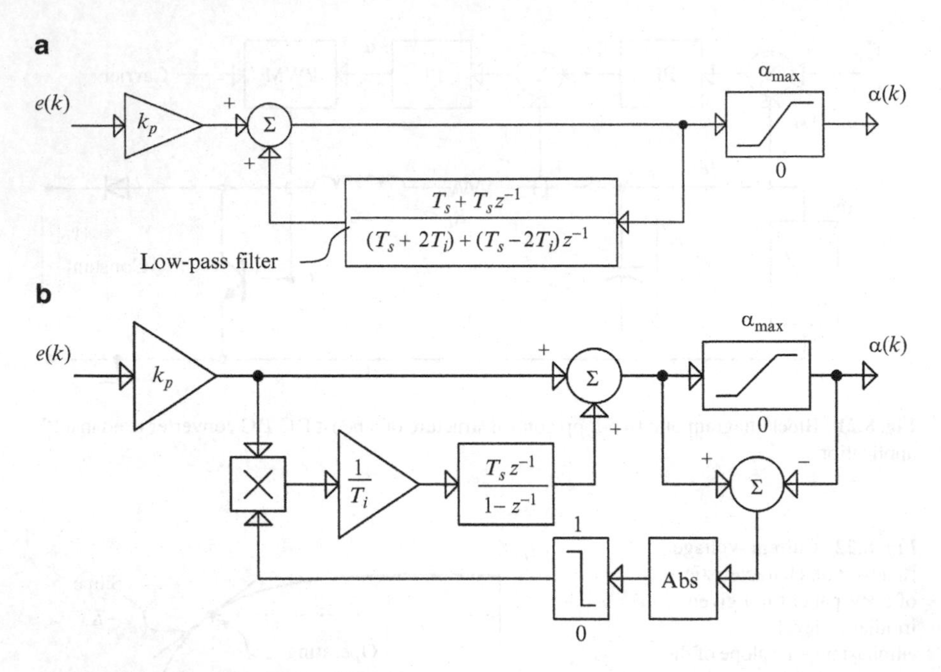

Fig. 8.23 Implementation of a digital PI controller allowing limitation of the output

the inner loop be at least five times faster than the outer one. To this end, computation relations detailed in Sect. 8.3 may be used. Developing these as done in the previous section, the transfer function of a digital PI controller turns out to be

$$H_{C\_PI}\left(z^{-1}\right) = k_{pz} \cdot \frac{1 - az^{-1}}{1 - z^{-1}}, \tag{8.32}$$

where $k_{pz} = k_p(1 + T_s/(2T_i))$ and $a = \dfrac{1 - T_s/(2T_i)}{1 + T_s/(2T_i)}$, with k_p and T_i being proportional gain and integral time constant of the original continuous controller, respectively. Implementation of the current digital PI controller according to relation (8.32) must ensure the possibility of limiting the control input, that is, the duty ratio α, within the interval $(0,\alpha_{max})$; to this end, a diagram like the one in Fig. 8.23a can be employed, where the proportional-integral action is equivalently represented by a first-order low-pass filter having T_i as time constant in a positive-feedback scheme (Åström and Hägglund 1995).

Another – more popular – way of implementing a digital PI controller having k_p and T_i as proportional gain and integral time constant, respectively, is shown in Fig. 8.23b; here the numerical version has been obtained by discretizing the integrator's transfer function through Euler's forward method, that is, by replacing variable s with $\dfrac{T_s z^{-1}}{1 - z^{-1}}$.

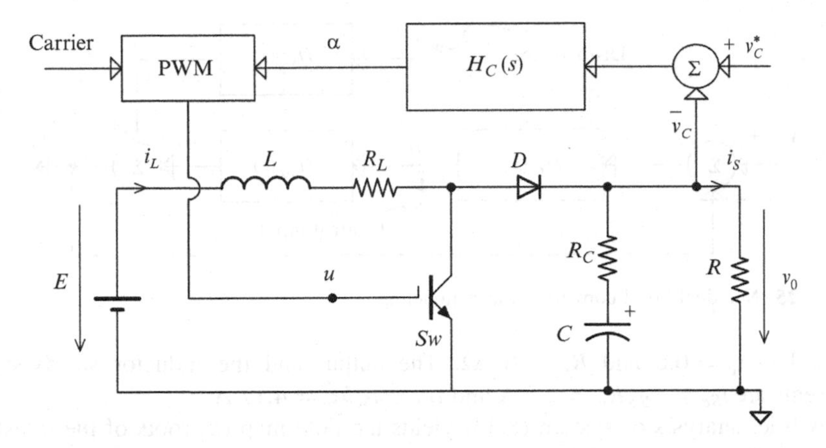

Fig. 8.24 Boost converter with losses: output voltage regulation loop

8.6 Case Studies

8.6.1 *Boost Converter Output Voltage Direct Control by Lead-lag Control*

Let us consider a nonideal boost power stage that supplies a variable resistive load. The rated power is 60 W, the input voltage is 15 V without significant variation. The inductor has $L = 0.5$ mH and $R_L = 0.1\ \Omega$; the capacitor has $C = 1000\ \mu$F, its equivalent series resistance being $R_C = 10\ \text{m}\Omega$. The PWM frequency is sufficiently high such as to allow the continuous-conduction mode (ccm) for the quasi-totality of the operating range. The goal is to design an output voltage lead-lag controller that regulates the circuit output voltage at 25 V.

The circuit closed-loop diagram is given in Fig. 8.24.

The system averaged model and its linearized version around a steady-state operating point have already been computed in Problem 4.3 from Chap. 4. To facilitate the discussion the boost circuit transfer functions are given again in relation (8.33) (input voltage E being considered constant).

$$\begin{cases} H_{v_0\alpha}(s) = \dfrac{R_e(CR_Cs+1)\left(v_{0e}\alpha_e' - i_{Le}R_L - i_{Le}Ls\right)}{CL(R_e+R_C)s^2 + \left[\alpha_e'^2R_eR_CC + CR_L(R_e+R_C) + L\right]s + \alpha_e'^2R_e + R_L}, \\ H_{v_0i_S}(s) = \dfrac{R_e(CR_Cs+1)(R_L+Ls)}{CL(R_e+R_C)s^2 + \left[\alpha_e'^2R_eR_CC + CR_L(R_e+R_C) + L\right]s + \alpha_e'^2R_e + R_L}. \end{cases} \tag{8.33}$$

Figure 8.25 presents the linearized plants' structure.

The operating point at which the control is designed is the point corresponding to the full load (i.e., 60 W). By considering that the control objective is achieved, the corresponding operating point is characterized by $E = 15$ V, $v_{0e} = 25$ V, $\alpha_e = 0.4$,

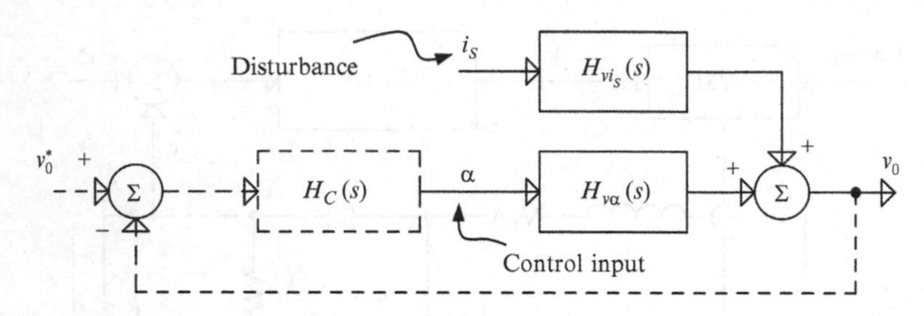

Fig. 8.25 Nonideal boost converter linear modeling

$\alpha_e^{'} = 1 - \alpha_e = 0.6$ and $R_e = 10\ \Omega$. The output and the inductor steady-state currents are $i_{Se} = v_{Se}/R_e = 2.5$ A and $i_{Le} = i_{Se}/\alpha_e^{'} = 4.17$ A.

A brief analysis of system (8.33) yields the pole map (as roots of the transfer function denominators). The time constant

$$T = \sqrt{\frac{CL(R_e + R_C)}{\alpha_e^{'2} R_e + R_L}} \tag{8.34}$$

and the damping coefficient

$$\zeta = \frac{1}{2T} \cdot \frac{\alpha_e^{'2} R_e R_C C + R_L(R_e + R_C)C + L}{\alpha_e^{'2} R_e + R_L} \tag{8.35}$$

computed at the chosen operating point lead to $T = 1.2$ ms and $\zeta = 0.18$, respectively.

By analyzing the control input channel (the first relation of (8.33)) one notes the presence of two zeros. One is situated in the left-half plane, at $-1/T_2$, with $T_2 = R_C C$, and the other one is in the right-half plane, at $1/T_1$, with

$$T_1 = \frac{i_{Le} L}{v_{0e}\alpha_e^{'} - i_{Le} R_L}.$$

Note that the latter zero induces a nonminimum-phase behavior, which becomes more significant (T_1 larger) for higher loads (high i_{Le}). At the chosen operating point one obtains $T_1 = 0.14$ ms and $T_2 = 10$ µs. Note that in this case the T_2 -related zero can practically be neglected in the control design. If the output capacitor is large (e.g., when an ultracapacitor is being used) this zero should be taken into account as its effect cannot be neglected any longer.

The DC gain over the control channel is given by

$$K_1 = R_e \frac{v_{0e}\alpha_e^{'} - i_{Le} R_L}{\alpha_e^{'2} R_e + R_L},$$

a relation which gives $K_1 = 39.4$ V for the selected operating point.

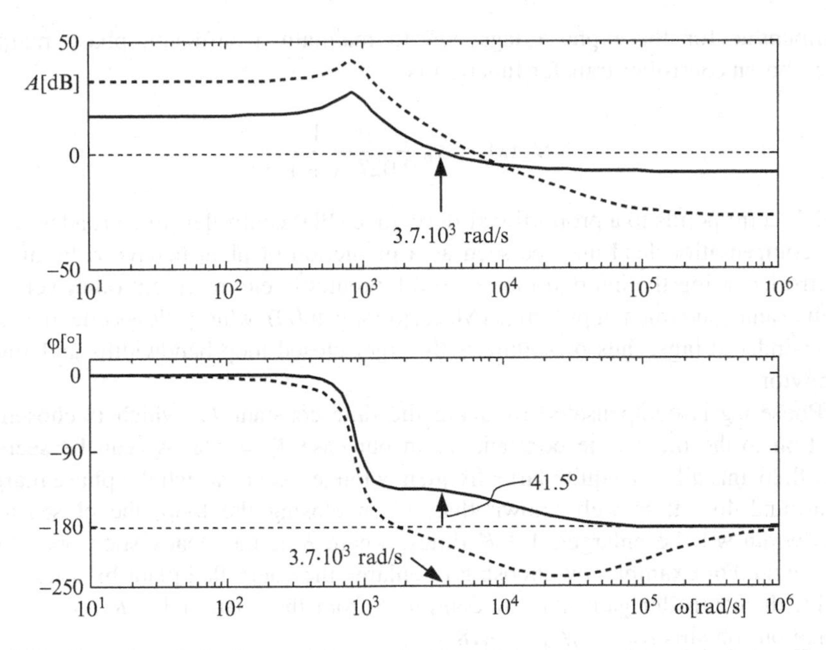

Fig. 8.26 Open-loop Bode characteristics: *dashed-line plots* correspond to original plant $H_{v\alpha}(s)$, and *solid-line plots* represent open-loop plant $H_C(s) \cdot H_{v\alpha}(s)$

Now, let us analyze the disturbance channel. There are two zeros placed in the left-half complex plane. The least important is described by the time constant $T_2 = 10$ μs and the other by $T_3 = L/R_L$, which gives $T_3 = 5$ ms. The DC gain of the disturbance channel is given by the relation

$$K_2 = \frac{R_e R_L}{\alpha_e'^2 R_e + R_L},$$

which gives $K_2 = 0.27\ \Omega$ for the selected operating point. One may easily note that the control channel gain is far larger that the disturbance channel gain. Obviously, this means that the disturbance effect over the output voltage will be compensated with a quite reduced control effort (duty ratio variation).

Now, following the remarks in Sect. 8.2, one may choose a PI controller that affects less the open-loop phase characteristic and offers a larger phase margin. Nevertheless, one should assume a nonzero steady-state output voltage error. Also, as the damping is very small, the phase characteristic evolves very abruptly towards –180°. Moreover, at higher frequencies the phase characteristic evolves towards –270° because of the presence of the right-half-plane zero. These aspects are reflected in the plant's Bode plots, represented in Fig. 8.26 by dashed lines. Therefore, a lead (derivative) component is necessary in the controller in order to compensate for these phase lags and to maintain a sufficient phase margin. The chosen controller transfer function is

$$H_C(s) = K_p \frac{T_d s + 1}{0.02 T_d s + 1},$$

which corresponds to a proportional-derivative (PD) controller. In a broader sense, the compensation lead may be seen as a prediction of plant behavior. Predictive controllers using the internal model, Smith predictor, etc. represent other versions of the same fundamental principle (Maciejowsky 2000), which allows one to cancel plant-induced lags, thus providing both larger closed-loop bandwidths and stable behavior.

Phase lag is compensated by using the time constant T_d, which is chosen in relation to the main time constant, T; in our case $T_d = T/2$. As can be seen in Fig. 8.26, this allows a quite large frequency range, one in which the phase margin is around 40°. It is well known that, when closing the loop, the closed-loop bandwidth will be enlarged 1 + K times, where K is the steady-state open-loop total gain. For example, by choosing to enlarge the controlled plant by a factor of eight, the controller gain may by computed from the relation $1 + K_p K_1 = 8$ and hence one obtains $K_p = 7/K_1 = 0.18$.

Provided with the above information, one can plot the open-loop Bode diagrams – i.e., of the transfer $H_C(s) \cdot H_{v_0\alpha}(s)$ – for the selected operating point (see the solid-line plots in Fig. 8.26).

Of course one may further increase the controller gain K_p, but beyond a certain value this will eventually lead to undesirable reduction in the phase margin. As the main objective is load variation rejection, one may want to analyze the closed-loop behavior of the plant having disturbance as input and output voltage as output (see Fig. 8.25).

Note that with respect to the original plant $H_{v_0 i_S}(s)$ (represented by dashed line in Fig. 8.27), the closed-loop steady-state gain (solid line) has been significantly reduced, namely from -11 to -29 dB. Also, the maximum gain has been reduced from +10 to around -10 dB, and this maximum has been displaced to a higher frequency. This means that, even if not zeroed, the output voltage error is made sufficiently small in order to fulfill reasonable requirements. Note also that the closed-loop damping has been significantly increased; therefore, the output voltage will vary without ringing.

In order to assess the closed-loop system time behavior, the nonlinear averaged model of the boost circuit together with its control has been implemented in MATLAB®-Simulink®. Due to nonzero steady-state error, the voltage reference must be a priori fixed to a higher value than the regulation target, $v_{0e} = 25$ V. For the present case study this value has been fixed at $v_0^* = 27.45$ V. The output voltage v_e will drop to v_{0e} and – as the following results will show – variations around this value as the load varies will be sufficiently small.

Figure 8.28a shows the closed-loop behavior at unit steps of load resistance around the operating point corresponding to the full load. The voltage variations

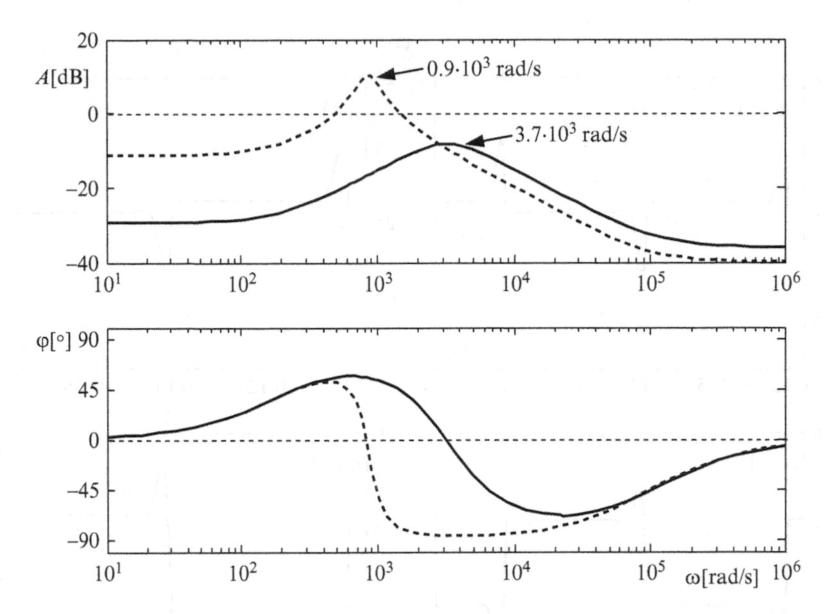

Fig. 8.27 Closed-loop Bode characteristics: original plant $H_{vi_S}(s)$ is drawn with *dashed line* and closed-loop plant is represented with *solid line*

are under 0.4 %. If this simulation is repeated at any other operating point, the variations will be even smaller. The voltage steady-state value depends slightly on the load resistance. Note that the duty ratio steady-state value slightly differs from the a priori stated value, α_e, because of losses present in the circuit model. Figure 8.28b shows the closed-loop behavior at large steps of load, from full load to 25 % load and vice versa. The output voltage variations are larger, at about 2 % and the response differs from the step-up to step-down case because the system is nonlinear.

The proposed controller has a quite straightforward analog implementation, which is depicted in Fig. 8.29.

8.6.2 Boost Converter Output Voltage Direct Control by Pole Placement

Let us consider the same boost power stage (with the same parameters) as the Case Study in Sect. 8.6.1. As it has been shown, the capacitor equivalent series resistor induces an influence at very high frequencies; it will therefore be neglected in this case study.

Knowing that the maximum output voltage is $v_{C\max} = 25$ V, it is required that the output voltage tracks – with zero steady-state error – the desired reference. The control structure to be employed is the one introduced in Sect. 8.4 and is given in Fig. 8.30.

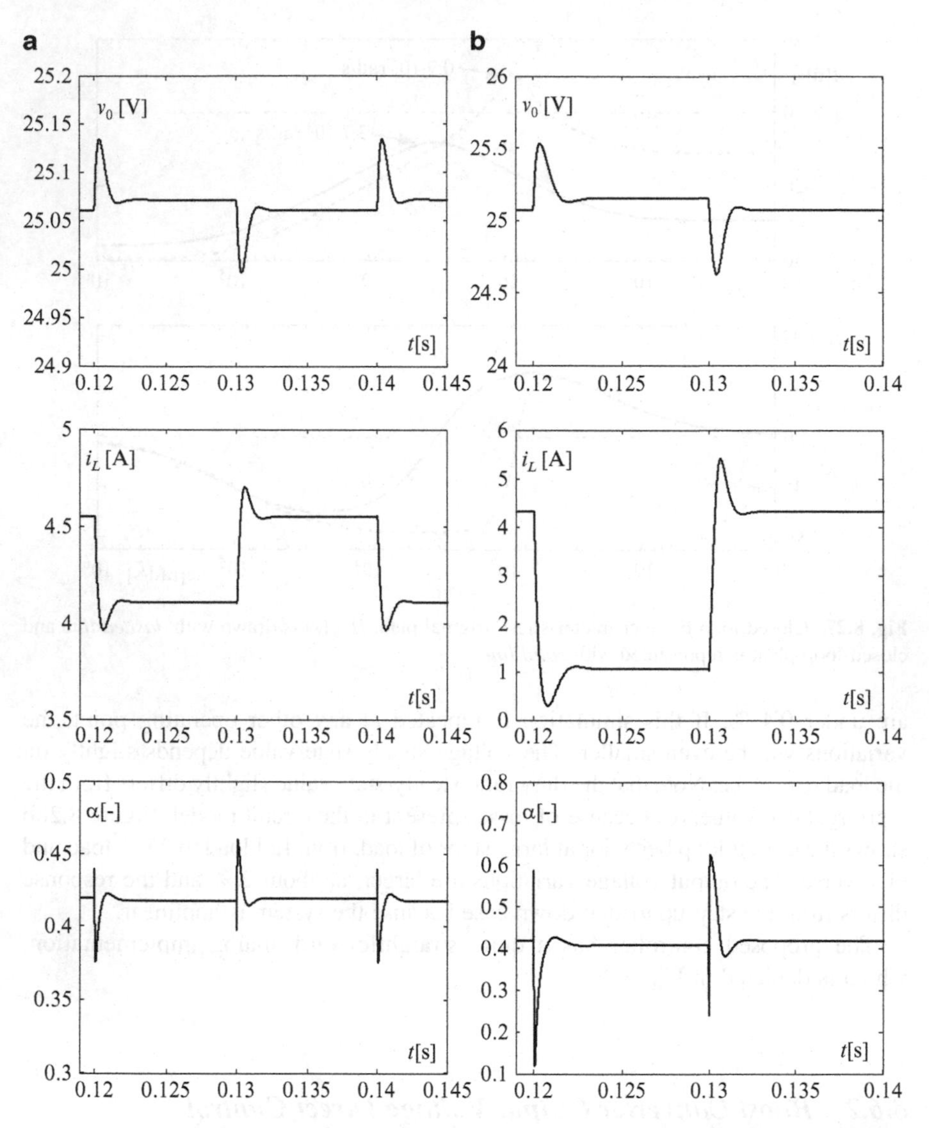

Fig. 8.28 MATLAB®-Simulink® results of a lead-lag-controlled nonideal boost converter: (**a**) behavior at a load resistance unit step variation around the selected operating point ($R_e = 10\ \Omega$); (**b**) behavior at a load variation from full load to 25 % load and backwards

As in the previous case study, the same full-load (60 W) operating point is chosen as steady state for design purposes. By considering the state vector $\mathbf{x} = \begin{bmatrix} \widetilde{i_L} & \widetilde{v_C} \end{bmatrix}^T$, and the input vector $u = \begin{bmatrix} \widetilde{\alpha} & \widetilde{i_S} \end{bmatrix}$, the circuit's linearized model in variations around the selected operating point is given by the relation $\dot{x} = \mathbf{Ax} + \mathbf{Bu}$, where

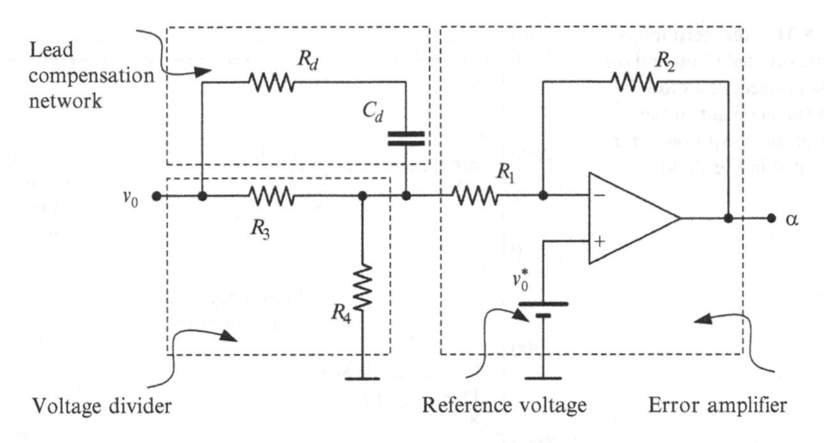

Fig. 8.29 Voltage controller diagram: the compensation network has $T_d = R_d C_d$ and the error amplifier outputs the duty ratio that will further be subject to pulse-width modulation

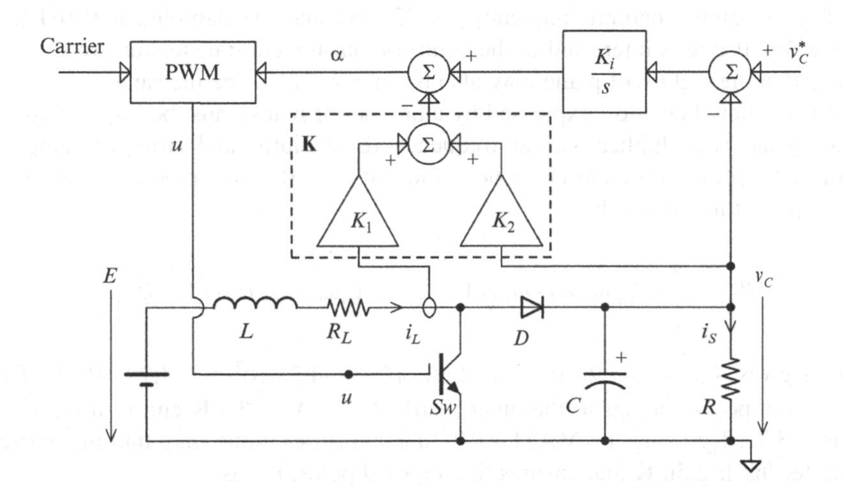

Fig. 8.30 Boost converter control using compensation by state feedback

$$\mathbf{A} = \begin{bmatrix} -\frac{R_L}{L} & -\frac{\alpha_e'}{L} \\ \frac{\alpha_e'}{C} & -\frac{1}{R_e C} \end{bmatrix} \text{ and } \mathbf{B} = \begin{bmatrix} \frac{v_{Ce}}{L} \\ -\frac{i_{Le}}{C} \end{bmatrix}. \tag{8.36}$$

One may compute the plant poles either by computing the roots of the equation $\det(s\mathbf{I}_2 - \mathbf{A}) = 0$ or by using dedicated CAD software (e.g., `eig` function in MATLAB®). For the selected steady-state operating point, this gives

$$p_{1,2} = -200 \pm j \cdot 1200.$$

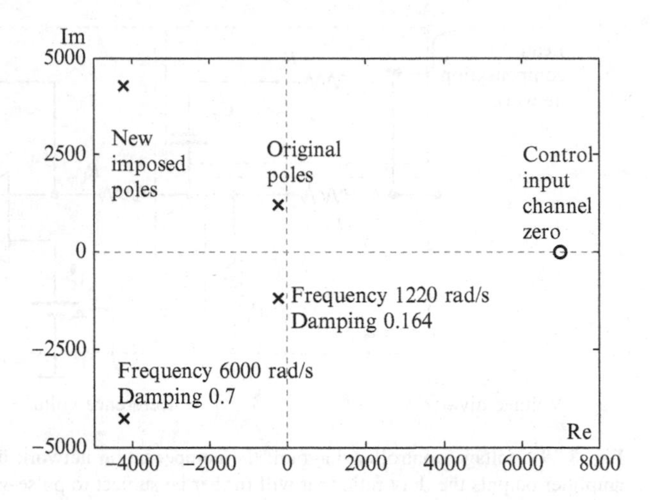

Fig. 8.31 Pole-zero maps for the original plant and for the state-feedback closed-loop (*inner*) plant in the case of the boost converter presented in Fig. 8.30

The associated natural frequency is 1220 rad and the damping is 0.164 (see Fig. 8.31). If one is interested in the control-input-related transfer function, a zero situated in the right-half plane may also be emphasized – see the same figure.

Now, a new behavior, expressed by a new set of poles, must be imposed to the inner plant, with higher natural frequency (bandwidth) and stronger damping. If these latter are chosen as $\omega_n = 6000$ rad and $\zeta_n = 0.7$, the imposed closed-loop set of poles turns out to be

$$P_n = \left[-\zeta_n\omega_n + j \cdot \omega_n\sqrt{1-\zeta_n^2} \quad -\zeta_n\omega_n - j \cdot \omega_n\sqrt{1-\zeta_n^2} \right],$$

which gives $P_n = \left[-4.2 \cdot 10^3 + j \cdot 4.28 \cdot 10^3 \quad -4.2 \cdot 10^3 + j \cdot 4.28 \cdot 10^3 \right]$. This new set of poles belongs to the inner matrix $\mathbf{A}_{0i} = \mathbf{A} - \mathbf{B} \cdot \mathbf{K}$ and is also plotted in Fig. 8.31. By using the MATLAB® function `acker` one may find the desired state feedback gain $\mathbf{K}$ that ensures the imposed poles, P_n, as

$$\mathbf{K} = \left[K_1 \quad K_2 \right] = \left[0.24 \quad 0.51 \right].$$

The inner plant transfer function may be computed as

$$H_{0i}(s) = \mathbf{C}(s\mathbf{I}_2 - \mathbf{A}_{0i})^{-1}\mathbf{B},$$

with output matrix $\mathbf{C} = [0 \quad 1]$ (emphasis on output voltage v_C). This gives

$$H_{0i}(s) = 1.62 \cdot \frac{-1.43 \cdot 10^{-3} \cdot s + 1}{2.8 \cdot 10^{-8} \cdot s^2 + 2.3 \cdot 10^{-4} \cdot s + 1}.$$

In regards to the same Fig. 8.31 that depicts the $H_{0i}(s)$ pole-zero map, one may observe that the right-half-plane zero of the plant has not changed.

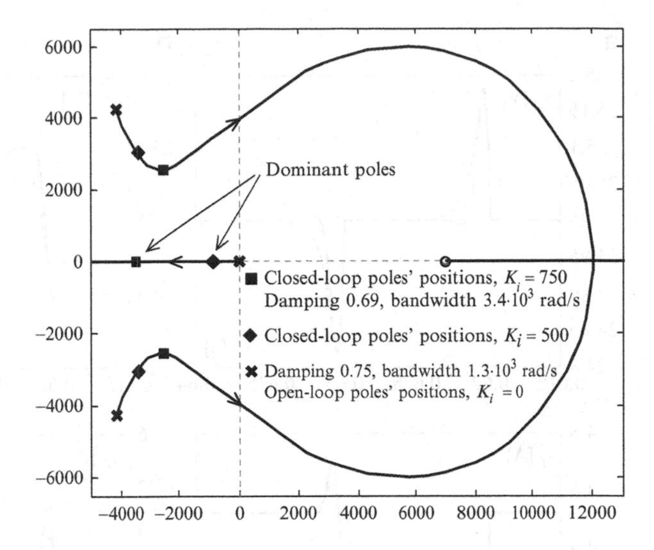

Fig. 8.32 Root locus for system described by (8.37) and position of poles for different values of controller gain K_i

Now, as analyzed before (see Sect. 8.4), the open-loop outer plant has the transfer function

$$H_{co}(s) = K_i \cdot \frac{H_{0i}(s)}{s}, \tag{8.37}$$

where the controller gain, K_i, still needs to be chosen. To this end, the root locus method is used, as the system (8.37) is of third order. Figure 8.32 shows the root locus of the system (8.37). This controller gain is chosen such that the real pole to remain the dominant one (i.e., the complex poles' natural frequency to not decrease below the real pole frequency) and the closed-loop damping remain at reasonable values. The poles' positions for two such values of K_i may be seen on Fig. 8.32.

In order to assess the closed-loop system time behavior, the nonlinear averaged model of the boost circuit together with its control has been implemented in MATLAB®-Simulink®. The evolution of the main variables in the regulation mode (load step variations at fixed output voltage reference) are given in Fig. 8.33a, while those concerning the tracking mode (reference step variations at fixed load) are shown in Fig. 8.33b. The dashed plots are those corresponding to the controller gain $K_i = 500$ and the solid ones correspond to gain $K_i = 750$ (see Fig. 8.32 for the corresponding closed-loop bandwidth and damping coefficients).

Note that the voltage error is nullified in 1–3 ms and its steady-state value is zero. In regulation mode, its maximum value is about 0.7 %. Note also that the nonminimum-phase behavior has not been cancelled (the right-half-plane zero influence is still visible – see Fig. 8.33b).

If the closed-loop dynamic response is not satisfactory, one may reiterate the problem solving by imposing another set of poles for the inner dynamic (at higher natural frequency) and choosing higher outer controller gain K_i in order to ensure even higher bandwidth with satisfactory damping.

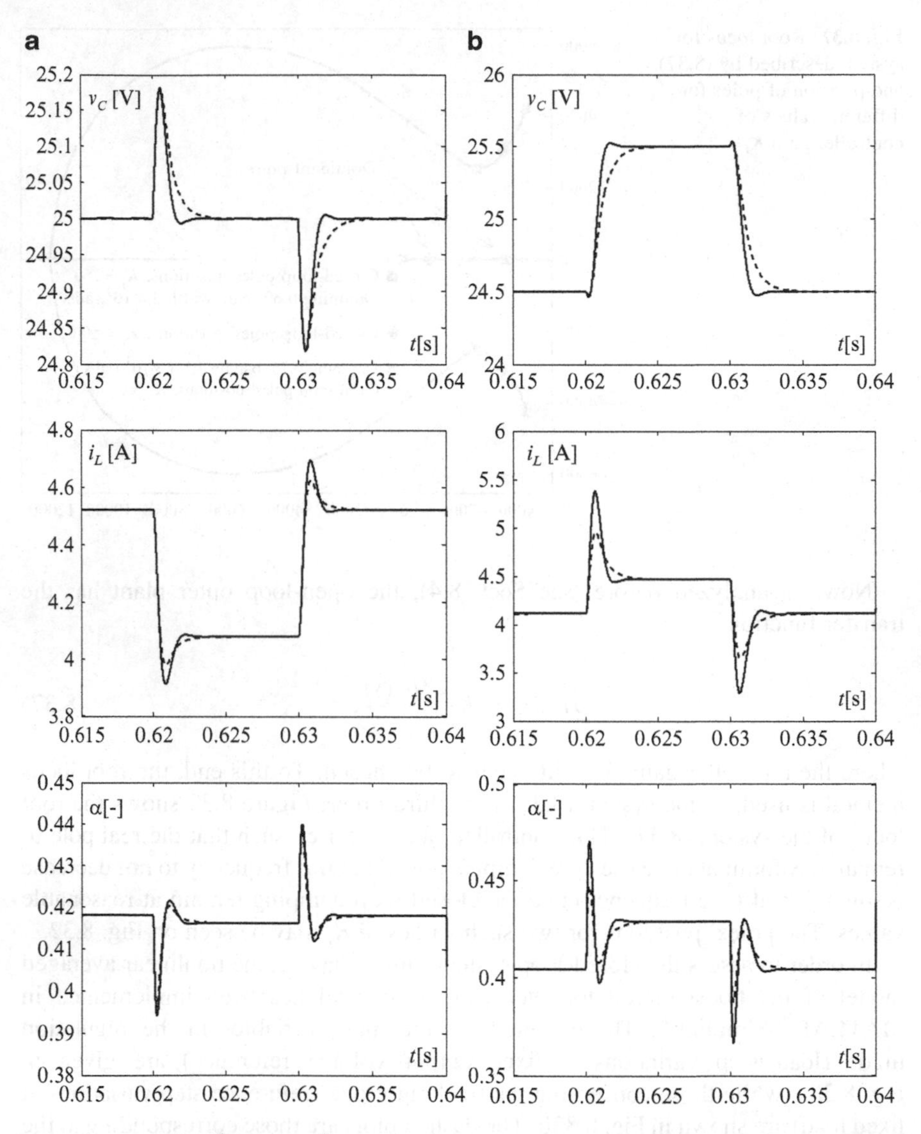

Fig. 8.33 MATLAB®-Simulink® results concerning closed-loop behavior of boost converter controlled by pole-placement compensation (see Fig. 8.30): (**a**) regulation mode – unit step load variations; (**b**) tracking mode – unit step voltage reference variations

In operating points corresponding to smaller loads, the nonminimum-phase effect and voltage error peak values will be even smaller. However, more tests must be done for assessing circuit operation in dcm, as the plant model changes significantly (see Chap. 6).

Also, depending on circuit power, the inductor current must be limited because, depending on inner dynamics tuning, current peaks in dynamic regime may reach dangerous values. This may be done by significantly reducing state feedback action when inductor current reaches its maximum limit (Matavelli et al. 1993).

8.7 Conclusion

This chapter has presented some linear control design techniques used for DC-DC power electronic converters. These approaches assume that converter inputs are pulse-width modulated and the converters operate in ccm. As a general remark, most DC-DC converters exhibit second-order weakly damped dominant dynamic, whereas a particular class – converters having boost capability – also exhibits nonminimum-phase behavior. The simultaneous presence of these two phenomena may induce significant stability problems; this is why they are not easily dealt with by classical control design. The three methods analyzed in this chapter are based upon somehow more elaborate algorithms, very useful very useful for situations where converters have two dynamic states and the goal is to control both the value of the output voltage and its dynamics.

Output voltage direct control is a method that relies on open-loop loop shaping and aims primarily at ensuring sufficient stability (phase) margin; meanwhile it tries to guarantee satisfactory dynamic performance. The pole-placement method allows both stability and dynamic performance to be ensured by means of an inner state feedback, embedded within a regulation outer loop. In this way, the strongly oscillatory pole pair giving the dominant dynamic may be relocated at convenient damping and bandwidth. While the latter two methods do not explicitly allow the inductor current to be controlled, a third method does so, relying upon a two-loop control structure. In this case, the basic assumption – of clearly separable dynamics within the two nested loops – may sometimes be fulfilled at the border, especially in boost-type power converters which exhibit nonminimum-phase behavior.

Linear controllers are designed for a certain operating point, and the system is variant with the load and the input supply; therefore the control design must comply with certain robustness requirements that ensure acceptable performance over the entire operating range and also in dcm. In the case when a single controller cannot yield satisfactory behavior, gain scheduling techniques and linear parameter varying approach may be successfully used (Shamma and Athans 1990).

These control techniques may also be successfully employed in converters with AC stages. For example, the cascade control structure is quite likely to be used, since a natural separation of the dynamics within these converters may be emphasized: the inner loop is built over the AC filter and the outer loop over the DC filter.

Based upon converter linearized averaged models, state observers may be built (Leung et al. 1991; Friedland 2011; Rytkonen and Tymerski 2012) in order to estimate the variations of some nonmeasurable or difficult-to-measure states, especially in the case of high-order converters. Design of such observers is especially time critical, considering the very fast dynamics of internal variables.

Problems

Problem 8.1 Cascaded-loop control of Watkins–Johnson converter

The converter in Fig. 8.34 has input supply voltage $E = 24$ V (constant), inductance $L = 0.33$ mH, parasitic series resistance $R_L = 0.1\ \Omega$ and output capacitor $C = 2700\ \mu$F. This converter supplies a mainly resistive load R and has a rated power of 350 W. The control input is PWM and the converter is supposed to operate in ccm. The output voltage must follow a variable reference, which may vary between $-3E$ and $+E$. To this end, design a cascaded control structure like the one detailed in Sect. 8.3. The following points must be addressed:

(a) obtain the converter's averaged model and study its steady-state behavior;
(b) obtain the linearized model of the inner current plant around a conveniently chosen operating point and compute the parameters of a PI controller for this plant starting from an imposed set of performances (bandwidth and damping);
(c) obtain the linearized model of the outer voltage plant around the same operating point, then compute the parameters of a PI controller for this plant based upon some imposed performance that ensures clear separation of dynamics within the two control loops;
(d) discuss how the plant's parameters vary when the operating point changes and propose solutions for dealing with this.

Solution

(a) The averaged model of the converter is

$$\begin{cases} L\dot{i_L} = (2\alpha - 1)E - \alpha v_C - R_L i_L \\ C\dot{v_C} = \alpha i_L - v_C/R. \end{cases} \tag{8.38}$$

In steady state (current derivative zeroed in the first equation of (8.38)) and by neglecting inductor losses (i.e., $R_L = 0$), one obtains that

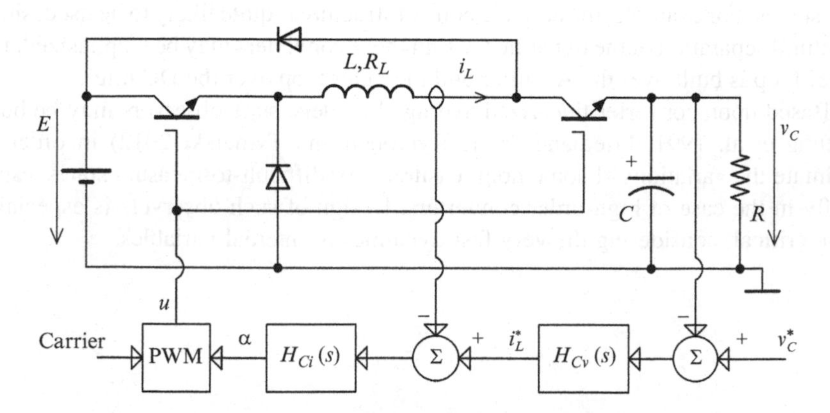

Fig. 8.34 Two-loop cascaded control of a Watkins–Johnson converter

$$v_{Ce} = E \cdot (2\alpha_e - 1)/\alpha_e,$$

a relation that describes the steady-state behavior and shows that the output voltage varies within the $(-\infty, E]$, being zero for $\alpha_e = 0.5$.

(b) Supposing that the control loop renders the voltage v_C slowly variable, this latter may be considered constant (at its reference v_C^*) from the current variation point of view. Therefore, the first equation of (8.38) becomes

$$L\frac{di_L}{dt} = (2\alpha - 1)E - \alpha v_C^* - R_L i_L.$$

By expressing the latter equation in variations around a quiescent operating point, one obtains the transfer function of the inner plant:

$$H_{i_L\alpha}(s) = \frac{\tilde{i}_L(s)}{\tilde{\alpha}(s)} = \frac{2E - v_C^*}{R_L} \cdot \frac{1}{L/R_L \cdot s + 1}. \tag{8.39}$$

By employing a PI controller one obtains a closed-loop inductor current behavior described by a transfer function like the one in (8.18), where $K_C = (2E - v_C^*)/R_L$ and $T_C = L/R_L = 3.3$ ms. If it is supposed that $v_C^* \in [-2E, E]$, the plant gain varies between E/R_L and $4E/R_L$ (according to relation (8.39)). The design operating point is the one corresponding to the smallest plant gain while at full load: $v_{Ce} \equiv v_C^* = -2E = -48$ V, $\alpha_e = 0.25$ (inductor losses neglected), $K_C = E/R_L = 240$ A, $R_e = 6.6\ \Omega$ and $i_{Le} = -29$ A.

One imposes the inner closed-loop performances: $T_{0C} = T_C/5 = 0.66$ ms and a damping of 0.8. Equations (8.19) give controller parameters values $T_{iC} = 0.9$ ms and $K_{pC} = 0.007\ \text{A}^{-1}$. Note that, as the output voltage reference changes, from $-2E$ to E, the steady-state gain K_C decreases by a factor of four and the closed-loop bandwidth decreases by a factor of two. A prefilter with the time constant T_{iC} is inserted on the current reference in order to compensate the inner loop zero.

(c) Now, concerning the outer voltage loop, by zeroing the dynamic in the first equation of (8.38) one obtains ($v_{Ce} \equiv v_C^*$)

$$\alpha = \frac{E + R_L i_L}{2E - v_C^*},$$

which is replaced in the second equation of (8.38), leading after linearization to

$$\tilde{v}_c(s) = \frac{(E + 2R_L i_{Le})R_e}{2E - v_C^*} \cdot \frac{1}{CR_e s + 1} \cdot \tilde{i}_L - \frac{R_e}{CR_e s + 1} \cdot \tilde{i}_S,$$

where $\tilde{i}_S = -v_C^*/R_e^2 \cdot \tilde{R}$ is load current variation due to load variation $\tilde{R}$. As a consequence, the transfer function corresponding to the inductor-current-to-output-voltage channel is a first-order one

$$H_{vi_L}(s) = \frac{(E + 2R_L i_{Le})R_e}{2E - v_C^*} \cdot \frac{1}{CR_e s + 1},$$

whose parameters – gain and time constant – depend on the operating point.

By using the same procedure as done for the inner loop, one may design the outer loop PI controller – see relations (8.20). In this case $K_V = \frac{(E+2R_L i_{Le})R_e}{2E-v_C^*}$ and $T_V = R_e C$. For the same operating point as in the inner loop, one obtains $K_V = 1.45$ V/A and $T_V = 17.8$ ms. One imposes the outer closed-loop performances: $T_{0V} = T_V/5 = 3.6$ ms and a damping of 0.8. Equations (8.20) give $K_{pV} = 4.83$ A/V and $T_{iV} = 5$ ms. Note that the ratio between the inner loop bandwidth and that of the outer loop is almost 5.4, which is satisfactory because it ensures the dynamics within the two nested loops are clearly separable. The voltage reference can be passed through a prefilter with time constant in order to compensate for the outer loop zero.

At this time the control design may be considered complete for the chosen operating point and one may proceed to closed-loop simulations.

(d) As regards change of steady-state operating point, this induces some degradation of the control performance. In our case variations of both load resistance and voltage reference represent causes of such degradation. Relations (8.18) and (8.19) show that for constant output voltage reference, closed-loop damping is slightly reduced for different load resistances R_e with respect to the imposed damping (0.8). Even more severe is the influence of the voltage reference change from $-2E$ to E. The outer plant steady-state gain K_V increases by a factor of four and the outer closed-loop bandwidth increases by two. Meanwhile, the inner closed-loop bandwidth is halved. This leads to the ratio between inner and outer bandwidth being reduced to only 1.35, which is not sufficient to guarantee clear dynamics separation within the two nested loops. Obviously, this will lead to unwanted global behavior.

To avoid this kind of problem, one solution is to adopt an adaptive control that modifies in real time the controllers' proportional gains, for example, by using a lookup table. Alternatively, one may use the following adaptation relations that alter the nominal proportional gain values:

$$K_{pC}^{new} = K_{pC}\frac{4E}{2E - v_C^f}, \quad K_{pV}^{new} = K_{pV}\frac{2E - v_C^f}{4E},$$

where K_{pC} and K_{pV} correspond to the precomputed values (in the chosen design operating point) and v_C^f is the quiescent output voltage and may be obtained either by suitably filtering the capacitor voltage or by filtering the voltage reference (e.g., one may use the voltage prefilter output).

The reader is invited to solve the following problems.

Problem 8.2 Buck DC-DC Converter Output Voltage Tracking

A buck converter (see Fig. 2.10 in Chap. 2) having $L = 0.5$ mH, $C = 680$ μF, $E = 72$ V and a rated load value $R = 10$ Ω must output a controlled voltage between 15 and 48 V. The minimum load value that still corresponds to ccm is considered to be 1000 Ω.

It is required to design a voltage-mode (direct) control structure based upon a proportional-integral controller that ensures the largest bandwidth possible and a damping coefficient between 0.65 and 0.85, all by maintaining a phase margin of 30° for the entire output voltage operating range.

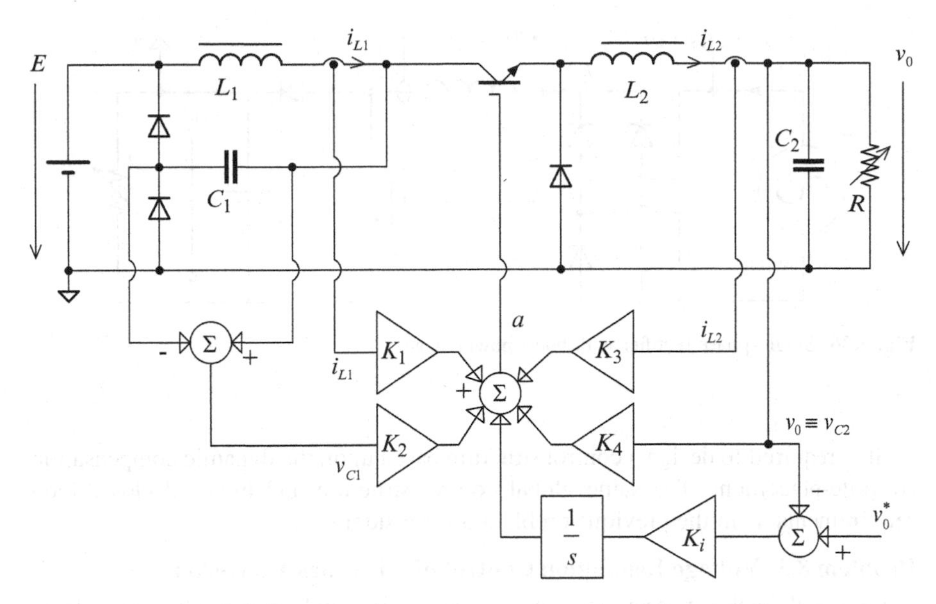

Fig. 8.35 Control structure of the quadratic buck DC-DC converter; PWM and gate driver omitted

The controller must be implemented using a digital signal processor having a sampling rate of 20 kHz. Obtain the discrete transfer function of the controller and its associated difference equation.

Problem 8.3 Quadratic Buck DC-DC Converter Control Using Dynamic Compensation by Pole-placement

Figure 8.35 shows a quadratic buck power stage (also presented in Fig. 4.38 at the end of Chap. 4).

The converter has the following circuit parameters: $L_1 = 0.5$ mH, $L_2 = 0.5$ mH, $C_1 = 470$ μF, $C_2 = 1000$ μF, $E = 100$ V and a rated load value $R = 5$ Ω. The control scope is to maintain a desired constant output voltage, $v_{C2}^* = 15$ V.

For this four-state converter, design a control structure based upon the dynamic compensation by pole-placement. Imposed control performance for the inner loop includes placement of the closed-loop dominant second-order dynamic at a bandwidth five times larger than the open-loop one and suitable damping coefficients (e.g., 0.8) for both second-order dynamics. Further, the integral gain in the outer regulation loop will be chosen by using the root locus method so as to ensure the maximal closed-loop bandwidth.

Problem 8.4 Ćuk Converter using Dynamic Compensation by Pole-placement

For the Ćuk converter presented in Fig. 3.17 at the end of Chap. 3, desired constant output voltage $v_{C2}^* = 50$V is aimed at. The circuit parameters are: $L_1 = 0.5$ mH, $L_2 = 0.5$ mH, $C_1 = 220$ μF, $C_2 = 1000$ μF, $E = 100$ V and a rated load value $R = 2$ Ω.

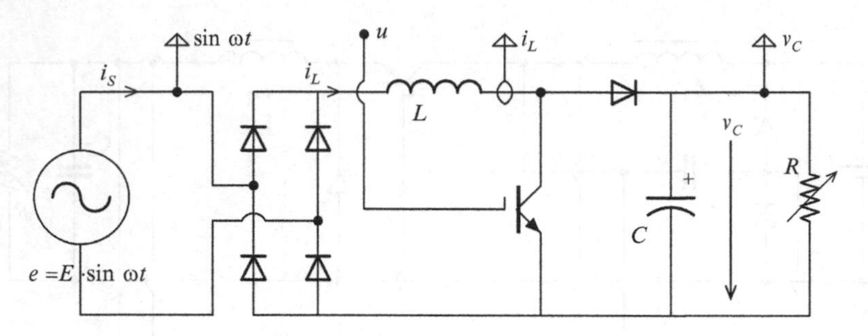

Fig. 8.36 Single-phase rectifier with boost power stage

It is required to design a control structure based upon the dynamic compensation by pole-placement. The same global control structure and imposed closed-loop requirements as in the previous problem are considered.

Problem 8.5 Voltage Regulation Control of a Flyback Converter

A lossless flyback converter operating in continuous-conduction mode is considered (see Fig. 4.24 of Chap. 4), which must output constant voltage $v_C^* = 3.3$ V on a constant load $R = 10\,\Omega$, while its supply voltage varies between $0.5v_C^*$ and $1.5v_C^*$. The circuit parameters are $L = 0.22$ mH and $C = 2200$ μF. Address the following points:

(a) Design a two-loop control structure (averaged current-mode control) that ensures a 3-kHz-bandwidth of the voltage closed loop for the entire operating range.
(b) Discretize the controllers obtained in part a and compute their digital counterparts for the sampling frequency $f_s = 40$ kHz.

Problem 8.6 Power Factor Correction Using Boost DC-DC Converter Control

Let us consider the case of a resistive load R supplied by means of a diode voltage rectifier and a boost configuration, as presented in Fig. 8.36. Provided that parameters L, C and E are known and measures of inductor current i_L and output voltage v_C are available, the global control goal is to design a control structure ensuring that the load voltage v_C is regulated at the reference value v_C^*, meanwhile reducing the harmonic content of the current absorbed from the grid i_S and ensuring operation at unit power factor.

The design will be based upon a cascaded control structure having averaged current i_L as inner variable and averaged voltage v_C as outer variable (Rossetto et al. 1994). Figure 8.37 presents this control structure.

Address the following issues:

(a) explain how current reference i_L^* is computed (where I_S is the amplitude of grid current i_S);
(b) obtain the inner plant linearized model and choose a suitable controller structure;
(c) knowing that the PWM switching frequency is $f = 10$ kHz and $\omega = 2\pi \cdot 50$ rad/s, suggest a pertinent way of imposing the inner closed-loop bandwidth, then compute the inner-loop controller;

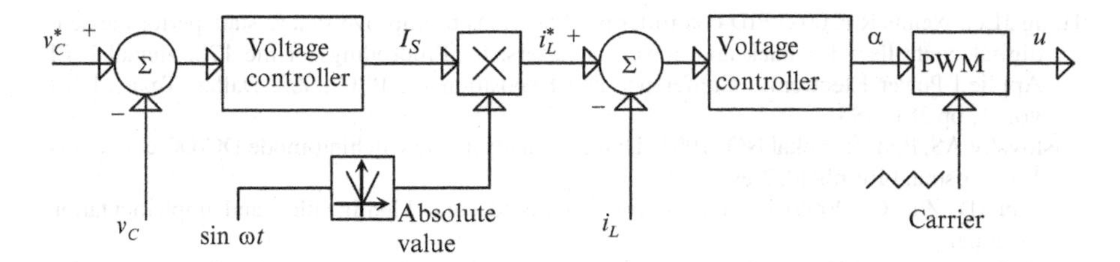

Fig. 8.37 Cascaded control structure ensuring unit-power-factor operation of the circuit in Fig. 8.36

(d) obtain the outer plant linearized model and choose a suitable controller structure;
(e) by requiring that the outer closed-loop bandwidth be one-tenth that of the inner loop, compute the outer-loop controller;
(f) build and simulate in MATLAB®-Simulink® the block diagram corresponding to the nonlinear averaged model of the uncontrolled circuit for the following set of parameters: $L = 2$ mH, $C = 1000$ μF, $E = 300$ V and load R taking values in the interval [100 Ω, 10 kΩ];
(g) by freezing the voltage v_C at its reference value $v_C^* = 500$ V and choosing the full-load operating point ($R = 100$ Ω) build and simulate the inner control loop – with the controller computed in part c – then validate the imposed dynamical performance;
(h) build the outer-loop block diagram – with the controller computed in part e, simulate its behavior and validate the closed-loop performance for the full-load operating point;
(i) move the entire controlled system around another operating point, e.g., $R = 10$kΩ, without modifying the previously computed controllers and simulate its behavior, then assess the control performance degradation;
(j) propose an adaptation law for the outer-loop controller.

References

Algreer M, Armstrong M, Giaouris D (2011) Adaptive control of a switch mode DC-DC power converter using a recursive FIR predictor. IEEE Trans Ind Appl 74(5):2135–2144

Arikatla V, Qahouq JAA (2011) DC-DC power converter with digital PID controller. In: Proceedings of the 26th annual IEEE Applied Power Electronics Conference and Exposition – APEC 2011. Fort Worth, Texas, USA, pp 327–330

Aström KJ, Hägglund T (1995) PID controllers: theory, design and tuning, 2nd edn. Instrument Society of America, Research Triangle Park

Ceangă E, Protin L, Nichita C, Cutululis NA (2001) Theory of control systems (in French: Théorie de la commande des systèmes). Technical Publishing House, Bucharest

d'Azzo JJ, Houpis CH, Sheldon SN (2003) Linear control system analysis and design with MATLAB, 5th edn. Marcel-Dekker, New York

Dixon L (1991) Average current mode control of switching power supplies. Unitrode power supply design seminar manual, pp C1-1–C1-14. Available at http://www.ti.com/lit/ml/slup075/slup075.pdf. Sept 2013

Erikson RW, Maksimović D (2001) Fundamentals of power electronics, 2nd edn. Kluwer, Dordrecht

Friedland B (2011) Observers. In: Levine WS (ed) The control handbook–control system advanced methods. CRC Press, Taylor & Francis Group, Boca Raton, pp 15-1–15-23

Hung JLG, Nelms R (2002) PID controller modifications to improve steady-state performance of digital controllers for buck and boost converters. In: Proceedings of the 17th annual IEEE Applied Power Electronics Conference and Exposition – APEC 2002. Dallas, Texas, USA, vol. 1, pp 381–388

Kislovsky AS, Redl R, Sokal NO (1991) Dynamic analysis of switching-mode DC/DC converters. Van Nostrand Reinhold, New York

Landau ID, Zito G (2005) Digital control systems: design. Identification and Implementation, Springer

Lee FC, Mahmoud MF, Yu Y (1980) Application handbook for a standardized control module (SCM) for DC to DC converters, vol I, NASA report NAS-3-20102. TRW Defense and Space Systems Group, Redondo Beach

Leung FHF, Tam PKS, Li CK (1991) The control of switching DC-DC converters – a general LQR problem. IEEE Trans Ind Electron 38(1):65–71

Levine WS (ed) (2011) The control handbook—control system advanced methods. CRC Press, Taylor & Francis Group, Boca Raton

Maciejowsky JM (2000) Predictive control with constraints. Prentice-Hall, Upper Saddle River

Matavelli P, Rossetto L, Spiazzi G, Tenti P (1993) General-purpose sliding-mode controller for DC/DC converter applications. In: Proceedings of the 24th annual IEEE Power Electronics Specialists Conference – PESC 1993. Seatle, Washington, USA, pp 609–615

Mathworks (2012) Control system toolbox – For use with MATLAB®. User's guide version 4.2. Available at http://www.mathworks.fr/fr/help/control/getstart/root-locus-design.html. Sept 2013

Mitchell DM (1988) Switching regulator analysis. McGraw-Hill, New York

Morroni J, Corradini L, Zane R, Maksimović D (2009) Adaptive tuning of switched-mode power supplies operating in discontinuous and continuous conduction modes. IEEE Trans Power Electron 24(11):2603–2611

Peng H, Maksimović D (2005) Digital current-mode controller for DC-DC converters. In: Proceedings of the 20th annual IEEE Applied Power Electronics Conference and Exposition – APEC 2005. Austin, Texas, USA, pp 899–905

Philips NJL, Francois GE (1981) Necessary and sufficient conditions for the stability of buck-type switched-mode power supplies. IEEE Trans Ind Electron Control Instrum 28(3):229–234

Prodić A, Maksimović D (2000) Digital PWM controller and current estimator for a low power switching converter. In: Proceedings of the 7th workshop on Computers in Power Electronics – COMPEL 2000. Blacksburg, Virginia, USA, pp 123–128

Rossetto L, Spiazzi G, Tenti P (1994) Control techniques for power factor correction converters. In: Proceedings of Power Electronics, Motion and Control – PEMC 1994. Warsaw, Poland, pp 1310–1318

Rytkonen F, Tymerski R (2012) Modern control regulator design for DC-DC converters. Course notes. Available at http://web.cecs.pdx.edu/~tymerski/ece451/Cuk_Control.pdf. Sept 2013

Santina M, Stubberud AR (2011) Discrete-time equivalents of continuous-time systems. In: Levine WS (ed) The control handbook—control system fundamentals. CRC Press, Taylor & Francis Group, Boca Raton, pp 12-1–12-34

Shamma JS, Athans M (1990) Analysis of gain scheduled control for nonlinear plants. IEEE Trans Autom Control 35(8):898–907

Sobel KM, Shapiro EY, Andry AN Jr (2011) Eigenstructure assignment. In: Levine WS (ed) The control handbook—control system advanced methods. CRC Press, Taylor & Francis Group, Boca Raton, pp 16-1–16-20

Stefani RT (1996) Time response of linear time-invariant systems. In: Levine WS (ed) The control handbook. CRC Press, Boca Raton, pp 115–121

Tan L (2007) Digital signal processing: fundamentals and applications. Academic, San Diego

Verghese GC, Ilic-Spong M, Lang JH (1986) Modeling and control challenges in power Electronics. In: Proceedings of the 25th IEEE Conference on Decision and Control – CDC 1986. Athens, Greece, vol. 25, pp 39–45

Wen Y (2007) Modelling and digital control of high frequency DC-DC power converters. Ph.D. thesis, University of Central Florida

Chapter 9
Linear Control Approaches for DC-AC and AC-DC Power Converters

This chapter addresses linear techniques for controlling the average (low-frequency) behavior of power electronic converters that include not only DC but also AC power stages. Without being exhaustive, the chapter focuses on the most-used control methods applied to somewhat popular single-phase and three-phase PWM- or full-wave-operated power electronic converters (mostly voltage source inverters). This chapter deals with control structures designed using both the rotating frame – relying upon *dq* models developed in Chap. 5 – and the natural AC stationary frame for different grid-connected and stand-alone applications. A case study, a number of examples and a set of problems related to comprehensive applications are also included.

9.1 Introductory Issues

The main control goal of AC-based converters depends on the converter role, specifying how the power is to be conveyed to/from the AC grid. For example, if an AC isolated load needs to be supplied, the output voltage amplitude and frequency must be regulated. On the contrary, if the converter conveys power between DC and AC sources, the DC-side voltage regulation may be the main control goal (see, for example, Sect. 8.3.2 from Chap. 8). As in Chap. 8, all these major goals include specifications that suppose the suitable changing of the converter dynamic behavior in terms of steady-state error, bandwidth and damping.

Generally speaking, converter models should be developed according to the stated control goal. As it has been explained in detail in Chap. 5, generalized averaged model (GAM) provides a mean of obtaining large-signal converter models that capture their most significant dynamic features. Further, these models (see, for example, Eq. (5.63) or (5.81)) may be linearized around a typical operating point in order to allow the application of linear control methods.

Using phase-locked-loop (PLL)-based synchronization with the AC grid, the *dq* modeling derived from the GAM performs an amplitude demodulation that

S. Bacha et al., *Power Electronic Converters Modeling and Control: with Case Studies*, Advanced Textbooks in Control and Signal Processing, DOI 10.1007/978-1-4471-5478-5_9,

transforms the original AC variables into DC variables. This makes possible the application of the methods used for DC-DC converters (already presented in Chap. 8) to converters that have AC stages. Also, this allows the separate control of active and reactive power components exchanged with the AC grid.

As in the previous chapter, these methods – such as loop shaping or pole placement – may be applied within a suitable control paradigm. The specific structure of these converters including both DC and AC stages with clearly separable dynamics makes useful the two-loop cascaded control structure presented in Sect. 8.3 of Chap. 8. In this context, the main control goal refers to the slower DC variables and will affect the outer control loop, whereas the inner control loop deals with the faster AC variables.

To conclude, in the case where the rotating *dq* frame is used, the control law (or a significant part of it) is applied to the *dq* DC variables, and supplementary transforms (amplitude demodulation/modulation) are employed in order to synchronously transform AC variables into DC ones. However, in some cases the control of AC variables in the original stationary frame may be preferred. In such cases resonant controllers that employ generalized integrators may be effectively used with enhanced performance. Combined *dq*-stationary control structures may also be envisioned for overall AC-DC converters control (Bose 2001; Blaabjerg et al. 2006; Etxeberria-Otadui et al. 2007).

Remarks. The remarks concerning closed-loop robustness in Chap. 8 for linear control of DC-DC converters remain valid for the case of AC-based converters. As in the previous chapter, for the sake of simplicity, the brackets denoting averaged values will be dropped and notations with subscript e are assumed to denote steady-state values.

9.2 Control in Rotating *dq* Frame

This approach refers to the single- or three-phase AC/DC (or DC/AC) converters operated by pulse-width modulation (PWM) and uses the models developed in Chap. 5.

Let us begin by analyzing the case of a three-phase rectifier whose diagram is given in Fig. 9.1. The DC voltage on the variable load R is to be regulated with certain dynamic performances, meanwhile extracting sinusoidal line currents in phase with grid voltages (thus ensuring unit power factor).

The use of the *dq* model facilitates the control of active and reactive power components as they can be expressed in relation to the components of voltage and currents (Bose 2001):

$$\begin{cases} P = v_d i_d + v_q i_q \\ Q = v_d i_q - v_q i_d. \end{cases} \tag{9.1}$$

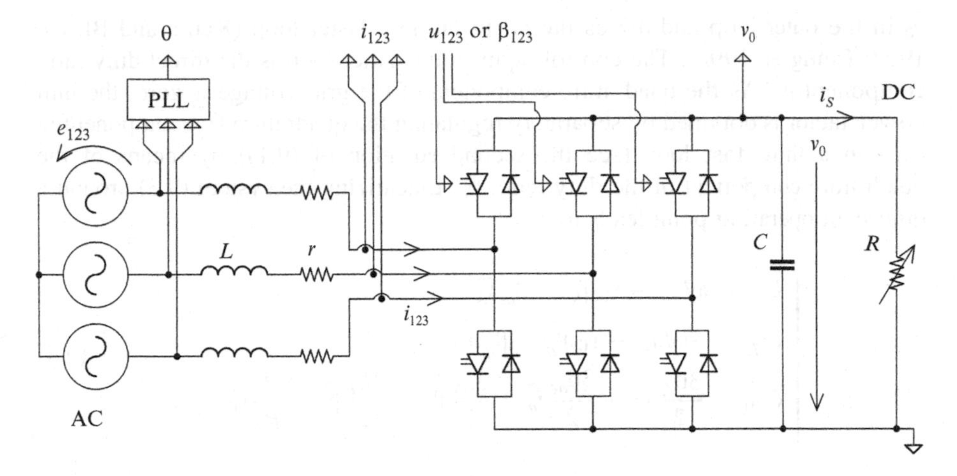

Fig. 9.1 Three-phase rectifier diagram

Now, by employing the GAM technique of Chap. 5 one obtains the state-space model of the circuit in Fig. 9.1 in the form

$$\begin{cases} L\dot{i_k} = e_k - v_0 \cdot u_k + \dfrac{v_0}{3}\displaystyle\sum_{k=1}^{3} u_k \quad k = 1, 2, 3, \\ C\dot{v_0} = \displaystyle\sum_{k=1}^{3} i_k \cdot u_k - \dfrac{v_0}{R}, \end{cases} \tag{9.2}$$

where the switching functions u_k, $k = 1, 2, 3$, take values in the set $\{0;1\}$.

Further, by considering symmetrical currents and by ignoring the AC component of voltage v_0, one can obtain the dq averaged state-space model of the circuit in Fig. 9.1 as (see Eq. (5.81) in Chap. 5):

$$\begin{cases} \dot{i_d} = \omega i_q - \dfrac{v_0}{L}\beta_d + \dfrac{E}{L} \\ \dot{i_q} = -\omega i_d - \dfrac{v_0}{L}\beta_q \\ \dot{v_0} = \dfrac{3}{2C}\left(i_d\beta_d + i_q\beta_q\right) - \dfrac{v_0}{RC}. \end{cases} \tag{9.3}$$

The state vector is $\left[i_d \quad i_q \quad v_0\right]^T$ and the control input vector is $\left[\beta_d \ \beta_q\right]$. A brief analysis reveals the bilinearity of the system and the existence of cross coupling between its equations (Blasko and Kaura 1997).

Suppose now that the DC capacitor is sufficiently high so as to induce significantly larger DC voltage inertia with respect to inductor current components. In this case one may want to employ a cascaded control structure which regulates voltage

v_0 in the outer loop and drives the i_d current in a faster loop (Kaura and Blasko 1997; Lindgren 1998). The control input in this latter loop is the direct duty ratio component β_d. As the quadrature component of the grid voltage is zero, the unit power factor is obtained by separately regulating the quadrature (i_q) component at zero in a third fast loop (see the second equation of (9.1)) by means of the quadrature component of the duty ratio, β_q. Linearizing the system (9.3) around a quiescent operating point leads to

$$\begin{cases} L\dot{\tilde{i_d}} = \omega L\widetilde{i_q} - v_{0e}\widetilde{\beta_d} - \beta_{de}\widetilde{v_0} \\ L\dot{\tilde{i_q}} = -\omega L\widetilde{i_d} - v_{0e}\widetilde{\beta_q} - \beta_{qe}\widetilde{v_0} \\ C\dot{\widetilde{v_0}} = \dfrac{3\beta_{de}}{2}\widetilde{i_d} + \dfrac{3\beta_{qe}}{2}\widetilde{i_q} + \dfrac{3i_{de}}{2}\widetilde{\beta_d} + \dfrac{3i_{qe}}{2}\widetilde{\beta_q} - \dfrac{1}{R_e}\widetilde{v_0} - \widetilde{i_S}, \end{cases} \tag{9.4}$$

where $\widetilde{i_S} = -\left(v_{0e}/R_e^2\right) \cdot \widetilde{R}$ is the load current variation.

Supplementary assumptions are necessary for control system design. Suppose that the current loops are very fast in comparison to the voltage v_0 loop. In this case one may consider $v_{0e} \equiv v_0^*$ constant and $\widetilde{v_0}$ having low-frequency content with respect to $\widetilde{i_d}$ and $\widetilde{i_q}$. This means that in the two first equations of (9.4) the last terms may represent low-frequency disturbances that can be easily rejected by the current controllers. In the same two equations the first (coupling) terms are high-frequency disturbances and may be zeroed by a decoupling structure (Teodorescu et al. 2006; Zmood et al. 2001). In this way, the current controllers may be independently designed for each channel d and q starting from the first two equations of (9.4), taking into account that the current plants are, practically, integrators. Figure 9.2 sketches the envisioned current control structures.

In order to alleviate low variations of v_0 (ramp-like), PI current controllers may be employed. As the control inputs β_d and β_q are intrinsically limited (between 0 and 1), antiwindup structures should be used in order to ensure proper behavior at sudden loads.

Now, in order to design the voltage controller one may suppose that the i_q control loop maintains $i_q = i_q{}^* \equiv 0$ irrespectively of perturbations acting on the q channel. Therefore, one may use a simplified reduced-order model:

$$\begin{cases} L\dot{\tilde{i_d}} = -v_{0e}\widetilde{\beta_d} - \beta_{de}\widetilde{v_0} \\ C\dot{\widetilde{v_0}} = \dfrac{3\beta_{de}}{2}\widetilde{i_d} + \dfrac{3i_{de}}{2}\widetilde{\beta_d} - \dfrac{1}{R_e}\widetilde{v_0} - \widetilde{i_S}, \end{cases} \tag{9.5}$$

whose second equation shows that the voltage v_0 can be driven by means of current i_d, as also shown in Fig. 9.3. For its part, this latter current is controlled by a very fast loop; therefore, the voltage plant "sees" $i_d \equiv i_d{}^*$ as a control input.

Note that control input i_d induces a supplementary high-frequency disturbance due to duty ratio β_d variations, through the branch having i_{de} as gain. In the very fast current closed loop, this variable has a very sharp derivative-like response to a step

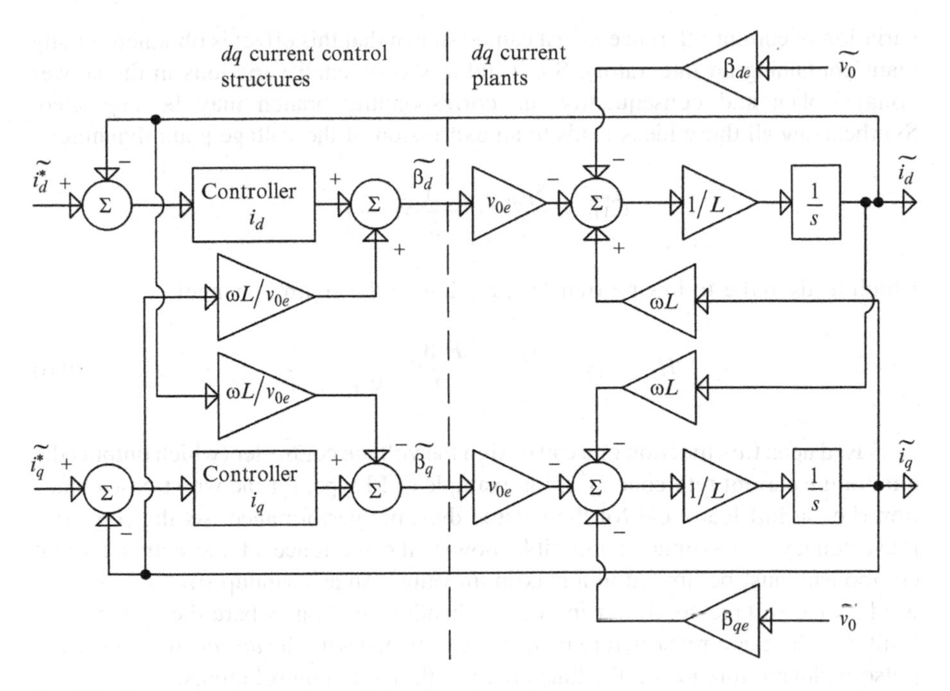

Fig. 9.2 *dq* current plants and associated control structures

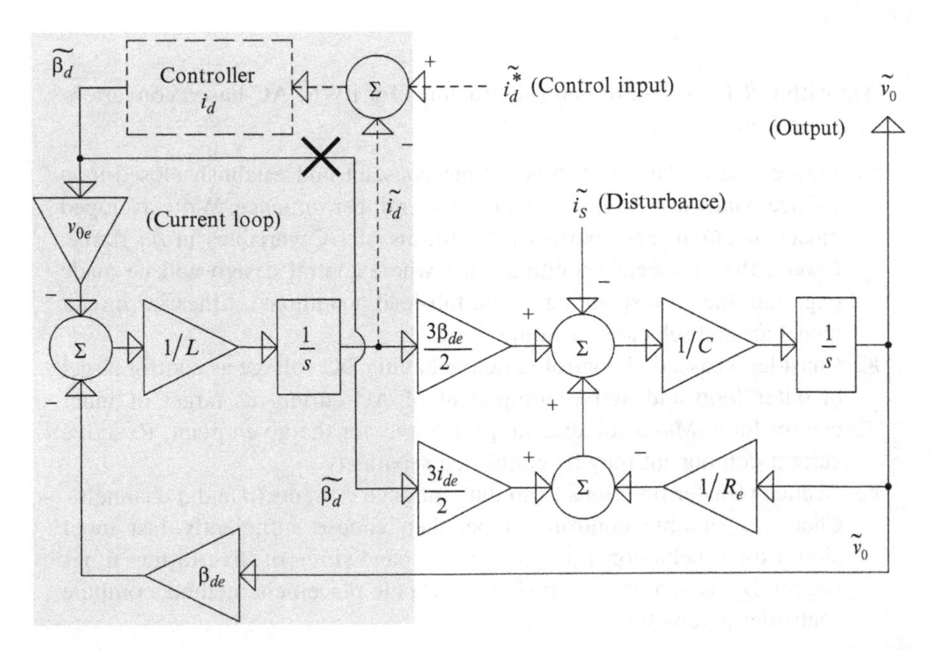

Fig. 9.3 Voltage plant – control of DC voltage by means of current i_d; current loop is indicated with a *gray* shading

variation of current reference $i_d{}^*$ (it can be shown that this effect is obtained for any plant containing an integrator). So, it induces very small variations in the slower voltage plant and consequently the corresponding branch may be neglected. Synthesizing all these ideas leads to an expression of the voltage plant dynamic:

$$C\dot{\widetilde{v_0}} = \frac{3\beta_{de}}{2}\widetilde{i_d} - \frac{1}{R_e}\widetilde{v_0} - \widetilde{i_s},$$

which leads to the following transfer function on the control channel:

$$H_{i_d \to v_0}(s) = \frac{\widetilde{v_0}}{\widetilde{i_d}} = \frac{3R_e\beta_{de}}{2} \cdot \frac{1}{R_e C s + 1}. \tag{9.6}$$

Based upon this function one can design the voltage controller (which outputs the quadrature current reference $i_d{}^*$), for example of PI type, for the worst-case condition (i.e., at full load) and for the desired dynamic performance. As the converter must convey a maximum admissible power, the reference of the active current component must be limited to a maximum value. An antiwindup structure may be used on the voltage controller in order to handle situations where the system is in limitation. The control structure is further completed with the *dq-abc* transform and pulse-width modulation on the thus obtained β_d and β_q control inputs.

Given the analysis within the above presented example, one may state the main steps of a generic design algorithm for the control of PWM AC-based converters in the *dq* frame.

Algorithm 9.1 Design of control structures for PWM AC-based converters in the *dq* frame

#1 Ensure that voltage has largest time constant and establish closed-loop voltage bandwidth based on imposed overall performance. Write averaged model of circuit and express it by means of DC variables in *dq* frame. Choose the quiescent operating point where control design will be made (e.g., the one corresponding to the full-load conditions). Linearize circuit model around selected operating point.

#2 Consider a cascaded control structure having DC voltage as control target of outer loop and active component of AC current as target of inner control loop. Make suitable simplifications for the given plant. Reactive current component may be controlled separately.

#3 Deduce transfer functions from duty ratios to currents (*d* and *q* channels). Choose a suitable controller type, then choose sufficiently fast inner closed-loop behavior (i.e., a current bandwidth of five-to-ten times larger). By using either loop-shaping or pole-placement method, compute controller parameters.

(continued)

Algorithm 9.1 (continued)

#4 Deduce transfer function from active current component to DC voltage. Ignore voltage ripple and reduce initial closed-loop bandwidth in order to avoid rejecting these variations. Use of a bandstop filter on main components of voltage variations may be effective in this sense. Deduce transfer function from active current component to output voltage. By using either loop-shaping or pole-placement method, compute controller parameters to satisfy the overall performance requirement.

#5 Change quiescent operating point (e.g., on a light load condition) and evaluate robustness of the obtained control law.

#6 If necessary, consider an adaptation of outer controller parameters in order to either obtain a smooth start-up or to preserve desired dynamic performance on entire operating range.

Remark The cascaded control structure is not the sole control paradigm that one can apply in the *dq* frame for this kind of converter A full-state feedback combined with an integral DC voltage controller could also be a good solution (in the sense provided by Sect. 8.4 in Chap. 8). A combined configuration involving partial-state feedback (e.g., for currents only) within a two-loop cascaded control structure may also be envisioned.

9.2.1 Example of a Grid-Connected Single-Phase DC-AC Converter

The PWM-operated power electronic converter of Fig. 9.4 conveys power from an unregulated DC current supply (e.g., a PV array) to a single-phase electrical grid. This DC power source outputs a variable current which depends on the primary resource and on terminal voltage values.

The design of the associated control structure must allow the available DC power to be injected into the grid. The converter must also provide a certain level of reactive power.

The control structure should be based on a two-loop cascade whose controllers are designed based upon the circuit linearized model by using Algorithm 9.1.

The control must establish the power balance between DC and AC parts in order to achieve its goal. A built-in imbalance sensor is the capacitor voltage v_0: it will increase to dangerous levels if the power supplied to the DC-link is larger than the power drawn; otherwise, it will decrease, eventually threatening the normal operation of the inverter. It is clear that both regimes must be avoided by maintenance of a sufficient level of DC voltage v_0. Therefore, the initial requirement of power

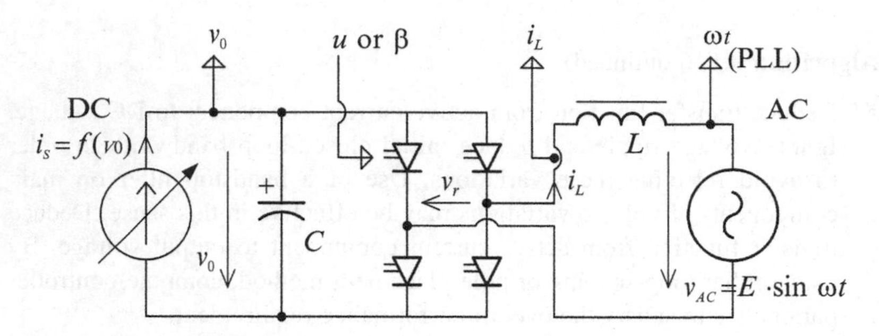

Fig. 9.4 Grid-connected single-phase inverter

balancing can be translated into the regulation of DC voltage v_0 to a constant value, thus ensuring the proper operation of the converter.

As the circuit also operates with AC variables, control actions also target the envelopes (amplitudes) of these variables (Roshan 2006). These are new DC variables that can be obtained by amplitude demodulation action, i.e., by dq transformation. In Chap. 5 it was shown that a generic AC variable y may be expressed by a sum of two components: $y = y_d \sin \omega t + y_q \cos \omega t$. The new dq frame is also useful for separately controlling the active power (d channel) and the reactive power (q channel). Suppose that the primary DC current source has an output characteristic of the form

$$i_S = f(v_0) = I_0 - \frac{v_0}{R},$$

where I_0 is a slow-variable parameter depending on the primary resource and R is the dynamic resistance of the source (the inverse of the characteristic $i_s - v_0$ slope), which is supposed to be rapidly variable.

Recall that the general averaged model of this circuit expressed in the synchronously rotating dq frame is given by (see Eq. (5.81) in Chap. 5):

$$\begin{cases} \dfrac{di_d}{dt} = \omega i_q + \dfrac{E}{L} - \dfrac{v_0}{L} \cdot \beta_d \\ \dfrac{di_q}{dt} = -\omega i_d - \dfrac{v_0}{L} \cdot \beta_q \\ \dfrac{dv_0}{dt} = \dfrac{1}{2C}\left(i_d \beta_d + i_q \beta_q\right) - \dfrac{v_0}{CR} + I_0, \end{cases} \tag{9.7}$$

with $\begin{bmatrix} \beta_d & \beta_q \end{bmatrix}^T$ as input vector and $\begin{bmatrix} i_d & i_q & v_0 \end{bmatrix}^T$ as state vector. Model parameters – grid pulsation ω, DC-link capacitor C, grid-side inductor L – are also visible in Fig. 9.4. Of course, current i_d must be negative in order to establish the DC-link balance.

The linearized version of this model around a quiescent operating point (whose variables are denoted by index e) is presented in (9.8), with $\widetilde{i_S}$ being the variation of the DC-link input current. In the first two equations of (9.8) the first terms on the right-hand side are high-frequency perturbations and should be compensated by a decoupling action as in the previous example. The second terms are low-frequency perturbations (v_0 variations are slow, ramp-like) and can be rejected by a suitably chosen control action.

$$\begin{cases} L\dot{\widetilde{i_d}} = \omega\widetilde{i_q} - \beta_{de}\widetilde{v_0} - v_{0e}\widetilde{\beta_d} \\ L\dot{\widetilde{i_q}} = -\omega\widetilde{i_d} - \beta_{qe}\widetilde{v_0} - v_{0e}\widetilde{\beta_q} \\ C\dot{\widetilde{v_0}} = \frac{\beta_{de}}{2}\widetilde{i_d} + \frac{i_{de}}{2}\widetilde{\beta_d} + \frac{\beta_{qe}}{2}\widetilde{i_q} + \frac{i_{qe}}{2}\widetilde{\beta_q} - \frac{\widetilde{v_0}}{R_e} + \widetilde{i_S}\,. \end{cases} \tag{9.8}$$

As a consequence, the transfer functions for both d and q channels represent integrators, as follows:

$$H_{\beta_d \to i_d}(s) = H_{\beta_q \to i_q}(s) = \frac{-v_{0e}}{Ls} = \frac{1}{T_c s},$$

where $T_c = -L/v_{0e}$. As a result, the current controllers may be either proportional or PI-type if a good rejection of perturbation v_0 is intended. In this latter case, if the controller is described by $H_c(s) = K_{pc}\left(1 + \frac{1}{T_{ic}s}\right)$, the closed-loop transfer function is $H_{0c}(s) = \frac{T_{ic}s+1}{\frac{T_{ic}T_c}{K_{pc}}s^2 + T_{ic}s + 1}$. Controller parameters are computed as usual, by imposing inner closed-loop bandwidth $1/T_{0c}$ and damping ξ_c:

$$K_{pc} = 2\xi_c T_c / T_{0c}\,, \quad T_{ic} = 2\xi_c T_{0c}$$

Remark. The inner closed-loop bandwidth is related to the outer loop's and should be five–ten times larger. Damping is usually taken between 0.7 and 0.85. The inner loop may need a prefilter $\frac{1}{T_{ic}s+1}$ in order to cancel the influence of the zero from the closed-loop transfer function.

Now, concerning the outer control loop, a thorough analysis of the current-controlled linearized voltage plant may be made – see Fig. 9.5 and the third equation of (9.8).

The i_d current is the control input that drives the voltage outer loop; it can be controlled within the significantly faster inner loop. As has been discussed before, the current plant contains an integrator. Hence, the variable β_d is the output of a derivative. Supposing that the current reference $i_d{}^*$ has a step variation, β_d will then have a narrow spike-like shape (due to the larger bandwidth of the inner loop). This variation of β_d does not, practically speaking, affect the value of v_0 within the outer loop. This allows one to neglect the direct influence of variable β_d on the DC voltage v_0. Because of its similar structure, the same argumentation applies to the q channel, so neither β_q nor β_d affect v_0. The i_q input may be considered a low-frequency disturbance if it is driven by a sufficiently slower outer loop aiming at controlling the reactive power; thus, it can also be neglected. Under these assumptions, the sole significant disturbance over the i_d-controlled voltage (outer) plant is the input current variation i_S. Figure 9.6 shows that

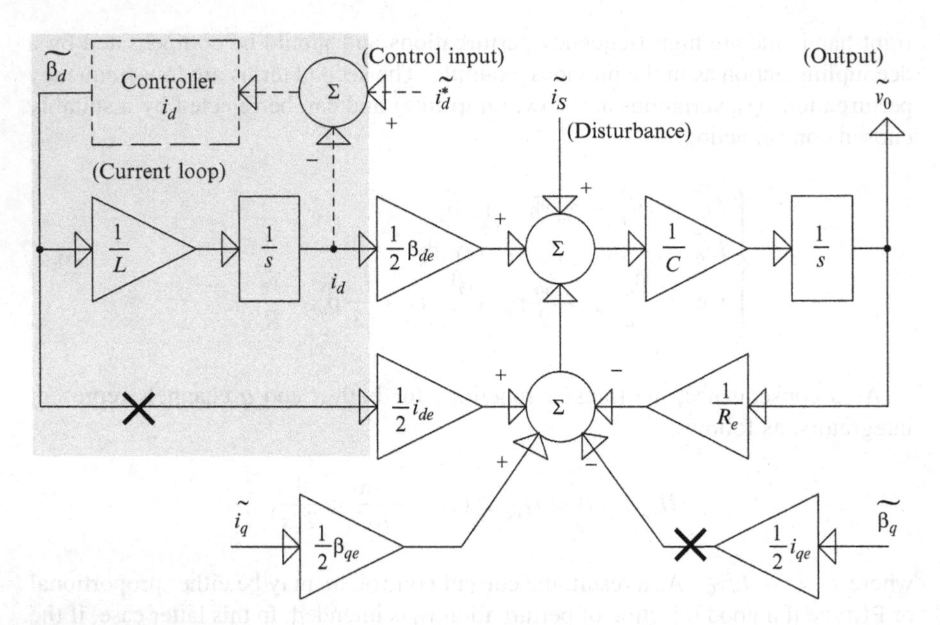

Fig. 9.5 Influences of various variables over DC voltage; i_d current loop is represented on *gray* background

Fig. 9.6 Simplified outer (DC voltage) plant

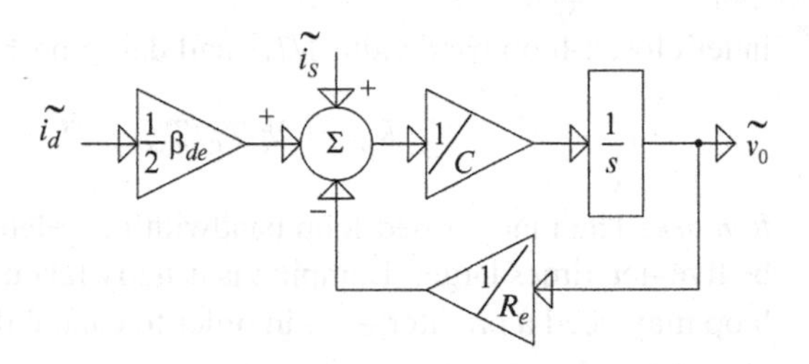

this plant is basically reduced to a first-order system having $0.5R_e\beta_{de}$ as gain and R_eC as time constant, both varying with the operating point (R_e is variable).

A PI controller $K_{pv}\left(1+\frac{1}{T_{iv}s}\right)$ may be chosen in order to regulate the output of this structure without steady-state error. The closed-loop transfer function is

$$H_{0v}(s) = \frac{T_{iv}s+1}{\frac{T_{iv}C}{0.5K_{pv}\beta_{de}}s^2 + \left(1+\frac{1}{0.5K_{pv}\beta_{de}R_e}\right)T_{iv}s+1}, \tag{9.9}$$

and by imposing outer closed-loop bandwidth $1/T_{0v}$ and damping ξ_v, one obtains the controller parameters by identification:

$$K_{pv} = \frac{2\xi_v CR_e - T_{0v}}{0.5\beta_{de}T_{0v}R_e}, \quad T_{iv} = T_{0v}\frac{2\xi_v CR_e - T_{0v}}{CR_e}$$

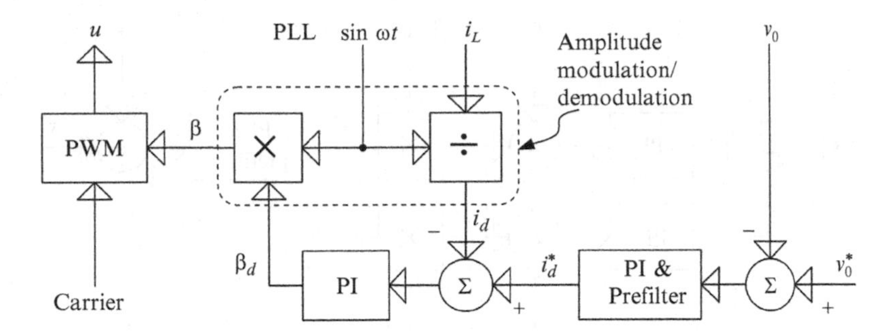

Fig. 9.7 Control diagram of the grid-connected single-phase inverter in Fig. 9.4

The first equation from (9.7) shows that the steady-state value of the duty ratio on the d channel is practically constant at $\beta_{de} = E/v_0$ (and slightly variable in the case of inductor losses). This further implies that the closed-loop bandwidth does not vary with the operating point (R_e), as Eq. (9.9) shows. However, the associated damping varies from the imposed value ξ_v as operating point changes, but this variation is quite reduced for large values of controller gain K_{pv}.

Remember that outer loop time constant T_{0v} should be imposed five–ten times the magnitude of T_{0c} (inner loop), otherwise undesired effects such as nonminimum-phase behavior or intensive ringing may be obtained. The outer-loop controller output represents the reference for the inner loop ($i_d{}^*$). The control structure needs a phase-locked loop (PLL) that provides the phase of the electrical grid so the variables may be computed in the dq frame according to $\beta = \beta_d \sin \omega t$ and $i_L = i_d \sin \omega t$ (see Chap. 5). Figure 9.7 depicts the control diagram that must be coupled with the circuit diagram in Fig. 9.4 for achieving closed-loop operation.

Some further improvements can be brought to this structure, for example, a supplementary filter (e.g., notch filter) on the DC-link voltage may be used to alleviate control structure excitation by the AC component (at twice ω) present in the variable v_0. The division block may be replaced by a product and a low-pass filter (as shown in Sect. 5.5.2 of Chap. 5). By displacing the product block ahead of the current controller, this structure can further be put into a more suitable form (which reduces the computation burden and avoids division), as can be seen in Fig. 9.8. Instead of comparing the d current components, one compares the inductor currents in the inner loop. The control on the q channel is similar, the inner loop reference results as $i_L = i_q \cos \omega t$. The slow-variable reference $i_q{}^*$ will adjust the reactive power supplied to the electric grid (see (9.1)).

Further, robustness study may be necessary (concerning mainly the damping of the outer voltage loop) in order to ensure satisfactory dynamics of the closed-loop system for the entire operating range.

In conclusion, two-loop cascaded control structures for AC-DC converters have been considered and designed in this section. All controllers process variables within the synchronously rotating dq frame. But, as will be detailed next, it is possible to synthesize controllers directly in the stationary frame (Malesani and Tomasin 1993;

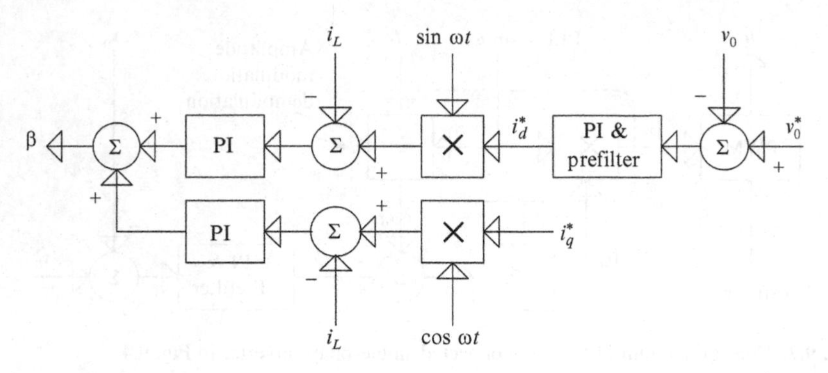

Fig. 9.8 Control diagram of grid-connected single-phase inverter from Fig. 9.4, redrawn and completed with q channel

Kazmierkowski and Malesani 1998) or to design hybrid control structures that use variables from both stationary and dq frames (Timbuş et al. 2009).

9.3 Resonant Controllers

This section also addresses the control of PWM-operated AC-based converters, but does so differently than in the previous section; here the control of stationary-frame AC variables is mainly addressed.

AC variable control in the stationary frame supposes a control loop that acts directly on the AC circuit, without intermediary amplitude modulation-demodulation (or dq transform).

9.3.1 Necessity of Resonant Control

Let us consider a single-phase grid-connected inverter, similar to the one described in Fig. 9.4. The transistor bridge is controlled such that it applies voltage $v_r = \beta \cdot v_0$ on the grid circuit, and therefore the dynamic of the averaged inductor current (injected to the grid) obeys the following equation:

$$L\dot{i_L} = v_r - v_{AC} - i_L r = \beta v_0 - v_{AC} - i_L r,$$

where L is inductance, r is inductor parasitic resistance and $v_{AC} = E \sin \omega_0 t$ is the voltage grid. If the amplitude of the grid voltage is considering invariant and the DC voltage constant at a desired value $v_0 \equiv v_{DC}$, the resulting associated transfer function (on the control channel) is

$$H_P(s) = \frac{\widetilde{i_L}}{\widetilde{\beta}} = \frac{v_{DC}}{Ls + r}.$$

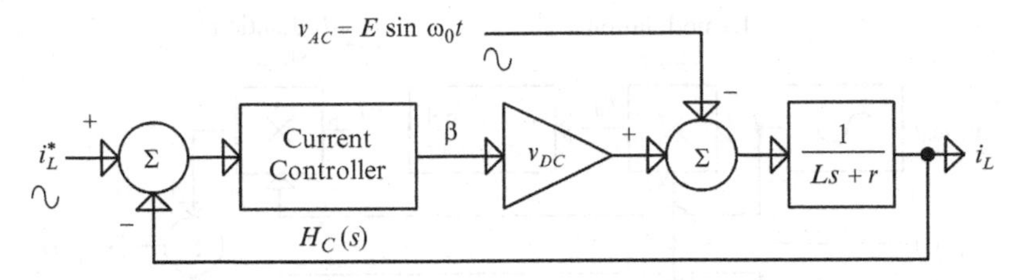

Fig. 9.9 AC current control loop for converter in Fig. 9.4

The inductor current control structure is schematized in Fig. 9.9. Supposing the current controller in this figure to be PI, with transfer function of the form $H_C(s) = K_p + K_i/s$, the closed-loop transfer function (control channel) is

$$H_C(s) = \frac{K_p s + K_i}{\frac{L}{v_{DC}} s^2 + \left(K_p + \frac{r}{v_{DC}}\right) s + K_i}.$$

The closed-loop frequency response is perfect at frequencies where the gain is one and the phase shift is zero. The expression for the closed loop gain is

$$H_0(\omega) = \sqrt{\frac{K_p^2 \omega^2 + K_i^2}{\left(K_p + \frac{r}{v_{DC}}\right)^2 \omega^2 + \left(K_i - \frac{L}{v_{DC}} \omega^2\right)^2}}$$

and it has unit value in DC (at $\omega = 0$). Also, depending on chosen parameters, it is possible for resonance to occur so that a second point of unit gain exists, namely at the frequency (Etxeberria-Otadui 2003)

$$\omega = \frac{1}{L} \sqrt{2 v_{DC} (K_i L - K_p r) - r^2}.$$

The closed-loop phase lag has the expression

$$\varphi = \arctan\left(\frac{K_p \omega}{K_i}\right) - \arctan\left(\omega \frac{v_{DC} K_p + r}{v_{DC} K_i - L\omega^2}\right),$$

which is zero only in DC (at $\omega = 0$). Therefore, the only frequency for which the closed-loop system from Fig. 9.9 employing a PI controller has an ideal frequency response is the zero frequency. This means that if the reference signal has AC components (contains nonzero frequencies) it is not possible to cancel the steady-state error in terms of amplitude and phase. Moreover, this error increases with the frequency. It is possible to further reduce the steady-state error by employing a very high controller gain, but there are two major limitations that rule out this solution. The first one is the control input saturation induced by a finite value of the DC-link voltage. Second, a high controller gain leads to an increase of the closed-loop bandwidth and a decrease of the phase margin, eventually leading further to system instability.

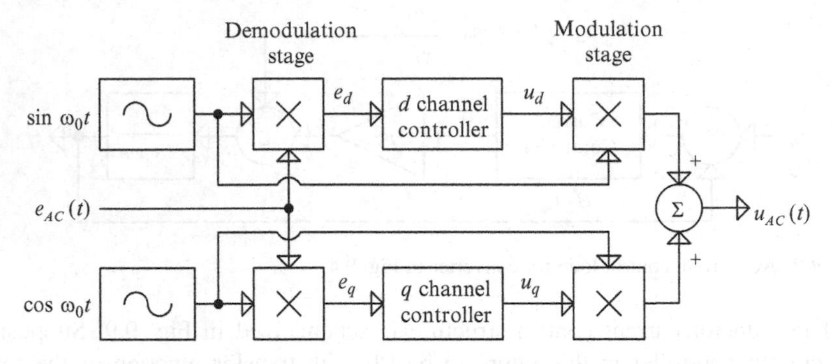

Fig. 9.10 AC variable control by demodulation-modulation (Zmood and Holmes 2003)

To conclude, as the reference is an AC variable having nonzero frequency, this kind of control loop cannot be properly built using standard PI controllers; therefore other solutions should be envisioned (Sato et al. 1998; Esselin et al. 2000; Fukuda and Yoda 2001).

9.3.2 Basics of Proportional-Resonant Control

As has already been described in Sect. 9.2, control in the rotating frame supposes amplitude demodulation of the AC variables, control of the envelopes (DC variables) on d and q channels, then amplitude modulation in order to obtain the AC control input(s). This approach is illustrated in Fig. 9.10, where e_{AC} represents the AC error signal and u_{AC} is the AC control input signal.

The error is demodulated through multiplication with sine and cosine of the grid phase (electrical angle), thus one signal centered on the DC component and another centered on the $2\omega_0$ component (ω_0 being the grid frequency) are obtained. If PI controllers in Fig. 9.10 are used, then one can obtain zero steady-state error for the DC component. Controller outputs are further modulated and by summing them up one obtains the AC control signal, u_{AC}. Note that the expression of u_{AC} does not contain the $2\omega_0$ signal as its components cancel each other by summing. The transfer function of the PI controller used in the dq frame is, as usual

$$H_{DC}(s) = K_p + \frac{K_i}{s}. \tag{9.10}$$

Now, by analyzing the diagram in Fig. 9.10 the time-domain expression of the control input u_{AC} is

$$u_{AC}(t) = [(e_{AC} \cos \omega_0 t) \otimes h_{DC}(t)] \cos \omega_0 t + [(e_{AC} \sin \omega_0 t) \otimes h_{DC}(t)] \sin \omega_0 t, \tag{9.11}$$

where $h_{DC}(t)$ is the controller impulse response and $\otimes$ denotes the convolution product. The final aim is to obtain a controller whose transfer function, denoted by $H_{AC}(s)$, has the same frequency response with the one given by Eq. (9.11), without using the demodulation-modulation action (Zmood and Holmes 2003). Such a system may be represented in the s domain by the equation

$$U_{AC}(s) = H_{AC}(s) \cdot E_{AC}(s), \tag{9.12}$$

where $E_{AC}(s)$ and $U_{AC}(s)$ are the Laplace images of the error and the control input, respectively. In the time domain, this is expressed as

$$u_{AC}(t) = e_{AC}(t) \otimes h_{AC}(t),$$

where $h_{AC}(t)$ is the impulse response of the equivalent AC controller. In Eq. (9.11) one makes the notations $f_1(t) = h_{DC}(t) \otimes [e_{AC}(t) \cdot \cos \omega_0 t]$ and $f_2(t) = h_{DC}(t) \otimes [e_{AC}(t) \cdot \sin \omega_0 t]$, respectively, taking account of the convolution product being commutative. Their Laplace images may be computed as follows:

$$F_1(s) = L\{h_{DC}(t) \otimes (e_{AC}(t) \cdot \cos \omega_0 t)\} = H_{DC}(s) \cdot L\{e_{AC}(t) \cdot \cos \omega_0 t\}.$$

By using the Laplace transform property of frequency shifting, one further obtains

$$F_1(s) = \frac{1}{2} H_{DC}(s) \cdot [E_{AC}(s + j\omega_0) + E_{AC}(s - j\omega_0)]. \tag{9.13}$$

In the same way one shows that

$$F_2(s) = \frac{j}{2} H_{DC}(s) \cdot [E_{AC}(s + j\omega_0) - E_{AC}(s - j\omega_0)]. \tag{9.14}$$

The Laplace transform of the first term in Eq. (9.11) gives

$$\begin{aligned} L_1(s) &= \mathcal{L}\{[(e_{AC}(t) \cdot \cos \omega_0 t) \otimes h_{DC}(t)] \cdot \cos \omega_0 t\} = \mathcal{L}\{f_1(t) \cdot \cos \omega_0 t\} \\ &= \frac{1}{2}[F_1(s + j\omega_0) + F_1(s - j\omega_0)]. \end{aligned} \tag{9.15}$$

The Laplace transform of the second term in Eq. (9.11) gives

$$\begin{aligned} L_2(s) &= \mathcal{L}\{[(e_{AC}(t) \cdot \sin \omega_0 t) \otimes h_{DC}(t)] \cdot \sin \omega_0 t\} = \mathcal{L}\{f_2(t) \cdot \sin \omega_0 t\} \\ &= \frac{j}{2}[F_2(s + j\omega_0) - F_2(s - j\omega_0)]. \end{aligned} \tag{9.16}$$

The Laplace image of the controller output is $U_{AC}(s) = L_1(s) + L_2(s)$; therefore, by combining Eqs. (9.15), (9.16) and (9.11) one obtains

$$U_{AC}(s) = \frac{1}{2}[F_1(s + j\omega_0) + F_1(s - j\omega_0)] + \frac{j}{2}[F_2(s + j\omega_0) - F_2(s - j\omega_0)]. \tag{9.17}$$

By replacing the values $F_1(s)$ and $F_2(s)$ as given by Eqs. (9.13) and (9.14) into Eq. (9.17) and by using (9.12) one finally obtains

$$H_{AC}(s) = \frac{1}{2}[H_{DC}(s + j\omega_0) + H_{DC}(s - j\omega_0)]. \tag{9.18}$$

An alternative to (9.18), when the reference signal bandwidth is small in comparison to the reference frequency itself, is to use the low-pass-to-band-pass filter transformation method developed in filter synthesis. In this way the equivalent AC controller transfer function that results is

$$H_{AC}(s) = K_p + \frac{2K_i s}{s^2 + \omega_0{}^2}. \tag{9.19}$$

Note that a resonant (second) term appears in this equation; for this reason this kind of controller is commonly called *proportional-resonant* (PR).

Remember that an integrator provides infinite gain at zero frequency (in DC). Conversely, the second term from (9.19) provides infinite gain at a certain frequency ω_0 – the resonance frequency – which may be different from zero, thus allowing zero steady-state error to be obtained at ω_0. This is why this term is also called *generalized integrator*. The gain of the generalized integrator is in a relation of inverse proportionality with the selectivity (quality factor) of the generalized integrator (resonant term), as can be seen in Fig. 9.11.

From a control theory point of view, for zeroing the steady-state error (canceling the disturbance effect in the steady-state regime) it is necessary to introduce in the controller the internal model of the exogenous signal as Laplace transform (Francis and Wonham 1976; Fukuda and Imamura 2005). Thus, for example, if the exogenous is a step variation, then a pole at the origin will be introduced, which is equivalent to the insertion of a delta distribution in the Bode characteristic at $\omega = 0$. The distinction of this situation is that the frequency characteristic is affected in the entire band. If the exogenous is a ramp, a double delta distribution is introduced at $\omega = 0$. Following the same reasoning, in the case of a harmonic exogenous (supposing either the tracking of a harmonic reference or rejection of a harmonic disturbance) a resonant component producing a delta distribution at the resonance frequency must be introduced in the controller. This action solves the problem of the steady-state regime (nullifying the steady-state error).

Further, one must solve the closed-loop dynamic performance problem by using either a state-space or frequency-domain approach. The latter is used more often in power electronics control. The frequency-based model (Nyquist locus or Bode characteristic) must be corrected according to dynamic performance requirements.

The PI controller is a particular case of PR, used when the delta distribution from the frequency characteristic is placed at $\omega = 0$. In this case the frequency characteristic is significantly affected in the entire frequency range, imposing the use of classical tuning methods by root locus (modulus and symmetry methods for the

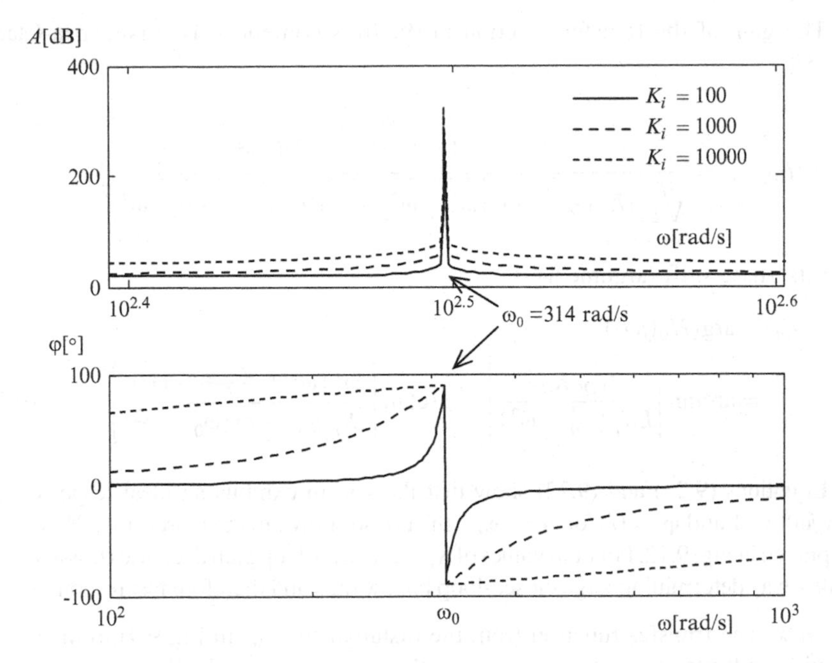

Fig. 9.11 Bode characteristics of the proportional-resonant controller for different integral gains K_i

single and the double pole, respectively) in the medium frequency domain. Now, concerning the PR controller the problem is similar, but the frequency characteristic is modified only locally, around the resonance frequency. The associated design methods are frequency-based (e.g., loop shaping and root locus – see `rltool` in MATLAB®) and obey the same design principles as happens in the case of PI controllers. However, there are some aspects specific to the power electronic converters: nonsinusoidal periodical exogenous (whose rejection requires multiresonant controllers), frequency variation (which supposes adapting the internal model poles), etc. Note that these aspects do not exist for controllers having poles in the origin.

Now, suppose that in Fig. 9.9 the current controller has the transfer function given in Eq. (9.19) (i.e., is a PR one). In this case the closed-loop transfer function is

$$H_0(s) = \frac{H_{AC}(s)H_P(s)}{1 + H_{AC}(s)H_P(s)}$$

$$= \frac{\dfrac{v_{DC}}{L}\left(K_p s^2 + 2K_i s + K_p \omega_0^2\right)}{s^3 + \dfrac{1}{L}\left(K_p v_{DC} + r\right)s^2 + \left(\omega_0^2 + \dfrac{2K_i v_{DC}}{L}\right)s + \dfrac{\omega_0^2}{L}\left(K_p v_{DC} + r\right)}. \tag{9.20}$$

The gain of the transfer function in (9.20) is computed as (Etxeberria-Otadui 2003):

$$|H_0(j\omega)| = \frac{\frac{v_{DC}}{L}\sqrt{K_p^2\left(\omega_0^2-\omega^2\right)^2+4K_i^2\omega^2}}{\sqrt{\frac{1}{L^2}\left(K_p v_{DC}+r\right)^2\left(\omega_0^2-\omega^2\right)^2+\omega^2\left(\omega_0^2+\frac{2K_i v_{DC}}{L}-\omega^2\right)^2}}, \tag{9.21}$$

and its phase shift (argument) is:

$$\begin{aligned}\varphi_0 &= \arg(H_0(j\omega)) \\ &= \arctan\left[\frac{2v_{DC}K_i\omega}{LK_p\left(\omega_0^2-\omega^2\right)}\right] - \arctan\left[\frac{\omega L\left(\omega_0^2+\frac{2K_i v_{DC}}{L}-\omega^2\right)}{\left(K_p v_{DC}+r\right)\left(\omega_0^2-\omega^2\right)}\right].\end{aligned} \tag{9.22}$$

Equations (9.21) and (9.22) show that the system exhibits an ideal response, i.e., $|H_0(j\omega)| = 1$ and $\varphi = 0°$ for $\omega = \omega_0$. This is also shown in the closed-loop Bode plots displayed in Fig. 9.12. For high values of K_p, the closed-loop gain decreases less abruptly, in this way determining large closed-loop bandwidths and therefore fast responses.

Remark The transfer function from the disturbance (v_{AC} in Fig 9.9) to the output provides infinite attenuation at $\omega = \omega_0$, thus rejecting completely the disturbance.

9.3.3 Design Methods

9.3.3.1 Loop Shaping

The most straightforward design method is the one based on loop shaping (Zmood and Holmes 2003; Yuan et al. 2002). The open-loop transfer function is

$$H_{OL}(s) = \left(K_p + K_i\frac{2s}{s^2+\omega_0^2}\right)\frac{v_{DC}}{Ls+r}$$

and the associated frequency characteristics are given in Fig. 9.13. First, let us consider that the effect of the resonant term affects the characteristic only locally, i.e., only around the frequency ω_0 (this supposes reasonably small values of gain K_i), so its presence does not significantly change the cut-off frequency, ω_{OL}. Under this assumption the simplified open-loop transfer function is the same as for a first-order system having a proportional controller – see Fig. 9.13:

$$H_{OL}^S(s) = \frac{K_p v_{DC}}{Ls+r}.$$

This case is treated classically; the desired settling time imposes the closed-loop bandwidth or equivalently the cut-off frequency value, ω_{OL}. This can be adjusted by means of the controller gain, K_p, as follows (see also Fig. 9.13).

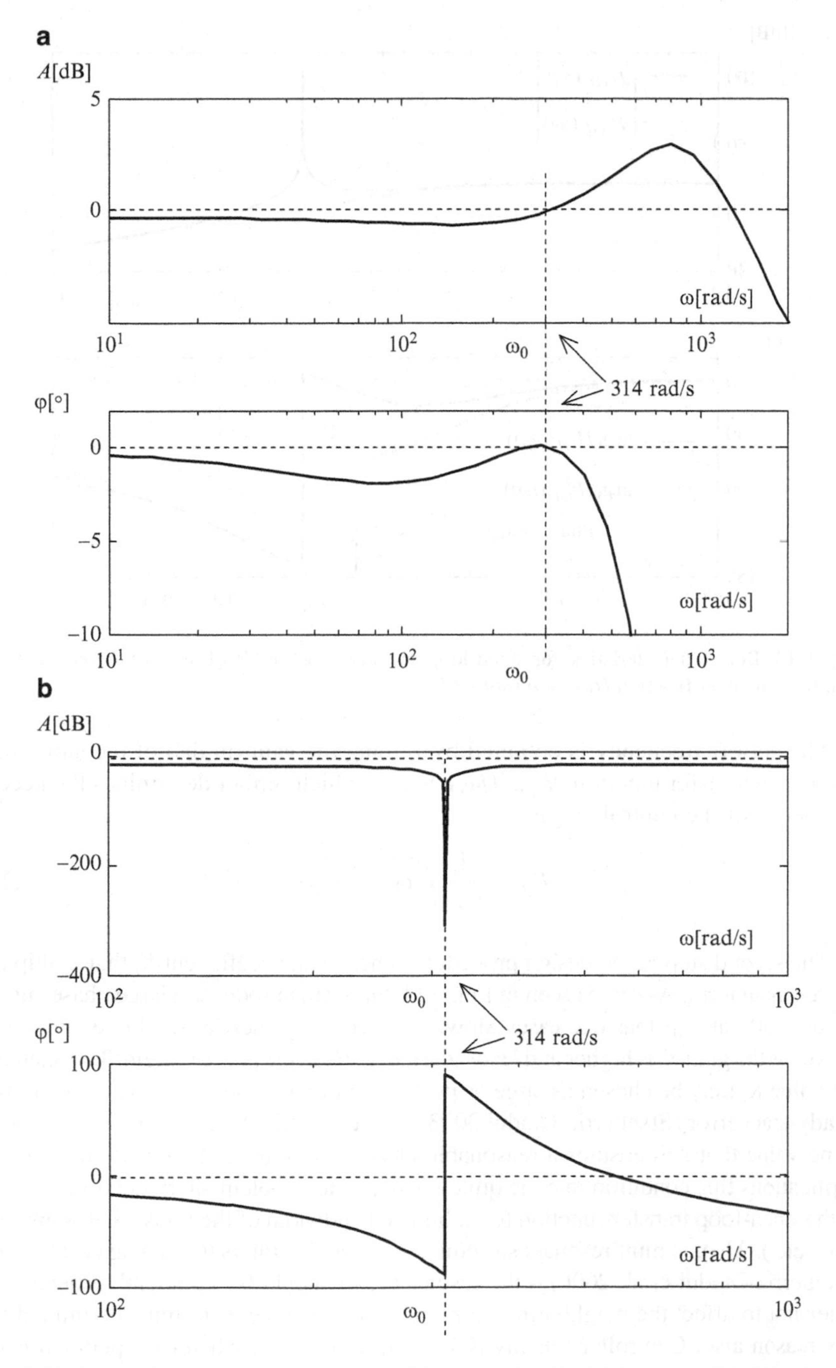

Fig. 9.12 Closed-loop Bode characteristics with proportional-resonant controller: (**a**) control-input-to-output transfer channel; (**b**) disturbance-to-output transfer channel

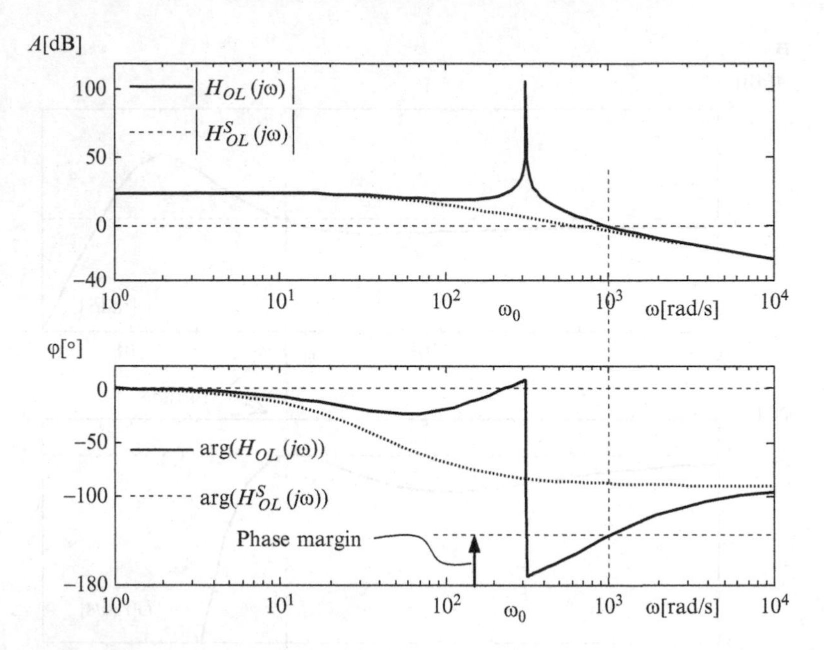

Fig. 9.13 Bode characteristics for open-loop transfer function $H_{OL}(j\omega)$ (*solid line*) and for simplified transfer function $H_{OL}{}^{S}(j\omega)$ (*dotted line*)

The cut-off frequency is obtained by zeroing the gain (in dB) of the simplified open-loop transfer function: $|H_{OL}{}^{S}(j\omega_{OL})| = 1$, which further determines the necessary value of the controller gain:

$$K_p = \frac{1}{v_{DC}} \sqrt{\omega_{OL}^2 L^2 + r^2} \tag{9.23}$$

The second step of the design procedure concerns the coefficient K_i that multiplies the resonant term. As can be seen in Fig. 9.11, this term introduces a large phase lag (of about $-90°$ at ω_0) that decreases slowly as frequency increases. This evolution is slower as the gain K_i is higher and leads to a reduction in the phase margin. To conclude, the value K_i may be chosen as large as possible in order to obtain rapid zeroing of the steady-state error (Etxeberria-Otadui 2003). However, this choice is upper bounded by some value that still ensures a reasonable phase margin (e.g., 40°). Note that in real applications this condition may be quite restrictive as supplementary delays intervene in the open-loop transfer function (e.g., limited bandwidth of the transducer, sampling time, etc.). Also, in multiresonant structures – e.g., active filters (Gaztañaga et al. 2005; Etxeberria-Otadui et al. 2006) – the resonant term should be sufficiently selective in order not to affect the neighboring harmonics, so the value of K_i must be limited for this reason also. Controller selectivity may be a drawback when the operation is on power grids having variable frequency (e.g., for weak grids or microgrids). This issue may be solved by using adaptive PR controllers (Timbuş et al. 2006).

9.3.3.2 Pole Placement

Controller tuning aims at ensuring a suitable dynamic performance of the controlled variable evolution towards a sinusoidal steady-state regime (and not to the DC steady-state). This dynamic regime – in terms of response time and shape – is essentially characterized by the poles' distribution of the closed-loop transfer function (9.20). The characteristic polynomial is

$$P(s) = s^3 + \frac{1}{L}\left(K_p v_{DC} + r\right)s^2 + \left(\omega_0^2 + \frac{2K_i v_{DC}}{L}\right)s + \frac{\omega_0^2}{L}\left(K_p v_{DC} + r\right), \qquad (9.24)$$

and a suitable placement of its roots may suppose, for example, one real and one pair of complex-conjugated. In this case there are only two degrees of freedom (K_p and K_i) in the control design and three parameters to configure (natural frequency and damping of complex-conjugated poles, and frequency of the real pole, respectively). The design may use the root locus method (e.g., `rltool` in MATLAB®); basic requirements are to ensure reasonable closed-loop damping and sufficiently large closed-loop bandwidth. The best way to validate closed-loop performance is to visualize the error signal evolution.

9.3.3.3 Naslin Polynomial Method

Controller parameters may also be computed by interpreting the closed-loop characteristic polynomial (9.24) as a third-order Naslin polynomial.

Naslin polynomials – also known as *normal polynomials with adjustable damping* – can be introduced starting by an analogy with second-order systems (Naslin 1968). Let us consider a second-order linear system described by the transfer function

$$H_0(s) = \frac{a_0'}{a_0 + a_1 s + a_2 s^2},$$

whose damping coefficient depends on the denominator's polynomial coefficients, according to $4\zeta^2 = a_1^2/(a_0 a_2)$, and gain is in general different from one ($a_0' \neq a_0$).

Definition of a damping coefficient for an nth-order system can be given in an analogous way; let

$$H_0(s) = \frac{a_0'}{a_0 + a_1 s + \cdots + a_n s^n} \qquad (9.25)$$

be the system's transfer function. One defines the so-called *characteristic ratios*

$$\alpha_1 = \frac{a_1^2}{a_0 a_2}, \quad \alpha_2 = \frac{a_2^2}{a_1 a_3}, \quad \cdots, \quad \alpha_n = \frac{a_n^2}{a_{n-1} a_{n+1}},$$

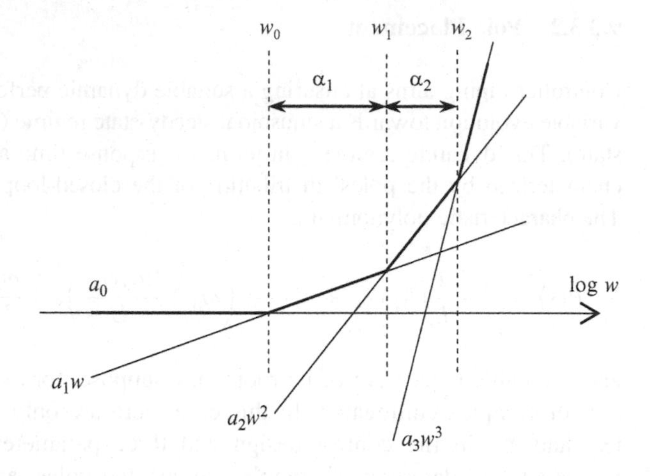

Fig. 9.14 Geometrical interpretation of characteristic pulsations and of characteristic ratios by means of a gain curve (Naslin 1968)

and the *characteristic pulsations*

$$w_0 = \frac{a_0}{a_1}, \quad w_1 = \frac{a_1}{a_2}, \quad \cdots, \quad w_n = \frac{a_n}{a_{n+1}}. \tag{9.26}$$

Simple algebra proves that

$$\alpha_1 = \frac{w_1}{w_0}, \quad \alpha_2 = \frac{w_2}{w_1}, \quad \cdots, \quad \alpha_n = \frac{w_n}{w_{n-1}}. \tag{9.27}$$

One can give a geometrical interpretation to characteristic ratios and pulsations on the gain curve of the H_0 characteristic polynomial, as depicted in Fig. 9.14.

In this figure lines corresponding to gain of successive terms $a_i s^i$ are represented; their envelope is an approximation of the gain curve. The characteristic pulsations are the abscissas of folding points of the gain curve, whereas the segments' lengths are given by the characteristic ratios. Imposing a sufficiently large value for characteristic ratios will ensure weak resonance (Naslin 1968). For its part, weak resonance ensure small errors of approximated curve in relation to real gain curve around folding frequencies. In conclusion, a good approximation is ensured by imposing characteristic ratios to be sufficiently large or, in other words, segments to be sufficiently long.

The above remarks allow a polynomial family to be defined whose members have the characteristic ratios with the same value denoted by α, which is related to damping. By requiring that α to be greater than a certain chosen value, one ensures stability and damping, all by guaranteeing transients that are described by a single dominant mode.

As consequence, a polynomial of this family is entirely described by its order n, the damping coefficient α and one of its characteristic pulsations. For example, if giving the first characteristic pulsation, w_0, the other ones result as depending of damping α, according to Eq. (9.27):

$$w_1 = \alpha w_0, \quad w_2 = \alpha^2 w_0, \quad \cdots, \quad w_n = \alpha^n w_0 . \tag{9.28}$$

Further, by giving the value of coefficient a_0, the other polynomial coefficients can be computed by using Eqs. (9.28) and (9.26):

$$a_1 = \frac{a_0}{w_0}, \quad a_2 = \frac{a_0}{\alpha w_0^2}, \quad a_3 = \frac{a_0}{\alpha^3 w_0^3}, \quad \ldots. \tag{9.29}$$

Equations (9.29) allow the general writing of Naslin polynomials as

$$P_N(s) = a_0 \left(1 + \sum_{n=1}^{N} \frac{s^n}{\alpha^{\frac{n(n-1)}{2}} w_0^n} \right). \tag{9.30}$$

Form (9.30) can be used in a control design procedure when the closed-loop transfer function is to have a Naslin polynomial as characteristic polynomial. Closed-loop dynamical performance can thus be configured by choosing two parameters, namely coefficient α and pulsation w_0.

In the case $n = 3$, the Naslin polynomial is (see Kazmierkowski et al. 2002)

$$P_N(s) = a_0 \left(1 + \frac{s}{w_0} + \frac{s^2}{\alpha w_0^2} + \frac{s^3}{\alpha^3 w_0^3} \right).$$

It is easy to verify that $s = -\alpha w_0$ is a root of the above polynomial; this remark helps us write successively developments in Eq. (9.31). The last relation of (9.31) allows one to identify the imposed closed-loop dynamics as third order; there are two complex-conjugated poles and a real pole, all of them being placed at the same frequency, αw_0. As a consequence, the closed-loop system has $\omega_{CL} = \alpha w_0$ as folding frequency, which depends on α and on the first characteristic pulsation, w_0.

$$\begin{aligned} P_N(s) &= \frac{a_0}{\alpha^3 w_0^3} \left(\alpha^3 w_0^3 + \alpha^3 w_0^2 \cdot s + \alpha^2 w_0 \cdot s^2 + s^3 \right) \\ &= \frac{a_0}{\alpha^3 w_0^3} \left[\left(s^3 + \alpha^3 w_0^3 \right) + \alpha^2 w_0 \cdot s(s + \alpha w_0) \right] \\ &= \frac{a_0}{\alpha^3 w_0^3} \left[(s + \alpha w_0)\left(s^2 - \alpha w_0 \cdot s + \alpha^2 w_0^2 \right) + \alpha^2 w_0 \cdot s(s + \alpha w_0) \right] \\ &= \frac{a_0}{\alpha^3 w_0^3} (s + \alpha w_0)\left(s^2 + \alpha w_0 (\alpha - 1) s + \alpha^2 w_0^2 \right). \end{aligned} \tag{9.31}$$

The damping coefficient of the complex pair also is related to the value of α, namely

$$\zeta = \frac{\alpha - 1}{2}. \tag{9.32}$$

Equation (9.32) shows that α must be larger than 1 for the closed loop to be stable, whereas values between 1 and 3 lead to complex-conjugated poles. Moreover, resonance is present if α is between 1 and $1 + \sqrt{2}$. Values of α larger than 3 render factorization of $(s^2 + \alpha w_0(\alpha - 1)s + \alpha^2 w_0^2)$ possible; therefore, the closed loop will have two real poles instead of a complex pair.

The case of a third-order closed-loop system is the one described by Eq. (9.24). Thus, if imposing that polynomial

$$P(s) = s^3 + \frac{1}{L}(K_p v_{DC} + r)s^2 + \left(\omega_0^2 + \frac{2K_i v_{DC}}{L}\right)s + \frac{\omega_0^2}{L}(K_p v_{DC} + r),$$

where ω_0 is frequency of sinusoidal signal to track and K_p and K_i are PI resonant controller parameters, to be a conveniently tuned Naslin polynomial

$$P_N(s) = a_0\left(1 + \frac{s}{w_0} + \frac{s^2}{\alpha w_0^2} + \frac{s^3}{\alpha^3 w_0^3}\right),$$

then the following relations hold after coefficient identification and some algebra:

$$\begin{cases} a_0 = \alpha^3 w_0^3 \\ \alpha^3 w_0^2 = \omega_0^2 + \dfrac{2K_i v_{DC}}{L} \\ \alpha^2 w_0 = \dfrac{1}{L}(K_p v_{DC} + r) \\ w_0 = \dfrac{\omega_0}{\sqrt{\alpha}}, \end{cases} \tag{9.33}$$

which finally leads to PI resonant controller parameters being determined as

$$K_p = \frac{\alpha\sqrt{\alpha}\omega_0 L - r}{v_{DC}}, \quad K_i = \frac{\omega_0^2 L(\alpha^2 - 1)}{2v_{DC}}. \tag{9.34}$$

From last relation of (9.33) it turns out that the closed-loop bandwidth ω_{CL} also depends on the frequency to track, ω_0:

$$\omega_{CL} = \alpha w_0 = \omega_0\sqrt{\alpha}. \tag{9.35}$$

For reasons of stability α must be greater than 1. Therefore the closed-loop bandwidth can theoretically be imposed at least equal to the frequency to track ω_0; but in this case the damping is null. That is resonance is infinite (see relation (9.32)). As the matter of fact, since the frequency to track ω_0 is given, there is a single degree of freedom when the closed-loop performance is imposed, i.e., α, which means that bandwidth and damping cannot be imposed separately. According to Eqs. (9.32) and (9.35), respectively, both damping ζ and bandwidth ω_{CL} increase with α (see also Fig. 9.15).

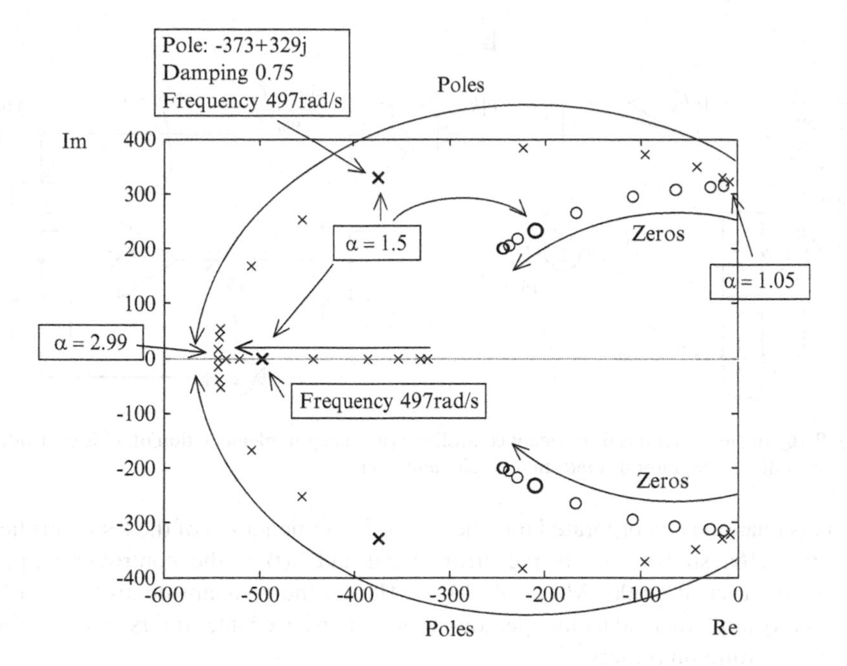

Fig. 9.15 Poles and zeros positions as α increases

Note that for values of α larger than 3 the closed-loop system becomes overdamped (having three different real poles); $\omega_0\sqrt{\alpha}$ no longer represents its bandwidth.

Once the dynamical performance is set, controller parameters result according to (9.34).

9.3.4 Implementation Aspects

It is well known that any linear time-invariant system may be represented by analog diagram containing integrators (energy accumulations) equal in number to the system's order. Starting from the generalized integrator transfer function

$$\frac{Y(s)}{U(s)} = \frac{s}{s^2 + \omega_0^2},$$

with $u(t)$ and $y(t)$ being generic input and output, respectively, one obtains the input-output relation as $Y(s) \cdot (s^2 + \omega_0^2) = s \cdot U(s)$. Multiplying this equation with $1/s^2$ gives

$$Y(s) = \frac{1}{s}\left(U(s) - Y(s)\frac{\omega_0^2}{s}\right). \tag{9.36}$$

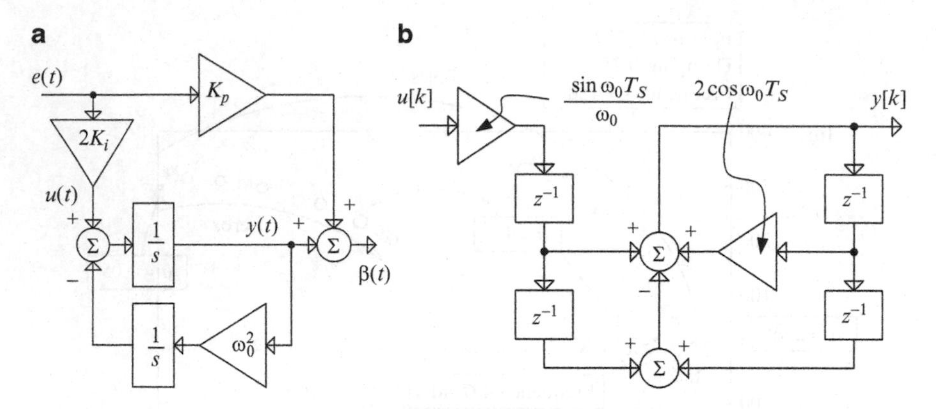

Fig. 9.16 Implementations of resonant controllers: (**a**) analog implementation of a PR controller; (**b**) generalized integrator discrete-time implementation

This relation is incorporated into the analog implementation of the PR controller, as Fig. 9.16a shows ($e(t)$ is the error signal and $\beta(t)$ is the controller output) (Teodorescu et al. 2006; Ma et al. 2011). Hence, the resonant controller can be physically implemented using operational-amplifier-based integrators and inverting or noninverting amplifiers.

Another solution for analog implementation is based upon Eq. (9.19), where the generalized integrator is replaced by a band-pass filter with a very high quality factor.

The discrete-time form of the generalized integrator, suitable to DSP implementation, can be obtained through the Z transform (Oppenheim et al. 1996; Yepes 2011). This form depends on the discretization method used. The "zero-order-hold" method supposes that the system to be discretized is connected in series with a zero-order hold element. The discrete transfer function is

$$H(z^{-1}) = \mathcal{Z}\left\{\mathcal{L}^{-1}\left[\frac{1-e^{sT_s}}{s}H(s)\right]\right\} = \left(1-z^{-1}\right)\cdot \mathcal{Z}\left\{\mathcal{L}^{-1}\left[\frac{H(s)}{s}\right]\right\},$$

where relation $z = e^{sT_S}$ has been used, with T_S being the sampling time. Then, by employing the generalized integrator transfer function (the second term of Eq. (9.19)), one obtains

$$H(z^{-1}) = (1-z^{-1})\cdot \mathcal{Z}\left\{\mathcal{L}^{-1}\left[\frac{1}{s^2+\omega_0^2}\right]\right\}.$$

By consulting tables giving the correspondence between Laplace and z transforms, one may find the generalized integrator discrete transfer function as

$$H\left(z^{-1}\right) = \frac{\left(1-z^{-1}\right)}{\omega_0}\cdot\frac{z\sin\omega_0 T_S}{z^2-2z\cos\omega_0 T_S+1},$$

Or, equivalently,

$$H(z^{-1}) = -\frac{\frac{\sin \omega_0 T_S}{\omega_0} \cdot z^{-1} + \frac{\sin \omega_0 T_S}{\omega_0} \cdot z^{-2}}{z^{-2} - 2\cos \omega_0 T_S \cdot z^{-1} + 1}. \tag{9.37}$$

By adopting the notation $y(kT_S) \equiv y[k]$ for the generic discrete-time signal y, the corresponding difference equation is

$$y[k] = \frac{\sin \omega_0 T_S}{\omega_0}(u[k-1] - u[k-2]) + 2\cos \omega_0 T_S \cdot y[k-1] - y[k-2] \tag{9.38}$$

and the associated diagram is presented in Fig. 9.16b.

Now, let us denote by $\theta = \omega_0 T_S$ the normalized digital frequency. If the Tustin discretization method is used, one substitutes $s = \frac{2}{T_S}\frac{1-z^{-1}}{1+z^{-1}}$ in $H(s) = \frac{s}{s^2+\omega_0^2}$. The generalized integrator transfer function is then

$$H(z^{-1}) = \frac{\frac{2T_S}{\theta^2+4} - \frac{2T_S}{\theta^2+4}z^{-2}}{z^{-2} - \frac{8-\theta^2}{4+\theta^2}z^{-1} + 1}, \tag{9.39}$$

and then the corresponding difference equation is

$$y[k] = \frac{2T_S}{4+\theta^2}(u[k] - u[k-2]) + \frac{8-\theta^2}{4+\theta^2}y[k-1] - y[k-2]. \tag{9.40}$$

Note that for reasonably small normalized frequency θ, the denominator of the transfer function from (9.39) approximates the one from Eq. (9.37). As Tustin's method does not map the frequency domain exactly, a prewarping action may be necessary (Tan 2007; Teodorescu et al. 2006). If Euler's forward method is used, one must substitute $s = \frac{1-z^{-1}}{T_S z^{-1}}$ in $H(s) = \frac{s}{s^2+\omega_0^2}$. The result is

$$H(z^{-1}) = \frac{T_S z^{-1} - T_S z^{-2}}{(1+\theta^2)z^{-2} - 2z^{-1} + 1}, \tag{9.41}$$

which corresponds to the following difference equation:

$$y[k] = T_S u[k-1] + T_S u[k-2] + 2y[k-1] - (1+\theta^2)y[k-2]. \tag{9.42}$$

Note that Eqs. (9.37), (9.39) and (9.41) give similar frequency responses in the domain of interest. Difference Eqs. (9.40) and (9.42) produce slightly different diagrams with respect to the one presented in Fig. 9.16b.

Other methods such as impulse-invariant method (Tan 2007) or direct pole-zero placement in the z plane are also available for designing the generalized integrator.

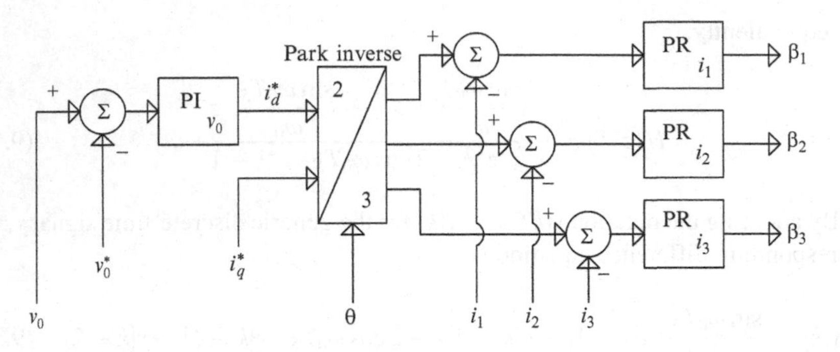

Fig. 9.17 Two-loop cascaded control structure in mixed *dq*-stationary frame dedicated to controlling the circuit of Fig. 9.1

The latter method is based upon discretizing the continuous plant and imposing a suitable closed-loop discrete-pole configuration. The predictive version of the resulting controller leads to the so-called *dead-beat control* ensuring a remarkably fast closed-loop dynamic (Timbuş et al. 2009).

9.3.5 *Use of Resonant Controllers in a Hybrid dq-Stationary Control Frame*

Let us reconsider the problem formulated at the beginning of Sect. 9.2, consisting of regulating the DC voltage of the circuit from Fig. 9.1 while controlling the reactive power exchanged with the three-phase grid. The proposed cascaded control structure that works exclusively in the synchronously rotating *dq* frame may be modified in order to accommodate PR current (inner) controllers that operate in the stationary (*abc*) frame. Thus, a combined *dq*-stationary control structure – as depicted in Fig. 9.17 – may be effectively used.

This structure supposes the outer loop control working in the *dq* frame, that is, references concerning the *d* and *q* current components issue from the active and power requirements (i.e., the necessary values that ensure the power balance). Further, these references are transformed in three-phase AC variables (inverse Park transform) by using the PLL-based grid electrical angle. The inner control structure contains three independent PR-driven control loops that track these AC references. Finally, one may state the main steps of a generic design algorithm for control in the hybrid *dq-abc* frame of PWM AC-based converters.

Algorithm 9.2 Design of control structures in the hybrid *dq-abc* frame

#1 Ensure that dynamics of circuit are separable such that cascaded control loop is suitable (i.e., DC variable has largest time constant).

#2 Consider a cascaded control structure having DC variable as target of outer loop and AC variables as target of inner control loop. Establish outer closed-loop bandwidth based on the imposed overall performance and

(continued)

Algorithm 9.2 (continued)

inner-loop performance that ensures complete frequency separation. Outer loop(s) operate(s) in *dq* frame and inner loop(s) work in stationary *abc* frame.

#3 Write down averaged model of circuit and express it by means of DC variables in *dq* frame (all high-frequency variations must be neglected). Consider only equations related to outer-loop dynamics. Choose the quiescent operating point where control design will be made (e.g., the one corresponding to full-load conditions) and linearize circuit model around selected operating point.

#4 Consider that inner loop is ideal (much faster than outer loop) and make suitable simplifications for given plant. Reactive power may be controlled separately. Deduce transfer function(s) that describe (simplified) dynamics of outer plant. Consider a suitable controller that may achieve in closed loop the performances established at #2. By using either loop-shaping or pole-placement method, compute controller parameters so as to satisfy these performances.

#5 Write down only the dynamic equations related to inner-loop dynamics in stationary frame, while considering DC outer variables as constant (this is possible at their setpoints). Make suitable simplifications for inner plant. For each phase write down inner plant transfer function. By using either loop-shaping, Naslin polynomials or pole-placement method, compute suitable proportional-resonant controllers that satisfy performances imposed at #2.

#6 If necessary, consider an adaptation of outer controller parameters in order to either obtain a smooth start-up or to preserve the desired dynamic performance on entire operating range.

9.3.6 Example of a Grid-Connected Three-Phase Inverter

The PWM-operated power electronic converter of Fig. 9.18 is powered by an unregulated DC current supply (e.g., a PV array) and supplies a strong three-phase electrical grid. The DC power source outputs a variable current as a function of its primary resource and of the DC voltage.

A control structure is to be conceived that allows the available DC power to be injected into the grid. The converter must also provide a certain level of reactive power.

The control structure should be a combined *dq*-stationary two-loop cascade and its controllers designed by using the circuit linearized model in the framework of Algorithm 9.2.

The circuit *dq* model is (see Eq. (5.80) in Chap. 5)

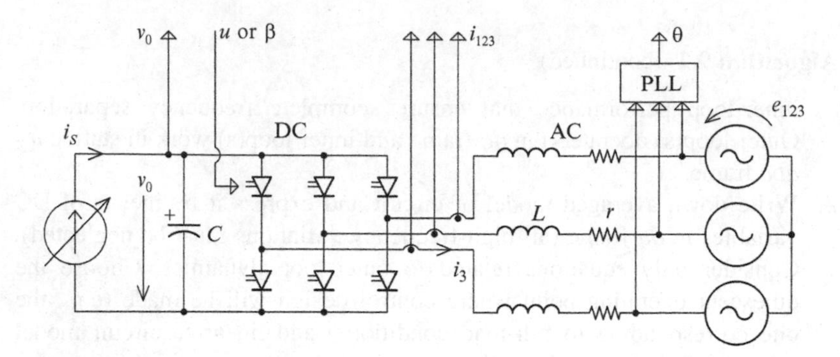

Fig. 9.18 Grid-connected three-phase inverter diagram

$$
\begin{cases}
\dfrac{di_d}{dt} = \omega i_q - r i_d + \dfrac{E}{L} - \dfrac{v_0}{2L} \cdot \beta_d \\
\dfrac{di_q}{dt} = -\omega i_d - r i_q - \dfrac{v_0}{2L} \cdot \beta_q \\
\dfrac{dv_0}{dt} = \dfrac{3}{4C} \left(i_d \beta_d + i_q \beta_q \right) - \dfrac{v_0}{CR} + I_0,
\end{cases}
\tag{9.43}
$$

where ω is the grid pulsation, C is the DC-link capacitor, and L and r are the grid-side inductor parameters. Also, it is supposed that the primary DC current source has an output characteristic of the form

$$
i_S = f(v_0) = I_0 - \frac{v_0}{R},
$$

where I_0 is a slowly-variable parameter depending on the primary resource and R is the dynamic resistance of the source (the inverse of the characteristic $i_S - v_0$ slope). First, suppose that the DC-link capacitor (at the rated circuit power) induces sufficiently slow dynamics with respect to the inductor current dynamic so that a two-loop cascaded control structure is pertinent. Secondly, suppose that the level of reactive power to be injected into the grid is low in relation to the active power and its reference varies slowly. In this case the influence of term $i_q \beta_q$ on the right-hand side of the third equation of (9.43) can be neglected (it may be seen as a disturbance). Under these assumptions this equation may be linearized around a quiescent operating point as

$$
C \frac{dv_0}{dt} = \frac{3}{4} \beta_{d0} i_d - \frac{v_0}{R_e} - \frac{3}{4} \beta_{q0} i_q,
\tag{9.44}
$$

corresponding to a structure like the one depicted in Fig. 9.6. Note that the last term is seen as disturbance. It follows that the control transfer function on the d channel (which represents here the outer plant) is

$$H_{i \to v}(s) = \frac{0.75\beta_{d0}}{Cs + 1/R_e}. \quad (9.45)$$

Regulation of DC voltage (outer loop) can be achieved by means of a PI controller whose parameters are designed according to the procedure detailed in Sect. 9.2.1.

As in the previous cases (see Sect. 9.2), this controller outputs the necessary current $i_d{}^*$ in order to maintain the DC-link power balance. Also, by modifying the value of $i_q{}^*$, one varies the reactive power injected into the grid (for simplicity, the regulation of this latter current is omitted here).

Now, consider that the two references $i_d{}^*$ and $i_q{}^*$ are transformed in three-phase AC variables, $i_{123}{}^*$, by using the Park inverse transform and a PLL-supplied electrical angle θ. The three inner control structures must ensure the tracking of these AC current references.

For any phase k, the inner plant has the following (current) averaged dynamic, given essentially by inductor inertia (see Eq. (9.2)):

$$L\frac{di_k}{dt} = e_k - \frac{\beta_k v_0}{2} - ri_k + \frac{v_0}{6}\sum_{k=1}^{3}\beta_k \quad k = 1, 2, 3, \quad (9.46)$$

with v_0 being the DC-link voltage value (supposed constant at the desired value).

Supposing a symmetrical grid voltage (even in dynamic conditions), then $\sum_{k=1}^{3} e_k = 0$ at any moment. Moreover, as the neutral is isolated, the sum of the line currents is also zero, $\sum_{k=1}^{3} i_{Lk} = 0$. Then, by summing all the three Eq. (9.46) (for all values of k), one obtains the identity

$$\frac{v_0}{2}\sum_{k=1}^{3}\beta_k = \frac{v_0}{6}\sum_{k=1}^{3}\beta_k,$$

which – for a reasonable operating point at $v_0 \neq 0$ – is valid only for $\sum_{k=1}^{3}\beta_k = 0$. Therefore, for a symmetrical undistorted condition, Eq. (9.46) can be rewritten as

$$L\frac{di_k}{dt} = e_k - \frac{\beta_k v_0}{2} - ri_k \quad k = 1, 2, 3. \quad (9.47)$$

By considering the grid voltages as disturbances, the control transfer function of a generic phase k (inner plant) is a linear time-invariant first-order system:

$$H_{\beta \to i}(s) = \frac{0.5v_0}{Ls + r}. \quad (9.48)$$

An inner control structure on each of the three grid phases can now be computed based upon use of a PR controller, whose transfer function is given by Eq. (9.19). For each phase, the open-loop transfer function is

$$H_{OL}(s) = \left(K_p + K_i \frac{2s}{s^2 + \omega_0^2}\right) \frac{0.5v_0}{Ls + r}, \tag{9.49}$$

where parameters K_p and K_i must yet be computed by trading off bandwidth and stability requirements (see also Fig. 9.13).

Now, by considering that the effect of the resonant term affects the characteristic only around frequency ω_0, the simplified open-loop transfer is

$$H_{OL}^S(s) = \frac{0.5K_p v_0}{Ls + r}.$$

Note that the open-loop cut-off frequency ω_{OL} is practically identical to the closed-loop system bandwidth. Usually the latter is imposed to be two–ten times larger than the original plant bandwidth, r/L (depending on how much one wants to "hurry" the system). In this example, let us consider $\omega_{OL} = 5 \cdot r/L$. Then one may compute the gain K_p by using Eq. (9.23):

$$K_p = \frac{1}{0.5v_0}\sqrt{\omega_{OL}^2 L^2 + r^2}.$$

Also, by neglecting the influence of inductor resistance r in the above equation, one may obtain an approximate value of the above gain as $K_p = \frac{L\omega_{OL}}{0.5v_0} = \frac{10r}{v_0}$.

Equivalently, one may write the closed-loop transfer function (always by neglecting the influence of the resonant term) as:

$$H_{CL}^S(s) = \frac{\dfrac{0.5K_p v_0}{0.5K_p v_0 + r}}{\dfrac{L}{0.5K_p v_0 + r}s + 1}.$$

This function has almost unit gain and the closed-loop bandwidth is $\frac{0.5K_p v_0 + r}{L}$. By equating this quantity with ω_{OL}, one obtains $K_p = \frac{8r}{v_0}$.

All these computations provide values for the proportional gain, K_p which are very close one to each other. For small resistance r at tenths of ohms and large values of v_0 at hundreds of volts, one obtains quite small values for K_p on the order of mΩ/V.

At this moment one may consider that this parameter has been computed and proceed to choose the controller parameter K_i that multiplies the resonant term. This value must be chosen as large as possible in order to obtain fast zeroing of the steady-state error, but as it introduces a large phase lag around ω_0, it must be limited to a reasonable value that ensures a sufficiently large phase margin. This latter quantity must be computed in radians as

$$\gamma = \pi - |\arg(H_{OL}(\omega_{OL}))|,$$

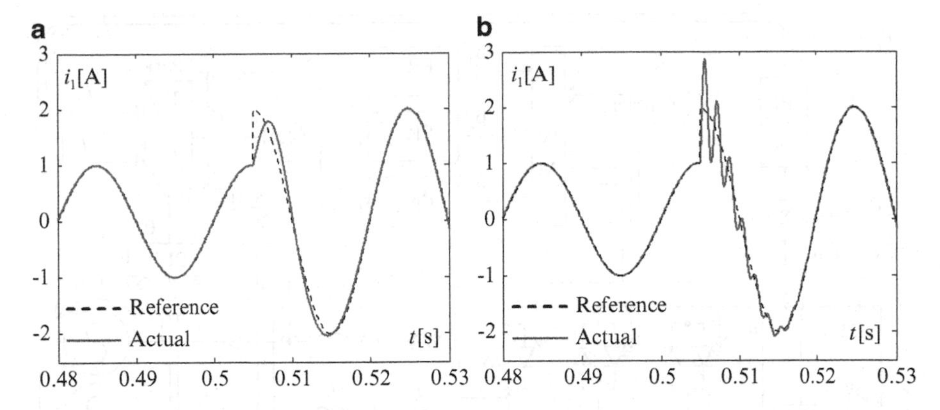

Fig. 9.19 Different inner loop responses at step variation of reference's amplitude for three-phase inverter controlled in hybrid *dq-abc* frame: (**a**) with phase margin of about 60°; (**b**) with phase margin of about 10°

and, by using Eq. (9.49), one obtains

$$\gamma = \pi - \arctan\frac{2K_i\omega_{OL}}{K_p\left(\omega_{OL}^2 - \omega_0^2\right)} - \arctan\frac{\omega_{OL}L}{r} \approx 1.77 - \arctan\frac{2K_i\omega_{OL}}{K_p\left(\omega_{OL}^2 - \omega_0^2\right)}.$$

By forcing this expression to be $\gamma = 0.7$ rad (which means a phase margin of about 40°), one may compute the value of K_i.

Remarks Due to inherent filtering, in real applications supplementary lags are likely to appear in the above phase margin expression, thus reducing the choice range of integral gain K_i Figure 9.19 shows system response (in terms of AC current) with sufficient versus insufficient phase margin.

The phase margin may also be chosen by a trial-and-error approach using interactive tools provided by control-oriented computer-aided design software like MATLAB® (e.g., `sisotool`).

The complete control structure embedding the inner and outer control loops is given in Fig. 9.20.

9.4 Control of Full-Wave Converters

Full-wave operated DC-AC converters are used in a wide variety of applications. Due to their quite reduced flexibility, they are modeled and controlled differently according to each particular case. For example, in the case of insulated load (either off-grid inverters or electronic ballasts driving high-intensity discharge lamps), the control input may be the frequency of the square wave (Yin et al. 2002). In grid-connected rectifiers, the frequency is fixed by the grid and the converter may be driven by the

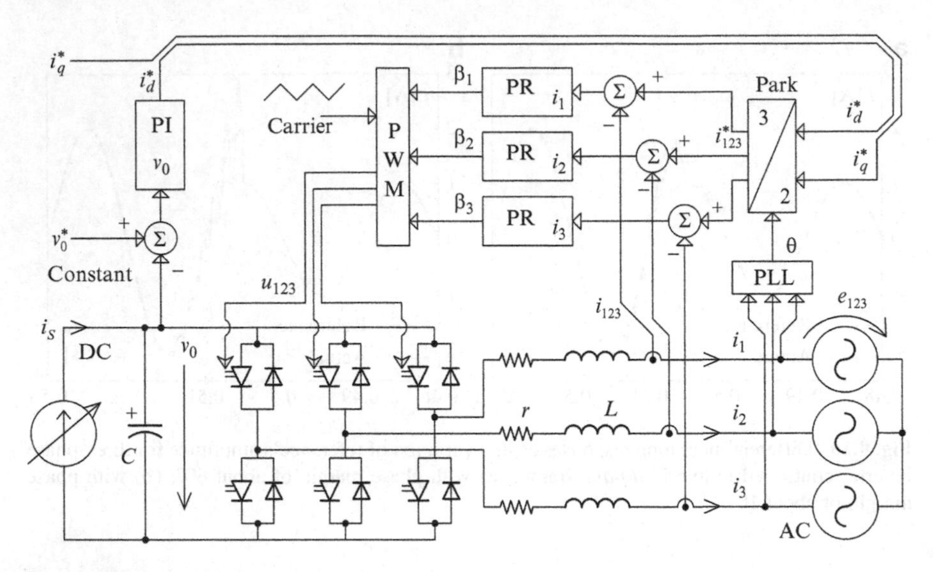

Fig. 9.20 Three-phase inverter controlled in the hybrid *dq-abc* frame

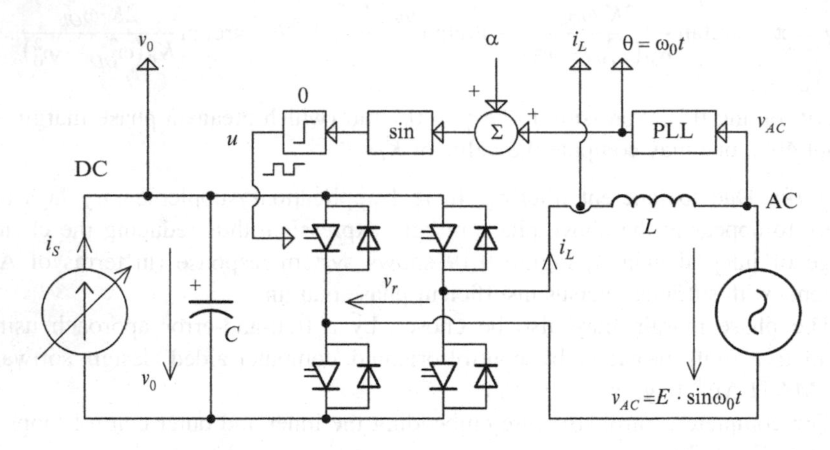

Fig. 9.21 Full-wave grid-connected single-phase inverter

phase lag of the square wave with respect to the grid voltage phase (Petitclair et al. 1996). In the case of resonant power supplies, the phase lag between control inputs of the two DC-AC converters surrounding the resonant tank may be utilized as a global control input (Bacha et al. 1994). Choosing different control inputs leads to different small-signal models and hence to using different control methods.

Next, a single-phase full-wave operated grid-connected inverter is considered – see Fig. 9.21. Its power structure is the same as in the PWM case (see Fig. 9.4), but switching function u is obtained in a different manner. The primary objective is to regulate the DC voltage and a control structure that accomplishes this goal is to be designed.

The states are chosen as usual, inductor current and capacitor voltage. Recall the generalized averaged model from Chap. 5, where the sense of the current corresponds to the rectifier mode (only the first harmonic of the AC variables and the average of the DC variables are considered):

$$\begin{cases} \dfrac{d\langle i_L\rangle_1}{dt} = -j\omega_0\langle i_L\rangle_1 + \dfrac{1}{L}\left(\langle v_{AC}\rangle_1 - \langle v_0 \cdot u\rangle_1\right) \\ \dfrac{d\langle v_0\rangle_0}{dt} = \dfrac{1}{C}\left(\langle i_L \cdot u\rangle_1 - \langle i_S\rangle_0\right), \end{cases} \tag{9.50}$$

where ω_0 is grid frequency, E is grid voltage amplitude, C is DC-link capacitor, L is AC inductance and i_S is current supplied/drawn by unregulated DC source/sink. The 0.5-duty-ratio rectangular wave control input u is supplied with a phase lag α with respect to grid voltage phase $\omega_0 t$ (see Chap. 5 for quick reference).

This model can be rewritten in terms of active and reactive components of the current inductor current i_L (i_d and i_q), as well as of the DC-link voltage average, $\langle v_0\rangle_0$ – see the developments in Chap. 5. For simplicity the averaging brackets will be further dropped from variable notation.

$$\begin{cases} \dfrac{di_d}{dt} = \omega_0 i_q + \dfrac{E}{L} - v_0\dfrac{4}{\pi L}\cos\alpha \\ \dfrac{di_q}{dt} = -\omega_0 i_d - v_0\dfrac{4}{\pi L}\sin\alpha \\ \dfrac{dv_0}{dt} = -\dfrac{2}{\pi C}\left(i_d\cos\alpha + i_q\sin\alpha\right) + \dfrac{i_S}{C}. \end{cases} \tag{9.51}$$

In the state-space model of Eq. (9.51), the control input is the phase lag α. Note that the model is nonlinear – it uses harmonic functions of α. Moreover, it has only one input and hence does not allow separate control of inductor current components i_d and i_q (as is the case of PWM-controlled converters). The active and reactive AC power components cannot be separately controlled.

Consider now a typical steady-state operating point of this model described by the values α_e, i_{Se}, v_{0e}, i_{de} and i_{qe}. The power exchanged with the DC source is $v_{0e}i_{Se}$ or $i_{Se}{}^2R_e$, where $R_e = v_{0e}/i_{Se}$ is the equivalent DC resistance in the quiescent operating point. As the inverter operation implies a nonnegligible DC power transfer, the current $i_{Se} = v_{0e}/R_e$ is also important (R_e has a small value). This operating point satisfies the following system of equations:

$$\begin{cases} \omega_0 L i_{qe} + E - \dfrac{4v_{0e}}{\pi}\cos\alpha_e = 0 \\ -\omega_0 L i_{de} - \dfrac{4v_{0e}}{\pi}\sin\alpha_e = 0 \\ -\dfrac{2}{\pi}\left(i_{de}\cos\alpha_e + i_{qe}\sin\alpha_e\right) - \dfrac{v_{0e}}{R_e} = 0. \end{cases} \tag{9.52}$$

The third relation of (9.52) shows that the expression of the current injected into the DC-link,

$$i_{de} \cos \alpha_e + i_{qe} \sin \alpha_e = \frac{\pi}{2} i_S,$$

is an important quantity for the chosen control input α_e. One multiplies the first equality of (9.52) by sin α_e and the second by cos α_e; then, by subtracting the second equation from the first, one obtains $\omega_0 L(i_{qe} \sin \alpha_e + i_{de} \cos \alpha_e) = -E \sin \alpha_e$; therefore

$$i_{qe} \sin \alpha_e + i_{de} \cos \alpha_e = -\frac{E}{\omega_0 L} \sin \alpha_e,$$

meaning that the value of sin α_e is also large. That is, the voltage equilibrium requires a sufficiently large value of phase lag α_e (around $\pi/2$) in order to compensate for the DC current i_{Se}.

The second equality of (9.52) gives $i_{de} = -\frac{4v_{0e}}{\pi\omega_0 L} \sin \alpha_e$, meaning that a large DC power requires a large value of the active power be drawn/supplied from/to the grid (note that $P = E \cdot i_d$).

To conclude, in the case where the AC-DC coupling acts as rectifier/inverter, the DC voltage balance may be achieved by imposing the right value of i_d (i.e., the DC voltage controller will output the active current reference). Thus, a cascaded control structure is envisioned, having an outer loop that deals with the DC voltage (v_0) control and an inner loop that performs the control of currents.

Note that if the AC-DC coupling acts as a STATCOM, then resistance R_e is large, DC power is not important and phase lag α_e is around zero (cos α_e is large).

Further, the small-signal state-space model developed around the above-defined quiescent operating point is

$$\begin{cases} L\dfrac{d\widetilde{i_d}}{dt} = \omega_0 L\widetilde{i_q} - \dfrac{4\cos\alpha_e}{\pi}\widetilde{v_0} + \dfrac{4v_{0e}\sin\alpha_e}{\pi}\widetilde{\alpha} \\ L\dfrac{d\widetilde{i_q}}{dt} = -\omega_0 L\widetilde{i_d} - \dfrac{4\sin\alpha_e}{\pi}\widetilde{v_0} - \dfrac{4v_{0e}\cos\alpha_e}{\pi}\widetilde{\alpha} \\ C\dfrac{d\widetilde{v_0}}{dt} = -\dfrac{2\cos\alpha_e}{\pi}\widetilde{i_d} - \dfrac{2\sin\alpha_e}{\pi}\widetilde{i_q} - \dfrac{\widetilde{v_0}}{R_e} + \dfrac{2\left(i_{de}\sin\alpha_e - i_{qe}\cos\alpha_e\right)}{\pi}\widetilde{\alpha} - \widetilde{i_S}, \end{cases} \tag{9.53}$$

where $\widetilde{i_S}$ is the DC source current variation around the steady-state value $i_{Se} = v_{0e}/R_e$. Note that the current components i_d and i_q can be measured using a phase-locked loop as detailed in Sect. 9.2.1. Since the angle α influences simultaneously the channels d and q, a single current may be controlled in the inner structure. If, for example, a loop of i_d is built (for reasons described in the previous paragraph), the current i_q remains uncontrolled and one must ensure that its dynamic is stable. However, one may impose

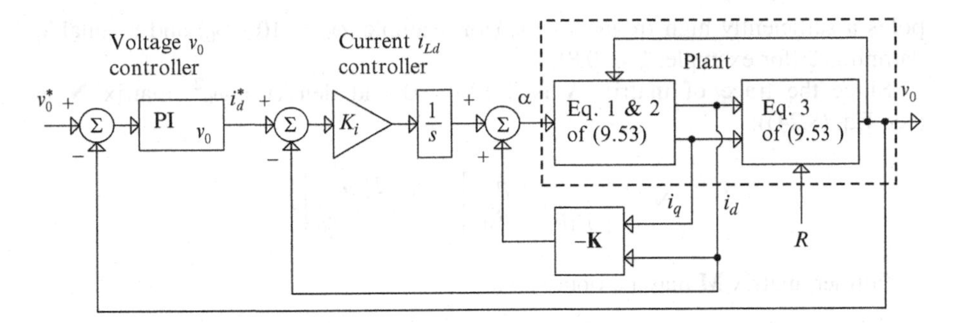

Fig. 9.22 Three-loop DC voltage control structure in which innermost loop is a partial-state feedback

a convenient dynamic to the pair (i_d, i_q) by using a partial-state feedback control structure (see Fig. 9.22). For this purpose, a simplified reduced-order current (inner) plant is considered, by neglecting the variations of the DC voltage v_0; its model is

$$\begin{cases} L\frac{d\widetilde{i_d}}{dt} = \omega_0 L\widetilde{i_q} + \frac{4v_{0e}\sin\alpha_e}{\pi}\widetilde{\alpha} \\ L\frac{d\widetilde{i_q}}{dt} = -\omega_0 L\widetilde{i_d} - \frac{4v_{0e}\cos\alpha_e}{\pi}\widetilde{\alpha}. \end{cases} \tag{9.54}$$

The design of the state feedback for the system (9.54) may be done by using either dedicated software (e.g., function `acker` in MATLAB®) or the procedure presented in Sect. 8.4.1 of Chap. 8. According to the developments in the cited chapter, one imposes the closed-loop poles pair as $\mathbf{P} = [p_1 \quad p_2]^T$, where $p_{1,2} = -\zeta\omega_n \pm j\omega_n\sqrt{1-\zeta^2}$. Then, the analytical solution for the state feedback computation may be obtained using the procedure developed in Sect. 8.2.1. as (see Eq. (8.25)):

$$\mathbf{K}^T = [k_1 \quad k_2]^T = \mathbf{M}^{-1} \cdot \mathbf{N}, \tag{9.55}$$

where matrices **M** and **N** are given by Eq. (8.24) involving state matrix **A** and input matrix **B** of system (9.54). These matrices are, respectively:

$$\mathbf{A} = \begin{bmatrix} 0 & \omega_0 \\ -\omega_0 & 0 \end{bmatrix}, \mathbf{B} = \begin{bmatrix} \frac{4v_{0e}}{\pi}\sin\alpha_e \\ -\frac{4v_{0e}}{\pi}\cos\alpha_e \end{bmatrix}.$$

The eigenvalues of matrix **A** are the solutions of equation det(s**I** − **A**) = 0, which further gives $s^2 + \omega_0{}^2 = 0$. This means that the poles of system (9.54) are placed on the imaginary axis at the grid frequency ω_0, the damping being zero due to the fact that inductor resistance has been neglected. One imposes to the closed-loop system poles a sufficiently high frequency ω_n (for example, $\omega_n = 10 \cdot \omega_0$) and a suitable damping, ζ (for example, $\zeta = 0.8$).

Since the trace of matrix **A** is $\mathrm{Tr}(\mathbf{A}) = 0$ and $\det(\mathbf{A}) = \omega_0{}^2$, matrix **N** is (see Eq. (8.24)):

$$\mathbf{N} = \begin{bmatrix} -(p_1 + p_2) \\ p_1 p_2 - \omega_0^2 \end{bmatrix} = \begin{bmatrix} 2\zeta\omega_n \\ \omega_n^2 - \omega_0^2 \end{bmatrix}.$$

Further, matrix **M** may be obtained as

$$\mathbf{M} = \frac{4v_{0e}}{\pi} \begin{bmatrix} \sin\alpha_e & -\cos\alpha_e \\ -\omega_0 \cos\alpha_e & -\omega_0 \sin\alpha_e \end{bmatrix}$$

having $-\frac{4v_{0e}\omega_0}{\pi}$ as determinant; its inverse is computed as

$$\mathbf{M}^{-1} = -\frac{\pi}{4v_{0e}\omega_0} \begin{bmatrix} -\omega_0 \sin\alpha_e & \cos\alpha_e \\ \omega_0 \cos\alpha_e & \sin\alpha_e \end{bmatrix}.$$

Equation (9.55) providing the gains of the state feedback leads in this case to

$$\begin{cases} k_1 = -\dfrac{\pi}{4v_{0e}\omega_0}\left[-2\zeta\omega_0\omega_n \sin\alpha_e + \left(\omega_n^2 - \omega_0^2\right)\cos\alpha_e\right] \\ k_2 = -\dfrac{\pi}{4v_{0e}\omega_0}\left[2\zeta\omega_0\omega_n \cos\alpha_e + \left(\omega_n^2 - \omega_0^2\right)\sin\alpha_e\right]. \end{cases} \tag{9.56}$$

The partial-state feedback structure intended for the innermost control loop may be seen in Fig. 9.22. In order to achieve the current i_d control, this structure must be completed with an intermediate control loop based upon an integrator (one considers the same framework as in Sect. 8.4.1).

The integrator gain K_i is chosen by means of root locus method so as to obtain a good closed-loop bandwidth (in terms of i_d tracking) and a suitable damping, ζ_C. For example, one may take K_i such that all three poles of the i_d closed-loop have the same frequency, $\omega_C = 1/T_C$ – see Fig. 9.23.

In this context, the i_d closed-loop transfer function can be described as

$$H_i(s) = \frac{1}{(T_C s + 1)\left(T_C^2 s^2 + 2\zeta_C T_C s + 1\right)}. \tag{9.57}$$

Note that a large value of ω_C will ease the design of the outermost control loop (i.e., of the DC voltage v_0). The plant of this loop has the transfer function given by Eq. (9.57).

Besides the PI component that ensures zero steady-state error, a suitable DC voltage controller must also have an anticipative component to ensure a reasonable phase margin (the third-order transfer function (9.57) induces large phase lag). To this end, one may use either a PI controller with a separate lead-lag transfer function

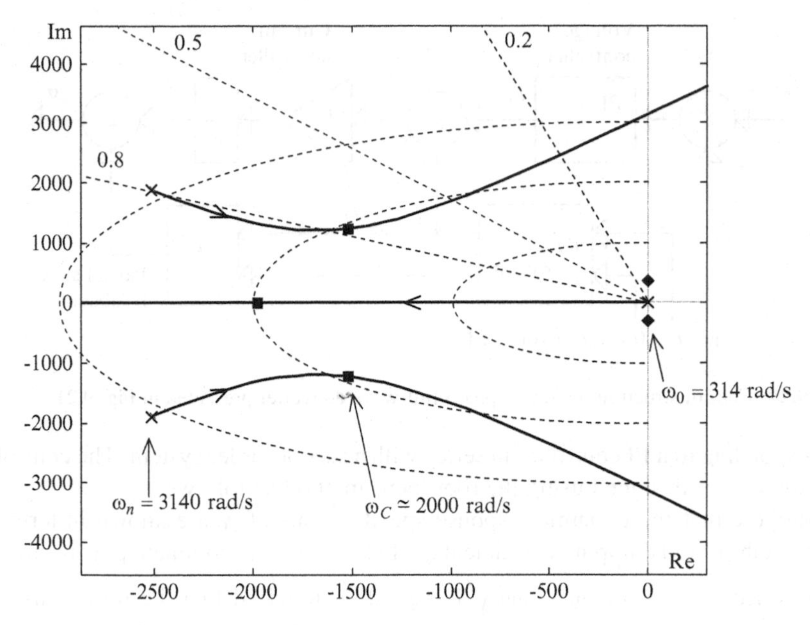

Fig. 9.23 Poles positions for different systems in example

or, directly, a PID controller. In the latter case, let us consider a PID controller with the transfer function:

$$H_{Cv}(s) = K_v \frac{(T_{C1}s + 1)(T_v s + 1)}{T_v s}. \tag{9.58}$$

Note that the term $(T_C s + 1)$ in the outer plant (see the denominator in Eq. (9.57)) varies with the operating point (**B** depends on α_e). If imposing $T_{C1} = T_C$, this term will be compensated exactly, but only in the design operating point (e.g., at full load). Otherwise, the transfer function $\frac{T_{C1}s+1}{T_C s+1}$ in the outer open loop corresponds to a lead-lag system which is likely to introduce small alterations in the initially imposed dynamic performance. A sensitivity analysis is thus necessary in each particular case.

In conclusion, consider that the open-loop system is described by the transfer function

$$H_{ol_v}(s) = \frac{K_v(T_v s + 1)}{T_v s(T_C^2 s^2 + 2\zeta_C T_C s + 1)}, \tag{9.59}$$

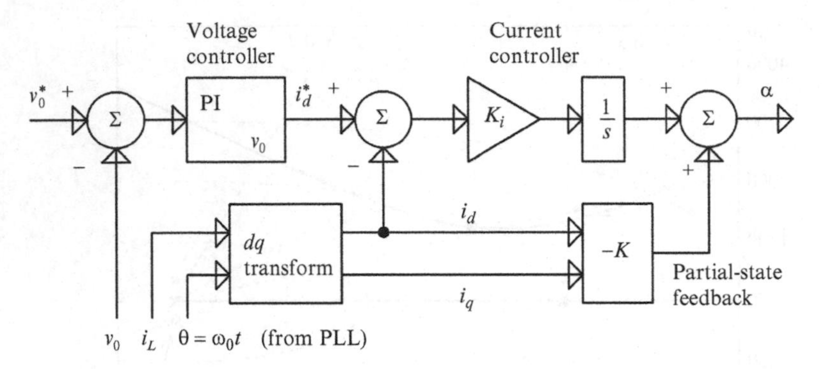

Fig. 9.24 Control structure for single-phase full-wave converter presented in Fig. 9.21

corresponding to a PI controller in series with a second-order system. The controller parameters are designed using the root locus method as follows.

Suppose that the dynamic response specifications of v_0 are known in terms of bandwidth ω_V and damping coefficient ζ_V. In this way, the dominant pole position is determined as $p = -\zeta_V\omega_V + j\omega_V\sqrt{1-\zeta_V^2}$. It is required that the root locus pass through pole p, corresponding to the imposed bandwidth and overshoot. The computation of controller parameters follows a quite laborious procedure requiring complex analysis tools (d'Azzo et al. 2003); alternatively, one can use dedicated software tools like, `rltool` from Control Toolbox of MATLAB®.

To conclude, the control structure contains three control loops, as can be seen in Fig. 9.22. The innermost control loop consists in fact in a pole placement using a partial-state feedback; in this way desired dynamics of currents i_d and i_q are imposed. The intermediate loop relies upon using an integral controller aiming at tracking the required active current i_d achieves power balance. Finally the outermost control loop is in charge of the DC voltage regulation, namely by employing a PID controller. The two latter control loops are designed by using root locus method. The global control structure to be coupled with the converter in Fig. 9.21 is described in Fig. 9.24.

9.5 Case Study: *dq*-Control of a PWM Three-Phase Grid-Tie Inverter

A three-phase voltage-source inverter which supplies an ideal AC grid is considered (see Fig. 9.25).

It receives energy from a controlled DC current source and must inject the entire available DC power to the AC grid. The output reactive power must be zero (i.e., the grid currents are in anti-phase with the corresponding grid voltages).

Inductor parameters are $L = 2.2$ mH and $r = 0.1\ \Omega$, and capacitor parameters are $C = 4700\ \mu$F and $R_C = 47$ kΩ. The DC source rated power is 2025 W, its rated voltage is $v_0{}^* = 450$ V. The grid amplitude is $E = 180$ V (with RMS value being 127 V) and its frequency is $\omega_0 = 50$ Hz.

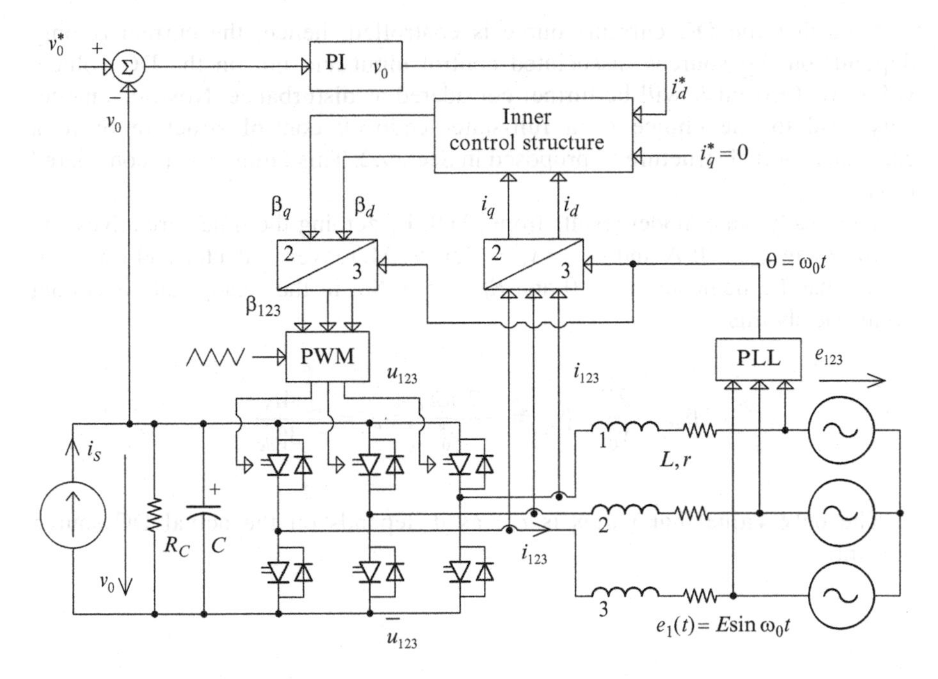

Fig. 9.25 Three-phase voltage-source grid-connected inverter connected with associated *dq*-frame control structure (inner control structure still to be detailed)

The control structure in the *dq* frame is to be designed so the stated control goal can be achieved. In the overall dynamic performance is included the requirement that DC voltage variations not exceed 1 % and its transient last for a maximum 0.25 s at unit step variation of the DC source voltage. Next, Algorithm 9.1 is used for design purpose.

9.5.1 System Modeling

Modeling follows essentially the steps presented in Sect. 9.2. But unlike the case in the cited section, the switching functions are in the set $\{-1; 1\}$. Hence, the three-phase control inputs β_{123} belong to the $[-1, 1]$ interval and, by considering that a current measure in the *dq* frame is available, the circuit averaged model that results is (see Eq. (5.80) of Chap. 5):

$$\begin{cases} L\dot{i_d} = \omega L i_q - \dfrac{v_0}{2}\beta_d + E - r i_d \\ L\dot{i_q} = -\omega L i_d - \dfrac{v_0}{2}\beta_q - r i_q \\ C\dot{v_0} = \dfrac{3}{4}\left(i_d\beta_d + i_q\beta_q\right) - \dfrac{v_0}{R_C} + i_S. \end{cases} \tag{9.60}$$

Note that the DC current source is controlled; hence, the current i_S only depends on the sources' associated control input and not on the DC voltage value, v_0. Current i_S will be further considered as disturbance. Now, this model may lead to the choice of a full-state feedback control structure or to a cascaded control structure as proposed in Sect. 9.2. This latter case is considered next.

The steady-state model results from (9.60) by zeroing the time derivatives and by imposing $i_{qe} = 0$ A and $v_{0e} = v_0^* = 450$ V. Moreover, inductor and capacitor losses may be neglected ($r = 0$ and $R_C = \infty$). So, in the steady-state operating point one obtains

$$\beta_{de} = \frac{2E}{v_0^*}, \quad \beta_{qe} = -\frac{2\omega L i_d}{v_0^*}, \quad i_{de} = -\frac{4 i_S}{3\beta_{de}}.$$

The only value that varies is i_{de}, as it depends on the actual DC source current.

9.5.2 Comments on the Adopted Control Structure

The proposed control structure – in the sense of Sect. 9.2 – is shown in Fig. 9.25. Control of the power injected to the grid supposes control of the inductor currents. Thus, the inner control loop(s) refer(s) to the inductor current. Note that the primary control loop is the power balance in the DC-link (the entire DC power supplied by the DC source must be evacuated to the AC circuit). As the DC voltage acts as a transducer of the power unbalance, this requirement is equivalent to maintaining the DC-link voltage constant at the rated value. As in the case of single-phase voltage-source converters (see Sect. 9.2.1), the DC-link output current depends mainly on the grid active current component i_d. This further implies that the voltage regulation acts on the active current reference i_d. The reactive power injected to the grid is proportional to the current component i_q. Controllers' outputs are altered in order to achieve decoupling between the d and q channels. PI controllers are used for each of the controlled variable (i_d, i_q and v_0). Prefilters are necessary in order to compensate for the zeros induced by the use of these types of controllers. These remarks issue from Sect. 9.2 and lead to the control structure depicted in Fig. 9.25.

9.5.3 Design of the Inner Loop (Current) Controllers

The linearized model around a quiescent operating point is

$$\begin{cases} L\dot{\widetilde{i_d}} = \omega L\widetilde{i_q} - \frac{v_{0e}}{2}\widetilde{\beta_d} - \frac{\beta_{de}}{2}\widetilde{v_0} - r\widetilde{i_d} \\ L\dot{\widetilde{i_q}} = -\omega L\widetilde{i_d} - \frac{v_{0e}}{2}\widetilde{\beta_q} - \frac{\beta_{qe}}{2}\widetilde{v_0} - r\widetilde{i_q} \\ C\dot{\widetilde{v_0}} = \frac{3\beta_{de}}{4}\widetilde{i_d} + \frac{3\beta_{qe}}{4}\widetilde{i_q} + \frac{3i_{de}}{4}\widetilde{\beta_d} + \frac{3i_{qe}}{4}\widetilde{\beta_q} - \frac{1}{R_C}\widetilde{v_0} + \widetilde{i_S}\,. \end{cases} \tag{9.61}$$

Consider that in the currents' equations the DC voltage is constant, as maintained at the desired value by the outer control loop ($v_{0e} \equiv v_0{}^*$). The current (inner) plant becomes

$$\begin{cases} L\dot{\widetilde{i_d}} = \omega L\widetilde{i_q} - \frac{v_{0e}}{2}\widetilde{\beta_d} - r\widetilde{i_d} \\ L\dot{\widetilde{i_q}} = -\omega L\widetilde{i_d} - \frac{v_{0e}}{2}\widetilde{\beta_q} - r\widetilde{i_q}\,. \end{cases} \tag{9.62}$$

In the first equation of (9.62) $\widetilde{i_q}$ is considered perturbation (which is further significantly reduced by the d-q decoupling structure). The same applies for $\widetilde{i_d}$ in the second equation of (9.62). Therefore, both d and q channels may be described by the transfer function

$$H_C(s) = -\frac{v_0^*}{2r}\cdot\frac{1}{rLs+1} = K_C\cdot\frac{1}{T_Cs+1}. \tag{9.63}$$

Using the parameter values, the gain of $H_C(s)$ results $K_C = 2250$ A and the time constant is $T_C = 22$ ms. As the current control loop is built by using a PI controller $H_{PIC}(s) = K_{pC}\left(1 + \frac{1}{T_{iC}s}\right)$, the closed-loop transfer function is

$$H_{0C}(s) = \frac{H_{PIC}(s)\cdot H_C(s)}{1 + H_{PIC}(s)\cdot H_C(s)} = \frac{T_{iC}s+1}{\frac{T_CT_{iC}}{K_CK_{pC}}s^2 + T_{iC}\left(\frac{1}{K_CK_{pC}}+1\right)s+1}.$$

By imposing a closed-loop transfer function of the form

$$H_{0C}(s) = \frac{T_{iC}s+1}{T_{0C}^2s^2 + 2\zeta_CT_{0C}s+1}$$

with imposed damping coefficient ζ_C and time constant T_{0C}, one obtains via identification

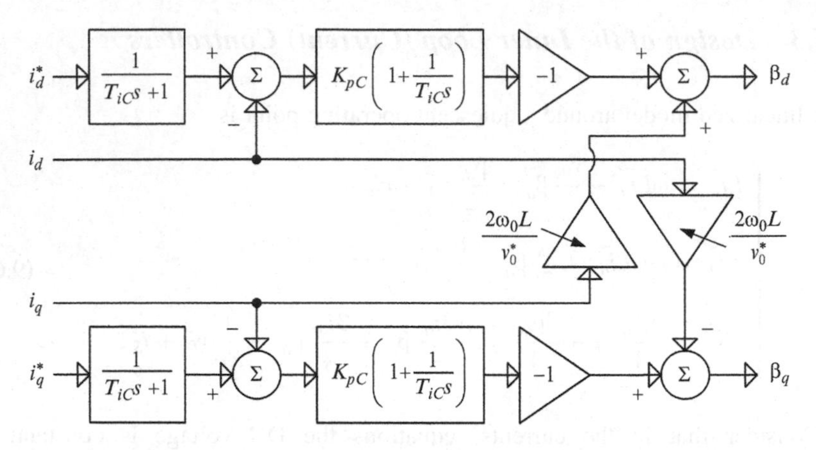

Fig. 9.26 Inner control structure (i_q prefilter may not be necessary if $i_q{}^*$ remains constant)

$$T_{iC} = 2\zeta_C T_{0C} - \frac{T_{0C}^2}{T_C}, \quad K_{pC} = \frac{T_C T_{iC}}{K_C T_{0C}^2}. \tag{9.64}$$

One imposes a closed-loop dynamic ten times faster than the original current plant, $T_{0C} = T_C/10 = 2.2$ ms, and a damping $\zeta_C = 0.7$. Using (9.64), one obtains

$$T_{iC} = 2.9 \text{ ms}, \quad K_{pC} = 5.8 \text{ mA}^{-1}.$$

The inner control structure is detailed in Fig. 9.26.

Note that the value of K_{pC} is quite small as the loop has already a high gain given by K_C (for $T_{0C} = T_C/10$, the total loop gain must be $K_{pc} \cdot K_C \simeq 10$). As Eq. (9.63) applies to both channels d and q, the PI parameters are the same for both currents i_d and i_q.

As regards the prefilters on the current references, both have the transfer function

$$H_{PF}(s) = \frac{1}{T_{iC}s + 1}.$$

Note that, under the assumption that the inductor parameters L and r are constant, the plant is invariant and the closed loop does not change its parameters with the operating point.

9.5.4 Simulations Results Concerning the Inner Loop

The converter nonlinear averaged model in the stationary frame given in (9.65) is implemented in MATLAB®-Simulink®, where the AC three-phase control inputs β_{123} belong to $[-1; 1]$ (see Chap. 5). Direct and inverse Park transforms are employed in order to convert the three-phase variables to dq variables and vice versa. For this purpose, an ideal PLL providing the grid phase is used.

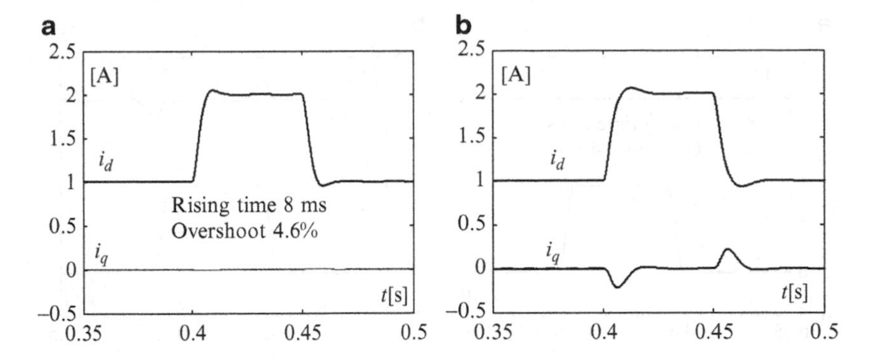

Fig. 9.27 i_d and i_q variations in response to unit step of active current reference, i_d^*: (**a**) with *dq* decoupling structure being active; (**b**) without *dq* decoupling structure

$$\begin{cases} L\dot{i}_k = e_k - \dfrac{v_0 \cdot \beta_k}{2} + \dfrac{v_0}{6}\displaystyle\sum_{k=1}^{3}\beta_k, \quad k = 1,2,3 \\ C\dot{v}_0 = \dfrac{1}{2}\displaystyle\sum_{k=1}^{3} i_k \cdot \beta_k - \dfrac{v_0}{R_C} + i_S, \end{cases} \tag{9.65}$$

In order to test the current controllers, the integrator of the second equation from (9.65) is "frozen" at the value $v_0 = v_0^* = 450$ V. In this way, the gain of the inner plant (see Eq. (9.63)) remains constant irrespective of the DC voltage imbalance.

Figure 9.27a shows the active current i_d response at unit step reference i_d^*; one can see that the settling time and overshoot correspond to the imposed dynamic performance. Figure 9.27b shows that the i_d dynamic is slightly altered if the decoupling structure is missing. Also, in this case the change in i_d induces a quite significant transient in i_q.

Figure 9.28a shows the reactive current i_q response at unit step reference i_q^*; here again the settling time and overshoot meet the imposed dynamic performance. Figure 9.28b shows that the i_q dynamic is slightly altered if the decoupling structure is missing. Also, in this case the change in i_q induces a quite significant transient in i_d. Figure 9.29 presents the system response to a large step of active current reference i_d^*, while current i_q remains constant. The system is linear and invariant as the large-signal variation has the same dynamic performances as in the case described in Fig. 9.27a.

Note that the grid current envelopes vary according to the shape of current i_d, as shown in Fig. 9.29.

Negative values of the current i_d correspond to injection of active power in the AC grid. The corresponding control efforts are described by the evolutions of β_d and β_q (which are the control inputs in the *dq* frame) in Fig. 9.30. Note that β_d is quite large, its steady-state ideal value being $\beta_{de} = 2E/v_0^* \simeq 0.8$ (inductor losses being neglected). Conversely, β_q is very small, corresponding to the steady-state ideal value of $\beta_{qe} = -\ 2\omega_0 L i_d / v_0^* \simeq 0.03$ (inductor losses being neglected).

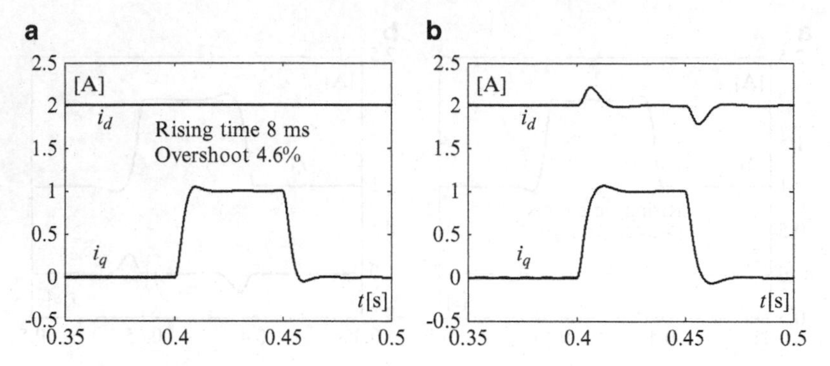

Fig. 9.28 i_d and i_q variations at unit step of the active current reference $i_q{}^*$: (a) with dq decoupling structure being active; (b) without the dq decoupling structure

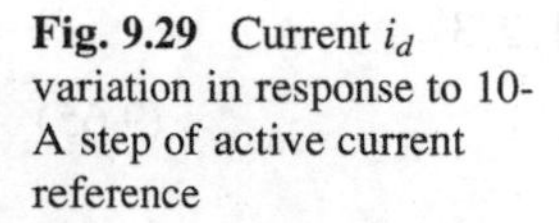

Fig. 9.29 Current i_d variation in response to 10-A step of active current reference

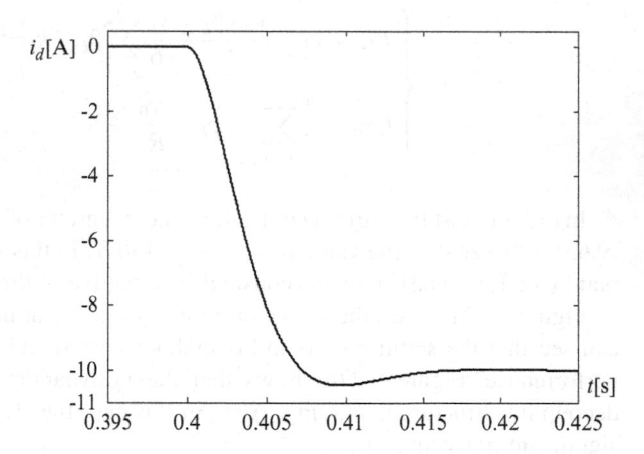

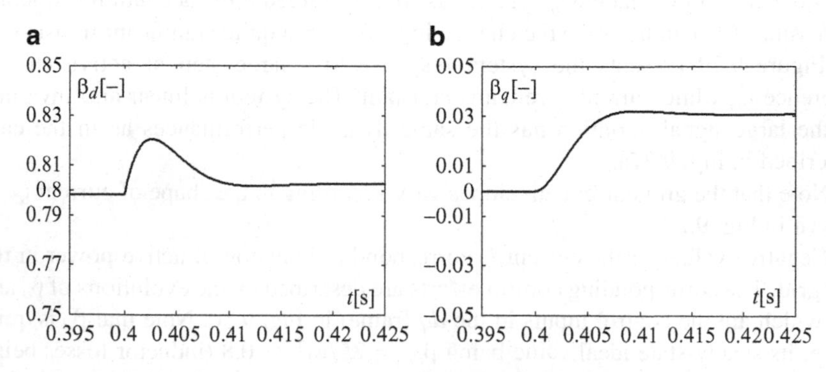

Fig. 9.30 i_d and i_q variations at unit step of the active current reference, $i_q{}^*$: (a) with dq decoupling structure being active; (b) without dq decoupling structure

9.5.5 Design of the Outer Loop (Voltage) Controller

The outer control loop is designed by considering that the separation of modes in the plant is effective. The i_d current dynamic being neglected in this loop (i.e., $\widetilde{i_d} \equiv \widetilde{i_d^*}$) and i_q being regulated at zero, the third equation of the linearized model (9.61) becomes

$$C\dot{\widetilde{v_0}} = \frac{3\beta_{de}}{4}\widetilde{i_d^*} + \frac{3i_{de}}{4}\widetilde{\beta_d} - \frac{1}{R_C}\widetilde{v_0} + \widetilde{i_S}. \tag{9.66}$$

As has been detailed earlier in this chapter (see, for example Sect. 9.2.1), the influence of the inner control input $\widetilde{\beta_d}$ can be neglected in the outer loop design: as the inner system is practically an integrator, β_d has a very narrow derivative-like evolution at steps of $i_d{}^*$ and has no influence on dv_0/dt. Consequently one may write that

$$C\dot{\widetilde{v_0}} = \frac{3\beta_{de}}{4}\widetilde{i_0^*} - \frac{1}{R_C}\widetilde{v_0} + \widetilde{i_S}. \tag{9.67}$$

Hence, the transfer function from the active current component to the DC voltage v_0 is

$$H_V(s) = \frac{\widetilde{v_0}}{\widetilde{i_0^*}} = \frac{3R_C\beta_{de}}{4} \cdot \frac{1}{CR_Cs + 1}. \tag{9.68}$$

For the chosen parameter values, the gain of $H_V(s)$ gives $K_V = 28000$ V/A and time constant $T_V = 221$ s. The voltage control loop is built using a PI controller $H_{PIV}(s) = K_{pV}\left(1 + \frac{1}{T_{iV}s}\right)$. The same principle as used for the inner loop (see Sect. 9.5.3) applies here. The voltage closed-loop transfer function

$$H_{0V}(s) = \frac{T_{iV}s + 1}{\frac{T_VT_{iV}}{K_VK_{pV}}s^2 + T_{iV}\left(\frac{1}{K_VK_{pV}} + 1\right)s + 1},$$

is imposed to be equal with

$$H_{0V}(s) = \frac{T_{iV}s + 1}{T_{0V}^2s^2 + 2\zeta_VT_{0V}s + 1}.$$

Imposing damping coefficient ζ_V and the time constant T_{0V}, one obtains via identification,

$$T_{iV} = 2\zeta_VT_{0V} - \frac{T_{0V}^2}{T_V}, \quad K_{pV} = \frac{T_VT_{iV}}{K_VT_{0V}^2}. \tag{9.69}$$

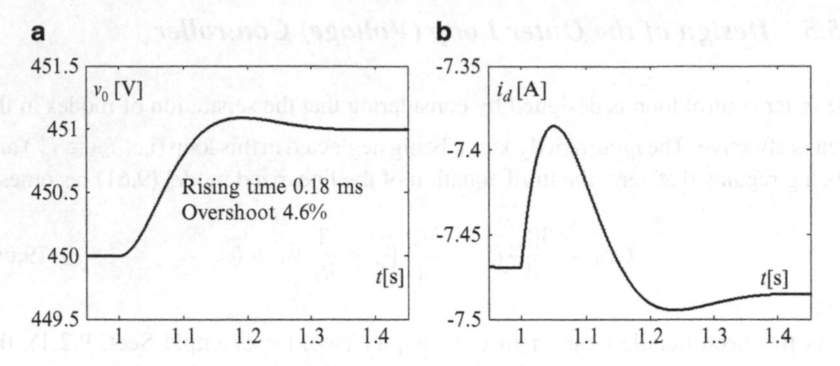

Fig. 9.31 (**a**) DC voltage variation at unit step of its reference $v_0{}^*$; (**b**) associated control effort $i_d{}^* \approx i_d$

Note that the outer plant time constant is extremely large. In fact, as the resistor R_C is high, the plant acts much more as an integrator. In this case, one must impose a sufficiently small closed-loop time constant in order to achieve a good voltage response (small variations around the setpoint). Let us consider, for example, $T_{0V} = 47$ ms (that is 4700 times faster than the original plant) and a damping $\zeta_V = 0.7$. Using (9.69), one obtains

$$T_{iV} = 65.8 \text{ ms}, \quad K_{pV} = 0.23 \text{ A/V}. \tag{9.70}$$

If one wants to vary the voltage setpoint, a prefilter must be inserted on the voltage reference:

$$H_{PF}(s) = \frac{1}{T_{iV}s + 1}.$$

Note that, for purposes concerning regulatory control, β_{de} is constant and the plant described by Eq. (9.68) is invariant.

9.5.6 Simulations Results Concerning the Outer Loop

For purposes of testing the outer control the same setup as in Sect. 9.5.4 has been used, this time with the full system (9.65). Also, the previously designed inner current loops have been used in order to render effective the voltage controller command.

Figure 9.31 shows the DC voltage evolution at unit step of its reference (in the neighborhood of the rated value of v_0) when at full load (i.e., $i_S = 4.5$ A) and using

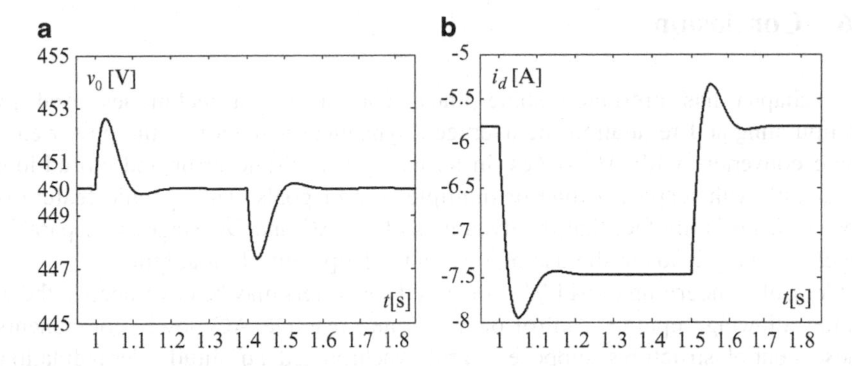

Fig. 9.32 (**a**) DC voltage v_0 variation at unit step variation of the DC source current i_S around rated value; (**b**) associated control effort $i_d^* \approx i_d$

the controller with parameters given by (9.70). The response has about 5 % overshoot (corresponding to a damping of 0.7) and a rising time of 0.18 s corresponding to the imposed closed-loop time constant $T_{0V} = 0.47$ s. The control effort is evident in Fig. 9.31b, where one can see that $i_{de} = -4i_S/3\beta_{de} \simeq 7.5$ A at full load. The dynamic regime lasts for about 0.4 s, which may be too long for certain applications.

Next, a new set of dynamic requirements has been adopted, namely $\zeta_V = 0.7$ and $T_{0V} = 25$ ms. Using Eq. (9.69), one obtains

$$T_{iV} = 35\ \text{ms},\quad K_{pV} = 0.44\ \text{A/V}, \tag{9.71}$$

so the controller is behaving more aggressively, as can be seen in Fig. 9.32a. This figure shows the voltage evolution at unit step of DC source current in the neighborhood of the full load regime (i.e., from 3.5 to 4.5 A and back). Note that the steady-state voltage error is zero, the dynamic regime takes less time with respect to the previous case (about 0.2 s) and the voltage deviation is also smaller (about 0.6 %). The associated control effort is seen in Fig. 9.32b (the steady-state active current is $i_{de} = -4i_S/3\beta_{de} \simeq 5.8$ A for $i_S = 3.5$ A).

The control structure must be completed with limiters of the control inputs in each loop. Besides the inherent limitations in the inner loops (the absolute value of β_d and β_q cannot exceed unit), the current reference, i_d^* must also be limited, for safety reasons. Antiwindup structures that disable the controller integrator during limitation regimes may prove a good solution in order to achieve smooth behavior in heavy loads or in start-up conditions. In this latter situation, the gains of both inner and outer plants are smaller than the rated, and the dynamic behavior is far from the one imposed by design. In this case the online adjustment of the controller's parameters (adaptive control) may also be considered in order to achieve good closed-loop behavior.

9.6 Conclusion

This chapter has presented some linear control design techniques used for manipulating and regulating the averaged dynamical behavior of the power electronic converters with AC stages, in terms of steady-state error and closed-loop bandwidth with respect to one or multiple control goals. The specific feature of these systems is the fact that the dynamics of the AC and DC stage are separable, which naturally leads to the use of a cascaded-loop control paradigm.

Controls concerning most PWM-operated converters may be developed in the *dq* frame, allowing separate control of active and reactive AC power components. These control structures suppose a grid-synchronized amplitude demodulation-modulation that transforms AC variables into DC ones and makes possible the application of control paradigms and methods used for DC-DC converters (see Chap. 8). These are very useful for the large majority of AC-DC converters having two dynamic states.

Stationary-frame control structures may also be devised. Generalized integrators (resonance elements) are able to introduce very large gain at certain frequencies (which, in general, are multiples of the grid frequency), thus leading to zero steady-state errors during tracking of sinusoidal references. Combined *dq*-stationary control structures may be used for the overall control of an AC-DC converter.

The control of full-wave-operated converters has also been visited. The dynamics of such controllers may be manipulated through the pole-placement method, and output variable regulation using integral or PI controllers may be achieved.

This chapter treats into detail the building of control laws associated with multiple types of converters, the accent being on both components of the control structure and computation of controller parameters. Although approximate, these control structures, based upon linearized converter models, may be very effective, especially when the circuit operating point does not significantly vary. When this is not the case, the intrinsic robustness of proportional-integral-based control structures may be effectively exploited. As control inputs are limited to certain values (by virtue of structure or safety reasons) antiwindup structures must be associated with this kind of controller.

These linear controllers are designed to meet control requirements for a certain operating point. As converter models are generally variant with load and input supply, control design must comply with certain robustness requirements that ensure acceptable performance over the entire operating range. Gain scheduling or other adaptive techniques may be used whenever a single controller cannot yield satisfactory behavior for the entire operating range. Even if the application examples given in this chapter concern mostly the control of voltage-source inverters, the techniques presented within may be easily "exported" to current-source inverter control.

Discussion of the use of discrete-time controllers is not the main purpose of this chapter; however, their use may be effective in AC-DC converters either by discretizing previously obtained continuous-time controllers or by modeling plants in discrete-time and then performing a direct synthesis of the discrete-time controller.

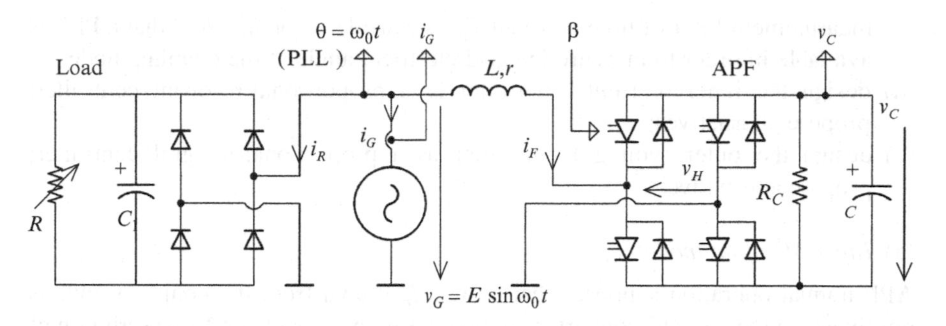

Fig. 9.33 Single-phase shunt active power filter diagram

Also, control methods and principles presented in this chapter can be easily applied to other types of applications related to energy quality such as STATCOMs and active power filters (Buso et al. 1998), AC drives (Cárdenas and Peña 2004), multilevel converters (Vazquez et al. 2009) or renewable energy systems (Andreica Vallet et al. 2011).

Problems

Problem 9.1 Control of a single-phase shunt active power filter

Figure 9.33 shows an ideal single-phase grid supplying a nonlinear load (diode rectifier and resistive load R – on the left side of the figure). Its current i_R is heavily distorted – it contains a rich spectrum with significant high-order harmonics content. The circuit also contains an active power filter (APF), the fully controlled rectifier having capacitor C on the DC side (right side of figure). Based upon the energy accumulated in this capacitor, the APF must draw from the common node a supplementary current i_F that compensates the high-order harmonics introduced by the nonlinear load. In such way, the total current drawn from the grid, i_G, remains sinusoidal (Miret et al. 2009). An essential aspect of the proper circuit operation is that the DC voltage v_C (on capacitor C) must be maintained at a certain level, v_C^*. Circuit parameters may be seen in Fig. 9.33.

Design the control structure so it ensures the above presented functionalities of the APF. Separation of modes between AC and DC variables is considered to be fulfilled; therefore a cascaded-loop control structure may be envisioned. In more detail, the following is required:

(a) model the averaged inner AC current dynamic (plant) provided that the output of the outer plant (v_C) is regulated to a constant value;
(b) model the outer DC voltage dynamic provided that the inner plant is much faster;
(c) propose a cascaded-loop control structure for the plants in part a and b that fulfills the above control goals. An indirect APF control that aims at controlling grid current i_G is envisioned. Also, by considering that the load R induces a parasitic high frequency (R contains switching devices), a supplementary

measurement filter of time constant T_{fC} is considered on i_G. Note that a PLL is available in order to measure the grid electrical angle at the coupling node;

(d) design the inner (current) controller as a proportional-resonant controller; propose alternatives;

(e) design the outer (voltage) controller as a proportional-integral controller; propose alternatives.

(a) *Inner AC plant modeling*

APF normal operation supposes that current i_F drawn from the coupling node is nonlinear, so the AC-DC converter *dq* model is no more suitable for description of its dynamical behavior. Therefore, stationary-frame modeling must be used in order to describe the AC-side behavior of the APF. Nevertheless, the model states are the variables that describe the energy accumulations in the circuit: inductor current i_F and capacitor voltage v_C.

The APF averaged current dynamic is described by

$$L\frac{di_F}{dt} = v_{AC} - v_H - r \cdot i_F = v_{AC} - \beta \cdot v_C - r \cdot i_F, \tag{9.72}$$

because the switched voltage is $v_H = \beta \cdot v_C$ (see AC-DC coupling equations in Chap. 5). Further,

$$Ls \cdot i_F + r \cdot i_F = -\beta \cdot v_C + E \cdot \sin \omega_0 t.$$

If one considers that capacitor voltage varies very slowly and remains almost constant at $v_C{}^*$ due to control action (according to control goals), the APF current is

$$i_F = \underbrace{\frac{-v_C^*}{r} \cdot \frac{1}{L/r \cdot s + 1}}_{\text{Control}} \cdot \beta + \underbrace{\frac{E}{r} \cdot \frac{1}{L/r \cdot s + 1}}_{\text{Disturbance}} \cdot \sin \omega_0 t. \tag{9.73}$$

In the coupling node one may write:

$$i_G = i_F + i_R;$$

therefore, the AC current plant may be described by the diagram in Fig. 9.34.

Remark As has been stated in the requirements, the indirect APF control primary goal supposes control of grid current i_G as a sinusoidal variable, in phase with voltage grid v_{AC} (this is why i_G is taken as the output of the inner plant in Fig 9.34). Further, the outer plant must output the proper reference for this current, $i_G^* = I_G^*$ sin $\omega_0 t$, which fulfills the secondary control goal – the regulation of the DC voltage v_C. Variable v_C acts as a sensor for the power unbalancing between the grid, the nonlinear load and the APF. Therefore, the outer control loop must provide the

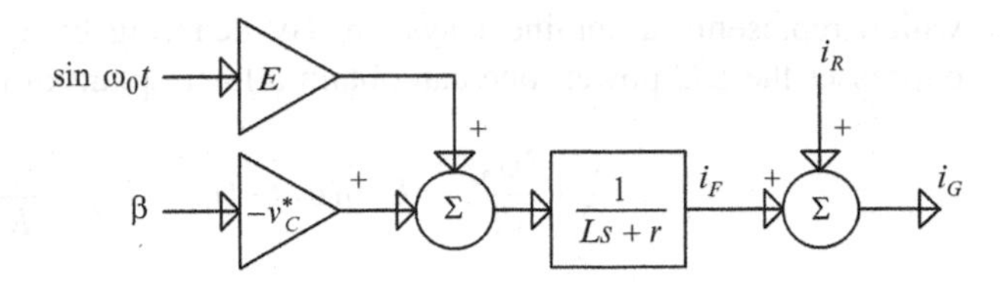

Fig. 9.34 Diagram of AC current plant

proper value of the current amplitude $I_G{}^*$, one that maintains the power balance irrespectively of load R variations.

(b) *Outer DC plant modeling*

The DC plant is described by the equation

$$C\frac{dv_C}{dt} = i_{DC} - \frac{v_C}{R_C} = \beta i_F - \frac{v_C}{R_C} \tag{9.74}$$

because $i_{DC} = \beta \cdot i_F$ (see Chap. 5). Suppose that the AC current plant is very fast with respect to the DC voltage plant (the mode separation condition is satisfied). That is, in the voltage control one considers that the current controlled in the inner loop varies instantaneously, i.e., $i_G \equiv i_G^*$ (in fact i_G has a much larger bandwidth). Also, this allows zeroing the inductor current dynamics in Eq. (9.72):

$$0 = E \sin \omega_0 t - \beta v_C,$$

the inductor losses being neglected, meaning that $v_H \simeq E \sin \omega_0 t$. This is natural because, when one wants to control the current drawn/injected from/into the node, the averaged value of voltage v_H must be slightly lower/higher than the coupling node's voltage. This shows that at a typical operating point where $v_C \neq 0$ the duty ratio is almost sinusoidal (although current i_F has higher-order harmonics):

$$\beta = \frac{E}{v_C} \cdot \sin \omega_0 t.$$

Equation (9.74) then becomes successively:

$$C\frac{dv_C}{dt} = \frac{E \sin \omega_0 t}{v_C} i_F - \frac{v_C}{R_C} = \frac{E \sin \omega_0 t}{v_C}\left(i_G^* - i_R\right) - \frac{v_C}{R_C},$$

and, by multiplying with v_C one obtains

$$Cv_C\frac{dv_C}{dt} = E \sin \omega_0 t \cdot \left(i_G^* - i_R\right) - \frac{v_C^2}{R_C},$$

or, equivalently,

$$\frac{C}{2} \cdot \frac{dv_C^2}{dt} = E \sin \omega_0 t \cdot \left(i_G^* - i_R\right) - \frac{v_C^2}{R_C},$$

which represents a nonlinear system. By denoting by $x_{DC} = {v_C}^2$ a variable that expresses the DC power, one can obtain a linear plant of the form

$$\frac{C}{2} \cdot \frac{dx_{DC}}{dt} = E \sin \omega_0 t \cdot \left(i_G^* - i_R\right) - \frac{x_{DC}}{R_C}. \tag{9.75}$$

Further, one develops the currents in (9.75). The grid current is $i_G = I_G \sin \omega_0 t$, its amplitude having been denoted by I_G. The load current i_R has multiple harmonics denoted by Σ_k

$$i_R = \underbrace{I_{R0} \cdot \sin \omega_0 t}_{\text{Fundamental}} + \underbrace{\sum_{k=1}^{\infty} I_{Rk} e^{j(k\omega_0 t + \varphi_k)}}_{\text{Superior-order harmonics}} = I_{R0} \cdot \sin \omega_0 t + \Sigma_k,$$

where I_{R0} is the amplitude of the fundamental, I_{Rk} is that of the kth-order harmonic and φ_k is the phase lag of the kth-order harmonic. Equation (9.75) becomes

$$\frac{C}{2} \cdot \frac{dx_{DC}}{dt} = E \sin \omega_0 t \cdot I_G^* \sin \omega_0 t - E \sin \omega_0 t \cdot I_{R0} \sin \omega_0 t - E \sin \omega_0 t \cdot \Sigma_\kappa - \frac{x_{DC}}{R_C}$$

or

$$\left(\frac{C}{2} s + \frac{1}{R_C}\right) \cdot x_{DC} = \left(EI_G^* - EI_{R0}\right) \cdot \sin^2 \omega_0 t - E \sin \omega_0 t \cdot \Sigma_\kappa.$$

Applying the trigonometric identity $\sin^2 \omega_0 t = (1 - \cos 2\omega_0 t)/2$ one obtains

$$\left(\frac{C}{2} s + \frac{1}{R_C}\right) \cdot x_{DC} = \underbrace{\frac{E}{2}\left(I_G^* - I_{R0}\right)}_{\text{DC component}} - \underbrace{\left[\frac{E}{2}\left(I_G^* - I_{R0}\right) \cos 2\omega_0 t - E \sin \omega_0 t \cdot \Sigma_\kappa\right]}_{\text{High-frequency component}}. \tag{9.76}$$

The last term in Eq. (9.76) contains only AC high-frequency components that induce only "local" variations and have no influence on the long-term evolution DC voltage, v_C. Only the DC component will be considered further:

$$x_{DC} = \underbrace{\frac{ER_C}{2} \cdot \frac{1}{\frac{CR_C}{2} s + 1}}_{\text{Control channel}} \cdot I_G^* - \underbrace{\frac{ER_C}{2} \cdot \frac{1}{\frac{CR_C}{2} s + 1}}_{\text{Disturbance channel}} \cdot I_{R0}. \tag{9.77}$$

Note that system (9.77) is linear in x_{DC}; therefore, when building the outer DC voltage loop one must compute and use $x_{DC} = {v_{DC}}^2$ as a feedback. Equation (9.77) shows that a proper value of I_G^* with respect to I_{R0} (depending on load R) may drive the variable x_{DC} (and therefore the value of v_{DC}) at a desired level.

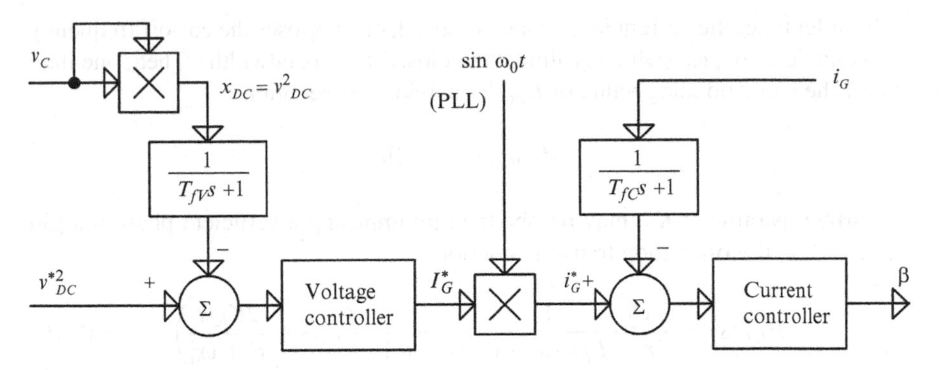

Fig. 9.35 Diagram of control structure for active power filter in Fig. 9.33

(c) *The control structure*

The above remarks and modeling actions suggest that a control structure could be the one presented in Fig. 9.35 (to be coupled with the circuit diagram from Fig. 9.33 for closed-loop operation).

As already stated, the outer controller outputs the necessary current amplitude to be drawn from the grid in order to establish the power balance. The PLL provides the grid phase in order to form a proper current reference – involving zero reactive power – for the inner loop. The insertion of a supplementary filter on the DC capacitor voltage is used in order to diminish high-frequency voltage variations (due to high-frequency current components – see Eq. (9.76)). Without this filter the squaring operation of the v_{DC} measure will introduce high-amplitude, high-frequency components, which will further affect controller output $I_G{}^*$ and hence introduce current distortions.

(d) *Inner (AC current) loop design*

Figure 9.34 and Eq. (9.73) provide the basis for designing the current controller, which may be solved in the spirit of Sects. 9.3.2 and 9.3.6. As the plant is a first-order system and the loop reference is sinusoidal, $i_G{}^* = I_G{}^* \sin \omega_0 t$, a proportional-resonant controller could be a good solution (see the internal model principle). Let us consider its transfer function

$$H_{PR}(s) = K_{pC} + \frac{2K_{iC}}{s^2 + \omega_0^2}.$$

Next, the loop-shaping method will be used in order to determine parameters K_{pC} and K_{iC}. In a first time, the influence of the generalized integrator is neglected; this allows one to determine K_{pC} when the open-loop bandwidth is imposed (see Sects. 9.3.2. and 9.3.6). The open-loop simplified transfer function is

$$H_{OL}^S(s) = \frac{-K_{pC} v_C^*}{r} \cdot \frac{1}{L/r \cdot s + 1} \cdot \frac{1}{T_{fC}s + 1}. \tag{9.78}$$

In order to set the current loop response speed, one imposes the cut-off frequency at a suitable value, ω_C (which is almost the closed-loop bandwidth). Then, one may obtain the corresponding value of K_{pC} by solving the equation

$$\left|H_{OL}^{S}(j\omega_C)\right| = 0.$$

Further, parameter K_{iC} may be obtained by imposing a sufficient phase margin (e.g., 40°) to the open-loop transfer function

$$H_{OL}(s) = \frac{-v_C^*}{r} \cdot \frac{1}{L/r \cdot s + 1} \cdot \frac{1}{T_{fC}s + 1} \cdot \left(K_{pC} + \frac{2K_{iC}s}{s^2 + \omega_0^2}\right), \tag{9.79}$$

that is,

$$\gamma = \pi - \left|\arg(H_{OL}(j\omega_C))\right| = 0.7 \text{ rad}.$$

According to Eq. (9.78), the latter relation further gives

$$\arctan\left(\frac{2K_i\omega_C}{K_p\left(\omega_C^2 - \omega_0^2\right)}\right) + \arctan\left(\frac{\omega_C L}{r}\right) + \arctan\left(\omega_C T_{fC}\right) = \pi - 0.7,$$

which enables computation of integral gain K_{iC}.

Hysteretic or sliding-mode controllers may be used as an alternative for the current controller; these solutions ensure sufficiently large closed-loop inner bandwidth with acceptable control efforts and, therefore, tracking of the sinusoidal reference with small time lags.

(e) *Outer (DC voltage) loop design*

Separation of modes being fulfilled, that is, ω_C being much higher than projected voltage bandwidth, ω_V, the current loop may be considered without dynamic, and the voltage plant transfer is given by Eq. (9.77). Provided the voltage reference is constant and the plant is essentially a first-order filter, a proportional-integral voltage controller is considered:

$$H_{PI}(s) = K_{pV}\left(1 + \frac{1}{T_{iV}s}\right).$$

By making the notation $H_V(s) = \frac{ER_C}{2} \cdot \frac{1}{CR_C/2 \cdot s + 1}$ (see Eq. (9.77)), the closed-loop transfer function is written as

$$H_{CL}(s) = \frac{H_{PI}(s) \cdot H_V(s)}{1 + H_{PI}(s) \cdot H_V(s)}, \tag{9.80}$$

and hence results as a second-order system. Further, one imposes suitable closed-loop transfer function parameters in terms of damping and bandwidth. Damping

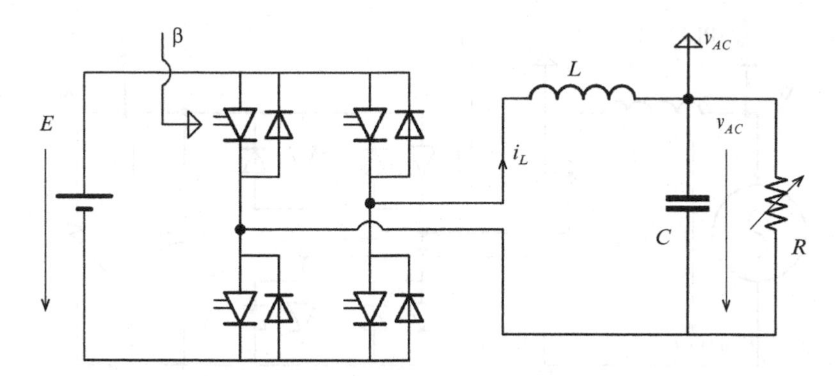

Fig. 9.36 Resonant control of a stand-alone PWM inverter

coefficient ζ_{0V} is usually taken between 0.7 and 0.9 and time constant $T_{0V} = 10/\omega_C$, that is, it is sufficiently large with respect to the inner loop time constant. Finally, proportional gain and integral time constant, K_{pV} and T_{iV} respectively, are computed by identifying the time constant and the damping of system (9.80) with T_{0V} and ζ_{0V}, as previously shown in Sects. 9.2 and 9.5.5.

Note that good dynamic performance is not mandatory for DC voltage regulation (it is required only that the voltage remain between certain limits of a quite large range). The steady-state error for the outer control loop is not critical in such an application; hence a simple proportional controller may alternatively be used.

The reader is invited to solve the following problems.

Problem 9.2 Resonant AC voltage control for a single-phase stand-alone PWM inverter
The off-grid inverter in Fig. 9.36 supplies a variable resistive load. Its parameters are $E = 300$ V, $L = 5$ mH and $C = 6.8$ μF. The rated load voltage is $v_{AC}{}^* = 127$ V RMS, the rated frequency is $f_0 = 50$ Hz and its resistance varies between $R_{max} = 1$ kΩ and $R_{min} = 25$ Ω (inductor and capacitor losses are neglected).

Design a control structure based upon a proportional-resonant (PR) controller that ensures the stabilization of the voltage amplitude on the AC load. PR controller parameters must be chosen such that the voltage closed-loop bandwidth is 20 kHz.

Problem 9.3 Averaged control of a current-source inverter used for induction heating
A current-source inverter – previously introduced in Sect. 3.2 of Chap. 3 and analyzed in Sect. 5.7 of Chap. 5 – is operated in full wave in the presence of a variable load. Its parameters are $L_f = 6$ mH, $r = 0.01$ Ω, $C = 6$ μF, $L = 100$ μH and $U_{d0} = 300$ V (see Fig. 3.3). Load resistance R varies between 10 Ω and 0.5 Ω. The general control goal is to control output voltage, the closed-loop bandwidth being $f_{AC} = 1$ kHz.

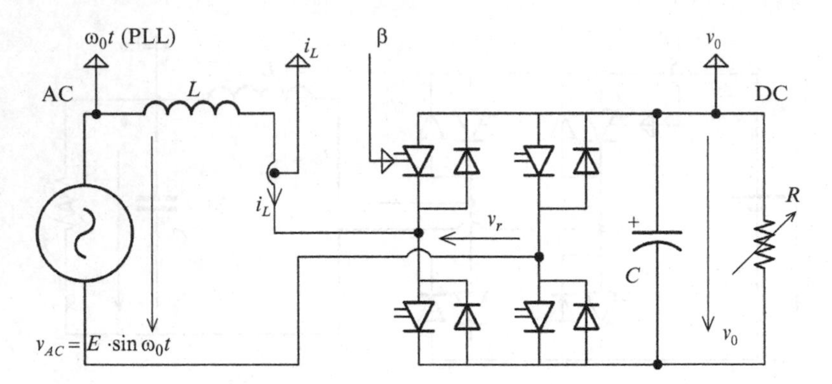

Fig. 9.37 Single-phase rectifier supplying a resistive variable load

By considering switching frequency as control input, address the following points.

(a) Deduce the linearized averaged model of the converter and plot the pole-zero map by using MATLAB® software.
(b) Conceive an inner partial-state feedback structure (by omitting the output voltage) and suitably place the system poles (the dominant time constant must be at most $10\pi \cdot f_{AC}$).
(c) Conceive the outer voltage loop by using an integral controller and choose the integral gain by using the root locus method (function `rltool` in MATLAB®).
(d) Simulate the converter response at step variations of load R, and compare the result with the dynamics predicted by the control design.

Problem 9.4 Control of a single-phase PWM rectifier
The rectifier inverter in Fig. 9.37 supplies a variable resistive load. Its parameters are $E = 340$ V, $\omega_0 = 100\pi$ rad/s, $L = 2.2$ mH and $C = 820$ μF. The rated load voltage is $v_0{}^* = 450$ V and the load resistance varies between $R_{max} = 20$ kΩ and $R_{min} = 100$ Ω (inductor and capacitor losses are neglected).

The primary control objective is to regulate DC voltage v_0 at its rated value. The outer closed-loop bandwidth must be 1 kHz. The secondary goal is to draw controllable reactive power Q within the limits $0 - 0.3P$, with P being the actual active power.

(a) Build a control structure in the dq frame that fulfills the stated goals.
(b) Build a control structure in the combined dq-stationary frame that fulfills the stated goals.
(c) Study the outer-loop robustness for the entire variation range of R (evaluate the variation of its performances).
(d) Simulate using the MATLAB®-Simulink® software the closed-loop response when R changes for control structures developed in parts a and b; compare the results with the dynamics predicted by the control design.

References

Andreica Vallet M, Bacha S, Munteanu I, Bratcu AI, Roye D (2011) Management and control of operating regimes of cross-flow water turbines. IEEE Trans Ind Electron 58(5):1866–1876

Bacha S, Brunello M, Hassan A (1994) A general large signal model for DC-DC symmetric switching converters. Electr Mach Power Syst 22(4):493–510

Blaabjerg F, Teodorescu R, Liserre M, Timbus AV (2006) Overview of control and grid synchronization for distributed power generation systems. IEEE Trans Ind Electron 53(5):1398–1409

Blasko V, Kaura V (1997) A new mathematical model and control of a three-phase AC-DC voltage source converter. IEEE Trans Power Electron 12(1):116–123

Bose BK (2001) Modern power electronics and AC drives. Prentice-Hall, Upper Saddle River

Buso S, Malesani L, Mattavelli P (1998) Comparison of current control techniques for active filter applications. IEEE Trans Ind Electron 45(5):722–729

Cárdenas R, Peña R (2004) Sensorless vector control of induction machines for variable-speed wind energy applications. IEEE Trans Energy Convers 19(1):196–205

d'Azzo JJ, Houpis CH, Sheldon SN (2003) Linear control system analysis and design with MATLAB, 5th edn. Marcel-Dekker, New York

Esselin M, Robyns B, Berthereau F, Hautier JP (2000) Resonant controller based power control of an inverter transformer association in a wind generator. Electromotion 7(4):185–190

Etxeberria-Otadui I (2003) On the power electronics systems dedicated to electrical energy distribution (in French: "Sur les systèmes d'électronique de puissance dédiés à la distribution électrique"). Ph.D. thesis, Grenoble Institute of Technology, Grenoble

Etxeberria-Otadui I, López de Heredia A, Gaztañaga H, Bacha S, Reyero MR (2006) A single synchronous frame hybrid (SSFH) multifrequency controller for power active filters. IEEE Trans Ind Electron 53(5):1640–1648

Etxeberria-Otadui I, Viscarret U, Caballero M, Rufer A, Bacha S (2007) New optimized PWM VSC control structures and strategies under unbalanced voltage transients. IEEE Trans Ind Electron 54(5):2902–2914

Francis BA, Wonham WM (1976) Internal model principle in control theory. Automatica 12 (5):457–465

Fukuda S, Imamura R (2005) Application of a sinusoidal internal model to current control of three-phase utility-interface converters. IEEE Trans Ind Electron 52(2):420–426

Fukuda S, Yoda T (2001) A novel current-tracking method for active filters based on a sinusoidal internal model. IEEE Trans Ind Appl 37(3):888–895

Gaztañaga H, Lopez de Heredia A, Etxeberria I, Bacha S, Roye D, Guiraud J, Reyero R (2005) Multi-resonant state feedback current control structure with pole placement approach. In: Proceedings of the European Conference on Power Electronics and Applications – EPE 2005. Dresden, Germany, pp 10–26

Kaura V, Blasko V (1997) Operation of a phase locked loop system under distorted utility conditions. IEEE Trans Ind Appl 33(1):58–63

Kazmierkowski MP, Malesani L (1998) Current control techniques for three-phase voltage-source PWM converters: a survey. IEEE Trans Ind Electron 45(5):691–703

Kazmierkowski MP, Malinowski M, Bech M (2002) Pulse width modulation techniques for three-phase voltage source converters. In: Kazmierkowski MP, Krishnan R, Blaabjerg F, Irwin JD (eds) Control in power electronics: selected problems, Academic Press series in engineering. Academic/Elsevier, San Diego/London, pp 89–160

Lindgren M (1998) Modelling and control of voltage source converters. Ph.D. thesis, Chalmers University of Technology, Göteborg

Ma L, Luna A, Rocabert J, Munoz R, Corcoles F, Rodriguez P (2011) Voltage feed-forward performance in stationary reference frame controllers for wind power applications. In: Proceedings of the 2011 international conference on power engineering, energy and electrical drives, POWERENG 2011. Torremolinos, Malaga, Spain, pp 1–5

Malesani L, Tomasin P (1993) PWM current control techniques of voltage source converters – A survey. In: Proceedings of the international conference on Industrial Electronics, Control and Instrumentation – IECON 1993. Maui, Hawaii, USA, vol 2, pp 670–675

Miret J, Castilla M, Matas J, Guerrero JM, Vasquez JC (2009) Selective harmonic-compensation control for single-phase active power filter with high harmonic rejection. IEEE Trans Ind Electron 56(8):3117–3127

Naslin P (1968) Practical technology and computation of controlled systems (in French: Technologie et calcul pratique des systèmes asservis), 3rd edn. Dunod, Paris, ch. 11

Oppenheim AV, Willsky AS, Nawab SH (1996) Signals and systems, 2nd edn. Prentice-Hall, Upper Saddle River

Petitclair P, Bacha S, Rognon JP (1996) Averaged modelling and nonlinear control of an ASVC (Advanced Static VAR Compensator). In: Proceedings of the 27th annual IEEE Power Electronics Specialists Conference – PESC 1996. Baveno, Italy, vol 1, pp 753–758

Roshan A (2006) A DQ rotating frame controller for single phase full-bridge inverters used in small distributed generation systems. M.Sc. thesis, Faculty of the Virginia Polytechnic Institute and State University, Blacksburg

Sato Y, Ishizuka T, Nezu K, Kataoka T (1998) A new control strategy for voltage-type PWM rectifiers to realize zero steady-state control error in input current. IEEE Trans Ind Appl 34(3):480–486

Tan L (2007) Digital signal processing: fundamentals and applications. Academic, San Diego

Teodorescu R, Blaabjerg F, Liserre M, Loh PC (2006) Proportional-resonant controllers and filters for grid-connected voltage-source converters. IEEE proceedings – Electric power applications 153(5):750–762

Timbuş AV, Ciobotaru M, Teodorescu R, Blaabjerg F (2006) Adaptive resonant controller for grid-connected converters in distributed power generation systems. In: Proceedings of the 21st annual IEEE Applied Power Electronics Conference and Exposition – APEC 2006. Dallas, Texas, pp 1601–1606

Timbuş A, Liserre M, Teodorescu R, Rodriguez P, Blaabjerg F (2009) Evaluation of current controllers for distributed power generation systems. IEEE Trans Power Electron 24(3):654–664

Vazquez S, Leon JI, Franquelo LG, Padilla JJ, Carrasco JM (2009) DC-voltage ratio control strategy for multilevel cascaded converters fed with a single DC source. IEEE Trans Ind Electron 56(7):2513–2521

Yepes AG (2011) Digital resonant current controllers for voltage source converters. Ph.D. thesis, University of Vigo, Vigo

Yin Y, Zane R, Glaser J, Erickson RW (2002) Small-signal analysis of frequency-controlled electronic ballasts. IEEE Trans Circuits Syst I Fundam Theory Appl 50(8):1103–1110

Yuan X, Merk W, Stemmler H, Allmeling J (2002) Stationary-frame generalized integrators for current control of active power filters with zero steady-state error for current harmonics of concern under unbalanced and distorted operating conditions. IEEE Trans Ind Appl 38(2):523–532

Zmood DN, Holmes DG (2003) Stationary frame current regulation of PWM inverters with zero steady-state error. IEEE Trans Power Electron 18(3):814–822

Zmood DN, Holmes DG, Bode GH (2001) Frequency-domain analysis of three-phase linear current regulators. IEEE Trans Ind Appl 37(2):601–610

Chapter 10
General Overview of Mathematical Tools Dedicated to Nonlinear Control

Specific features of power electronic converters as intrinsically nonlinear variable-structure dynamical systems and particular constraints of their operation require the use of some powerful tools provided by the theory of nonlinear control. These obviously appear, in general, more appropriate than linear approaches in dealing with nonlinearities and unavoidable uncertainties. Some nonlinear control approaches employed with good results for power electronic converters will be detailed in the next three chapters. This chapter aims at reviewing the most often utilized concepts and formal results specific to these approaches, as well as at identifying relations between them and some of their common characteristics.

10.1 Issues and Basic Concepts

This section will briefly cover mathematical definitions and developments related to some fundamental concepts employed in nonlinear control, such as Lie derivative, relative degree, normal form and zero dynamics. Their use aids apprehension of the power electronic converter dynamical behavior within a framework of the different nonlinear control formalisms.

10.1.1 Elements of Differential Geometry

Before introducing some elements of a Lie algebra – indispensable for obtaining the main results of nonlinear control applicable to power electronic converters – some preliminary notions, like switching function, vector field and Lie derivative, will first be discussed.

Let us recall the notion of *switching function* introduced in the first part of this book, dedicated to modeling of power electronic converters (see Chap. 3). Using

S. Bacha et al., *Power Electronic Converters Modeling and Control: with Case Studies*, Advanced Textbooks in Control and Signal Processing, DOI 10.1007/978-1-4471-5478-5_10,

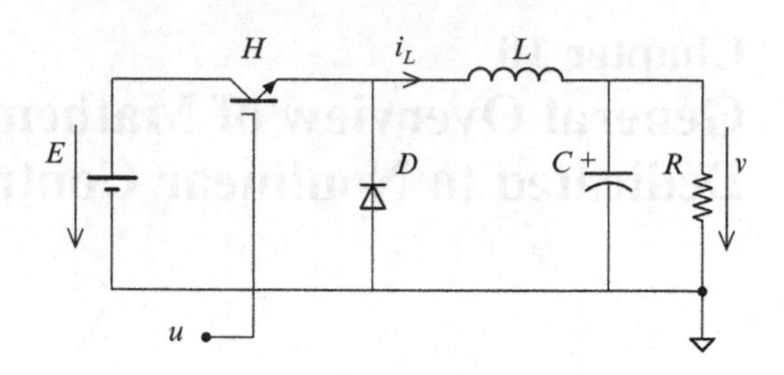

Fig. 10.1 Electrical circuit of a buck DC-DC converter

notations previously introduced with their usual meaning, any power converter can be described by a generally nonlinear mathematical model:

$$\frac{d\mathbf{x}}{dt} = \mathbf{f}(\mathbf{x}) + \mathbf{g}(\mathbf{x}) \cdot \mathbf{u}, \tag{10.1}$$

where $\mathbf{x}$ is the n-dimension column state vector. Function $\mathbf{u}$ controlling the switching from one configuration to another (structure change) is called a switching function. In the multivariable case this is represented as $\mathbf{u} = [\,u_1 \quad u_2 \quad \cdots \quad u_p\,]^T$, where each function u_i takes values within a discrete set $\{u_{i\,-};u_{i\,+}\}$. Function $\mathbf{u}$ varies according to how the switching surface has been chosen. Function $\mathbf{f}(\mathbf{x})$ from (10.1) is called *vector field*; function $\mathbf{g}(\mathbf{x})$ is a n-by-p matrix of vector fields. If $\mathbf{f}(\mathbf{x})$ and $\mathbf{g}(\mathbf{x})$ do not explicitly depend on time, it is said that they are *invariant* (or *autonomous*). Otherwise, it is said that they are *variant* (or *nonautonomous*). Vector fields within Eq. (10.1) are exclusively invariant.

In the linear case

$$\begin{cases} \mathbf{f}(\mathbf{x}) = \mathbf{A} \cdot \mathbf{x} \\ \mathbf{g}(\mathbf{x}) = \mathbf{B}, \end{cases} \tag{10.2}$$

where $\mathbf{A}$ is a n-by-n matrix and $\mathbf{B}$ is a n-by-p matrix. In the bilinear case

$$\begin{cases} \mathbf{f}(\mathbf{x}) = \mathbf{A} \cdot \mathbf{x} \\ \mathbf{g}(\mathbf{x}) = \mathbf{x} \cdot \mathbf{b}^T + \boldsymbol{\delta}, \end{cases} \tag{10.3}$$

where $\mathbf{A}$ is a n-by-n matrix, $\mathbf{b}$ is a p-dimension column vector and $\boldsymbol{\delta}$ is a n-by-p matrix. If taking the example of a buck converter as in Fig. 10.1, then by choosing the switching function u such that $u = 0$ if H is open and $u = 1$ if H is closed, the converter model results as:

$$\dot{\mathbf{x}} = \mathbf{f}(\mathbf{x}) + \mathbf{g}(\mathbf{x}) \cdot u,$$

where $\mathbf{x} = [\,i_L \quad v_C\,]^T$ and the vector fields correspond to the linear case

$$\mathbf{f}(\mathbf{x}) = \begin{bmatrix} -v_C/L \\ i_L/C - v_C/(RC) \end{bmatrix} = \underbrace{\begin{bmatrix} 0 & -1/L \\ 1/C & -1/(RC) \end{bmatrix}}_{\mathbf{A}} \cdot \mathbf{x},\ \mathbf{g}(\mathbf{x}) = \begin{bmatrix} E/L \\ 0 \end{bmatrix} = \mathbf{B}.$$

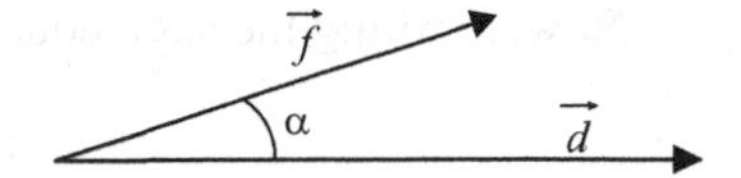

Fig. 10.2 Representation of vectors $\overline{d}$ and $\overline{f}$

In a similar way, having defined the switching function, one can identify *switching vector fields*. Thus, the vector field $\mathbf{g}(\mathbf{x})$ switches between

$$\mathbf{g}^{-}(\mathbf{x}) = \mathbf{g}(\mathbf{x}) \cdot u^{-} \text{ and } \mathbf{g}^{+}(\mathbf{x}) = \mathbf{g}(\mathbf{x}) \cdot u^{+},$$

whereas the vector field defined in (10.1) switches between

$$\mathbf{f}(\mathbf{x}) + \mathbf{g}(\mathbf{x}) \cdot u^{-} \text{ and } \mathbf{f}(\mathbf{x}) + \mathbf{g}(\mathbf{x}) \cdot u^{+}.$$

Next, the notion of Lie derivative will be introduced. Let a vector $\overline{f}$ be considered, which is applied in a given point by an angle of α in relation to the direction given by a unit vector $\overline{d}$, as shown in Fig. 10.2. Projecting vector $\overline{f}$ on the direction of $\overline{d}$ is equivalent to making the scalar (inner) product of the two vectors, denoted as $\overline{f} \bullet \overline{d}$ and defined as $\overline{f} \bullet \overline{d} = \overline{f}^{T} \cdot \overline{d}$.

The same applies when defining the time variation of a vector field $\mathbf{s}(\mathbf{x})$. Thus, one can write

$$\frac{d\mathbf{s}(\mathbf{x})}{dt} = \frac{\partial \mathbf{s}(\mathbf{x})}{\partial \mathbf{x}} \bullet \frac{d\mathbf{x}}{dt}. \tag{10.4}$$

Term $\partial \mathbf{s}(\mathbf{x})/\partial \mathbf{x}$ is known as the Jacobian of $\mathbf{s}(\mathbf{x})$. In the general case of a function $\mathbf{s} : \mathbb{R}^n \rightarrow \mathbb{R}^p$ its Jacobian is expressed as

$$\frac{\partial \mathbf{s}(\mathbf{x})}{\partial \mathbf{x}} = \begin{bmatrix} \frac{\partial s_1}{\partial x_1} & \cdots & \frac{\partial s_1}{\partial x_n} \\ \vdots & \ddots & \vdots \\ \frac{\partial s_p}{\partial x_1} & \cdots & \frac{\partial s_p}{\partial x_n} \end{bmatrix}. \tag{10.5}$$

If s is a scalar linear function, its Jacobian is a vector normal to the surface defined as $s(\mathbf{x}) = 0$. Notation

$$\mathbf{ds} = \begin{bmatrix} \frac{\partial s}{\partial x_1} & \cdots & \frac{\partial s}{\partial x_n} \end{bmatrix}$$

is adopted, which denotes the *gradient vector* of function s.

Now, recalling the state equation in the form

$$\dot{\mathbf{x}} = \mathbf{f}(\mathbf{x}) + \mathbf{g}(\mathbf{x}) \cdot \mathbf{u},$$

gives

$$\frac{d\mathbf{s}(\mathbf{x})}{dt} = \frac{\partial \mathbf{s}(\mathbf{x})}{\partial \mathbf{x}} \bullet \mathbf{f}(\mathbf{x}) + \left(\frac{\partial \mathbf{s}(\mathbf{x})}{\partial \mathbf{x}} \bullet \mathbf{g}(\mathbf{x}) \right) \cdot \mathbf{u}. \tag{10.6}$$

Equation (10.6) can be written more compactly as

$$\frac{d\mathbf{s}(\mathbf{x})}{dt} = \mathrm{L}_{\mathbf{f}(\mathbf{x})}\mathbf{s}(\mathbf{x}) + \mathrm{L}_{\mathbf{g}(\mathbf{x})}\mathbf{s}(\mathbf{x}) \cdot \mathbf{u}, \tag{10.7}$$

where term $\mathrm{L}_{\mathbf{f}(\mathbf{x})}\mathbf{s}(\mathbf{x})$ defined as

$$\mathrm{L}_{\mathbf{f}(\mathbf{x})}\mathbf{s}(\mathbf{x}) \frac{\partial \mathbf{s}(\mathbf{x})}{\partial \mathbf{x}} \bullet \mathbf{f}(\mathbf{x}) \tag{10.8}$$

is called the *Lie derivative* of $\mathbf{s}$ along vector field $\mathbf{f}$. Equation (10.7) may further be written as

$$\frac{d\mathbf{s}(\mathbf{x})}{dt} = \mathrm{L}_{\mathbf{f}(\mathbf{x})+\mathbf{g}(\mathbf{x})\cdot\mathbf{u}}\mathbf{s}(\mathbf{x}). \tag{10.9}$$

In the single-input–single-output case – functions u and $s(\mathbf{x})$ being scalars – the Lie derivative of $s(\mathbf{x})$ along the vector field $\mathbf{f}$ is computed as

$$\mathrm{L}_{\mathbf{f}}\mathbf{s} = \mathbf{ds} \bullet \mathbf{f} = \sum_{i=1}^{n} \left(\frac{\partial s}{\partial x_i} \cdot f_i(\mathbf{x}) \right), \tag{10.10}$$

where $\mathbf{f} = [f_1 \quad f_2 \quad \cdots \quad f_n]^T$.

Remark. Sometimes the term "Lie bracket" is used, which is defined as

$$[\mathbf{f}, \mathbf{g}] = \mathrm{L}_{\mathbf{g}}\mathbf{f} - \mathrm{L}_{\mathbf{f}}\mathbf{g}. \tag{10.11}$$

In the case of a linear system switching between two configurations – such that $\mathbf{f}_1 = \mathbf{A}_1\mathbf{x}$ and $\mathbf{f}_2 = \mathbf{A}_2\mathbf{x}$ – the Lie bracket of the vector field pair $\mathbf{f}_1$ and $\mathbf{f}_2$ is given by

$$[\mathbf{f}_1, \mathbf{f}_2] = (\mathbf{A}_1 \cdot \mathbf{A}_2 - \mathbf{A}_2 \cdot \mathbf{A}_1) \cdot \mathbf{x}. \tag{10.12}$$

Equation (10.12) contains the matrix expression $(\mathbf{A}_1 \cdot \mathbf{A}_2 - \mathbf{A}_2 \cdot \mathbf{A}_1)$, which gives the precision of the first-order approximation of the averaged model in the case of noncommutative matrices $\mathbf{A}_1$ and $\mathbf{A}_2$ (see Eq. (4.31) from Chap. 4).

10.1.2 Relative Degree and Zero Dynamics

SISO case

The single-input–single-output (SISO) nonlinear system of the form

$$\begin{cases} \dot{\mathbf{x}} = \mathbf{f}(\mathbf{x}) + \mathbf{g}(\mathbf{x}) \cdot u \\ y = h(\mathbf{x}) \end{cases} \tag{10.13}$$

is said to have *relative degree r* at a point $\mathbf{x}^0$ if $L_\mathbf{g}L_\mathbf{f}^k h(\mathbf{x}) = 0$ for all points $\mathbf{x}$ in a neighborhood of $\mathbf{x}^0$ and for all $k < r - 1$, and $L_\mathbf{g}L_\mathbf{f}^{r-1} h(\mathbf{x}^0) \neq 0$ (Isidori 1989).

A pragmatic approach for deducing the relative degree of a system is to compute time derivatives of the output variable y until the input u appears explicitly; the order of the corresponding time derivative is the relative degree. For example, in the case of output voltage regulation of a flyback converter, whose model has been deduced in Problem 4.1 of Chap. 4, the relative degree is $r = 1$ because the input u appears explicitly in the expression of the first derivative of the output voltage v_C (see Eq. (4.42)).

One can note that the notion of relative degree of a nonlinear system is a generalization of the same notion defined in the case of a linear system: the difference between the number of poles and the number of zeros in the associated transfer function. For a nonlinear system with the relative degree smaller than the system degree n ($r < n$), a concept that plays a role similar to the one of the zeros in the transfer function can be defined. It is *zero dynamics*, related to solving the *problem of zeroing the output*. This is formulated as follows: find all pairs consisting of an initial state $\mathbf{x}^0$ and of an input function $u^0(t)$, defined for all t in a neighborhood of $t = 0$, such that the corresponding output $y(t)$ of the system is identically zero for all t in a neighborhood of $t = 0$.

A partition of the state-space vector is possible, namely, into a vector grouping the r state variables that give the relative degree and a vector of the remaining $n - r$ state variables:

$$\boldsymbol{\xi} = \begin{bmatrix} z_1 & \cdots & z_r \end{bmatrix}^T, \quad \boldsymbol{\eta} = \begin{bmatrix} z_{r+1} & \cdots & z_n \end{bmatrix}^T,$$

where

$$\begin{cases} y(t) = z_1(t), \quad \dot{z_1} = z_2, \quad \dot{z_2} = z_3, \quad \cdots, \quad \dot{z_{r-1}} = z_r \\ \dot{z_r} = b(\boldsymbol{\xi}, \boldsymbol{\eta}) + a(\boldsymbol{\xi}, \boldsymbol{\eta})u \\ \dot{\boldsymbol{\eta}} = \mathbf{q}(\boldsymbol{\xi}, \boldsymbol{\eta}). \end{cases} \tag{10.14}$$

Equation (10.14) introduces the so-called *normal form* (Isidori 1989). If the output $y(t)$ has to be zero, then necessarily $\dot{z_1}(t) = \dot{z_2}(t) = \cdots = \dot{z_r}(t) = 0$, so $\boldsymbol{\xi}(t) = 0$ for all t. Hence, in order to zero the output, the system initial state must be set to $\boldsymbol{\xi}(0) = 0$, whereas $\boldsymbol{\eta}(0) = \boldsymbol{\eta}^0$ can be chosen arbitrarily. It follows that the problem of zeroing the output admits the solution

$$u(t) = -\frac{b(0, \boldsymbol{\eta}(t))}{a(0, \boldsymbol{\eta}(t))},$$

with $\boldsymbol{\eta}(t)$ being the solution of the differential equation

$$\dot{\boldsymbol{\eta}}(t) = \mathbf{q}(0, \boldsymbol{\eta}(t)), \quad \boldsymbol{\eta}(0) = \boldsymbol{\eta}^0. \tag{10.15}$$

Equation (10.15) represents the *zero dynamics* of the considered nonlinear system; they are called in this way because they describe the "internal" behavior of the system when input and initial conditions have been chosen in such a way as to constrain the output to remain identically zero (Isidori 1989).

MIMO case

Extensions of definitions of relative degree and zero dynamics formulated in the SISO case can quite naturally be made for multi-input–multi-output (MIMO) dynamical systems belonging to some special class, i.e., which have the number of inputs equal to that of outputs. These generalizations are briefly presented next (Isidori 1989). It is said that a dynamical system in the state-space representation

$$\begin{cases} \dot{\mathbf{x}} = \mathbf{f}(\mathbf{x}) + \sum_{i=1}^{m} \mathbf{g}_i(\mathbf{x}) \cdot u_i \\ y_1 = h_1(\mathbf{x}), \quad \cdots, \quad y_m = h_m(\mathbf{x}), \end{cases} \tag{10.16}$$

– where m is the common number of inputs and outputs, $\mathbf{f}(\mathbf{x})$, $\mathbf{g}_1(\mathbf{x}), \cdots, \mathbf{g}_m(\mathbf{x})$, are smooth vector fields and $h_1(\mathbf{x}), \cdots, h_m(\mathbf{x})$ are smooth functions, all defined on an open set of $\mathbb{R}^n$ – has a *vector relative degree* $\{r_1, \cdots, r_m\}$ at point $\mathbf{x}^0$ if two conditions are simultaneously met, namely:

- $\mathrm{L}_{g_j}\mathrm{L}_{\mathbf{f}}^k h_i(\mathbf{x}) = 0$ for all $1 \le j \le m$, for all $1 \le i \le m$, for all $k < r_i - 1$ and for all $\mathbf{x}$ in a neighborhood of $\mathbf{x}^0$; and
- the $m \times m$ matrix

$$\mathbf{A}(\mathbf{x}) = \begin{bmatrix} \mathrm{L}_{\mathbf{g}_1}\mathrm{L}_{\mathbf{f}}^{r_1-1}h_1(\mathbf{x}) & \cdots & \mathrm{L}_{\mathbf{g}_m}\mathrm{L}_{\mathbf{f}}^{r_1-1}h_1(\mathbf{x}) \\ \mathrm{L}_{\mathbf{g}_1}\mathrm{L}_{\mathbf{f}}^{r_2-1}h_2(\mathbf{x}) & \cdots & \mathrm{L}_{\mathbf{g}_m}\mathrm{L}_{\mathbf{f}}^{r_2-1}h_2(\mathbf{x}) \\ \vdots & \ddots & \vdots \\ \mathrm{L}_{\mathbf{g}_1}\mathrm{L}_{\mathbf{f}}^{r_m-1}h_m(\mathbf{x}) & \cdots & \mathrm{L}_{\mathbf{g}_m}\mathrm{L}_{\mathbf{f}}^{r_m-1}h_m(\mathbf{x}) \end{bmatrix} \tag{10.17}$$

is nonsingular at $\mathbf{x} = \mathbf{x}^0$.

The zero dynamics of such a nonlinear system are defined in close analogy with the results established for the SISO case, always related to solving the problem of zeroing the output. If the output $\mathbf{y}(t)$ has to be zero, then necessarily the system's initial state must be set to $\boldsymbol{\xi}(0) = 0$, whereas $\boldsymbol{\eta}(0) = \boldsymbol{\eta}^0$ can be chosen arbitrarily. It follows that the problem of zeroing the output admits the solution

$$\mathbf{u}(t) = -[\mathbf{A}(0, \boldsymbol{\eta}(t))]^{-1} \cdot \mathbf{b}(0, \boldsymbol{\eta}(t)),$$

with matrix $\mathbf{A}$ defined by (10.17) and

$$\mathbf{b}(\mathbf{x}) = \begin{bmatrix} L_{\mathbf{f}}^{r_1} h_1(\mathbf{x}) \\ L_{\mathbf{f}}^{r_2} h_2(\mathbf{x}) \\ \vdots \\ L_{\mathbf{f}}^{r_m} h_m(\mathbf{x}) \end{bmatrix}$$

and $\boldsymbol{\eta}(t)$ defining the *zero dynamics* as being the solution of the differential equation of the form

$$\dot{\boldsymbol{\eta}}(t) = \mathbf{q}(0, \boldsymbol{\eta}(t)) - \mathbf{p}(0, \boldsymbol{\eta}(t)) \cdot [\mathbf{A}(0, \boldsymbol{\eta}(t))]^{-1} \cdot \mathbf{b}(0, \boldsymbol{\eta}(t)), \quad \boldsymbol{\eta}(0) = \boldsymbol{\eta}^0. \quad (10.18)$$

10.1.3 Lyapunov Approach

In the Lyapunov approach a function $V(\mathbf{x},u,t)$ is called a *Lyapunov candidate function* if

$$\begin{cases} \alpha\|\mathbf{x}\| \leq V(\mathbf{x}, u, t) \leq \beta\|\mathbf{x}\| \\ V(\mathbf{x}, t) = 0 \ \text{ for } \mathbf{x} = 0, \end{cases}$$

with α and β being K-class functions. By a K-class function one understands a function defined in $\mathbb{R}^+$, continuous, monotone nondecreasing, null at the origin and indefinitely increasing with its argument.

Let us recall the following.

Theorem (Lyapunov stability) Given a Lyapunov function $V(\mathbf{x},t)$ and the system defined by $\mathbf{x} = \mathbf{f}(\mathbf{x}) + \mathbf{g}(\mathbf{x}) \cdot u$, its time derivative has the expression

$$\frac{d}{dt}V(\mathbf{x}, t) = L_{\mathbf{f}(\mathbf{x})}V(\mathbf{x}, t) + u \cdot g(\mathbf{x}) \cdot V(\mathbf{x}, t) + \frac{\partial}{\partial t}V(\mathbf{x}, t).$$

(a) If the time derivative of the Lyapunov function satisfies $\frac{d}{dt}V(\mathbf{x}, t) \leq 0 \ \forall \ \mathbf{x}, t$, then the considered system is stable but the equilibrium point can be different from the origin of the state space.
(b) If $\frac{d}{dt}V(\mathbf{x}, t) \leq -\gamma \cdot \|\mathbf{x}\| \ \forall \ \mathbf{x}, t$, with γ being a K-class function, then the considered system is stable and the zero equilibrium point (in the state space) is reached.

Although the way of finding a Lyapunov function in order to characterize a given nonlinear dynamical system from an energy viewpoint is not always a systematic one, the concept of Lyapunov function is crucial for stability analysis (Khalil 2011), as well as for designing control laws that represent the control goal as being equivalent with minimizing the energy of the system (the so-called Lyapunov design; Freeman and Kokotović 2011).

10.2 Overview of Nonlinear Control Methods for Power Electronic Converters

An overview of the content of the next three chapters is presented here; these chapters are respectively dedicated to three nonlinear control approaches applied to power electronic converters.

One possible classification of the approaches presented next is according to the type of control law provided: *continuous* and *discontinuous* (*variable-structure*) nonlinear control laws. Notions like relative degree and zero dynamics are widely used by all these approaches for characterizing the structural properties of power converters as dynamical systems.

Another common feature of these approaches is the fact that they yield control laws generally dependent on the system's operating point and/or sensitive to parameter variations. Adaptive approaches employing supplementary measurements or parameter estimations are necessary in almost all cases in order to prevent significant performance degradation of control quality when the operating point covers the entire range. Thus, the general control algorithms provided in each case can be rendered adaptive after being enriched with appropriately tuned on-line parameter estimators.

Continuous control laws
Chapter 11 is dedicated to the *feedback-linearization control approach*, which proposes to overcome undesired converters behavior due to their intrinsic nonlinearities and dependence of the operating point (Isidori 1989). Its result is a nonlinear state feedback that ensures a pure-integrator input-output behavior whose order is equal to the system relative degree in the SISO case. Linearization reduces control structure complexity and enhances plant robustness.

The *energy-based control approach* for power electronic converters is the subject of Chap. 12. It treats two control methods based on concepts of energy: so-called *stabilizing control*, employing Lyapunov-based control design methods (Sanders and Verghese 1992); and *passivity-based control*, exploiting certain specific structural properties for control purposes (Ortega et al. 1998, 2001). Energy processing described in terms of power flows proves to be a natural representation of the phenomena that govern power converter behavior. The basic idea of energy-based control relies upon adjusting the speed of the energy dissipation process that ensures convergence to the steady-state operation (Stanković et al. 2001). The concept of *energy in the increment* (Sanders 1989; Sanders and Verghese 1992) is of extreme importance in characterizing dynamical behavior and designing control laws in a comprehensive manner.

The most significant drawback exhibited by energy-based control laws is their complexity and dependence difficult to measure variable parameters, such as load characteristics. Power converter time-criticality worsens the situation. Energy-based control approaches may be combined with linear control techniques (Pérez et al. 2004) as well as with other nonlinear control methods, for example, feedback-linearization techniques (Sira-Ramírez and Prada-Rizzo 1992) or variable-structure control (Sira-Ramírez et al. 1996; Ortega et al. 1998) – in order to improve overall control performance.

Variable-structure control laws
The final chapter of this book, Chap. 13, approaches the topics of the *variable-structure–* or *sliding-mode – control* (Filippov 1960; Emelyanov 1967; Utkin 1972), which appears a reasonable way of robustly controlling systems whose structure switches between several configurations. Power converters offer a good application area for this class of control methods because they are described by differential equations with discontinuous right-hand sides (i.e., discontinuous inputs) (Sira-Ramírez and Silva-Ortigoza 2006; Tan et al. 2011). Notions like *sliding surface* and *equivalent control* are fundamental here. In this case, good control performance such as large bandwidth is ensured because the switching solution is obtained directly without needing any other form of supplementary modulation; the fastest closed-loop response is thus ensured. Switching control benefits from intrinsic robustness and thus exhibits reduced sensitivity to uncertainties likely to occur in power electronic converters when the operating point and/or operating mode changes.

References

Emelyanov SV (1967) Variable structure control systems (in Russian). Nauka, Moscow

Filippov AF (1960) Differential equations with discontinuous right hand side. Am Math Soc Transl 62:199–231

Freeman RA, Kokotović PV (2011) Lyapunov design. In: Levine WS (ed) The control handbook, 2nd edn. CRC Press/Taylor & Francis Group, Boca Raton, pp 49-1–49-14

Isidori A (1989) Nonlinear control systems, 2nd edn. Springer, Berlin

Khalil H (2011) Lyapunov stability. In: Levine WS (ed) The control handbook, 2nd edn. CRC Press/Taylor & Francis Group, Boca Raton, pp 43-1–43-10

Ortega R, Loría A, Nicklasson PJ, Sira-Ramírez H (1998) Passivity-based control of Euler-Lagrange systems. Springer, London

Ortega R, van der Schaft AJ, Mareels I, Maschke B (2001) Putting energy back in control. IEEE Control Syst Mag 21(2):18–33

Pérez M, Ortega R, Espinoza JR (2004) Passivity-based PI control of switched power converters. IEEE Trans Control Syst Technol 12(6):881–890

Sanders SR (1989) Nonlinear control of switching power converters. Ph.D. thesis, Massachusetts Institute of Technology

Sanders SR, Verghese GC (1992) Lyapunov-based control for switched power converters. IEEE Trans Power Electron 7(1):17–24

Sira-Ramírez H, Prada-Rizzo MT (1992) Nonlinear feedback regulator design for the Ćuk converter. IEEE Trans Autom Control 37(8):1173–1180

Sira-Ramírez H, Silva-Ortigoza R (2006) Control design techniques in power electronics devices. Springer, London

Sira-Ramírez H, Escobar G, Ortega R (1996) On passivity-based sliding mode control of switched DC-to-DC power converters. In: Proceedings of the 35th Conference on Decision and Control – CDC 1996. Kobe, Japan, pp 2525–2526

Stanković AM, Escobar G, Ortega R, Sanders SR (2001) Energy-based control in power electronics. In: Banerjee S, Verghese GC (eds) Nonlinear phenomena in power electronics: attractors, bifurcations, chaos and nonlinear control. IEEE Press, Piscataway, pp 25–37
Tan S-C, Lai Y-M, Tse C-K (2011) Sliding mode control of switching power converters: techniques and implementation. CRC Press/Taylor & Francis Group, Boca Raton
Utkin VA (1972) Equations of sliding mode in discontinuous systems. Autom Remote Control 2 (2):211–219

Chapter 11
Feedback-linearization Control Applied to Power Electronic Converters

Feedback linearization is a powerful instrument that transforms a generic nonlinear plant model into a linear one by using a nonlinear feedback that cancels the original plant nonlinearity. Usually the target plant is a pure integrator; hence it can be controlled (with zero steady-state error) by a simple proportional controller (Isidori 1989). However, this feature comes with a drawback: the linearized system is likely to be sensitive to parameters' variations and/or to the operating point, as the nonlinear feedback derives from the system model.

This control structure outputs a continuous control input that needs supplementary (e.g., PWM) modulation in order to be applied to power switches. For power electronic converters the feedback linearization may be conceived for both single-input–single-output (SISO) and multi-input–multi-output (MIMO) cases by using their averaged models, but in this chapter the focus is on the first type of method.

Corresponding to the primary control target, one may define the *direct* control in which the linearized dynamics correspond to the controlled variable, or *indirect* control, in which the linearized dynamics output a different variable (Sira-Ramírez and Silva-Ortigoza 2006). In this latter case, the control structure turns out to be more complicated because a supplementary outer control loop is required in order to achieve the primary control target (Jung et al. 1999; Song et al. 2009). Because in this case the outer loop is built on a (generally) nonlinear plant, its design may require either plant-approximated linearization (employing techniques amply presented in Chaps. 8 and 9 of this book) or the employment of a robust nonlinear control law (see, for example, Chaps. 12 and 13). In either case the system's zero dynamics represent a difficult issue to deal with, except when the system's relative degree is equal to the system's order – and zero dynamics do not exist; feedback linearization is called *exact* in such cases.

Using some of the mathematical tools presented in Chap. 10, this chapter introduces feedback linearization of a generic converter and the algorithm to be employed for control law design. Some examples, a case study and a problem with its solution complete the discussion of this subject. The reader is invited to solve several other problems at the end of this chapter.

S. Bacha et al., *Power Electronic Converters Modeling and Control: with Case Studies*, Advanced Textbooks in Control and Signal Processing, DOI 10.1007/978-1-4471-5478-5_11,

11.1 Basics of Linearization *via* Feedback

Theoretical foundations of feedback-linearization control have been proposed by Isidori (1989), in both SISO and MIMO cases; they are based on the differential geometric approach in the theory of nonlinear control systems. This section aims at offering a quick overview of this topic and specific solution tools.

11.1.1 Problem Statement

Linearization by feedback aims at finding the expression of the control law that is able to transform a nonlinear system into a linear and controllable one by means of feedback and coordinates transformation. The main advantage of such an approach is the possibility it allows for using further linear control design techniques, which are simple and intuitive and, at the same time, robust and easy to implement.

11.1.2 Main Results

This section presents briefly the main ideas regarding how the linearizing-feedback control problem is solved in the SISO case and in a special case of MIMO dynamical systems. This presentation is only intended to provide essential tools for application problems. Technical details, complete proofs and in-depth interpretations can be found in Isidori (1989).

Given a SISO nonlinear system described by the model

$$\begin{cases} \dot{\mathbf{x}} = \mathbf{f}(\mathbf{x}) + \mathbf{g}(\mathbf{x}) \cdot u \\ y = h(\mathbf{x}), \end{cases} \tag{11.1}$$

which has relative degree r at a point $\mathbf{x}^0$, where $r \leq n$, with n being the system's order. From Chap. 10 recall that $\mathrm{L}_\mathbf{f}$ is the Lie derivative of a function along a vector field $\mathbf{f}$. This means that $\mathrm{L}_{\mathbf{g}(\mathbf{x})}\mathrm{L}_{\mathbf{f}(\mathbf{x})}^{r-1}h_{(\mathbf{x})} \neq 0$ and the state feedback computed as

$$u = \frac{1}{\mathrm{L}_{\mathbf{g}(\mathbf{x})}\mathrm{L}_{\mathbf{f}(\mathbf{x})}^{r-1}h(\mathbf{x})}\left(-\mathrm{L}_{\mathbf{f}(\mathbf{x})}^{r}h(\mathbf{x}) + v\right) \tag{11.2}$$

transforms system (11.1) into a system whose input-output behavior is identical to that of an rth-order integrator having transfer function

$$H(s) = \frac{Y(s)}{V(s)} = \frac{1}{s^r},$$

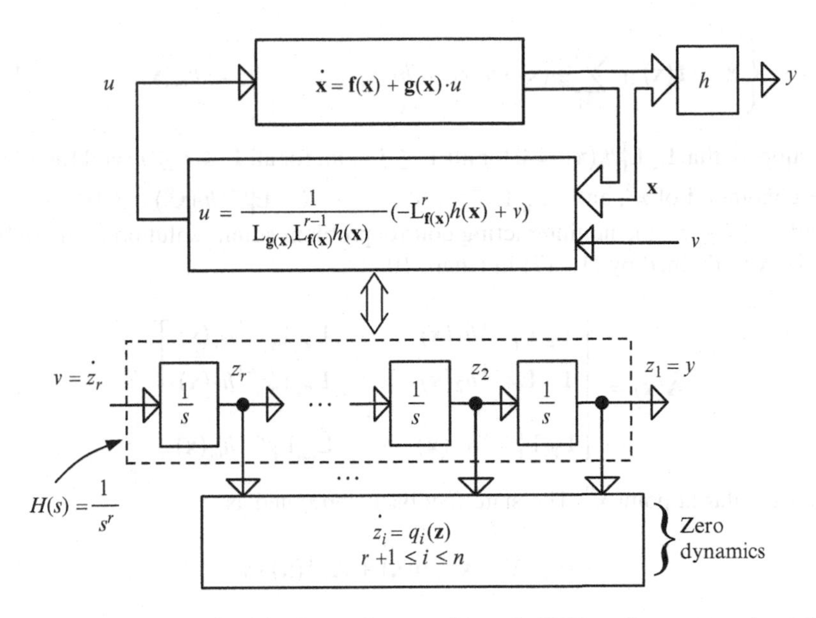

Fig. 11.1 Linearizing feedback leading to decomposition of a SISO nonlinear system into a linear subsystem and a nonlinear subsystem representing the zero dynamics

where $V(s)$ and $Y(s)$ are the Laplace images of signals v and y, respectively. Further adopting a change of coordinates $\mathbf{z} = \mathbf{F}(\mathbf{x})$ such that

$$\begin{cases} \dot{z_1} = z_2, \quad \dot{z_2} = z_3, \quad \cdots, \quad \dot{z_{r-1}} = z_r \\ \dot{z_r} = v, \quad \dot{z_{r+1}} = q_{r+1}(\mathbf{z}), \quad \cdots, \quad \dot{z_n} = q_n(\mathbf{z}), \quad y = z_1, \end{cases}$$

one can note that the initial system appears as a decomposition into an rth-order linear subsystem, the only one that determines the input-output behavior, and a possibly nonlinear system of dimension n–r, whose dynamics do not influence the output, as shown in Fig. 11.1. These latter dynamics are the so-called zero dynamics, defined in Sect. 10.1.2 of Chap. 10.

This latter subsystem plays in many circumstances a role similar to that of the zeros of the transfer function of a linear system. It corresponds to "internal" behavior that is unobservable in the output behavior.

Isidori (1989) showed that quite straightforward extensions of the results obtained for SISO systems are also valid for a special case of MIMO systems, that is, for systems having multiple inputs and the same number of outputs. In this case, an interesting problem is the use of feedback to obtain an equivalent system representing – at least from an input-output viewpoint – an aggregation of independent SISO channels. This is called the *noninteracting control problem.*

Let us resume the model of a MIMO nonlinear system with m inputs and m outputs presented in Eq. (10.16) in Chap. 10:

$$\begin{cases} \dot{\mathbf{x}} = \mathbf{f}(\mathbf{x}) + \sum_{i=1}^{m} \mathbf{g}_i(\mathbf{x}) \cdot u_i \, y_1 = h_1(\mathbf{x}), \quad \cdots \quad y_m = h_m(\mathbf{x}), \end{cases} \tag{11.3}$$

and suppose that $\mathrm{L}_{g_j}\mathrm{L}_{\mathbf{f}}^{k}h_i(\mathbf{x}) = 0$ for all $1 \le j \le m$, for all $1 \le i \le m$ and for all $\mathbf{x}$ in a neighborhood of $\mathbf{x}^0$, and $\left[\mathrm{L}_{\mathbf{g}_1}\mathrm{L}_{\mathbf{f}}^{r_i-1}h_i(\mathbf{x}^0) \quad \cdots \quad \mathrm{L}_{\mathbf{g}_m}\mathrm{L}_{\mathbf{f}}^{r_i-1}h_i(\mathbf{x}^0)\right] \neq [0 \quad \cdots \quad 0]$ for all $1 \le i \le m$, the noninteracting control problem admits solution if and only if matrix $\mathbf{A}(\mathbf{x})$ defined by (10.17) in Chap. 10,

$$\mathbf{A}(\mathbf{x}) = \begin{bmatrix} \mathrm{L}_{\mathbf{g}_1}\mathrm{L}_{\mathbf{f}}^{r_1-1}h_1(\mathbf{x}) & \cdots & \mathrm{L}_{\mathbf{g}_m}\mathrm{L}_{\mathbf{f}}^{r_1-1}h_1(\mathbf{x}) \\ \mathrm{L}_{\mathbf{g}_1}\mathrm{L}_{\mathbf{f}}^{r_2-1}h_2(\mathbf{x}) & \cdots & \mathrm{L}_{\mathbf{g}_m}\mathrm{L}_{\mathbf{f}}^{r_2-1}h_2(\mathbf{x}) \\ \vdots & \ddots & \vdots \\ \mathrm{L}_{\mathbf{g}_1}\mathrm{L}_{\mathbf{f}}^{r_m-1}h_m(\mathbf{x}) & \cdots & \mathrm{L}_{\mathbf{g}_m}\mathrm{L}_{\mathbf{f}}^{r_m-1}h_m(\mathbf{x}) \end{bmatrix}, \tag{11.4}$$

is nonsingular at point $\mathbf{x}^0$. The state feedback computed as

$$\mathbf{u} = -\mathbf{A}^{-1}(\mathbf{x}) \cdot \mathbf{b}(\mathbf{x}) + \mathbf{A}^{-1}(\mathbf{x}) \cdot \mathbf{v}, \tag{11.5}$$

where

$$\mathbf{b}(\mathbf{x}) = \begin{bmatrix} \mathrm{L}_{\mathbf{f}}^{r_1}h_1(\mathbf{x}) \\ \mathrm{L}_{\mathbf{f}}^{r_2}h_2(\mathbf{x}) \\ \vdots \\ \mathrm{L}_{\mathbf{f}}^{r_m}h_m(\mathbf{x}) \end{bmatrix}, \tag{11.6}$$

transforms system (11.3) into a linear system whose input-output behavior is described by the transfer function matrix of the form

$$\mathbf{H}(s) = \begin{bmatrix} \frac{1}{s^{r_1}} & 0 & \cdots & 0 \\ 0 & \frac{1}{s^{r_2}} & \cdots & 0 \\ \vdots & \vdots & \ddots & \vdots \\ 0 & 0 & \cdots & \frac{1}{s^{r_m}} \end{bmatrix}.$$

Matrix $\mathbf{A}$ given by (11.4) is also called a decoupling matrix because it plays a role in the separation of the individual input-output channels. Its nonsingularity at point $\mathbf{x}^0$ means that the MIMO nonlinear system (11.3) has the vector relative degree $\{r_1,\cdots,r_m\}$ at $\mathbf{x}^0$ (see definitions in Sect. 10.1.2 of Chap. 10). One can note that, similarly to the SISO case, if $r = r_1 + r_2 + \cdots + r_m$ is less than the system dimension n, an unobservable part is present in the closed-loop system, which has no influence on the outputs; this part corresponds to the zero dynamics. This decomposition is depicted in Fig. 11.2. The advantage of such an approach is

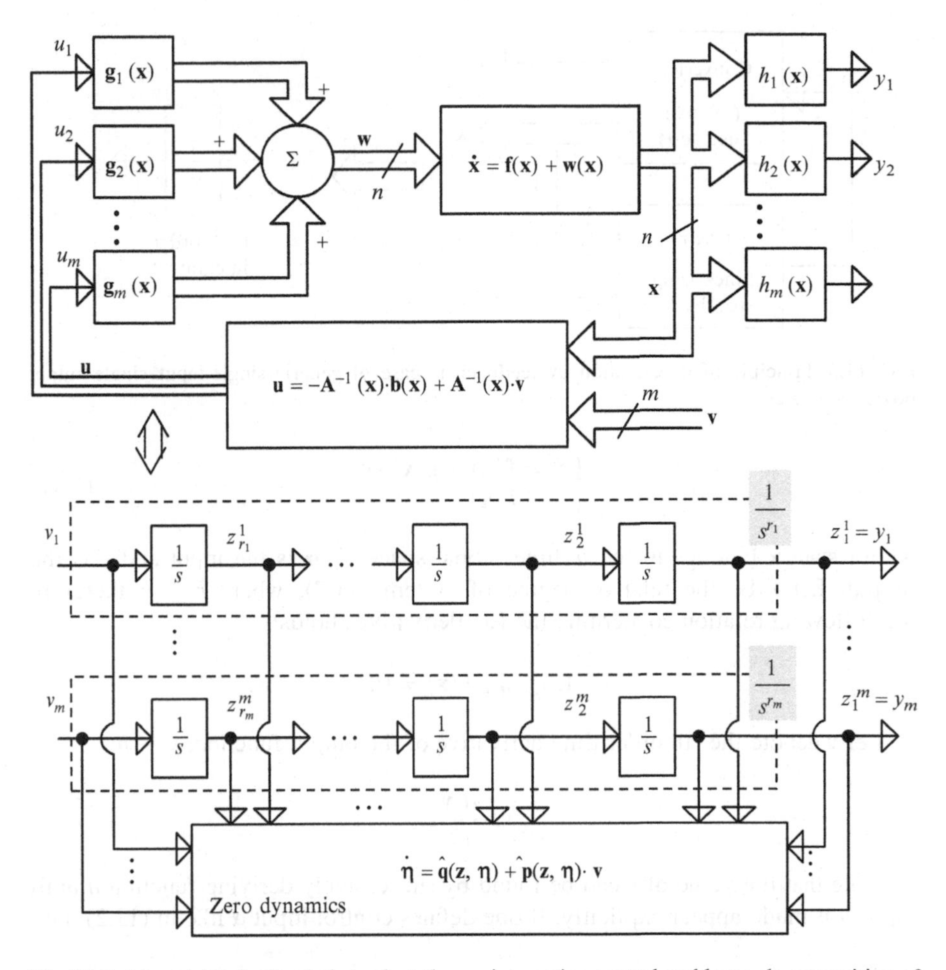

Fig. 11.2 Linearizing feedback that solves the noninteracting control problem – decomposition of a MIMO $m \times m$ nonlinear system into a set of linear subsystems and a nonlinear subsystem representing the zero dynamics

that each input v_i controls the corresponding output y_i throughout a chain of r_i integrators. The feedback of the form (11.5) that solves the noninteracting control problem is called *standard noninteractive feedback* (Isidori 1989).

11.2 Application to Power Electronic Converters

11.2.1 Feedback-Linearization Control Law Computation

Let us consider the averaged model of a single-input–single-output (SISO) power converter in its state-space bilinear representation (time argument being dropped):

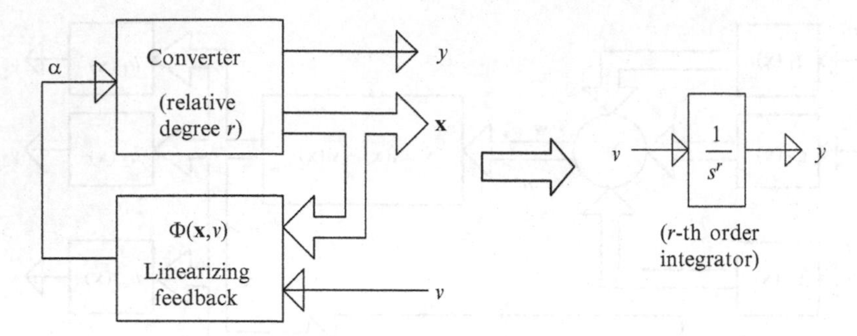

Fig. 11.3 Principle of linearization by feedback in case of generic single-input–single-output power converter

$$\begin{cases} \dot{\mathbf{x}} = \mathbf{f}(\mathbf{x}) + \mathbf{g}(\mathbf{x}) \cdot \alpha \\ y = h(\mathbf{x}), \end{cases} \tag{11.7}$$

where state $\mathbf{x}$ belongs to the n-dimensional space $\mathbf{X}$, α is the input and y is the output. Let r be the relative degree of system (11.7), where $r \leq n$; therefore the following relation concerning the Lie derivatives holds:

$$\mathrm{L}_{\mathbf{g}(\mathbf{x})}\mathrm{L}_{\mathbf{f}(\mathbf{x})}^{r-1}h(\mathbf{x}) \neq 0.$$

Let v denote the rth order time derivative of the output function, $y \equiv h$:

$$v \triangleq \frac{d^r y}{dt}.$$

Note that the value of r can be found by successively deriving function h until input u is made appear explicitly. If one defines control input α like in (11.2), i.e.,

$$\alpha \equiv \Phi(\mathbf{x}, v) \triangleq \frac{v - \mathrm{L}_{\mathbf{f}(\mathbf{x})}^{r}h(\mathbf{x})}{\mathrm{L}_{\mathbf{g}(\mathbf{x})}L_{\mathbf{f}(\mathbf{x})}^{r-1}h(\mathbf{x})}, \tag{11.8}$$

then it will play the role of a feedback-linearization control because it ensures that nonlinear system (11.7) will behave like an rth-order integrator on the transfer channel from v to y; this equivalence is depicted in Fig. 11.3. Relation (11.8) allows emphasizing a general feature of the linearizing control law as partial-state feedback, namely its dependence on system parameters. This further leads to the necessity of embedding parameter estimators into the control law's implementation (Sira-Ramirez et al. 1995).

In this way, output vector y can be controlled by means of a simple gain. Figure 11.4 offers details about how such control input ensures the dynamic equivalence with a linear system, where notations

$$F(\mathbf{x}) \triangleq \mathrm{L}_{\mathbf{f}(\mathbf{x})}^{r}h(\mathbf{x}), \quad G(\mathbf{x}) \triangleq \mathrm{L}_{\mathbf{g}(\mathbf{x})}\mathrm{L}_{\mathbf{f}(\mathbf{x})}^{r-1}h(\mathbf{x})$$

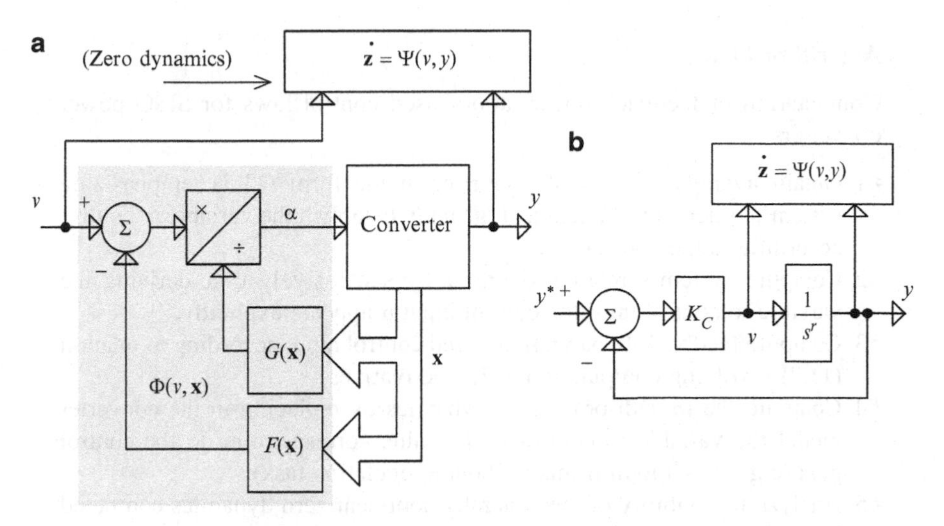

Fig. 11.4 Schematics of a feedback linearization control scheme allowing a generic single-input–single-output power electronic converter modeled as a nonlinear plant being controlled by means of a proportional controller: (**a**) model of the uncontrolled linearized-by-feedback system; (**b**) equivalent linear system controlled by a gain

have been adopted to detail the implementation of function $\Phi(\mathbf{x},v)$ (emphasized on a gray background in this figure). There are n–r variables denoting the so-called free-variable dynamics, or otherwise the *zero dynamics*; they have been denoted by $\dot{\mathbf{z}} = \Psi(v, y)$ in Fig. 11.4. Although these variables have no influence on the system's output, in any approach for finding control laws for the first r variables one must ensure that the zero dynamics are stable by suitably choosing the r variables to control, so the resulting zero dynamics are stable. In some cases it is possible to control the zero dynamics (Lee 2003); in particular, adding a supplementary degree of freedom in the control goal definition may allow stabilizing the eventually unstable zero dynamics (Petitclair 1997). In this way, problems like control input saturation can be avoided.

11.2.2 Pragmatic Design Approach

A synthesis of the above listed computation steps is presented as the algorithm below. Taking account that elementary theory of linearizing feedback control for SISO systems can be extended with minor changes to multi-input-multi-output (MIMO) systems (Isidori 1989), the algorithm below can serve as generalization basis to the case of MIMO power converters.

Algorithm 11.1.

Computation of feedback-linearization-based control laws for SISO power converters

#1 Obtain averaged model of converter in the form (11.7), emphasizing system's order n and functions **f**, **g** and h. Establish the variable to control according to the control goal.
#2 Compute system's relative degree r by successively time-deriving the variable to control until the control input α appears explicitly.
#3 Compute feedback-linearization-based control input according to relation (11.8) involving computation of Lie derivatives.
#4 Compute the $(n–r)$th-order zero dynamics by replacing in the converter model the variable to control by its value corresponding to the control goal (e.g., its setpoint if this is about a regulation task).
#5 Analyze the stability of the generally nonlinear zero dynamics computed in previous step.
#6 If zero dynamics prove unstable, choose one possibility: either adopt an indirect control method, that is, achieve the control goal established at Step #1 by means of one of the "free" variables, or stabilize the zero dynamics by means of enriching the control goal with a supplementary degree of freedom that allows stabilization.
#7 Select controller gain that ensures desired closed-loop performance of rth-order integrator equivalent system.
#8 Study sensitivity of control law to parameter variations. Design suitable parameter estimators.
#9 Perform closed-loop numerical simulations, eventually reiterating controller's gain selection and/or parameter estimators tuning for best performance.

11.2.3 Examples: Boost DC-DC Converter and Buck DC-DC Converter

In order to illustrate the application of Algorithm 11.1 to computing feedback linearization control input expressions, two examples will next be considered, namely the boost DC-DC converter – as a case for which the relative degree is inferior to the system's order – and the buck DC-DC converter supplied by a weak voltage source. The examples illustrate a case where the relative degree is equal to the system's order and therefore the zero dynamics are absent.

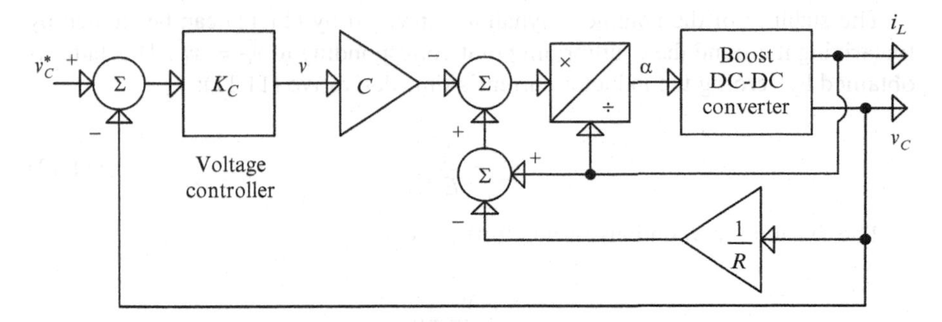

Fig. 11.5 Closed-loop schematics of feedback-linearization-based direct voltage control of boost DC-DC converter; zero dynamics corresponding to inductor current are unstable

Example 1. The boost converter electrical circuit is the one presented in Sect. 3.2.3 from Chap. 3; its averaged model for the continuous-conduction case is given by Eq. (3.11) from the same section:

$$\begin{cases} \dot{i_L} = -(1-\alpha)v_C/L + E/L \\ \dot{v_C} = (1-\alpha)i_L/C - v_C/(RC), \end{cases} \tag{11.9}$$

where the usual notations have been employed, with α being the average value of switching function u. It is required to regulate the output voltage, $y \equiv v_C$, its setpoint being denoted by v_C^*.

(a) *Output voltage control – direct approach*

Equation (11.9) shows that control input α appears explicitly in the time derivative of the variable to control, v_C; therefore the relative degree is $r = 1$. After adopting notation $v = \dot{y} \equiv \dot{v_C}$, the second equation from (11.9) allows the expression of the linearizing feedback to be obtained as

$$\alpha = \frac{i_L - v_C/R - C \cdot v}{i_L}. \tag{11.10}$$

Use of control input (11.10) allows the converter to be equivalent to an integrator – from input v to output y – which can therefore be controlled by a simple gain, denoted by K_C in Fig. 11.5, where the closed-loop diagram is shown. As a general remark, the dependence of the control law on system's parameters is noted.

Except for the first-order controllable dynamic a first-order zero dynamic also exists in this case. It is the dynamic of the other variable, the inductor current i_L and it can be emphasized by assuming the value of the controlled variable being equal to its setpoint, i.e., $v_C = v_C^*$, and substituting it into the converter dynamics (11.9). Note first that $v = 0$, so the corresponding control input value is $\alpha = 1 - v_C^*/(Ri_L)$, according to (11.10). With these values, the zero dynamic is further obtained from the first equation of (11.9):

$$\dot{i_L} = \frac{1}{L}\left(E - \frac{v_C^{*2}}{Ri_L}\right). \tag{11.11}$$

The stability of the nonlinear dynamic expressed by (11.11) can be studied by linearizing it around the equilibrium point corresponding to $v_C = v_C^*$. This latter is obtained by zeroing the inductor current's time derivative (11.11):

$$i_{Le} = \frac{v_C^{*2}}{ER}. \tag{11.12}$$

The dynamic in variations results further as

$$\dot{\tilde{i}}_L = \frac{v_C^{*2}}{RLi_{Le}^2}\tilde{i}_L,$$

which indicates that the zero dynamic is unstable because $\dot{\tilde{i}}_L \cdot \tilde{i}_L > 0$. In this case, one can decide to keep the control diagram in Fig. 11.5 and study the possibility of zero dynamic stabilization by changing the expression of the reference. Another possibility is to change the control method from the direct to the indirect one, that is, to control the inductor current instead of the capacitor voltage.

(b) *Inductor current control – indirect approach*

In this case the current setpoint, denoted by i_L^*, is given by expression (11.12), as it must correspond to the desired target voltage v_C^*. Equation (11.9) shows that control input α appears explicitly in the time derivative of the variable to control, i_L; therefore, the relative degree is $r = 1$ in this case also. After adopting notation $w = \dot{y} \equiv \dot{i}_L$, the first equation from (11.9) yields the expression of the linearizing control input as

$$\alpha = \frac{Lw - E + v_C}{v_C}. \tag{11.13}$$

Use of control input (11.13) allows the converter to be equivalent to an integrator – from input w to output y – which can be controlled by a simple gain like in the direct control method. This time again, the control law depends on system parameters. As regards the zero dynamic, it corresponds in this case to the capacitor voltage and it is stable, as is shown next.

The zero dynamic expression is obtained by assuming the value of the controlled variable to be equal to its setpoint, i.e., $i_L = i_L^*$, and substituting it into the converter dynamics (11.9). Because $w = 0$, the corresponding control input value is $\alpha = 1 - E/v_C$, according to (11.13). With these values, the zero dynamic is further obtained from the second equation of (11.9):

$$\dot{v_C} = \frac{1}{C}\left(i_L^*\frac{E}{v_C} - \frac{v_C}{R}\right). \tag{11.14}$$

Expression (11.14) corresponds to a nonlinear dynamic. Stability can be studied by linearization in this case also, but a simpler method consists in noting that, if

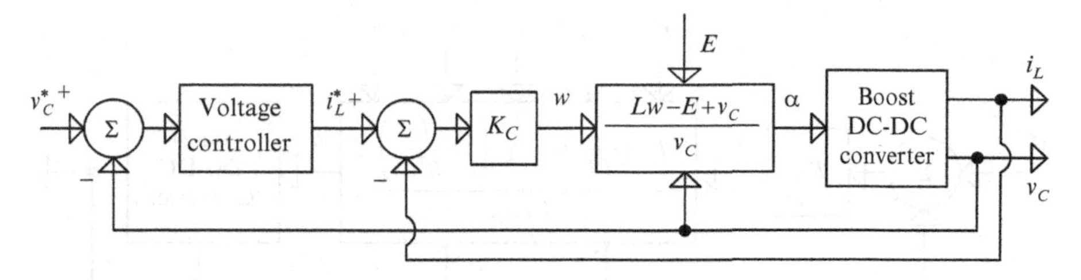

Fig. 11.6 Closed-loop schematics of a feedback-linearization-based indirect voltage control of a boost DC-DC converter; the zero dynamic corresponding to the capacitor voltage are controlled by an outer loop

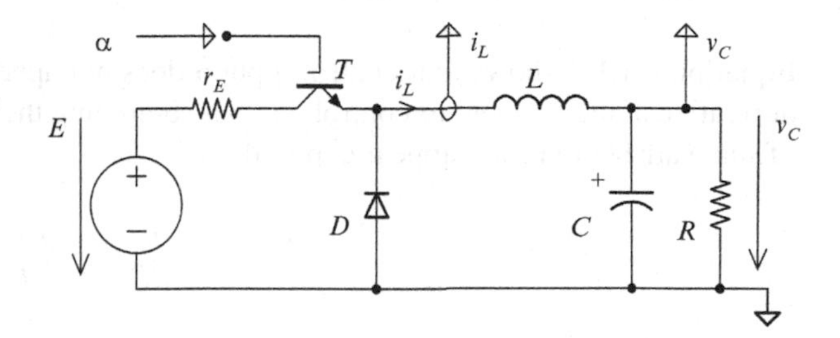

Fig. 11.7 Electrical circuit of a buck DC-DC converter supplied by a voltage source having nonnegligible internal resistance r_E

multiplying (11.14) by v_C, the dynamic of variable v_C^2 is obtained, which is a first-order linear and stable one:

$$\dot{v_C^2} = -\frac{v_C^2}{RC} + \frac{1}{C} i_L^* E. \tag{11.15}$$

Finally, the control diagram can be obtained by adding an outer voltage control loop to the one controlling the current, as shown in Fig. 11.6. Note that such a setup is valid for controlling either v_C^2, based upon relation (11.15), or v_C, based upon the linearized version of relation (11.15).

Example 2. The electrical circuit of a buck converter supplied by a weak voltage source is presented in Fig. 11.7, where r_E denotes the nonnegligible internal resistance of the voltage source and the other notations are the usual ones, being clear from the context. As in the previous example, the output voltage $y \equiv v_C$ is to be regulated, its setpoint being denoted by $\overset{*}{v_C}$.

The buck converter's averaged model for the continuous-conduction case is the following:

$$\begin{cases} \dot{i_L} = (E - r_E i_L) \cdot \alpha/L - v_C/L \\ \dot{v_C} = i_L/C - v_C/(RC). \end{cases} \tag{11.16}$$

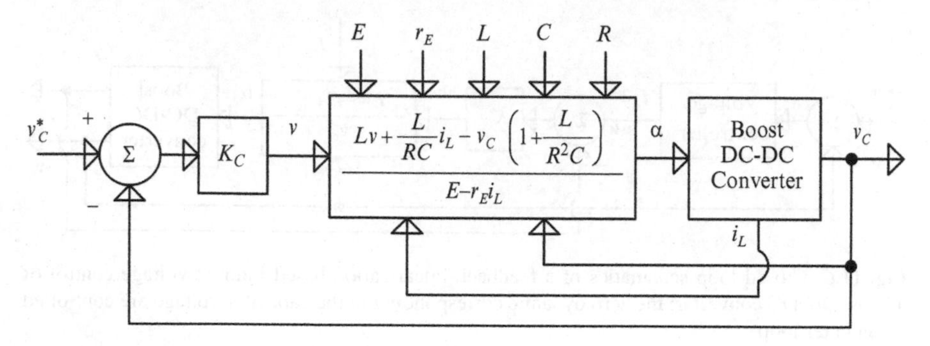

Fig. 11.8 Closed-loop schematics of a feedback-linearization-based voltage control of a buck DC-DC converter; zero dynamics are absent

Equation (11.16) shows that control input α does not appear explicitly in the time derivative of the variable to control, v_C, but computing the second-order derivative of this variable makes α appear explicitly:

$$\ddot{v_C} = \dot{i_L} - \frac{\dot{v_C}}{R} = \frac{E - r_E i_L}{L} \cdot \alpha - \frac{i_L}{RC} - v_C\left(\frac{1}{L} + \frac{1}{R^2C}\right). \tag{11.17}$$

In conclusion, the relative degree is $r = 2$, being equal to the system's order. This means that, unlike in the first example, in this case there is no zero dynamic. After adopting notation $v = \ddot{y} \equiv \ddot{v_C}$, Eq. (11.17) allows the expression of the linearizing control input being obtained as full-state feedback:

$$\alpha = \frac{Lv + Li_L/(RC) - v_C\left[1 + L/\left(R^2C\right)\right]}{E - r_E i_L}. \tag{11.18}$$

Control input (11.18) allows the converter to be equivalent to a double integrator – from input v to output y – which can further be controlled by a gain controller K_C. Figure 11.8 shows the closed-loop diagram, in which the linearizing control law's dependence on converter parameters is emphasized.

11.2.4 Dealing with Parameter Uncertainties

As it has been shown in the examples discussed above, the expression of the linearizing feedback able to transform the initial nonlinear converter into a partially linear and controllable system generally depends on converter parameters (see expression (11.13) in the case of indirect control of a boost converter and expression (11.18) obtained for the buck converter with a nonideal voltage source). This is in fact a common drawback of practically all nonlinear control methods for power converters. This means that exact linearization is valid only if one knows precisely the values of the parameters involved in feedback control law computation. Among these parameters are some that could be easily measured (like source voltage E in the above examples), as well as others that need to be estimated (the converter's functional parameters, like inductance L and capacity C or load resistance R).

Parameter estimation may sometimes be difficult to achieve, especially for time-critical systems like power converters, because one must ensure both a very high estimation convergence speed along with sufficient estimation precision (Petitclair and Bacha 1997). Before a decision is made that implementation of parameter estimators is unavoidable, a preliminary sensitivity analysis is necessary in order to assess the influence of each parameter's variation on the quality of linearization. Effective design of an estimator will be done if it turns out that the influence of the respective parameter is important. In this case an *adaptive* approach must be employed.

11.3 Case Study: Feedback-Linearization Control of a Flyback Converter

The case of a flyback converter providing isolated noninverting boost topology by means of a transformer is here considered; its electrical circuit is shown in Fig. 4.24 in the text of Problem 4.1 of Chap. 4 and repeated in Fig. 11.9.

For this converter the output voltage regulation to setpoint v_C^* is stated as the control problem; its solution by feedback linearization is illustrated next.

11.3.1 Linearizing Feedback Design

The flyback converter's averaged model results from Eq. (4.42) in Chap. 4 by replacing switching function u by duty ratio α; so

$$\begin{cases} L\dot{i_L} = -(1-\alpha)\dfrac{v_C}{n} + \alpha E \\ C\dot{v_C} = (1-\alpha)\dfrac{i_L}{n} - \dfrac{v_C}{R}, \end{cases} \tag{11.19}$$

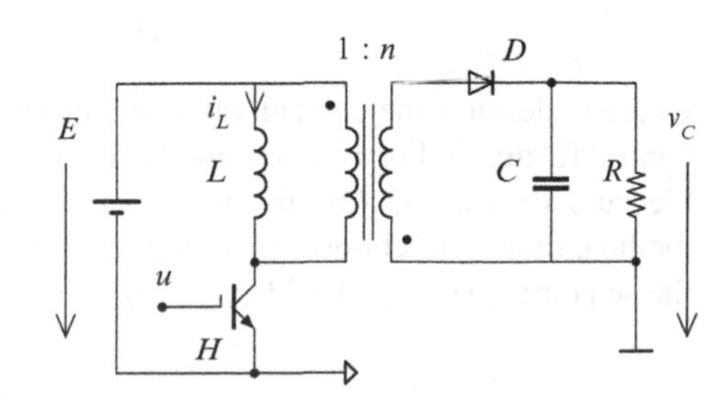

Fig. 11.9 Electrical circuit of flyback converter, including transformer model

where the usual notations have been used, with n being the transformer ratio, E the source voltage and duty ratio α the input.

In order to deduce the system's relative degree, as the output voltage is chosen as output variable $y \equiv v_C$, one first looks at the second equation from (11.19) and notes that the input α appears explicitly. Therefore, the relative degree is $r = 1$. Notation $v = \dot{y} \equiv \dot{v_C}$ is adopted, which allows the linearizing control input α to be computed based upon the second equation of (11.19) as

$$\alpha = -\frac{nCv - i_L + nv_C/R}{i_L}. \tag{11.20}$$

The result is that a first-order zero dynamic exists, in which the other variable, the inductor current i_L, is involved. One must further check whether this dynamic is stable or not; to this end, conditions for $y \equiv v_C^*$ (hence, $v \equiv \dot{y} = 0$) are considered. Expression (11.20) provides the corresponding expression of the control input:

$$\alpha_1 = \frac{i_L - nv_C^*/R}{i_L}, \tag{11.21}$$

which is further substituted in the first equation of (11.19) and yields

$$L\dot{i_L} i_L = i_L E - \frac{v_C^{*2}}{R} - \frac{nv_C^* E}{R},$$

or, otherwise,

$$\frac{1}{2} L\dot{i_L^2} = i_L E - \frac{v_C^{*2}}{R} - \frac{nv_C^* E}{R}. \tag{11.22}$$

Equation (11.22) describes the nonlinear first-order zero dynamic (of inductor current i_L), whose equilibrium point

$$i_{Le} = \frac{v_C^* \left(v_C^* + nE\right)}{RE} \tag{11.23}$$

is an unstable one. Indeed, if linearizing (11.22) around point (11.23) the following unstable model in variations results:

$$L i_{Le} \dot{\widetilde{i_L}} - E\widetilde{i_L} = 0,$$

where $\widetilde{i_L}$ denotes the current variation around i_{Le}. One can conclude that the direct control feedback-linearization method is not feasible as it is associated with unstable zero dynamics. At this point we will therefore prefer the indirect control method, that is, to choose the inductor current as output variable and regulate it at the setpoint given by (11.23), $i_L^* = i_{Le}$.

In the indirect-control case it is the first equation of (11.19) that will allow relative degree computation. Indeed, one can see that, like in the direct-control case, here also the relative degree is $r = 1$, because input α appears explicitly in the expression of the inductor current's first-order time derivative. Notation $v = \dot{y} \equiv \dot{i}_L$ allows the linearizing control input α to be computed based upon the first equation of (11.19) as

$$\alpha = \frac{Lv + v_C/n}{v_C/n + E}. \tag{11.24}$$

Now again a first-order zero dynamic exists, in which the capacitor voltage v_C is this time involved. In order to check whether this dynamic is stable or not, conditions for $y \equiv i_L^*$ (and therefore $v \equiv \dot{y} = 0$) are considered. According to expression (11.24), the corresponding expression of the control input is obtained as

$$\alpha_1 = \frac{v_C/n}{v_C/n + E}, \tag{11.25}$$

which is further substituted in the second equation of (11.19) and yields

$$C\dot{v_C} = \frac{E}{v_C + nE} i_L^* - \frac{1}{R} v_C.$$

After some simple manipulations, the expression of the zero dynamic can be written as

$$\frac{C}{2n}\dot{v_C^2} + \frac{v_C^2}{nR} + CE\dot{v_C} + \frac{E}{R} v_C = E\frac{i_L^*}{n}. \tag{11.26}$$

The zero dynamic expressed in (11.26) is a nonlinear one; the stability of its equilibrium point $v_{Ce} = v_C^*$ can be deduced after linearizing (11.26) around this point. The corresponding linear dynamic equation in variations results:

$$\frac{C}{n}\left(v_C^* + nE\right)\dot{\widetilde{v_C}} + \frac{2v_C^* + nE}{nR}\widetilde{v_C} = 0, \tag{11.27}$$

where $\widetilde{v_C}$ denotes the voltage variation around v_C^*. Equation (11.27) describes a stable dynamic. The conclusion is that the indirect control is feasible and can be used for output voltage regulation.

By summarizing the above results, the plant fed with the control input (11.24) behaves like an integrator from the inductor current viewpoint, whereas the other variable – the capacitor voltage – exhibits a stable zero dynamic. Note that control input (11.24) does not depend on load resistance R, nor on operating point, but only on system parameters L, E and n.

The original control task – of regulating the voltage at the setpoint v_C^* – can finally be achieved by a two-loop control structure as depicted in Fig. 11.10, in which both loops are driven by linear controllers. Indeed, because the plant in the

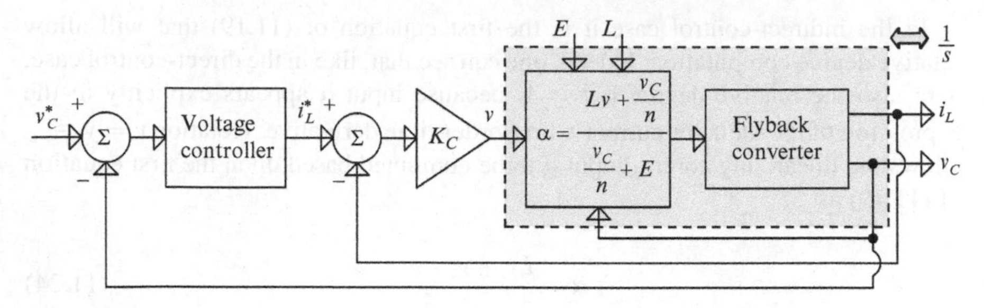

Fig. 11.10 Flyback converter output voltage regulation by using indirect control based upon feedback linearization

inner loop is an integrator, its controller can be chosen as a simple gain K_C. At its turn, the controller in the outer loop can be chosen in different ways; the simplest remains to use linearization and then linear design techniques, provided the desired operating point is stable.

The following basic parameter values are considered next for numerical simulation illustration: inductor inductance $L = 2$ mH, capacitor capacity $C = 860$ μF, source voltage $E = 9$ V, transformer ratio $n = 2$, rated load $R = 10\ \Omega$.

Figure 11.11a shows that the inner loop indeed behaves like an integrator when in open loop, as its output i_L is a ramp when its input $v \equiv \dot{i}_L$ is a step. Selection of gain K_C in the inner loop is obvious, as it represents the inverse of the desired closed-loop bandwidth; here, $K_C = 5000$, which ensures 1 ms of settling time in response to a 1 A step reference, as shown in Fig. 11.11b.

Figure 11.11c, d contain the corresponding time evolutions of capacitor voltage v_C and duty ratio α, respectively. One can note the change in the voltage steady-state value and quite large values of the duty ratio. It is also important to note the nonminimum-phase behavior of v_C; this remark will be useful for analyzing the outer loop dynamics.

11.3.2 Outer Loop Analysis

As regards the outer-loop controller, two possible ways of designing it are illustrated next. First, let us try to characterize the transfer between the inductor current and the capacitor voltage when the converter is fed by the linearizing feedback α given by (11.24). Simulations have already shown nonminimum-phase behavior (see Fig. 11.11c), which will be characterized analytically next. By replacing (11.24) in the second equation of (11.19) one obtains a nonlinear dependence of the capacitor voltage of the inductor current, namely:

$$\frac{C}{2n}\dot{v_C^2} + \frac{1}{nR}v_C^2 + CE\dot{v}_C + \frac{E}{R}v_C = \frac{E}{n}i_L - \frac{L}{2n}\dot{i_L^2},$$

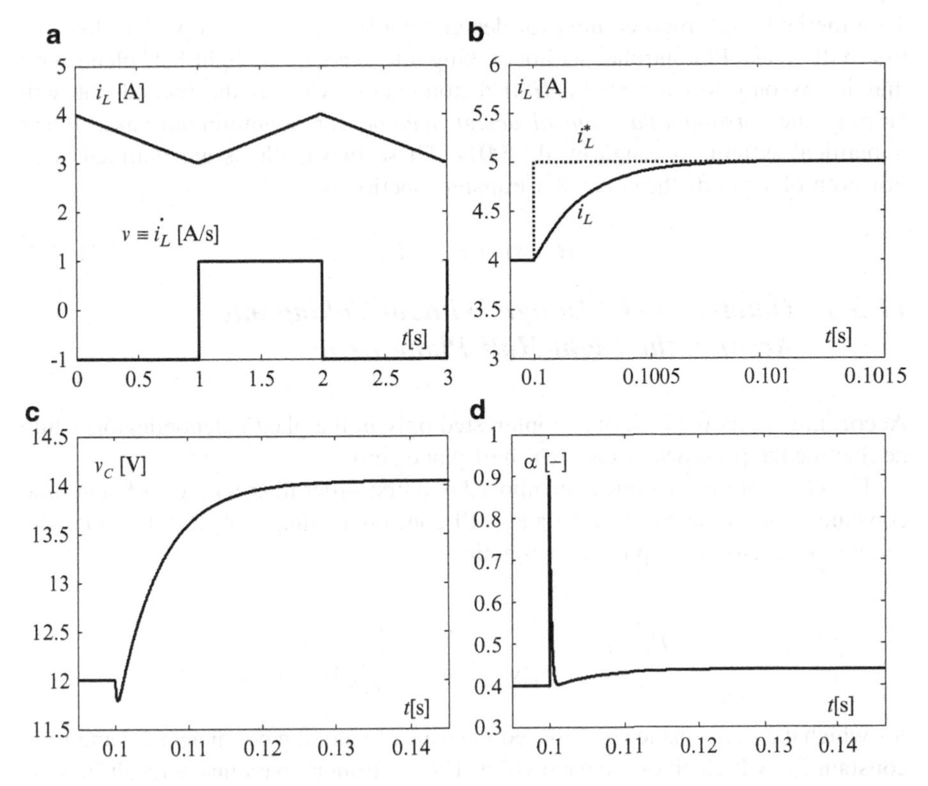

Fig. 11.11 Numerical tests regarding inner loop (inductor current) dynamic: (**a**) integrator-like open-loop behavior; (**b**) closed-loop current response to 1 A step reference applied to the system at operating point ($i_L = 4$ A, $v_C = 12$ V); (**c**) corresponding evolution of capacitor voltage emphasizing nonminimum phase; (**d**) corresponding evolution of duty ratio

which gives further, by linearization around the point (i_{Le}, v_C^*):

$$C\left(\frac{v_C^*}{n} + E\right)\dot{\tilde{v_C}} + \left(\frac{2v_C^*}{nR} + \frac{E}{R}\right)\tilde{v_C} = \frac{E}{n}\tilde{i_L} - \frac{Li_{Le}}{n}\dot{\tilde{i_L}}, \qquad (11.28)$$

where i_{Le} corresponds to voltage setpoint v_C^* and is given by (11.23). Relation (11.28) allows transfer function $H(s) = \frac{\tilde{v_C}(s)}{\tilde{i_L}(s)}$ to be computed as in (11.29), where $K = \frac{ER}{2v_C^* + nE}$, $T = \frac{RC(v_C^* + nE)}{2v_C^* + nE}$ and $T_{nmp} = \frac{Lv_C^*(v_C^* + nE)}{RE^2}$.

By analyzing (11.29), one can see that the nonminimum-phase behavior of the voltage v_C anticipated by simulation is confirmed by the presence of a right-half-plane zero in the transfer function. For the considered case, parameters of transfer function (11.29) are $K = 2.14\ \Omega$, $T = 6.1$ ms and $T_{nmp} = 0.88$ ms.

$$H(s) = K \cdot \frac{1 - T_{nmp}s}{1 + Ts}. \qquad (11.29)$$

Two methods are proposed next for designing a PI-type outer-loop controller: the first will yield a PI controller without taking into account the right-half-plane zero, that is, by only looking at the $H(s)$ denominator, whereas the second one will employ the *maximum-flat control design method* for nonminimum-phase linear dynamical systems (Ceangă et al. 2001). These two methods are outlined next. For both of methods the controller transfer function is

$$H_{PI}(s) = K_p + K_i/s. \tag{11.30}$$

11.3.3 Outer-Loop PI Design Without Taking into Account the Right-Half-Plane Zero

According to this method, one is interested only in the plant's denominator, while neglecting the presence of the right-half-plane zero.

The closed loop is therefore composed of a first-order filter with gain K and time constant T (see relation (11.29)) and a PI controller. Simple algebra provides the second-order closed-loop transfer function

$$H_{OC}(s) = \frac{\frac{K_p}{K_i}s + 1}{\frac{T}{K_i K}s^2 + \frac{K_p}{K_i}\left(1 + \frac{1}{K_p K}\right)s + 1},$$

for which the performance is imposed in terms of damping coefficient ζ_{0v} and time constant T_{0v}, which gives the bandwidth. The controller's parameters result further:

$$K_p = \frac{T}{KT_{0v}^2}\left(2\zeta_{0v}T_{0v} - \frac{T_{0v}^2}{T}\right), \quad K_i = \frac{T}{KT_{0v}^2}.$$

For the considered case, $\zeta_{0v} = 0.75$ and $T_{0v} = 2$ ms reflect a reasonable choice of the closed-loop behavior, corresponding to imposing an outer-loop dynamic that is ten times faster than that of the inner loop, with slight overshoot. Figure 11.12 shows the tracking performance of the outer loop around the steady-state point ($i_L = 4$ A, $v_C = 12$ V) in response to voltage step reference. The dynamic response in Fig. 11.12a shows that design requirements are met.

Figure 11.13 presents how disturbances resulting from load resistance variation are rejected by the outer loop. Here, a 1 Ω load variation in relation to the rated value of $R = 10$ Ω has been considered. One notes the same transient duration as seen in the tracking case (about 10 ms), as emphasized in Fig. 11.13a.

11.3.4 Outer-Loop PI Design While Taking into Account the Right-Half-Plane Zero

Unlike the above detailed method, the second design method takes into account the presence of the right-half-plane zero in the plant's transfer function and aims at

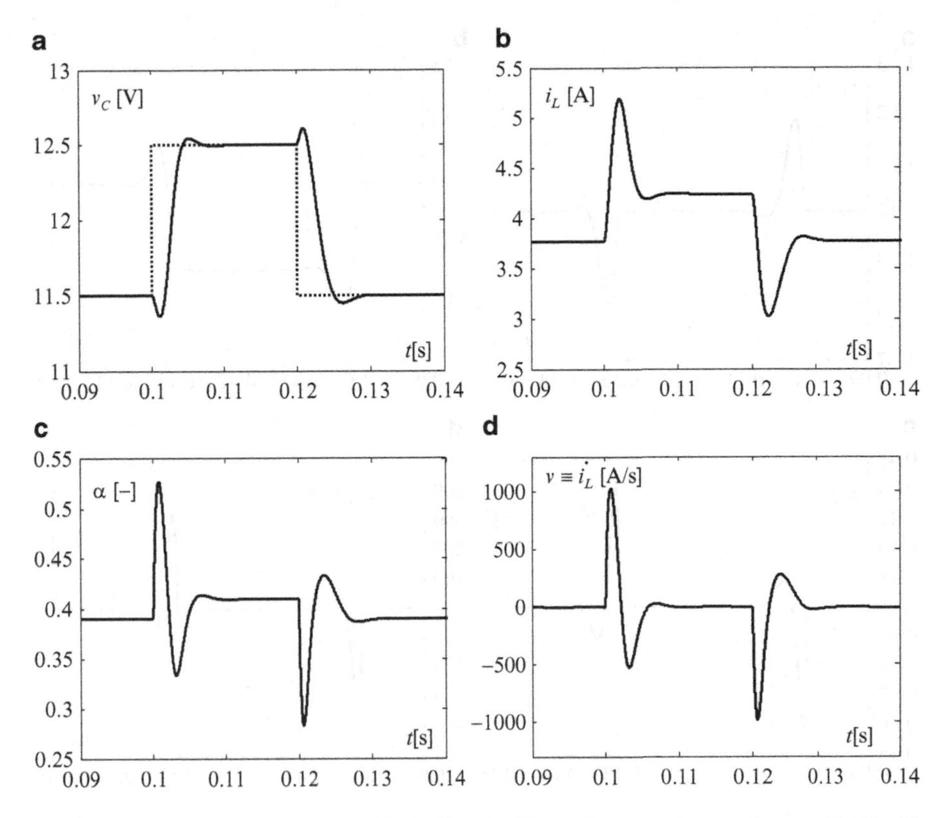

Fig. 11.12 Numerical simulation results showing tracking performance by outer loop equipped with "ordinary" PI controller – time evolutions of variables of interest to 1 V voltage reference variations: (**a**) capacitor voltage; (**b**) inductor current; (**c**) duty ratio; (**d**) inductor current first-order time derivative

alleviating its influence on the closed-loop time response allure. This influence can be noted only on the transfer channel from the reference to output, whereas the response to eventual disturbances due to load R variation is not influenced. This is why closed-loop performance will not be the same on both channels.

The closed-loop transfer function of plant (11.29) with PI controller (11.30) is:

$$H_0(s) = \frac{-\frac{K_p T_{nmp}}{K_i} s^2 + \left(\frac{K_p}{K_i} - T_{nmp}\right) s + 1}{\left(\frac{T}{K_i K} - \frac{K_p T_{nmp}}{K_i}\right) s^2 + \left(\frac{K_p}{K_i} - T_{nmp} + \frac{1}{K_i K}\right) s + 1}. \tag{11.31}$$

Transfer function (11.31) has to have unit gain for an arbitrarily large frequency domain, i.e., $|H_0(j\omega)| = 1$ for $\omega \leq \omega_0$, with ω_0 defining a conveniently chosen bandwidth. In this particular case, ω_0 must be at least one-fifth the inner loop's bandwidth. To facilitate computation the following notations are adopted:

$$a = -\frac{K_p T_{nmp}}{K_i}, \quad b = \frac{K_p}{K_i} - T_{nmp}. \tag{11.32}$$

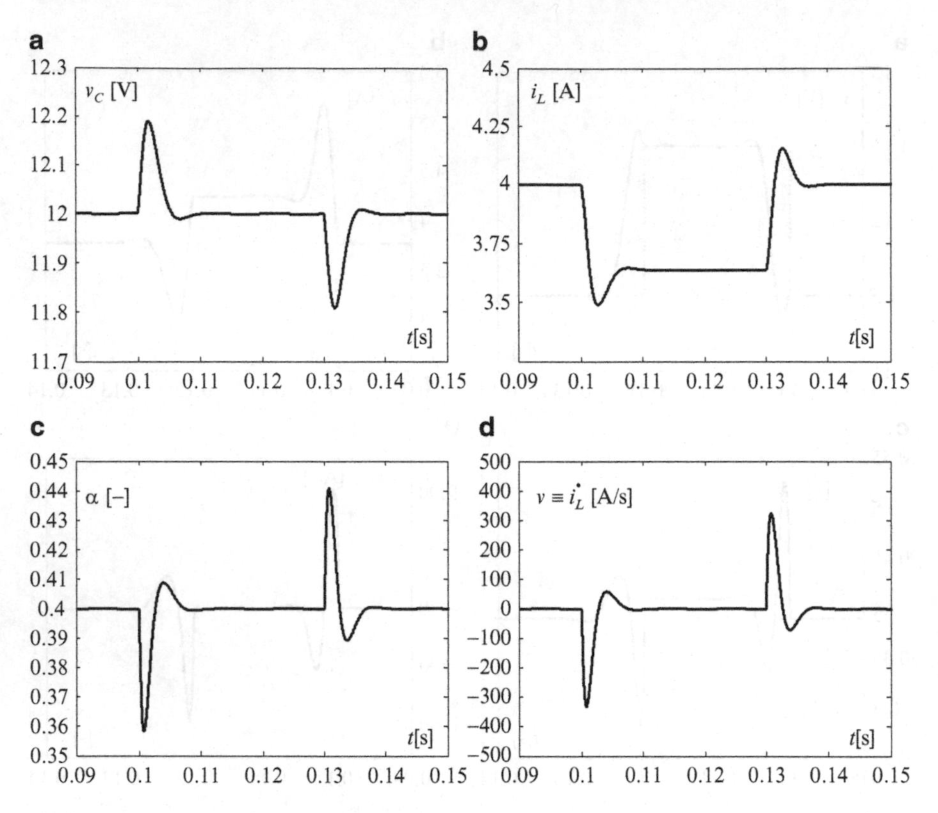

Fig. 11.13 Numerical simulation results showing disturbance rejection by outer loop equipped with an ordinary PI controller – time evolutions of interest variables to 1 Ω load variations (from rated $R = 10\ \Omega$ to 11 Ω and back): (**a**) capacitor voltage; (**b**) inductor current; (**c**) duty ratio; (**d**) inductor current first-order time derivative

In this way, the frequency response of the closed-loop plant is

$$H_0(j\omega) = \frac{1 - a\omega^2 + jb\omega}{1 - [a + T/(K_iK)]\omega^2 + j[b + 1/(K_iK)]\omega}$$

and its gain is

$$|H_0(j\omega)| = \frac{a^2\omega^4 + (b^2 - 2a)\omega^2 + 1}{[a + T/(K_iK)]^2\omega^4 + \left[(b + 1/(K_iK))^2 - 2(a + T/(K_iK))\right]\omega^2 + 1}. \tag{11.33}$$

Based upon Eq. (11.33), the aim of having $|H_0(j\omega)| = 1$ for $\omega \leq \omega_0$ is translated into requiring that the numerator's polynomial have the same coefficients as the denominator's, except for the coefficients of ω^4. Indeed, by requiring that $[a + T/(K_iK)]^2$ be close to a^2 one actually imposes the closed-loop bandwidth. This request

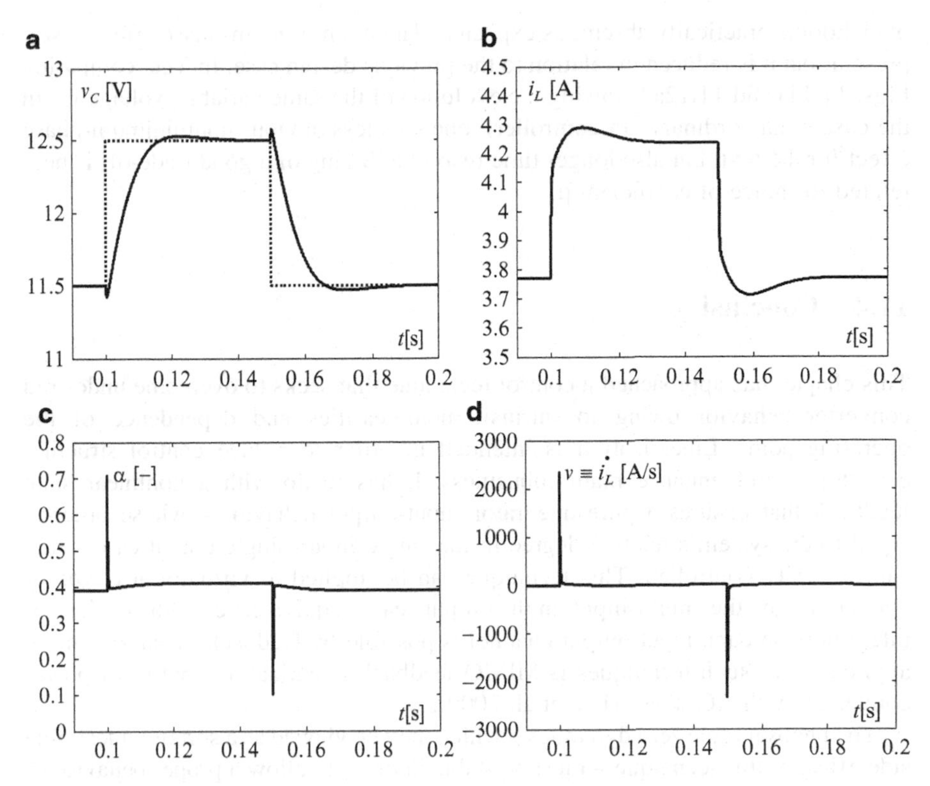

Fig. 11.14 Numerical simulation results showing tracking performance of outer loop equipped with maximum-flat PI controller; time evolutions of interest variables to 1 V voltage reference variations: (**a**) capacitor voltage; (**b**) inductor current; (**c**) duty ratio; (**d**) inductor current first-order time derivative

will be expressed as imposing that term $T/(K_iK)$ be put in relation to the absolute value of a by means of a positive coefficient μ, whose effective selection can be made clear by numerical simulation. To conclude, the following equations are obtained:

$$\frac{T}{K_iK} = \mu|a|, \quad b^2 - 2a = \left(b + \frac{1}{K_iK}\right)^2 - 2\left(a + \frac{T}{K_iK}\right),$$

which, taking into account notations (11.32) adopted for a and b, further provide computation relations for the controller's coefficients:

$$K_p = \frac{T}{\mu K T_{nmp}}, \quad K_i = \frac{2T + \mu T_{nmp}}{2K(T + T_{nmp})}. \tag{11.34}$$

Figure 11.14 presents results that validate the outer loop tracking performance when voltage controller is designed according to the latter method; they have been obtained for $\mu = 10$. In the voltage evolution (Fig. 11.14a) one notes that the overshoot is practically absent, as expected. The nonminimum-phase effect is still present, but it is reduced in relation to the previous design case. Indeed, comparing Figs. 11.14a and 11.12a (showing the evolution of the same variable, voltage v_C, in the case of an "ordinary" PI controller), one sees less obvious nonminimum-phase effect for the first, but also longer time response. Sizing of a good trade-off is here related to choice of coefficient μ.

11.4 Conclusion

This chapter has approached a control technique that seeks to overcome undesired converter behavior owing to intrinsic nonlinearities and dependence of the operating point. Linearization is intended in order to reduce control structure complexity and enhance plant robustness. It has to do with a nonlinear state feedback that ensures a pure-integrator input-output behavior – whose order is equal to the system's relative degree in the single-input–single-output case – that can be easily controlled. This technique can be applied to virtually any type of converter. In the multi-input–multi-output case equivalence with chains of integrators on each input-output channel is possible by feedback linearization; an application of such techniques is MIMO feedback-linearization control of power converters with AC stages (Lee et al. 2000).

The unobservable zero dynamics – which may be viewed as a sort of undesirable side effect of this technique – must be stable in order to allow a proper behavior of the linearized system. In some of cases – see, for example, converters exhibiting "boost" effect – this may not be possible with direct control, and so an indirect control approach is needed. This latter case involves the use of an outer control loop in order to achieve the primary control goal. Setting up the associated controller in this case may not be a trivial task, as the plant behavior exhibits nonminimum-phase effects (Jain et al. 2006).

In the dedicated literature one may find advanced solutions to this problem that employ robust and complex nonlinear methods as passive or sliding-mode control (Sira-Ramírez and Ilic-Spong 1989; Escobar et al. 1999; Matas et al. 2008). As has been shown, linearizing feedback may be sensitive to the converters' operating points and parameter variations, so adaptive approaches employing supplementary measures or parameter estimation may be necessary in order to preserve control quality for the entire converter operating range (Sira-Ramírez et al. 1997).

Problems

Problem 11.1 Feedback-linearization control of series-resonant converter output voltage

Let us consider the series-resonant power converter whose circuit diagram is shown in Fig. 11.15, where primary power source is voltage source E. Resonant filter bank has inductor L, capacitor C and equivalent resistance r_L.

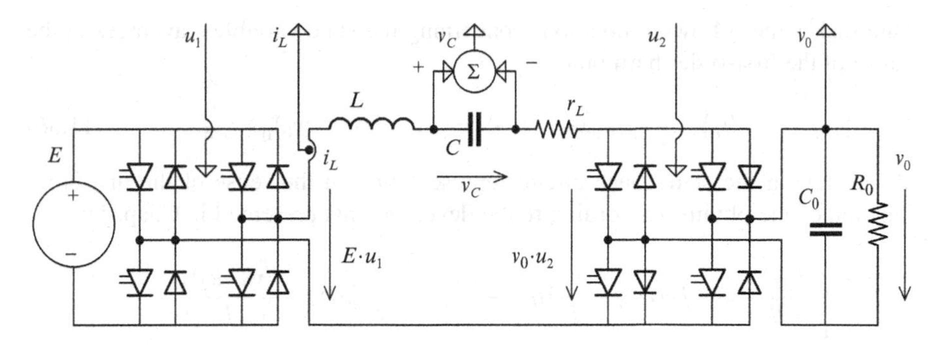

Fig. 11.15 Circuit diagram of series-resonant DC-DC converter

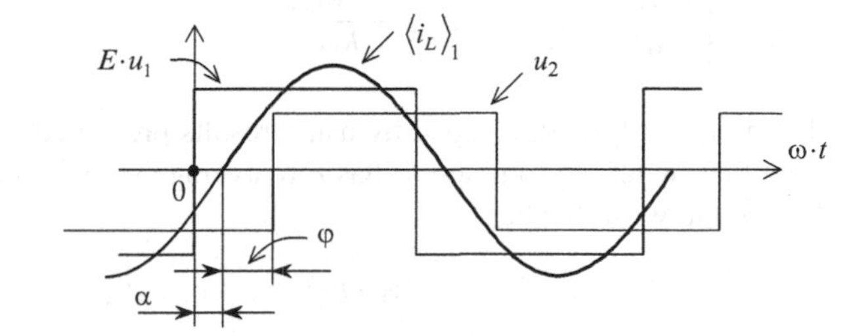

Fig. 11.16 Rectangular waveforms of switching signals in series-resonant converter case

Output voltage is filtered using capacitor C_0. Output resistance R_0 is considered constant and known. Signals u_1 and u_2 are rectangular waveforms of fixed frequency and 0.5 duty ratio that are switching the DC variables E and v_0. The input variable is the phase lag φ between these two switching signals, where u_1 is taken as phase origin.
Control of output voltage v_0 is the stated goal. For this purpose, the expression of the control input φ that linearizes the plant having v_0 as output is to be obtained.

Solution. Figure 11.16 presents waveforms of signals u_1 and u_2 and emphasizes the angle lag between them, φ, which serves as control input in this application. Current i_L is lagged by angle α in relation to signal u_1.
By considering as state variables the ones corresponding to energy accumulations in the circuit, the converter switched state-space model is given by (see Sect. 5.7.2)

$$\begin{cases} L\dfrac{di_L}{dt} = E \cdot u_1 - v_C - r_L i_L - v_0 \cdot u_2 \\ C\dfrac{dv_C}{dt} = i_L \\ C_0\dfrac{dv_0}{dt} = i_L u_2 - \dfrac{v_0}{R_0}. \end{cases} \tag{11.35}$$

One makes the following notations concerning the state variables' averages in the sense of the first-order harmonic:

$$\langle i_L \rangle_1 = x_1 + jx_2, \quad \langle v_C \rangle_1 = x_3 + jx_4, \quad \langle v_0 \rangle_0 = x_5 \,. \tag{11.36}$$

By averaging the state-space circuit representation in the sense of the first-order harmonic, one obtains (according to the developments presented in Chap. 5)

$$\begin{cases} \dfrac{d\langle i_L \rangle_1}{dt} = -j\omega \langle i_L \rangle_1 - \dfrac{r_L}{L}\langle i_L \rangle_1 - \dfrac{\langle v_C \rangle_1}{L} + \dfrac{E}{L}\langle u_1 \rangle_1 - \dfrac{\langle v_0 \cdot u_2 \rangle_1}{L} \\ \dfrac{d\langle v_C \rangle_1}{dt} = -j\omega \langle v_C \rangle_1 + \dfrac{\langle i_L \rangle_1}{C} \\ \dfrac{d\langle v_0 \rangle_0}{dt} = \dfrac{1}{C_0}\langle i \cdot u_2 \rangle_0 - \dfrac{\langle v_0 \rangle_0}{R_0 C_0} \,. \end{cases} \tag{11.37}$$

Term $\langle v_0 \cdot u_2 \rangle_1$ is developed by using results presented in Chap. 5. Thus, one can note that switching function u_2 has zero average value; therefore relation (5.19) can be used to write further:

$$\langle v_0 \cdot u_2 \rangle_1 = \langle v_0 \rangle_0 \cdot \langle u_2 \rangle_1, \tag{11.38}$$

where $\langle u_2 \rangle_1$ is expressed by using expression (5.23) of a rectangular zero-averaged switching function delayed by angle $\varphi + \alpha$:

$$\langle u_2 \rangle_1 = \frac{2}{\pi j} e^{-(\varphi+\alpha)} \,. \tag{11.39}$$

Further, by replacing notations (11.36) and expressions (11.38) and (11.39) in Eq. (11.37) and by separating the real and the imaginary parts, one obtains

$$\begin{cases} \dot{x_1} = -\dfrac{r_L}{L}x_1 + \omega x_2 - \dfrac{x_3}{L} - \mathrm{Re}\left(\dfrac{2}{j\pi L}\langle v_0 \rangle_0 e^{-j(\varphi+\alpha)}\right) \\ \dot{x_2} = -\omega x_1 - \dfrac{r_L}{L}x_2 - \dfrac{x_4}{L} - \mathrm{Im}\left(\dfrac{2}{j\pi L}\langle v_0 \rangle_0 e^{-j(\varphi+\alpha)}\right) \\ \dot{x_3} = \dfrac{x_1}{C} + \omega x_4 \\ \dot{x_4} = \dfrac{x_2}{C} - \omega x_3 \\ \dot{x_5} = \dfrac{4}{\pi C_0}\sqrt{x_1^2 + x_2^2}\cos\varphi - \dfrac{\langle v_0 \rangle_0}{R_0 C_0} \,. \end{cases} \tag{11.40}$$

The new control variable $w = d\langle v_0 \rangle_0 / dt = \dot{x_5}$ is introduced. One can note that the last equation from (11.40) contains an explicit expression of the control input; therefore, the relative degree is $r = 1$. This last equation may be rewritten as

$$w = \frac{4}{\pi C_0}\sqrt{x_1^2 + x_2^2}\cos\varphi - \frac{\langle v_0\rangle_0}{R_0 C_0},$$

which allows the angle φ to be computed as the linearizing control input

$$\varphi = \arccos\left(\frac{\pi}{4}\cdot\frac{wC_0 + x_5/R_0}{\sqrt{x_1^2 + x_2^2}}\right). \tag{11.41}$$

Knowing that $i_L(t) \approx 2(x_1 \cos \omega t - x_2 \sin \omega t)$ (see relation (5.29) from Chap. 5), the quantity $\sqrt{x_1^2 + x_2^2}$ can otherwise be expressed as $I_L/2$, where I_L represents the first-order harmonic's magnitude of current i_L. Therefore, the final expression of the linearizing feedback is

$$\varphi = \arccos\left(\frac{\pi}{2}\cdot\frac{vC_0 + \langle v_0\rangle_0/R_0}{I_L}\right). \tag{11.42}$$

Before declaring expression (11.42) as providing the linearizing feedback sought for, one must analyze the stability of the zero dynamics. One notes that the fourth-order dynamics composed of the first four equations from (11.40) must be analyzed. To this end, one puts $w = \dot{x_5} = 0$ and $x_5 \equiv \langle v_0\rangle_0^*$ in (11.41) to obtain the expression of the control input to be further replaced into these equations:

$$\varphi_0 = \arccos\left(\frac{\pi}{4R_0}\cdot\frac{\langle v_0\rangle_0^*}{\sqrt{x_1^2 + x_2^2}}\right). \tag{11.43}$$

The fourth-order nonlinear zero dynamics that result are:

$$\begin{cases} \dot{x_1} = -\dfrac{r_L}{L}x_1 + \omega x_2 - \dfrac{x_3}{L} - \operatorname{Re}\left(\dfrac{2}{j\pi L}\langle v_0\rangle_0^* e^{-j(\varphi_0+\alpha)}\right) \\ \dot{x_2} = -\omega x_1 - \dfrac{r_L}{L}x_2 - \dfrac{x_4}{L} - \operatorname{Im}\left(\dfrac{2}{j\pi L}\langle v_0\rangle_0^* e^{-j(\varphi_0+\alpha)}\right) \\ \dot{x_3} = \dfrac{x_1}{C} + \omega x_4 \\ \dot{x_4} = \dfrac{x_2}{C} - \omega x_3. \end{cases} \tag{11.44}$$

Stability of nonlinear dynamical system (11.44) will be analyzed by means of the Lyapunov approach. Notation $\mathbf{y} = \begin{bmatrix} x_1 & x_2 & x_3 & x_4 \end{bmatrix}^T$ is used to denote the state vector corresponding to the zero dynamics. The following Lyapunov candidate function is defined as a quadratic form of the state $\mathbf{y}$:

$$V(\mathbf{y}) = \frac{1}{2}\mathbf{y}^T\mathbf{Q}\mathbf{y}, \tag{11.45}$$

with **Q** being a symmetric positive defined matrix containing characteristic values of the energy accumulation elements in the circuit (inductor's inductance and capacitor's capacity, respectively):

$$\mathbf{Q} = \begin{bmatrix} L & 0 & 0 & 0 \\ 0 & L & 0 & 0 \\ 0 & 0 & C & 0 \\ 0 & 0 & 0 & C \end{bmatrix}.$$

Function (11.45) can be declared a Lyapunov function – or else, an energy function – if its time derivative is negative whatever the value of state **y** is. This is proved next. By time-deriving expression (11.45) one obtains

$$\frac{dV(\mathbf{y})}{dt} = Lx_1\dot{x_1} + Lx_2\dot{x_2} + Cx_3\dot{x_3} + Cx_4\dot{x_4}. \tag{11.46}$$

To simplify writing, notation

$$z = \frac{2}{j\pi L}\langle v_0\rangle_0^* e^{-j(\varphi_0+\alpha)} \tag{11.47}$$

is adopted. Equation (11.46) is further developed by using expressions of the state variables' time derivatives as given in (11.44); it becomes, after some simple algebra,

$$\frac{dV(\mathbf{y})}{dt} = -r_L x_1^2 - r_L x_2^2 - Lx_1\mathrm{Re}(z) - Lx_2\,\mathrm{Im}(z). \tag{11.48}$$

One must now get the expressions of Re(z) and Im(z). According to (11.47), variable z can be written

$$z = -\frac{2}{\pi L}\langle v_0\rangle_0^* \sin(\varphi_0 + \alpha) - j\frac{2}{\pi L}\langle v_0\rangle_0^* \cos(\varphi_0 + \alpha),$$

therefore

$$\mathrm{Re}(z) = -\frac{2}{\pi L}\langle v_0\rangle_0^* \sin(\varphi_0 + \alpha), \quad \mathrm{Im}(z) = -\frac{2}{\pi L}\langle v_0\rangle_0^* \cos(\varphi_0 + \alpha). \tag{11.49}$$

At this point, we will establish a connection between complex components of the inductor current's first-order harmonic, x_1 and x_2, and lag angle α. Developments presented in Sect. 5.5.1 of Chap. 5 – related to the extraction of real-time-varying signal from the components of the generalized averaged model – are used for this purpose. Thus, according to Eq. (5.33) and taking into account that $\alpha \in (0, \pi/2)$ is a lag angle (therefore $\alpha = -\psi$), the following relations hold:

$$\sin(-\alpha) = -\sin\alpha = \frac{x_1}{\sqrt{x_1^2 + x_2^2}}, \quad \cos(-\alpha) = \cos\alpha = -\frac{x_2}{\sqrt{x_1^2 + x_2^2}};$$

therefore the real and imaginary part of the first-order harmonic as a complex variable can respectively be expressed as

$$x_1 = -\sin\alpha \cdot \sqrt{x_1^2 + x_2^2}, \quad x_2 = -\cos\alpha \cdot \sqrt{x_1^2 + x_2^2}. \tag{11.50}$$

Finally, expressions (11.49) and (11.50) are substituted into Eq. (11.48) to provide the function V's time derivative in the form

$$\begin{aligned}\dot{V}(\mathbf{y}) = & -r_L x_1^2 - r_L x_2^2 \\ & -\frac{2}{\pi}\langle v_0\rangle_0^* \sqrt{x_1^2 + x_2^2} \cdot \sin\alpha \cdot \sin(\varphi_0 + \alpha) \\ & -\frac{2}{\pi}\langle v_0\rangle_0^* \sqrt{x_1^2 + x_2^2} \cdot \cos\alpha \cdot \cos(\varphi_0 + \alpha),\end{aligned}$$

or, equivalently,

$$\dot{V}(\mathbf{y}) = -r_L x_1^2 - r_L x_2^2 - \frac{2}{\pi}\langle v_0\rangle_0^* \sqrt{x_1^2 + x_2^2} \cdot \cos\varphi_0. \tag{11.51}$$

Expression (11.43) of the control input φ_0 is substituted into (11.51) to yield the final expression of $\dot{V}(\mathbf{y})$:

$$\dot{V}(\mathbf{y}) = -r_L x_1^2 - r_L x_2^2 - \frac{\langle v_0\rangle_0^{*2}}{2R_0} < 0. \tag{11.52}$$

Relation (11.52) shows that the energy function V defined as a quadratic form of the zero dynamics state vector (see Eq. (11.45)) is strictly decreasing with time, and this allows a conclusion that the zero dynamics are in this case stable.

Once Eq. (11.42) of the necessary phase lag has been found as linearizing feedback, it will be used for obtaining u_2. Indeed, u_2 is obtained by delaying u_1 by the phase lag φ given in (11.42).

The control block diagram, profiting from the feedback linearization, is given in Fig. 11.17, which shows that gain K_C is now sufficient for controlling output voltage v_0.

The following problems are left to the reader to solve.

Problem 11.2 Voltage regulation of buck-boost converter using feedback-linearization control

Given the electrical circuit of a buck-boost DC-DC converter supplying a resistance load R, presented in Fig. 4.12 from Chap. 4, and its model given by Eq. (4.34) from the same chapter:

$$\begin{cases} L\dot{i_L} = Eu - v_C(1-u) - r_L i_L \\ C\dot{v_C} = -i_L(1-u) - \dfrac{v_C}{R}, \end{cases}$$

where r_L is the inductor's resistance and the other notations preserve their usual meaning, address the following points.

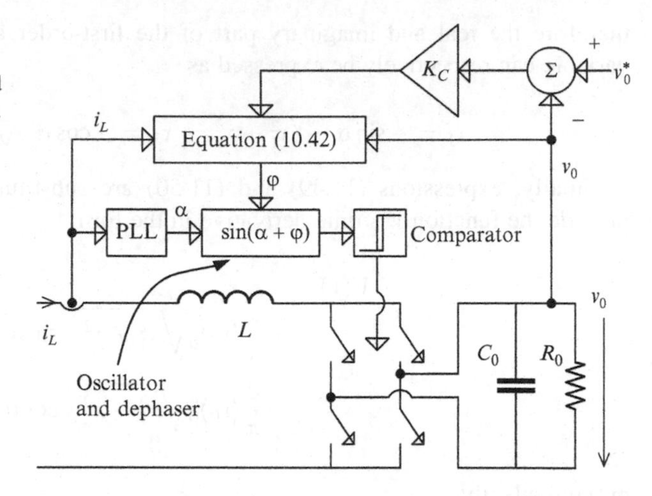

Fig. 11.17 Series-resonant converter output voltage control block diagram based upon feedback linearization and gain control

(a) Design a feedback-linearization control in order to regulate the output voltage, v_C. Based upon the zero dynamics analysis, decide whether a direct or an indirect control approach is suitable.
(b) For the numerical values $L = 3$ mH, $C = 1200$ μF, $r_L = 0.1$ Ω, $E = 12$ V and rated $R = 100$ Ω compute the linear controller that ensures a voltage closed-loop bandwidth of 50 rad/s.
(c) Implement the numerical simulation block diagram in Simulink® in order to validate the closed-loop behavior imposed in part b. Analyze the dynamical behavior in response to load step variations.

Problem 11.3 Reactive power control of STATCOM using linearizing feedback
The electrical circuit of a static synchronous compensator (STATCOM) operating in full wave is given in Fig. 11.18. Resistance R_S represents the inductor resistance on a phase and load resistance R includes power switch losses.

The reactive power control will be achieved by using the *dq* model in the sense of the first-order harmonic, namely by regulating the averaged value of the three-phase current q component at reference value i_q^*. Output voltage v_C is uncontrolled. The control input is α, the phase lag between the switching function u_1 and the first-phase grid voltage e_1, which is taken as phase origin.

(a) Using the approach presented in Chap. 5 – related to computation of the generalized averaged model (GAM) – prove that the first-order harmonic *dq* model of the STATCOM is the following (Petitclair et al. 1996):

$$\begin{cases} \dot{\langle i_q \rangle_0} = -\dfrac{R_S}{L_S}\langle i_q \rangle_0 - \omega\langle i_d \rangle_0 + \dfrac{2}{\pi L_S}\langle v_C \rangle_0 \sin\alpha \\ \dot{\langle i_d \rangle_0} = \omega\langle i_q \rangle_0 - \dfrac{R_S}{L_S}\langle i_d \rangle_0 - \dfrac{2}{\pi L_S}\langle v_C \rangle_0 \cos\alpha + \dfrac{E}{L_S} \\ \dot{\langle v_C \rangle_0} = -\dfrac{3}{\pi C}\langle i_q \rangle_0 \sin\alpha + \dfrac{3}{\pi C}\langle i_d \rangle_0 \sin\alpha - \dfrac{1}{RC}\langle v_C \rangle_0, \end{cases}$$

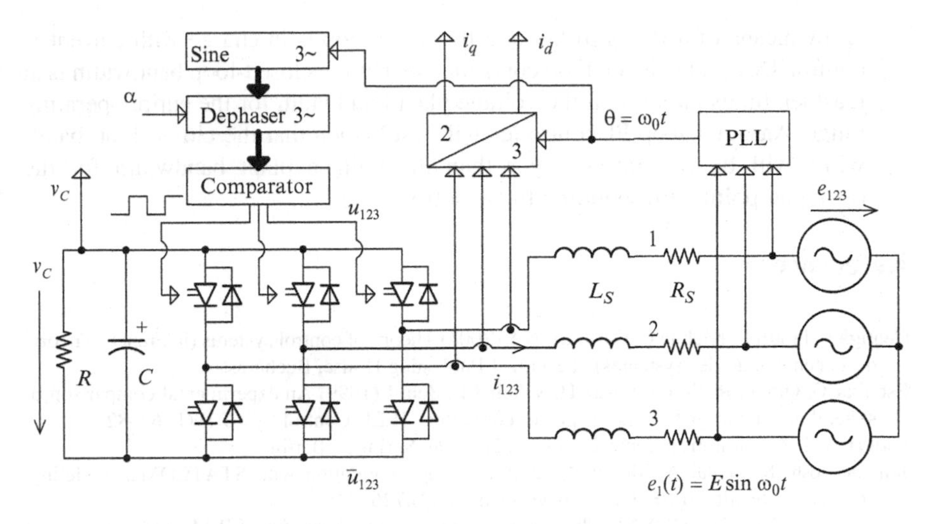

Fig. 11.18 Electrical circuit of static synchronous compensator (STATCOM)

where E is the grid voltage amplitude and ω is the grid voltage pulsation ($\omega = 2\pi \cdot 50$ rad/s).

(b) Obtain the expression of the linearizing feedback ensuring regulation of $\langle i_q \rangle_0$ at the value i_q^*.
(c) Analyze the zero dynamics stability by using the small-signal model.
(d) Provide the global reactive power control structure and emphasize the relation between the imposed value of the reactive power Q^* and i_q^*.

Problem 11.4 Watkins–Johnson DC-DC converter controlled by feedback linearization

For a Watkins–Johnson converter – whose electrical circuit was presented in Fig. 4. 37 in Chap. 4 and repeated in Fig. 8.34 in Problem 8.1 in Chap. 8 – it is required to regulate the output voltage at the setpoint v_C^*. Its averaged model, given by relation (8.38) of Chap. 8, is given hereafter:

$$\begin{cases} L\dot{i_L} = (2\alpha - 1)E - \alpha v_C \\ C\dot{v_C} = \alpha i_L - \dfrac{v_C}{R}, \end{cases}$$

where notations used have their usual meaning, with duty ratio α being the control input and R being the load resistance. The following issues must be addressed.

(a) Deduce the current value i_L^* that corresponds to the voltage setpoint v_C^*.
(b) Compute the expression of a feedback-linearization control law issued from an indirect control approach, that is, from regulating current i_L to the setpoint i_L^* computed in part a.
(c) Knowing that the output voltage operating range is $[-3E, E]$, design a two-loop control structure whose outer loop is dedicated to regulating the output voltage

v_C by means of a PI controller and the inner loop is in charge with current i_L control. Design the inner-loop controller so that the closed-loop bandwidth is at least ten times larger than the voltage plant bandwidth for the entire operating range. An outer loop PI controller will result such that the closed-loop bandwidth will be two times larger than the voltage plant bandwidth for the operating point corresponding to $v_C = 0$ V.

References

Ceangă E, Protin L, Nichita C, Cutululis NA (2001) Theory of control systems (in French: Théorie de la commande des systèmes). Technical Publishing House, Bucharest

Escobar G, Ortega R, Sira-Ramírez H, Vilain J-P, Zein I (1999) An experimental comparison of several nonlinear controllers for power converters. IEEE Control Syst J 19(1):66–82

Isidori A (1989) Nonlinear control systems, 2nd edn. Springer, Berlin

Jain A, Joshi K, Behal A, Mohan N (2006) Voltage regulation with STATCOMs: modeling, control and results. IEEE Trans Power Deliv 21(2):726–735

Jung J, Lim S, Nam K (1999) A feedback linearizing control scheme for a PWM converter-inverter having a very small DC-link capacitor. IEEE Trans Ind Appl 35(5):1124–1131

Lee T-S (2003) Input-output linearization and zero-dynamics control of three-phase AC/DC voltage-source converters. IEEE Trans Power Electron 18(1):11–22

Lee D-C, Lee G-M, Lee K-D (2000) DC-bus voltage control of three-phase AC/DC PWM converters using feedback linearization. IEEE Trans Ind Appl 36(3):826–833

Matas J, de Vicuña LG, Miret J, Guerrero JM, Castilla M (2008) Feedback linearization of a single-phase active power filter via sliding mode control. IEEE Trans Power Electron 23(1):116–125

Petitclair P (1997) Modelling and control of FACTS (Flexible Alternative Current Transmission System): application to STACOM (STATic Compensator) (in French: "Modélisation et commande de structures FACTS: Application au STACOM"). Ph.D. thesis, Grenoble Institute of Technology, France

Petitclair P, Bacha S (1997) Optimized linearization via feedback control law for a STATCOM. In: Proceedings of the 32nd IEEE Industry Applications Society annual meeting – IAS 1997. New Orleans, Louisiana, USA, vol 2, pp 880–885

Petitclair P, Bacha S, Rognon JP (1996) Averaged modelling and nonlinear control of an ASVC (Advanced Static Var Compensator). In: Proceedings of the 27th annual IEEE Power Electronics Specialists Conference – PESC 1996. Baveno, Italy, pp 753–758

Sira-Ramírez H, Ilic-Spong M (1989) Exact linearisation in switched-mode DC-to-DC power converters. Int J Control 50(2):511–524

Sira-Ramírez H, Silva-Ortigoza R (2006) Control design techniques in power electronics devices. Springer, London

Sira-Ramírez HJ, Rios-Bolivar M, Zinober ASI (1995) Adaptive input-output linearization for PWM regulation of DC-to-DC power converters. In: Proceedings of the American control conference – ACC 1995. Seattle, Washington, USA, vol 1, pp 81–85

Sira-Ramírez HJ, Rios-Bolivar M, Zinober ASI (1997) Adaptive dynamical input-output linearization of DC to DC power converters: a backstepping approach. Int J Robust Nonlinear Control 7:279–296

Song E, Lynch AF, Dinavahi V (2009) Experimental validation of nonlinear control for a voltage source converter. IEEE Trans Control Syst Technol 17(5):1135–1144

Chapter 12
Energy-Based Control of Power Electronic Converters

This chapter aims at presenting basic ideas and main insights of the energy-based control approach for power electronic converters. Thus, two control methods based on concepts of energy will be here detailed: the so-called *stabilizing control* – based upon Lyapunov control design methods – and *passivity-based control*, relying upon specific structural properties – for example, passivity and dissipativity – and on control methods that exploit these properties.

As is true for most engineering systems, power electronic converter modeling is closely related to control objectives. In this case, it is energy processing that must be described in terms of power flows, and this turns out to be a natural representation of the underlying phenomena. Further, the mathematical developments involved make use of some remarkable properties of these systems. The essence of energy-based control consists in taking advantage of the fact that power electronic converters reach their steady-state operation by dissipating energy, that is, in controlling the speed of this dissipation (Stanković et al. 2001).

The concept of incremental energy (Sanders 1989) is crucial for designing control laws for DC-DC converters in a comprehensive manner; this is why buck-boost and boost DC-DC converters serve in this chapter as benchmarks. But the energy-based control is not effective for DC-DC converters only; applications of the same general methodology to converters having AC stages (Komurcugil and Kukrer 1998; Escobar et al. 2001; Mattavelli et al. 2001), as well as to multilevel converters (Liserre 2006; Noriega-Pineda and Espinosa-Pérez 2007) have also been reported.

This chapter begins by introducing some basic concepts specific to the energy-based control approach. Stabilizing control is detailed in the nonlinear case, then in the linearized one and illustrated with an example. Next, modeling of power electronic converters in the Euler–Lagrange formalism is presented, with emphasis on the passivity property. The use of this property gives its name to the passivity-based control approach detailed further. After providing a general design methodology, the necessity of on-line parameter estimation is considered, which leads to adaptive versions of the general control algorithm. The buck-boost converter has been chosen as a case study to illustrate the passivity-based control approach.

S. Bacha et al., *Power Electronic Converters Modeling and Control: with Case Studies*, Advanced Textbooks in Control and Signal Processing, DOI 10.1007/978-1-4471-5478-5_12,

Connections between energy-based control methods and other nonlinear control approaches are briefly reviewed. A set of problems with solutions and some unsolved problems are provided at the end of this chapter.

12.1 Basic Definitions

Some basic notions will be here briefly reviewed, which are related to formalization of systems' fundamental property of processing energy, part of which is stored and the rest dissipated.

Let Σ be a system in its usual state-space representation

$$\Sigma : \begin{cases} \dot{\mathbf{x}} = \mathbf{f}(\mathbf{x}, \mathbf{u}) \\ \mathbf{y} = \mathbf{g}(\mathbf{x}, \mathbf{u}), \end{cases} \tag{12.1}$$

whose state $\mathbf{x}$ belongs to the n-dimensional space $\mathbf{X}$. It is said that system Σ is *dissipative* in relation to the energy flow $w(t)$ if there exists a nonnegative continuous function $H : \mathbf{X} \to \mathbb{R}$ satisfying

$$H(\mathbf{x}(t)) - H(\mathbf{x}(0)) \leq \int_0^t w(\tau)d\tau. \tag{12.2}$$

One can note that the right-hand side of the inequality in relation (12.2) has energy dimension because it is obtained as the time integral of an energy flow (or otherwise said power). The result is H as a function of the stored energy, representing the energy of state $\mathbf{x}$ at a given moment; thus, this function is called an *(energy) storage function*. The left-hand side of relation (12.2) then represents the stored energy. If supposing that flow $w(t)$ is an input for the system Σ, then (12.2) indicates that the energy of flow $w(t)$ is not entirely transferred to the system – a part of it is dissipated.

A lumped-parameter system connected with its environment by means of *port-power-conjugated variables* $\mathbf{u} \in \mathbb{R}^m$ and $\mathbf{y} \in \mathbb{R}^m$ (the product of which has power dimension), which is dissipative in relation with the energy flow $w(\mathbf{u},\mathbf{y}) = \mathbf{u}^T\mathbf{y}$, that is, for which

$$H(\mathbf{x}(t)) - H(\mathbf{x}(0)) \leq \int_0^t \mathbf{u}^T(\tau)\mathbf{y}(\tau)d\tau \tag{12.3}$$

holds, with the storage function satisfying $H(0) = 0$, is said to be *passive*. Relation (12.3) gives information about the stability of a passive system. Thus, if letting $\mathbf{u}^T = 0$, then, according to (12.3), $H(\mathbf{x})$ decreases starting from any trajectory of Σ, which shows that passive systems having a positive definite energy storage function are stable in the Lyapunov sense. Note also that zeroing the output ($\mathbf{y} = 0$), conserves the property is, which denotes stable zero dynamics.

12.2 Stabilizing Control of Power Electronic Converters

According to standard control goals in power electronics, one wants to maintain some variables – e.g., the output voltage – at a desired value $\mathbf{x}_d$, in spite of parametric disturbances typically induced by load and input voltage. Steady-state value $\mathbf{x}_d$ is in general known; next we will suppose in the first place that either the full state – or a subset of it – is available for measurements.

Let

$$\widetilde{\mathbf{x}} = \mathbf{x} - \mathbf{x}_d$$

denote the deviation of the real state $\mathbf{x}$ from the steady-state operating point $\mathbf{x}_d$. The stabilizing control approach makes use of quadratic forms of variable $\widetilde{\mathbf{x}}$,

$$V(\widetilde{\mathbf{x}}) = \frac{1}{2} \cdot \widetilde{\mathbf{x}}^T \cdot \mathbf{Q} \cdot \widetilde{\mathbf{x}}, \tag{12.4}$$

with $\mathbf{Q}$ being a symmetric positive definite matrix. Functions of type (12.4) give an image of how far the system's current operating point is placed in relation to system's equilibrium that expresses its minimum energy. Such quadratic forms are generally used in defining Lyapunov candidate functions for deriving control laws aimed at stabilizing the operation at $\mathbf{x}_d$.

In the case of power electronic converters, the most natural way of expressing the system's energy is by summing up its energy accumulations, that is,

$$V = \frac{1}{2}\left(\sum_{j=1}^{n_L} L_j \widetilde{i_{Lj}}^2 + \sum_{k=1}^{n_C} C_k \widetilde{v_{Ck}}^2\right), \tag{12.5}$$

where L_j are the inductances of its n_L inductors, C_k are the capacities of its n_C capacitors and $\widetilde{i_{Lj}}$ and $\widetilde{v_{Ck}}$ denote variations of inductor currents and capacitor voltages in relation to their respective steady-state values, i_{Ljd} and v_{Ckd}. Form (12.5) has been called the *energy in the increment* (Sanders and Verghese 1992). In this case, as the state vector $\mathbf{x}$ is composed of currents i_{Lj} and voltages v_{Ck}:

$$\mathbf{x} = \begin{bmatrix} i_{L1} & i_{L2} & \cdots & i_{Ln_L} & v_{C1} & v_{C2} & \cdots & v_{Cn_C} \end{bmatrix}^T, \tag{12.6}$$

the function given by (12.5) can obviously be expressed as a quadratic form of type (12.4), where matrix $\mathbf{Q}$ may be identified by its main diagonal containing the characteristic values of the power converter's accumulation elements; therefore, it can be configured like in (12.7). Lyapunov candidate functions of form (12.5) with matrix $\mathbf{Q}$ of form (12.7) are typically used for designing stabilizing control laws. Details of the design procedure in the nonlinear and in the linearized case are presented in the next two sections, respectively.

$$\mathbf{Q} = \begin{bmatrix} L_1 & & & & & \\ & \ddots & & & 0 & \\ & & L_{nL} & & & \\ & & & C_1 & & \\ & 0 & & & \ddots & \\ & & & & & C_{nc} \end{bmatrix}. \tag{12.7}$$

12.2.1 General Nonlinear Case

One may consider the general state-space description of a power electronic converter as a nonlinear system emphasizing the dynamic of the state vector $\mathbf{x}$ having the usual composition (12.6) and input $\mathbf{u}$, which is in the general case the vector of duty ratios. The steady-state vector $\mathbf{x}_d$ has the expression

$$\mathbf{x}_d = \begin{bmatrix} i_{L1d} & i_{L2d} & \cdots & i_{Ln_Ld} & v_{C1d} & v_{C2d} & \cdots & v_{Cn_Cd} \end{bmatrix}^T,$$

allowing the vector of deviations to be

$$\widetilde{\mathbf{x}} = \mathbf{x} - \mathbf{x}_d = \begin{bmatrix} \widetilde{i_{L1}} & \widetilde{i_{L2}} & \cdots & \widetilde{i_{Ln_L}} & \widetilde{v_{C1}} & \widetilde{v_{C2}} & \cdots & \widetilde{v_{Cn_C}} \end{bmatrix}^T. \tag{12.8}$$

It is always possible to obtain the mathematical model of the system in variations in the general nonlinear form

$$\dot{\widetilde{\mathbf{x}}} = \mathbf{f}(\widetilde{\mathbf{x}}) + \mathbf{g}(\widetilde{\mathbf{x}}) \cdot \mathbf{u}, \tag{12.9}$$

where nonlinear function $\mathbf{f}$ also depends on the steady-state values $\mathbf{x}_d$. Function

$$V(\widetilde{\mathbf{x}}) = \frac{1}{2} \cdot \widetilde{\mathbf{x}}^T \cdot \mathbf{Q} \cdot \widetilde{\mathbf{x}}, \tag{12.10}$$

with $\mathbf{Q}$ being symmetrical and positive definite, is chosen as a Lyapunov candidate function. Obviously, $V(\widetilde{\mathbf{x}}) > 0$. A suitable control input $\mathbf{u}$ may be found that ensures decreasing of the energy function $V(\widetilde{\mathbf{x}})$ over time. It follows that $\mathbf{u}$ results from imposing $dV(\widetilde{\mathbf{x}})/dt < 0$. Taking account of (12.8) and (12.9), one can then write:

$$\begin{aligned} \frac{dV(\widetilde{\mathbf{x}})}{dt} &= \frac{\partial V}{\partial \widetilde{\mathbf{x}}} \cdot \dot{\widetilde{\mathbf{x}}} = \frac{\partial V}{\partial \widetilde{\mathbf{x}}} \cdot (\mathbf{f}(\widetilde{\mathbf{x}}) + \mathbf{g}(\widetilde{\mathbf{x}}) \cdot \mathbf{u}) \\ &= \widetilde{\mathbf{x}}^T \mathbf{Q} \cdot \mathbf{f}(\widetilde{\mathbf{x}}) + \widetilde{\mathbf{x}}^T \mathbf{Q} \cdot \mathbf{g}(\widetilde{\mathbf{x}}) \cdot \mathbf{u}, \end{aligned} \tag{12.11}$$

where the equality $\partial V / \partial \widetilde{\mathbf{x}} = \widetilde{\mathbf{x}}^T \cdot \mathbf{Q}$ has been used. The time derivative of the Lyapunov function must be negative:

$$\frac{dV(\widetilde{\mathbf{x}})}{dt} < 0, \tag{12.12}$$

thus forcing by control action that the energy function to decrease. By combining relations (12.11) and (12.12), one obtains

$$\widetilde{\mathbf{x}}^T\mathbf{Q}\cdot\mathbf{f}(\widetilde{\mathbf{x}})+\widetilde{\mathbf{x}}^T\mathbf{Q}\cdot\mathbf{g}(\widetilde{\mathbf{x}})\cdot\mathbf{u}<0,$$

which shows that it is sufficient that both terms be negative for relation (12.12) to be met. Thus, it is sufficient to find matrix $\mathbf{Q}$ symmetrical and positive definite such that $\mathbf{Q}\cdot\mathbf{f}(\widetilde{\mathbf{x}})=-\mathbf{P}\cdot\widetilde{\mathbf{x}}$, where $\mathbf{P}$ is a suitably chosen symmetrical and positive semidefinite matrix, to ensure that the first term is negative. Once matrix $\mathbf{Q}$ has been found, it is then sufficient to define a control law

$$\mathbf{u}=-\lambda\cdot\mathbf{g}(\widetilde{\mathbf{x}})^T\mathbf{Q}\cdot\widetilde{\mathbf{x}}, \tag{12.13}$$

with λ being an appropriately chosen positive scalar, in order to ensure the negativity of the function V time derivative and hence the system's convergence to the origin, which is equivalent to accomplishing the control goal. Here λ is responsible for the convergence speed and its value can be finely tuned by numerical simulation. Equation (12.13) suggests that the stabilizing control solution is obtained as a nonlinear full-state feedback of the system in variations, thus depending on the steady-state operating point chosen as setpoint. One can also note that constraints on $\mathbf{u}$ – composed of averages of switching functions, commonly restricted to either [0,1] or [−1, 1] – suppose an appropriate choice of constant λ.

12.2.2 Linearized Case

The linearized case refers to the use of averaged models in the stabilizing control design; it allows using linear tools in the design and it is for this reason more intuitive. Averaging methodology was explained in Chap. 4 of this book. Nonlinear models are linearized around the steady-state point chosen as control target. Next, notation $\mathbf{x}_d$ of this point is maintained; let $\mathbf{u}_d$ be the input value corresponding to $\mathbf{x}_d$. In this way, linear models involving variations of variables result. Let us consider such a model in the well-known form

$$\dot{\widetilde{\mathbf{x}}}=\mathbf{A}\cdot\widetilde{\mathbf{x}}+\mathbf{B}\cdot\widetilde{\mathbf{u}}, \tag{12.14}$$

where $\widetilde{\mathbf{x}}=\mathbf{x}-\mathbf{x}_d$, $\widetilde{\mathbf{u}}=\mathbf{u}-\mathbf{u}_d$ and matrices $\mathbf{A}$ and $\mathbf{B}$ depend on the operating point $(\mathbf{x}_d,\mathbf{u}_d)$. One can choose a Lyapunov candidate function of the same form as in (12.10):

$$V(\widetilde{\mathbf{x}})=\frac{1}{2}\cdot\widetilde{\mathbf{x}}^T\cdot\mathbf{Q}\cdot\widetilde{\mathbf{x}}, \tag{12.15}$$

which is an energy function identifying the energy in the increment, but this time requiring only that matrix $\mathbf{Q}$ be symmetric and positive definite (that is, without imposing that it necessarily have the form given by (12.7)). Elements of $\mathbf{Q}$ will result from imposing that function V become a Lyapunov function. To this end, one requires that $dV(\widetilde{\mathbf{x}})/dt < 0$. The time derivative of V is computed by using Eqs. (12.14) and (12.15); it then gives:

$$\begin{aligned}\frac{dV(\widetilde{\mathbf{x}})}{dt} &= \frac{1}{2}\left(\dot{\widetilde{\mathbf{x}}}^T \cdot \mathbf{Q} \cdot \widetilde{\mathbf{x}} + \widetilde{\mathbf{x}}^T \cdot \mathbf{Q} \cdot \dot{\widetilde{\mathbf{x}}}\right) = \frac{1}{2}(\mathbf{A}\widetilde{\mathbf{x}} + \mathbf{B}\widetilde{\mathbf{u}})^T\mathbf{Q}\widetilde{\mathbf{x}} + \frac{1}{2} \cdot \widetilde{\mathbf{x}}^T\mathbf{Q}(\mathbf{A}\widetilde{\mathbf{x}} + \mathbf{B}\widetilde{\mathbf{u}}) \\ &= \frac{1}{2} \cdot \widetilde{\mathbf{x}}^T \cdot \left(\mathbf{A}^T\mathbf{Q} + \mathbf{Q}\mathbf{A}\right) \cdot \widetilde{\mathbf{x}} + \frac{1}{2}\left(\widetilde{\mathbf{u}}^T\mathbf{B}^T \cdot \mathbf{Q}\widetilde{\mathbf{x}} + \widetilde{\mathbf{x}}^T\mathbf{Q} \cdot \mathbf{B}\widetilde{\mathbf{u}}\right).\end{aligned} \tag{12.16}$$

Expression (12.16) can further be written by analyzing its last two terms. Thus, by using matrix transposition properties and the fact that $\mathbf{Q}$ is symmetric, i.e., $\mathbf{Q} = \mathbf{Q}^T$, one obtains that

$$\widetilde{\mathbf{u}}^T\mathbf{B}^T \cdot \mathbf{Q}\widetilde{\mathbf{x}} = (\mathbf{B}\widetilde{\mathbf{u}})^T \cdot \mathbf{Q}\widetilde{\mathbf{x}} = \left((\mathbf{Q}\widetilde{\mathbf{x}})^T \cdot \mathbf{B}\widetilde{\mathbf{u}}\right)^T = \left(\widetilde{\mathbf{x}}^T\mathbf{Q}^T \cdot \mathbf{B}\widetilde{\mathbf{u}}\right)^T = \left(\widetilde{\mathbf{x}}^T\mathbf{Q} \cdot \mathbf{B}\widetilde{\mathbf{u}}\right)^T;$$

therefore the last two terms of (12.16) are equal, since they are in fact scalars. It follows that relation (12.16) can finally be posed in the form

$$\frac{dV(\widetilde{\mathbf{x}})}{dt} = \frac{1}{2} \cdot \widetilde{\mathbf{x}}^T \cdot \left(\mathbf{A}^T\mathbf{Q} + \mathbf{Q}\mathbf{A}\right) \cdot \widetilde{\mathbf{x}} + \widetilde{\mathbf{x}}^T\mathbf{Q} \cdot \mathbf{B}\widetilde{\mathbf{u}}. \tag{12.17}$$

From relation (12.17) it results that it is sufficient to require both terms of $dV(\widetilde{\mathbf{x}})/dt$ be negative in order for the time derivative of V to be negative. In this way, elements of $\mathbf{Q}$ result by solving the Lyapunov equation

$$\mathbf{A}^T\mathbf{Q} + \mathbf{Q}\mathbf{A} = -\mathbf{P}, \tag{12.18}$$

with $\mathbf{P}$ being a symmetric positive semidefinite matrix suitably chosen. In order to obtain negative time derivative of the Lyapunov function expressed in (12.17) and taking account of (12.18), it results that the second term of (12.17) must also be negative:

$$\widetilde{\mathbf{x}}^T\mathbf{Q}\mathbf{B}\widetilde{\mathbf{u}} < 0, \tag{12.19}$$

In order to fulfill relation (12.19) it is sufficient that the control input sought be chosen as:

$$\widetilde{\mathbf{u}} = -\lambda \cdot \mathbf{B}^T\mathbf{Q} \cdot \widetilde{\mathbf{x}}, \tag{12.20}$$

with λ being a positive scalar playing the same role as in the nonlinear case (see Eq. 12.13). One can note that the control solution $\widetilde{\mathbf{u}}$ is obtained in the form of a linear full-state feedback for the system in variations, obviously depending on the setpoint $\mathbf{x}_d$, $\widetilde{\mathbf{u}} = \mathbf{K} \cdot \widetilde{\mathbf{x}}$, where matrix $\mathbf{K}$ is given by

$$\mathbf{K} = -\lambda \cdot \mathbf{B}^T\mathbf{Q}. \tag{12.21}$$

One notes that too large values of λ could negatively impact control input values, whose averages are confined to either [0,1] or [−1, 1]. Obviously, the total stabilizing control is obtained as $\mathbf{u} = \mathbf{u}_d + \widetilde{\mathbf{u}}$.

Guidelines for choosing λ may in this case result from analyzing the excursion of poles for the closed-loop system with control law (12.21) (Sanders and Verghese 1992), that is, the eigenvalues of matrix $\mathbf{A} + \mathbf{B} \cdot \mathbf{K}$, where $\mathbf{K}$ is the stabilizing feedback matrix gain given by (12.21). It is preferable that such analysis be performed numerically, even for small-order converters. Note that matrix $\mathbf{Q}$ also influences the distribution of closed-loop poles.

12.2.3 Stabilizing Control Design Algorithm

As a synthesis to the above presented issues, implementation of a stabilizing control approach for a given power electronic converter in the linearized case follows the general steps of Algorithm 12.1. Algorithm 12.2 summarizes the procedure to be followed in the nonlinear case.

12.2.4 Example: Stabilizing Control Design for a Boost DC-DC Converter

12.2.4.1 Basic Stabilizing Control Design

The design of a stabilizing control law will be here illustrated for the linearized case of a boost converter. Its large-signal bilinear model has been presented in Sect. 3.2.3 of Chap. 3:

$$\begin{bmatrix} \dot{i_L} \\ \dot{v_C} \end{bmatrix} = \begin{bmatrix} 0 & -1/L \\ 1/C & -1/(RC) \end{bmatrix} \cdot \begin{bmatrix} i_L \\ v_C \end{bmatrix} + \begin{bmatrix} 0 & 1/L \\ -1/C & 0 \end{bmatrix} \cdot \begin{bmatrix} i_L \\ v_C \end{bmatrix} \cdot u + \begin{bmatrix} E/L \\ 0 \end{bmatrix},$$

where notations have their usual meaning employed until now. As detailed in Sect. 4.5.3 of Chap. 4, notations $x_1 = \langle i_L \rangle_0$ and $x_2 = \langle v_C \rangle_0$ are adopted to represent the averaged values of the two state variables respectively. Suppose that the control task consists in regulating the averaged value of the output voltage x_2 to the value $v_{Cd} \equiv x_{2d}$. Noting by α_d the corresponding duty ratio value and taking into account that $x_{2d} = E/(1 - \alpha_d)$, one gets

$$\alpha_d = 1 - E/v_{Cd}. \tag{12.22}$$

Algorithm 12.1

Steps of stabilizing control design for power electronic converters by using linearized averaged model

#1. Write down the switched model of the converter and then obtain its averaged model.
#2. Linearize averaged model around chosen operating point $\mathbf{x}_d$, typically corresponding to control target if this involves a regulation task. Corresponding input value is denoted by $\mathbf{u}_d$.
#3. Arrange linearized model as a state model, allowing corresponding matrices **A** and **B** to be identified.
#4. Choose diagonal semidefinite positive matrix **P** and find positive definite matrix **Q** that verifies Lyapunov Eq. (12.18). Without loss of generality, matrix **Q** can be chosen of form (12.7), having on its main diagonal the characteristic values of energy accumulation elements of the circuit.
#5. Choose positive scalar λ having role of accelerating output error convergence to zero (at this step this choice is arbitrary; it will be clarified after numerical analysis of the closed-loop poles).
#6. Compute stabilizing control input variation $\widetilde{\mathbf{u}}$ according to relation (12.20).
#7. Perform a numerical analysis of linearized closed-loop system poles – i.e., of closed-loop matrix's $\mathbf{A} - \lambda \cdot \mathbf{B}^T\mathbf{Q}$ eigenvalues – for different values of λ. Resume selection of λ as a consequence of some suitable closed-loop pole placement.
#8. Compute total stabilizing control input $\mathbf{u} = \mathbf{u}_d + \widetilde{\mathbf{u}}$ and simulate the *nonlinear* closed-loop system numerically.
#9. Analyze robustness of obtained control to parameter variations. Design parameter estimators for most variable parameters – e.g., by using the approach detailed further in this chapter (Sect. 12.4.3) – to be embedded in closed-loop system.

Algorithm 12.2

Stabilizing control design procedure for power electronic converters by using the nonlinear model

#1. Write down the switched model of the converter. Choose operating point $\mathbf{x}_d$, typically corresponding to control target if this involves a regulation task.
#2. Obtain nonlinear model in variations around point $\mathbf{x}_d$ in form (12.9), emphasizing functions **f** and **g**.

#3. Find matrix $\mathbf{Q}$ such that $\mathbf{Q} \cdot \mathbf{f}(\widetilde{\mathbf{x}}) = -\mathbf{P} \cdot \widetilde{\mathbf{x}}$, after having suitably chosen matrix $\mathbf{P}$ symmetrical and positive semidefinite. As a hint, verify first that matrix $\mathbf{Q}$ chosen of form (12.7) – having on its main diagonal the characteristic values of energy accumulation elements of the circuit – meets the requirement.
#4. Choose positive scalar λ having the role of accelerating the output error convergence to zero (as in the linearized case, at this step this choice is arbitrary; it can be clarified further by numerical simulation).
#5. Compute stabilizing control input $\mathbf{u}$ according to relation (12.13).
#6. Perform a numerical simulation of nonlinear closed-loop system for different values of λ and select its value for most suitable closed-loop dynamic performance.
#7. Analyze robustness of obtained control to parameter variations. Design parameter estimators for most variable parameters to be embedded in closed-loop system.

In the same way, taking into account that $i_{Ld} \equiv x_{1d} = E/((1 - \alpha_d)^2 R)$, the corresponding desired value of the inductor current is

$$i_{Ld} \equiv x_{1d} = v_{Cd}^2/(ER). \tag{12.23}$$

The following notations are adopted: $\mathbf{x} = [x_1 \quad x_2], \mathbf{x}_d = [x_{1d} \quad x_{2d}], \widetilde{\mathbf{x}} = \mathbf{x} - \mathbf{x}_d$ and $\widetilde{\alpha} = \alpha - \alpha_d$. The linearized model is then obtained in the form

$$\dot{\widetilde{\mathbf{x}}} = \mathbf{A} \cdot \widetilde{\mathbf{x}} + \mathbf{B} \cdot \widetilde{\alpha}, \tag{12.24}$$

with matrices

$$\mathbf{A} = \begin{bmatrix} 0 & -(1-\alpha_d)/L \\ (1-\alpha_d)/C & -1/(RC) \end{bmatrix}, \quad \mathbf{B} = \begin{bmatrix} E/(L(1-\alpha_d)) \\ -E/((1-\alpha_d)^2 RC) \end{bmatrix}. \tag{12.25}$$

The next steps concern the choice of matrix $\mathbf{P}$, which must be symmetric and positive semidefinite and of positive constant λ. These choices will become clearer when the closed-loop system's poles are analyzed; let us for the moment suppose that $\mathbf{P}$ and λ are known.

Provided that matrix $\mathbf{P}$ is known, one further solves Lyapunov Eq. (12.18) in order to obtain $\mathbf{Q}$ as a symmetric and also positive definite matrix:

$$\mathbf{Q} = \begin{bmatrix} q_1 & 0 \\ 0 & q_2 \end{bmatrix}. \tag{12.26}$$

To this end, function `lyap` in MATLAB® can be used as a numerical solving tool. A simpler way to find matrix $\mathbf{Q}$ is to take it of the form (12.7). In this case, this corresponds to $\mathbf{Q}$ being chosen as

$$\mathbf{Q} = \begin{bmatrix} L & 0 \\ 0 & C \end{bmatrix}. \tag{12.27}$$

Further, once matrix $\mathbf{Q}$ is found, one can compute the stabilizing control solution based upon (12.20) as

$$\tilde{\mathbf{u}} = -\lambda \cdot \mathbf{B}^T \mathbf{Q} \cdot \tilde{\mathbf{x}}, \tag{12.28}$$

with suitably chosen positive scalar λ. By using expressions of matrices $\mathbf{B}$ and $\mathbf{Q}$ (relations (12.25) and (12.26)), expression (12.28) particularized to our case gives

$$\tilde{\alpha} = -\lambda \left[\frac{q_1 E}{L(1-\alpha_d)}(x_1 - x_{1d}) - \frac{q_2 E}{(1-\alpha_d)^2 RC}(x_2 - x_{2d}) \right], \tag{12.29}$$

where $x_{2d} \equiv v_{Cd}$ is the output voltage setpoint, α_d is given by (12.22) and $x_{1d} \equiv i_{Ld}$ is given by (12.25). If choosing matrix $\mathbf{Q}$ as (12.27), expression (12.29) becomes simpler:

$$\tilde{\alpha} = -\frac{\lambda E}{1-\alpha_d} \left[(x_1 - x_{1d}) - \frac{1}{R(1-\alpha_d)}(x_2 - x_{2d}) \right], \tag{12.30}$$

which can be put into form

$$\tilde{\alpha} = \underbrace{\begin{bmatrix} -\dfrac{\lambda E}{(1-\alpha_d)} & \dfrac{\lambda E}{R(1-\alpha_d)^2} \end{bmatrix}}_{\mathbf{K}} \cdot \tilde{\mathbf{x}}, \tag{12.31}$$

which makes appear the expression of the stabilizing feedback gain, $\mathbf{K}$. One can note that scalar λ is measured in $A^{-1}V^{-1}$. The block diagram of the closed-loop system based upon stabilizing control is given in Fig. 12.1.

Now, one can resume the problem of choosing λ. To this end, the closed-loop pole distribution must be analyzed, that is, the eigenvalues of the closed-loop state matrix $\mathbf{A} + \mathbf{B} \cdot \mathbf{K}$, with $\mathbf{K}$ being given by (12.31) corresponding to the matrix $\mathbf{Q}$ having been chosen in the form (12.27). The most convenient way to perform this analysis is numerically for each particular case, by using dedicated software tools, for example, MATLAB®.

Let us consider the following numerical instance for the considered boost converter: $L = 500\ \mu\text{H}$, $C = 1000\ \mu\text{H}$, $R = 10\ \Omega$, $E = 5$ V, $v_{Cd} = 15$ V. Relation (12.22) gives the corresponding value of the "desired" averaged duty ratio $\alpha_d = 0.66$. Figure 12.2 shows the excursion of the closed-loop system's poles in the complex plane for increasing values of λ, starting from 0.00015 $\text{A}^{-1} \cdot \text{V}^{-1}$.

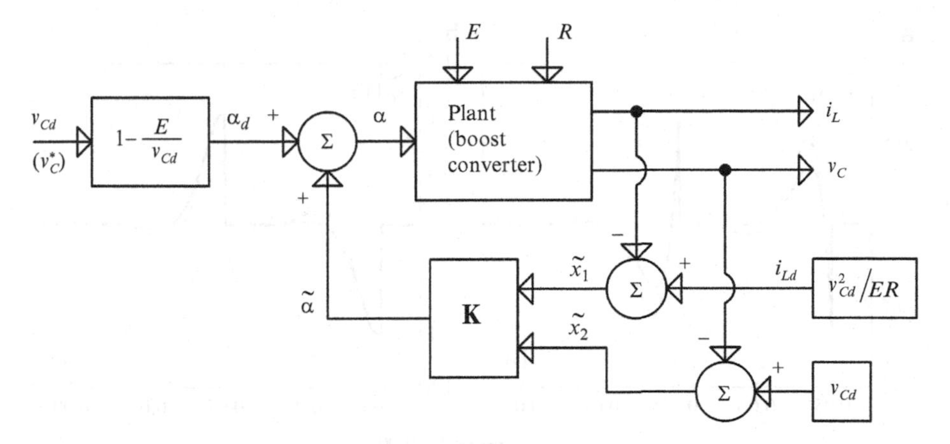

Fig. 12.1 Output voltage stabilizing control of a boost converter, relying upon a full-state-variation feedback of form (12.31)

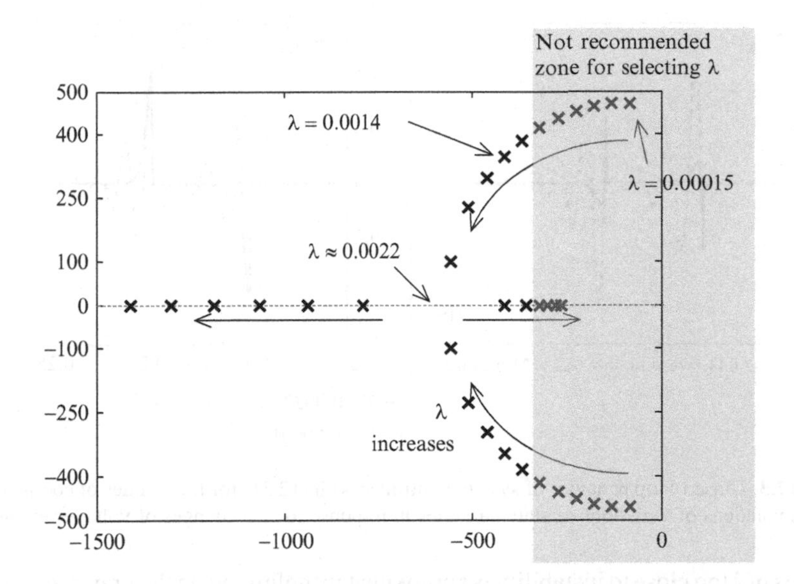

Fig. 12.2 Pole distribution of boost converter linearized model in variations controlled by stabilizing control law of form (12.31) for different values of parameter λ

One can note that values of λ placed around 0.0022 already determine in this case the nature of poles to change: they become real instead of complex-conjugated with a trend to be increasingly remote one to the other. It may happen that small values of λ to not ensure closed-loop stability (existence of two positive-real-part complex-conjugated poles), whereas very large values could lead to one of the poles to become unstable. Besides, values of λ that are too large could lead to control input saturation. Numerical simulation can be used in order to show that operation in the so-called saturated region is stable.

According to Fig. 12.2, $\lambda = 0.0014$ ensures in the considered case a closed-loop damping coefficient around 0.71 and a bandwidth of 534 rad/s. The zone including

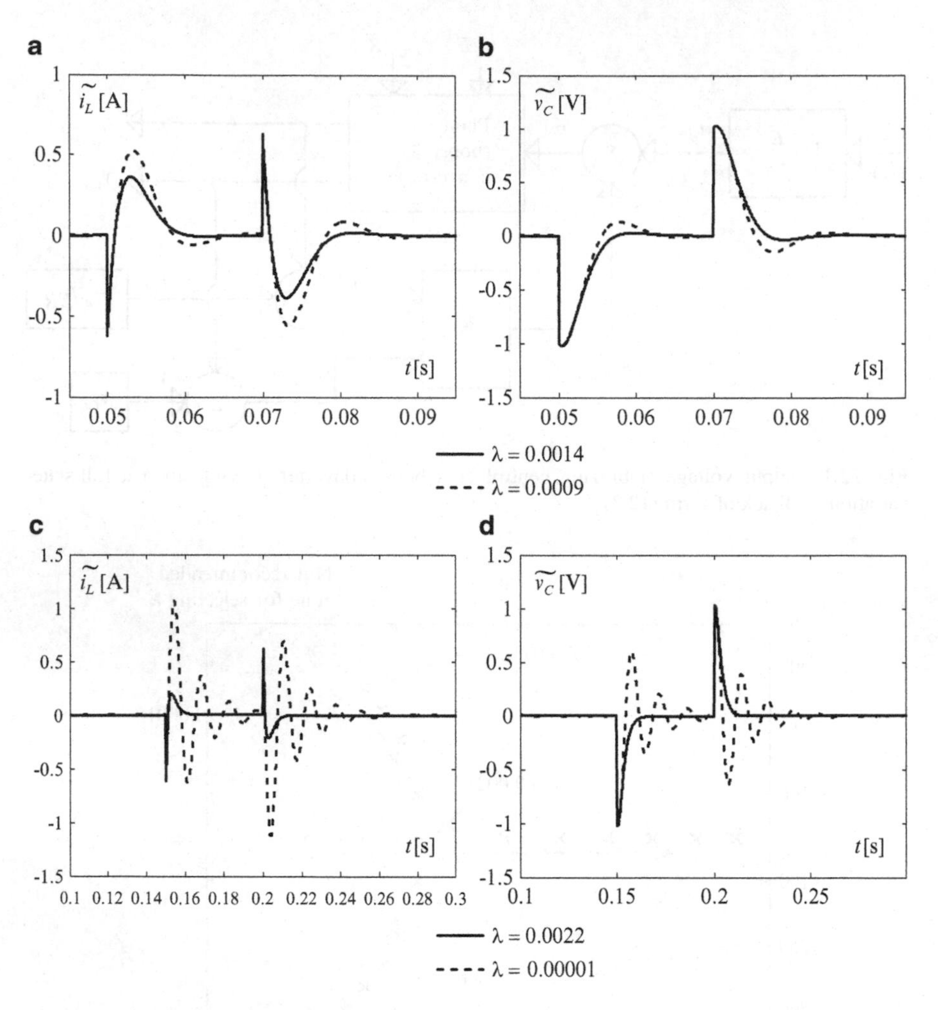

Fig. 12.3 Closed-loop behavior of system controlled with (12.31) for four values of coefficient λ: time evolutions of variations of state variables in response to step changes of voltage setpoint

values of λ too close to instability is approximately delimited on the gray background on Fig. 12.2. Another way of selecting gain λ is by requiring the closed-loop poles be real and equal; in this case, the corresponding value of λ would be around 0.0022.

12.2.4.2 Validation of Imposed Dynamic Performance

Simulations performed for different values of λ allow assessing the dynamic performance in response to step variations of the voltage reference v_{Cd}.

Figure 12.3 presents time evolutions of variation variables $\widetilde{i_L}$ and $\widetilde{v_C}$ in response to application of a step variation of ± 1 V in v_{Cd} around the steady point given by $v_{Cd} = 15$ V for two values of λ.

In Fig. 12.3a, b one can note that, for $\lambda = 0.0014$ (the curves represented by solid line in the figure), the settling time of both variables is coherent with the corresponding bandwidth (534 rad/s) and that the weak overshoot indeed reflects a damping coefficient around 0.7. Smaller values of λ correspond to weaker damping and smaller bandwidth (time evolutions for $\lambda = 0.0009$ are represented by dashed line in Fig. 12.3). Figure 12.3c, d show a similar comparison, namely for another two values of λ, one corresponding to real and almost equal poles ($\lambda = 0.0022$) and the other much smaller and determining oscillatory behavior ($\lambda = 0.00001$).

12.2.4.3 Control Input Saturation Issues

Let us now resume the problem of control input saturation on the considered example. Suppose that an event has taken place which can produce averaged control input saturation (this is the case, for example, of a sudden change in the voltage reference). The operation enters a region where the averaged control input is upper-saturated at the value α_{sat} (typically around 0.9). Consequently, the linearized system in variations (12.24) takes the form of an autonomous dynamical system, since $\widetilde{\alpha} = 0$:

$$\dot{\widetilde{\mathbf{x}_{sat}}} = \mathbf{A}_{sat} \cdot \widetilde{\mathbf{x}_{sat}}\,, \tag{12.32}$$

with state matrix

$$\mathbf{A}_{sat} = \begin{bmatrix} 0 & -(1-\alpha_{sat})/L \\ (1-\alpha_{sat})/C & -1/(RC) \end{bmatrix}. \tag{12.33}$$

In this case the Lyapunov function of form (12.15) has the time derivative of the form detailed in (12.16), but preserving the first term only:

$$\frac{dV(\widetilde{\mathbf{x}})}{dt} = \frac{1}{2}\widetilde{\mathbf{x}_{sat}}^{T}\left(\mathbf{A}_{sat}^{T}\mathbf{Q} + \mathbf{Q}\mathbf{A}_{sat}\right)\widetilde{\mathbf{x}_{sat}} \tag{12.34}$$

Choosing $\mathbf{Q}$ of particular form (12.27) allows rapid proof that $dV(\widetilde{\mathbf{x}})/dt < 0$. Thus, by substituting (12.27) and (12.33) into (12.34), one obtains

$$\frac{dV(\widetilde{\mathbf{x}})}{dt} = \frac{1}{2}\cdot\widetilde{\mathbf{x}_{sat}}^{T}\cdot\begin{bmatrix} 0 & 0 \\ 0 & -2/R \end{bmatrix}\cdot\widetilde{\mathbf{x}_{sat}} = -\frac{\widetilde{\mathbf{x}_{sat}}^{2}}{R} = -\frac{\widetilde{v_C}^{2}}{R} < 0 \tag{12.35}$$

The fact that the Lyapunov function is strictly decreasing when the system operates with saturated control input lets one expect that the closed-loop system will quickly leave the saturated region and enter the unsaturated one.

This is indeed the case, as simulations in Figs. 12.4 and 12.5 show. These two figures depict the case of applying a significant step change of the voltage setpoint, v_{Cd} – from 25 to 30 V – to the system at equilibrium ($\widetilde{i_L} = 0$, $\widetilde{v_C} = 0$). This is an

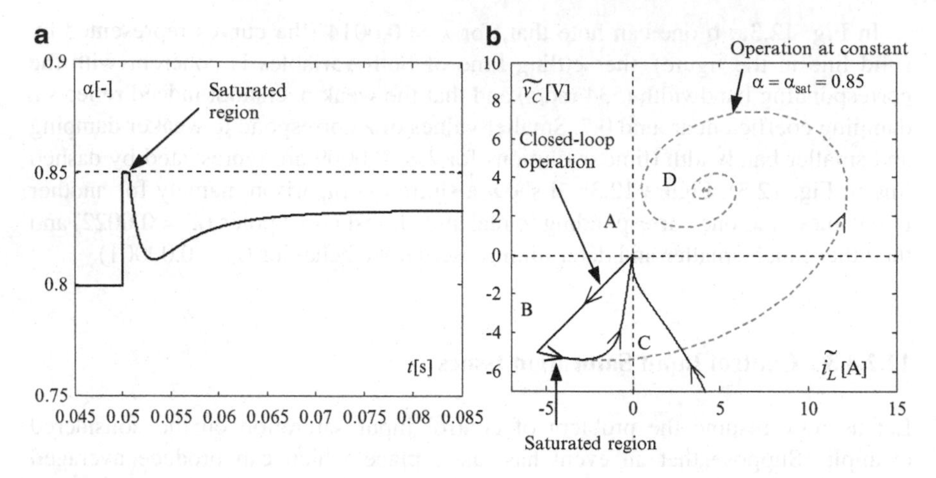

Fig. 12.4 Illustration of closed-loop system operation in saturated zone, followed by re-entering unsaturated zone: (**a**) duty ratio evolution; (**b**) state-space trajectory of system in variations

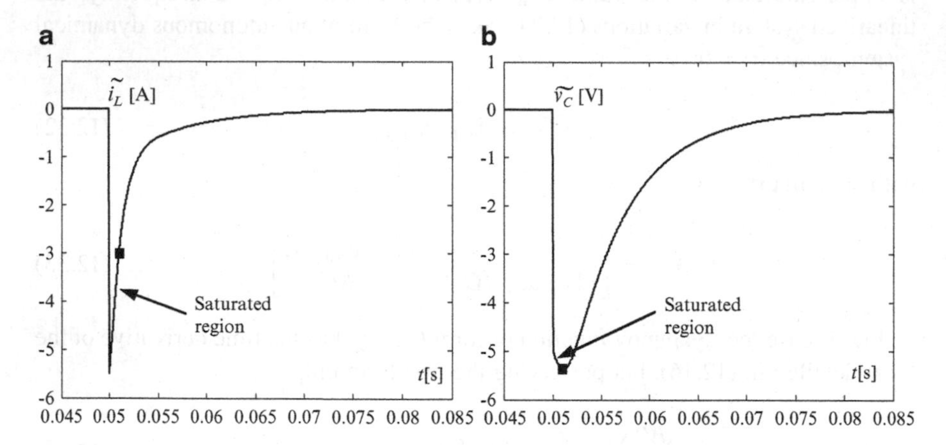

Fig. 12.5 Illustration of closed-loop linearized system operation in saturated zone, followed by re-entering unsaturated zone: (**a**) current variations; (**b**) voltage variations

event that determines operation in the saturated region for a quite short time interval, as Fig. 12.4a containing the evolution of the duty ratio shows. In this case, upper saturation has been chosen as $\alpha_{sat} = 0.85$. Figure 12.5a, b respectively present the time evolutions of the current and respectively voltage variations, $\widetilde{i_L}$ and $\widetilde{v_C}$. Figure 12.4b contains an interesting analysis performed in the phase plane $\left(\widetilde{i_L}, \widetilde{v_C}\right)$, where the closed-loop state trajectory, ABC, is shown to coincide for a time with CD, the system state trajectory that would be obtained if the control input were to continue to be saturated. Thus, segment AB corresponds to the time interval immediately after the event occurred. Segment BC describes the system evolution when the control input is saturated. One can note that the constant-α operation of the system is stable, but its equilibrium point (around D) is not the origin.

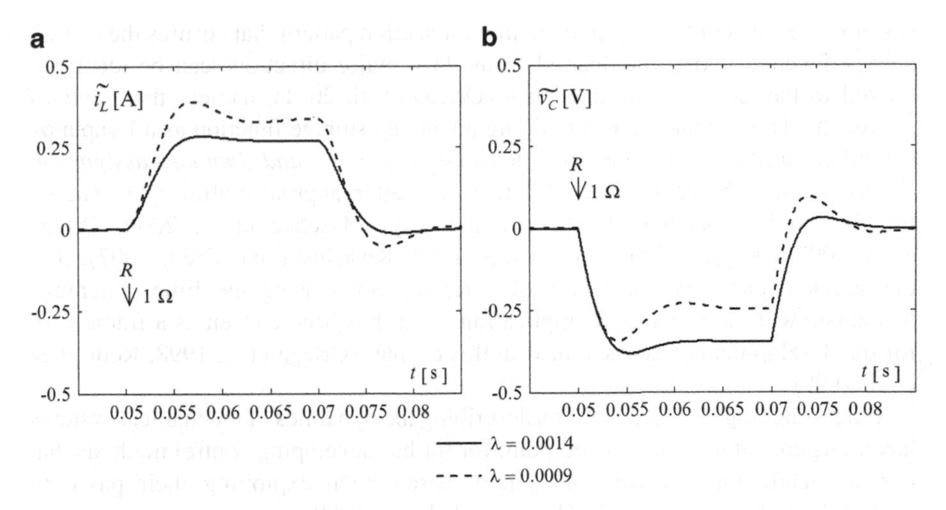

Fig. 12.6 Nonrejection of load perturbations by full-state-feedback-gain stabilizing control law (12.31): nonzero steady-state error on both current (**a**) and voltage (**b**) time evolutions

12.2.4.4 Adaptive Stabilizing Control

Expression (12.31) of the stabilizing control shows stabilization control involves a feedback gain which alone cannot ensure rejection of perturbations. Figure 12.6 presents simulation results obtained for 1 Ω step variations of the load R applied to the system at equilibrium, both positive and negative. Note that the "desired" value of the current – viewed here as something like an "internal" variable – also changes abruptly as a consequence of the load step change (see Eq. 12.23). One notes the presence of a nonzero steady-state error on both current and voltage time evolutions.

To conclude from the last set of simulation results, the necessity of employing parameter estimators – especially load estimators – is obvious for real-time implementation of stabilizing control. Details about how such observers can be designed to implement adaptive control laws will be given further in this chapter, namely related to passivity-based control (Sect. 12.4.3).

12.3 Approaches in Passivity-Based Control. Euler–Lagrange General Representation of Dynamical Systems

The action of designing a passivity-based controller has often been expressed somewhat metaphorically as an "energy-shaping" action. First, an energy function must be suitably defined for the considered system. Then one must find a dynamical system – i.e., the controller – and an interconnection pattern that ensures the overall energy function taking the desired form. Two major directions can be identified related to the passivity-based control (Ortega et al. 2001), namely the *standard* approach – based upon a priori defining an energy storage function as a Lyapunov candidate function – and the so-called *interconnection and damping assignment* (IDA) passivity-based control. Whereas the latter approach allows the energy function to be obtained in a systematic way (Maschke et al. 2000; Ortega et al. 2002; Ortega and García-Canseco 2004; Kwasinski and Krein 2007), it is the standard passivity-based control – relying upon using the Euler–Lagrange formalism with more intuitive application – that has been chosen as a framework for the developments discussed next in this chapter (Ortega et al. 1998; Rodríguez et al. 1999).

Euler–Lagrange (E-L) equations describing the dynamics of mechanical systems have long constituted the starting point for further developing control methods that aim at stabilizing this kind of system, based upon exploiting their passivity (Takegaki and Arimoto 1981; Ortega and Spong 1989). In a next step, one can deduce the generalization of the Euler–Lagrange formalism to any kind of passive dynamical system in order to design a general passivity-based control algorithm (Ortega et al. 1998).

12.3.1 Original Euler–Lagrange Form for Mechanical Systems

The Euler–Lagrange (E-L) representation of dynamical mechanical systems is (van der Schaft 1996)

$$\mathbf{M}(\mathbf{q}) \cdot \ddot{\mathbf{q}} + \mathbf{C}(\mathbf{q}\dot{\mathbf{q}}) \cdot \dot{\mathbf{q}} + \mathbf{g}(\mathbf{q}) = \tau, \tag{12.36}$$

where $\mathbf{q}$ represents the vector of generalized coordinates, $\mathbf{M}(\mathbf{q})$ the generalized inertias, $\mathbf{C}(\mathbf{q}, \dot{\mathbf{q}})$ the centrifugal and Coriolis force matrix, $\mathbf{g}(\mathbf{q})$ the gravitational force and τ the vector of generalized forces applied to the system. Relation (12.36) emphasizes the passivity of the $\tau \mapsto \dot{\mathbf{q}}$ input-to-output map, assuming that the associated potential energy has an absolute minimum. From the control viewpoint, examples of systems of form (12.36) are robot manipulators, for which the control inputs are as many as the states and the relative degree is one irrespective of the output to control. This is not the case for so-called *underactuated systems*, having free modes, as do, for example, rotating electrical machines or power electronic converters. Their passivity-based control requires adaptations of the original approach (Ortega and Espinosa 1993).

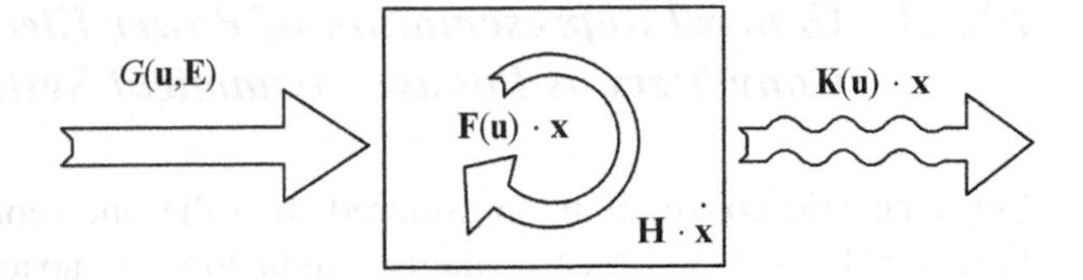

Fig. 12.7 Dynamical system that processes energy; identification of elements of formal representation (12.38) (Oyarbide and Bacha 2000)

12.3.2 Adaptation of Euler-Lagrange Formalism to Power Electronic Converters

A power electronic converter's dynamics can be cast into the following Euler–Lagrange representation (Oyarbide-Usabiaga 1998):

$$\mathbf{M} \cdot \ddot{\mathbf{q}} + \mathbf{C}(\mathbf{u}) \cdot \dot{\mathbf{q}} + \mathbf{K}(\mathbf{u}) \cdot \dot{\mathbf{q}} = \mathbf{G}(\mathbf{u}), \tag{12.37}$$

where $\mathbf{u}$ is the input vector. Note that dynamical systems of type (12.37) are not "naturally" suitable to be controlled by a passivity-based approach because the size of the control vector is almost always smaller than the system order and the terms containing the control input cannot be separated on the right-hand side of the differential equation.

An adaptation of the general passive-control formulation for power electronic converters has been proposed by Sira-Ramírez et al. (1997). This "modified" passivity-based control relies upon a suitable E-L representation of the system allowing its possible passivity being exploited in the sense of deducing the expression of a control input that stabilizes the system to an equilibrium point.

As a matter of fact, the E-L representation of power electronic converters which is presented next is general for a class of dynamical systems that exchange energy with their environment, a part of which is stored and the other part is dissipated. Such behavior is formally described by (Oyarbide-Usabiaga 1998):

$$\mathbf{H} \cdot \dot{\mathbf{x}} + \mathbf{F}(\mathbf{u}) \cdot \mathbf{x} + \mathbf{K}(\mathbf{u}) \cdot \mathbf{x} = \mathbf{G}(\mathbf{u}, \mathbf{E}), \tag{12.38}$$

where $\mathbf{x}$ is the n-dimensional state vector, $\mathbf{u}$ the m-dimensional input vector, $\mathbf{E}$ the vector of exogenous actions applied to the system (e.g., forces in the mechanical case, voltage or current sources in the electrical case) and $\mathbf{G}(\mathbf{u},\mathbf{E})$ models the way these actions are applied (i.e., the input energy). Matrices $\mathbf{H}$, $\mathbf{F}(\mathbf{u})$ and $\mathbf{K}(\mathbf{u})$ are related to the energy exchanged; thus, $\mathbf{H}$ is a positive definite matrix related to the energy being stored by different elements of the system (this is why it is called a storage function), $\mathbf{F}(\mathbf{u})$ takes into account the "internal" system's energy and $\mathbf{K}(\mathbf{u})$ is a positive semidefinite matrix providing information about the energy dissipation rate. Figure 12.7 gives an intuitive image of a system formalized according to (12.38).

A necessary condition for a system in form (12.38) to be passive is that its internal energy be zero, which is mathematically described by $\mathbf{x}^T \cdot \mathbf{F}(\mathbf{u}) \cdot \mathbf{x} = 0$. The validity of this assumption is related to the configuration of a passive system. This property is verified for power electronic converters, as shown further in the next section.

12.3.3 General Representation of Power Electronic Converters as Passive Dynamical Systems

Let a generic converter be considered, like the one represented symbolically in Fig. 3.1 of Chap. 3, which contains k inductors, m capacitors, n perfect switches, q voltage sources and p current sources. These elements are interconnected to define various configurations.

Let $\mathbf{E} = [E_1 \quad \cdots \quad E_q]^T$ be the vector of voltage sources, $\mathbf{I} = [I_1 \quad \cdots \quad I_p]^T$ the vector of current sources and $\mathbf{u} = [u_1 \quad \cdots \quad u_n]^T$ the vector giving the states of switches. Let also $\mathbf{I}_L = [i_{L1} \quad \cdots \quad i_{Lk}]^T$ be the vector of k inductor currents and $\mathbf{V}_C = [v_{C1} \quad \cdots \quad v_{Cm}]^T$ the vector of m capacitor voltages. This system has $k + m$ states, which are the inductor currents and the capacitor voltages; its state vector is:

$$\mathbf{x} = [\mathbf{I}_L \quad \mathbf{V}_C] = [i_{L1} \quad \cdots \quad i_{Lk} \quad v_{C1} \quad \cdots \quad v_{Cm}]^T. \tag{12.39}$$

Dynamics of each current and voltage result according to the influence of all voltage sources and all current sources weighted by the corresponding linear combinations of switch states. Thus, the dynamic of inductor current i results from representing the inductor by its equivalent Thévenin generator and can be written as

$$L_i \dot{i}_{Li} = \underbrace{\mathbf{T}_{SEL_i}(\mathbf{u}) \cdot \mathbf{E} + \mathbf{T}_{SIL_i}(\mathbf{u}) \cdot \mathbf{I} + \mathbf{T}_{IL_i}(\mathbf{u}) \cdot \mathbf{I}_L + \mathbf{T}_{VL_i}(\mathbf{u}) \cdot \mathbf{V}_C}_{\text{Thévenin voltage } i} - R_{Li}(\mathbf{u}) \cdot i_{Li}, \tag{12.40}$$

where $R_{Li}(\mathbf{u})$ is the Thévenin equivalent resistance and all $\mathbf{T}$ vectors have elements that are linear combinations of switch states $\mathbf{u}$. Thus, vectors $\mathbf{T}_{SEL_i}(\mathbf{u})$ and $\mathbf{T}_{SIL_i}(\mathbf{u})$ model the influence of the voltage sources and current sources, respectively, on the dynamic of inductor current i, whereas $\mathbf{T}_{IL_i}(\mathbf{u})$ and $\mathbf{T}_{VL_i}(\mathbf{u})$ represent the influence of all currents $\mathbf{I}_L$ and all voltages $\mathbf{V}_C$, respectively (all state variables), on the same dynamic. One can note that the ith element of $\mathbf{T}_{IL_i}(\mathbf{u})$ is null because the Thévenin voltage i is not related to current i_{Li}.

The same formalism is used for expressing the dynamic of capacitor voltage j, which may be represented by its equivalent Norton generator:

$$C_j \dot{v}_{Cj} = \underbrace{\mathbf{T}_{SEC_j}(\mathbf{u}) \cdot \mathbf{E} + \mathbf{T}_{SIC_j}(\mathbf{u}) \cdot \mathbf{I} + \mathbf{T}_{IC_j}(\mathbf{u}) \cdot \mathbf{I}_L + \mathbf{T}_{VC_j}(\mathbf{u}) \cdot \mathbf{V}_C}_{\text{Norton current } j} - \frac{v_{Cj}}{R_{Cj}(\mathbf{u})}, \tag{12.41}$$

where $R_{Cj}(\mathbf{u})$ is the Norton equivalent resistance and vectors $\mathbf{T}_{SEC_j}(\mathbf{u})$ and $\mathbf{T}_{SIC_j}(\mathbf{u})$ model the influence of the voltage sources and current sources, respectively, on the dynamic of capacitor voltage j, whereas $\mathbf{T}_{IC_j}(\mathbf{u})$ and $\mathbf{T}_{VC_j}(\mathbf{u})$ represent the influence of all currents $\mathbf{I}_L$ and all voltages $\mathbf{V}_C$, respectively, on the same dynamic. Similarly as above, one can note that the jth element of $\mathbf{T}_{VC_j}(\mathbf{u})$ is null because the Norton current j is not related to voltage v_{Cj}.

By grouping together all k dynamic equations of form (12.40) and all m dynamic equations of form (12.41), the dynamics of the generic converter can be represented in the Euler–Lagrange formalism (12.38), where the q voltage sources and the p current sources constitute the external effort:

$$\mathbf{H} \cdot \dot{\mathbf{x}} + \mathbf{F}(\mathbf{u}) \cdot \mathbf{x} + \mathbf{K}(\mathbf{u}) \cdot \mathbf{x} = \mathbf{G}(\mathbf{u}, \mathbf{E}, \mathbf{I}), \tag{12.42}$$

where $\mathbf{x}$ is the state vector given by (12.39) and the involved matrices are

$$\mathbf{H} = \begin{bmatrix} L_1 & & & & & \\ & \ddots & & & 0 & \\ & & L_k & & & \\ & & & C_1 & & \\ & 0 & & & \ddots & \\ & & & & & C_m \end{bmatrix}, \quad \mathbf{F}(\mathbf{u}) = \begin{bmatrix} \mathbf{T}_{IL_1}(\mathbf{u}) & \mathbf{T}_{VL_1}(\mathbf{u}) \\ \vdots & \vdots \\ \mathbf{T}_{IL_k}(\mathbf{u}) & \mathbf{T}_{VL_k}(\mathbf{u}) \\ \mathbf{T}_{IC_1}(\mathbf{u}) & \mathbf{T}_{VC_1}(\mathbf{u}) \\ \vdots & \vdots \\ \mathbf{T}_{IC_m}(\mathbf{u}) & \mathbf{T}_{VC_m}(\mathbf{u}) \end{bmatrix}, \tag{12.43}$$

$$\mathbf{K}(\mathbf{u}) = \begin{bmatrix} R_{L1}(\mathbf{u}) & & & & & \\ & \ddots & & & 0 & \\ & & R_{Lk}(\mathbf{u}) & & & \\ & & & \dfrac{1}{R_{C1}(\mathbf{u})} & & \\ & 0 & & & \ddots & \\ & & & & & \dfrac{1}{R_{Cm}(\mathbf{u})} \end{bmatrix}, \tag{12.44}$$

$$\mathbf{G}(\mathbf{u}, \mathbf{E}, \mathbf{I}) = \begin{bmatrix} \mathbf{T}_{SEL_1}(\mathbf{u}) \cdot \mathbf{E} + \mathbf{T}_{SIL_1}(\mathbf{u}) \cdot \mathbf{I} \\ \vdots \\ \mathbf{T}_{SEL_k}(\mathbf{u}) \cdot \mathbf{E} + \mathbf{T}_{SIL_k}(\mathbf{u}) \cdot \mathbf{I} \\ \mathbf{T}_{SEC_1}(\mathbf{u}) \cdot \mathbf{E} + \mathbf{T}_{SIC_1}(\mathbf{u}) \cdot \mathbf{I} \\ \vdots \\ \mathbf{T}_{SEC_m}(\mathbf{u}) \cdot \mathbf{E} + \mathbf{T}_{SIC_m}(\mathbf{u}) \cdot \mathbf{I} \end{bmatrix}, \tag{12.45}$$

where matrix **G(u,E,I)** defines the way the external sources **(E,I)** are applied to the generic converter, **F(u)** takes account of the configuration allowing the energy exchange between inductors and capacitors, **H** contains the terms related to energy storage inside of inductors and capacitors and **K(u)** represents the energy dissipation matrix.

In order for the Eq. (12.42) to correspond to an Euler–Lagrange representation, one has to verify the property $\mathbf{x}^T\mathbf{F}(\mathbf{u})\mathbf{x} = 0$. Whereas this property is usually associated in the literature with the matrix **F(u)** being antisymmetrical – i.e., $\mathbf{F}(\mathbf{u}) = -\mathbf{F}(\mathbf{u})^T$ – this is not necessary in the case of electric circuits. Indeed, this property can be verified by using certain properties of a Kirchhoff's network

(Penfield et al. 1970), namely that the subspaces of its currents and its voltages are orthogonal. In particular, the result of product $\mathbf{F(u)x}$ belongs to a subspace which is orthogonal to the state space $\mathbf{x}$. Moreover, the property $\mathbf{x}^T\mathbf{F(u)x} = 0$ is preserved under usual transforms, like the Park or Concordia transform or the one leading to the generalized averaged model (GAM) (Oyarbide-Usabiaga 1998).

12.3.4 Examples of Converter Modeling in the Euler–Lagrange Formalism

Next, two examples are presented in order to illustrate Euler–Lagrange modeling and invariance of passivity properties at the usual transforms. Namely, the case of a buck-boost converter is first considered, emphasizing the antisymmetry of matrix $\mathbf{F(u)}$; then the case of a voltage inverter and the invariance of properties of its Euler–Lagrange representation at the Park transform are presented.

The switched model of a buck-boost converter supplying a load R – whose electrical circuit is shown in Fig. 4.12 in Chap. 4 – is given by relation (4.34) of the same chapter:

$$\begin{cases} \dot{i_L} = \dfrac{1}{L}[Eu - v_C(1-u) - ri_L] \\ \dot{v_C} = \dfrac{1}{C}\left[-i_L(1-u) - \dfrac{v_C}{R}\right], \end{cases} \tag{12.46}$$

where notations preserve their usual meaning. There is a single voltage source, E, and no current sources. By noting by $\mathbf{x} = [i_L \quad v_C]^T$ the state vector composed of the inductor current i_L and the capacitor voltage v_C, Eq. (12.46) can be arranged to fit the Euler–Lagrange model (12.42), where the different matrices are

$$\mathbf{H} = \begin{bmatrix} L & 0 \\ 0 & C \end{bmatrix}, \quad \mathbf{K} = \begin{bmatrix} r & 0 \\ 0 & 1/R \end{bmatrix}, \quad \mathbf{G} = \begin{bmatrix} uE \\ 0 \end{bmatrix}, \quad \mathbf{F}(u) = \begin{bmatrix} 0 & (1-u) \\ -(1-u) & 0 \end{bmatrix}$$

and one can note that matrix $\mathbf{F}(u)$ is antisymmetrical.

The second modeling example consists of the case of a three-phase voltage inverter supplying a load R, whose circuit is shown in Fig. 5.28 in Chap. 5. Based upon the switched model of this converter – presented in relation (5.67) from the same chapter, where the current exchanged with an additional DC stage is $i_S = v_0/R$ – one can deduce its Euler–Lagrange representation (12.42), with

$$\begin{cases} \mathbf{H} = \begin{bmatrix} L & & & 0 \\ & L & & \\ & & L & \\ 0 & & & C \end{bmatrix} & \mathbf{K} = \begin{bmatrix} r & & & 0 \\ & r & & \\ & & r & \\ 0 & & & 1/R \end{bmatrix} \\ \mathbf{G} = \begin{bmatrix} e_1 \\ e_2 \\ e_3 \\ 0 \end{bmatrix} & \mathbf{F}(u) = \begin{bmatrix} 0 & 0 & 0 & (2u_1 - u_2 - u_3)/6 \\ 0 & 0 & 0 & (-u_1 + 2u_2 - u_3)/6 \\ 0 & 0 & 0 & (-u_1 - u_2 + 2u_3)/6 \\ -u_1/2 & -u_2/2 & -u_3/2 & 0 \end{bmatrix} \end{cases}$$

being the matrices involved and $\mathbf{x} = [i_1 \quad i_2 \quad i_3 \quad v_0]^T$ being the state vector. First, note that matrix $\mathbf{F}(u)$ is no longer antisymmetrical, but one can verify that the property $\mathbf{x}^T\mathbf{F}(\mathbf{u})\mathbf{x} = 0$ still holds, provided $i_1 + i_2 + i_3 = 0$. Next it is shown that the passivity properties are preserved in the new state space defined by the Park transform that conserves the exchanged energy and whose expression in a rotating frame dq is (see also relations (5.74) in Chap. 5)

$$\begin{bmatrix} f_q \\ f_d \end{bmatrix} = \sqrt{\frac{2}{3}} \cdot \begin{bmatrix} \cos\omega t & \cos(\omega t - 2\pi/3) & \cos(\omega t + 2\pi/3) \\ \sin\omega t & \sin(\omega t - 2\pi/3) & \sin(\omega t + 2\pi/3) \end{bmatrix} \cdot \begin{bmatrix} f_1 \\ f_2 \\ f_3 \end{bmatrix}.$$

The new state vector is $\mathbf{x} = [i_q \quad i_d \quad v_0]^T$. If the inverter is operated in full wave, with the control input being the phase lag between source voltages e_i and the corresponding voltages v_i, denoted by α, the generalized averaged model of this converter in the dq frame contains the following Euler–Lagrange matrices:

$$\begin{cases} \mathbf{H} = \begin{bmatrix} L & 0 & 0 \\ 0 & L & 0 \\ 0 & 0 & C \end{bmatrix} & \mathbf{K} = \begin{bmatrix} r & 0 & 0 \\ 0 & r & 0 \\ 0 & 0 & 1/R \end{bmatrix} \\ \mathbf{G} = \begin{bmatrix} E_q \\ E_d \\ 0 \end{bmatrix} & \mathbf{F}(u) = \begin{bmatrix} 0 & \omega L & -\sqrt{6}\sin\alpha/\pi \\ -\omega L & 0 & \sqrt{6}\cos\alpha/\pi \\ \sqrt{6}\sin\alpha/\pi & -\sqrt{6}\cos\alpha/\pi & 0 \end{bmatrix} \end{cases}$$

that fit into the form expressed in (12.42), with $\mathbf{F}(u)$ being antisymmetrical.

12.4 Passivity-Based Control of Power Electronic Converters

12.4.1 Theoretical Background

The passivity property of a system whose state model can be arranged in form (12.42) can be exploited for control purposes. Let $\mathbf{x}_d$ be the trajectory around which one wants to stabilize system (12.42) and $\widetilde{\mathbf{x}} = \mathbf{x} - \mathbf{x}_d$ be the state error in relation to the imposed trajectory. Then relation (12.42) can be rewritten as

$$\mathbf{H}\cdot\dot{\widetilde{\mathbf{x}}}+\mathbf{F}(\mathbf{u})\cdot\widetilde{\mathbf{x}}+\mathbf{K}(\mathbf{u})\cdot\widetilde{\mathbf{x}}=\mathbf{G}(\mathbf{u},\mathbf{E},\mathbf{I})\\ -\{\mathbf{H}\cdot\dot{\mathbf{x}}_d+\mathbf{F}(\mathbf{u})\cdot\mathbf{x}_d+\mathbf{K}(\mathbf{u})\cdot\mathbf{x}_d\}. \quad (12.47)$$

Suppose that a control input $\mathbf{u}_C$ zeroing the right side of (12.47) exists and can be computed; it therefore satisfies

$$\mathbf{G}(\mathbf{u}_C,\mathbf{E},\mathbf{I})-\{\mathbf{H}\cdot\dot{\mathbf{x}}_d+\mathbf{F}(\mathbf{u}_C)\cdot\mathbf{x}_d+\mathbf{K}(\mathbf{u}_C)\cdot\mathbf{x}_d\}=0. \quad (12.48)$$

Under these assumptions, the error dynamics are given by

$$\mathbf{H}\cdot\dot{\widetilde{\mathbf{x}}}+\mathbf{F}(\mathbf{u})\cdot\widetilde{\mathbf{x}}+\mathbf{K}(\mathbf{u})\cdot\widetilde{\mathbf{x}}=0. \quad (12.49)$$

By posing $V(\widetilde{\mathbf{x}})=1/2\cdot\widetilde{\mathbf{x}}^T\mathbf{H}\widetilde{\mathbf{x}}$ as the candidate Lyapunov function and by using the property $\mathbf{x}^T\mathbf{F}(\mathbf{u})\mathbf{x}=0$ that also holds for the state error, i.e., $\widetilde{\mathbf{x}}^T\mathbf{F}(\mathbf{u})\widetilde{\mathbf{x}}=0$, one proves the asymptotic stability of the considered system around the imposed trajectory $\mathbf{x}_d$ (Oyarbide-Usabiaga 1998; Oyarbide and Bacha 1999; Oyarbide et al. 2000).

The convergence towards $\mathbf{x}_d$ can be rendered faster if adding a supplementary term in expression (12.48), which allows computation of the control input $\mathbf{u}_C$. The role of this added term is to increase some of the elements of the dissipativity matrix $\mathbf{K}(\mathbf{u})$, that is, to increase the speed at which the energy gets out of the system. A diagonal matrix is thus defined as $\mathbf{Ki}=\mathrm{diag}(k_1,\cdots,k_n)$, with $k_j\geq 0$ and n being the system's order, which is called *damping injection matrix*. As a consequence, the new form of relation (12.48) is

$$\mathbf{G}(\mathbf{u}_C,\mathbf{E},\mathbf{I})-\{\mathbf{H}\cdot\dot{\mathbf{x}}_d+\mathbf{F}(\mathbf{u}_C)\cdot\mathbf{x}_d+\mathbf{K}(\mathbf{u}_C)\cdot\mathbf{x}_d\}+\mathbf{Ki}\cdot\widetilde{\mathbf{x}}=0, \quad (12.50)$$

which serves to compute the stabilizing control $\mathbf{u}_C$.

The problem of effectively computing $\mathbf{u}_C$ depends on whether the system is fully controllable – i.e., the size of the control input vector is the same as that of the state vector – or not. Power electronic converters belong to the second class, of so-called underactuated systems that exhibit free modes.

In order to compute the stabilizing control input $\mathbf{u}_C$, the nth order system (12.42) may be decomposed into two subsystems according to the state vector $\mathbf{x}$ being partitioned as $\mathbf{x}=\begin{bmatrix}\mathbf{x}_C & \mathbf{x}_F\end{bmatrix}$, where $\mathbf{x}_C$ contains the "controlled" states and $\mathbf{x}_F$ contains the free states. Let m be the size of $\mathbf{x}_C$; therefore, $n-m$ will be the number of free modes. Supposing that the dissipativity matrix $\mathbf{K}(\mathbf{u})$ is diagonal, the corresponding decomposition of system (12.42) is

$$\begin{cases}\mathbf{H}_C\cdot\dot{\mathbf{x}}_C+\mathbf{F}_C(\mathbf{u})\cdot\mathbf{x}_C+\mathbf{K}_C(\mathbf{u})\cdot\mathbf{x}_C+\mathbf{F}_{CF}(\mathbf{u})\cdot\mathbf{x}_F=\mathbf{G}_C(\mathbf{u},\mathbf{E},\mathbf{I})\\ \mathbf{H}_F\cdot\dot{\mathbf{x}}_F+\mathbf{F}_F(\mathbf{u})\cdot\mathbf{x}_F+\mathbf{K}_F(\mathbf{u})\cdot\mathbf{x}_F+\mathbf{F}_{FC}(\mathbf{u})\cdot\mathbf{x}_C=\mathbf{G}_F(\mathbf{u},\mathbf{E},\mathbf{I}),\end{cases}$$

where subscript C refers to variables related to controlled states $\mathbf{x}_C$ and subscript F refers to variables related to free states.

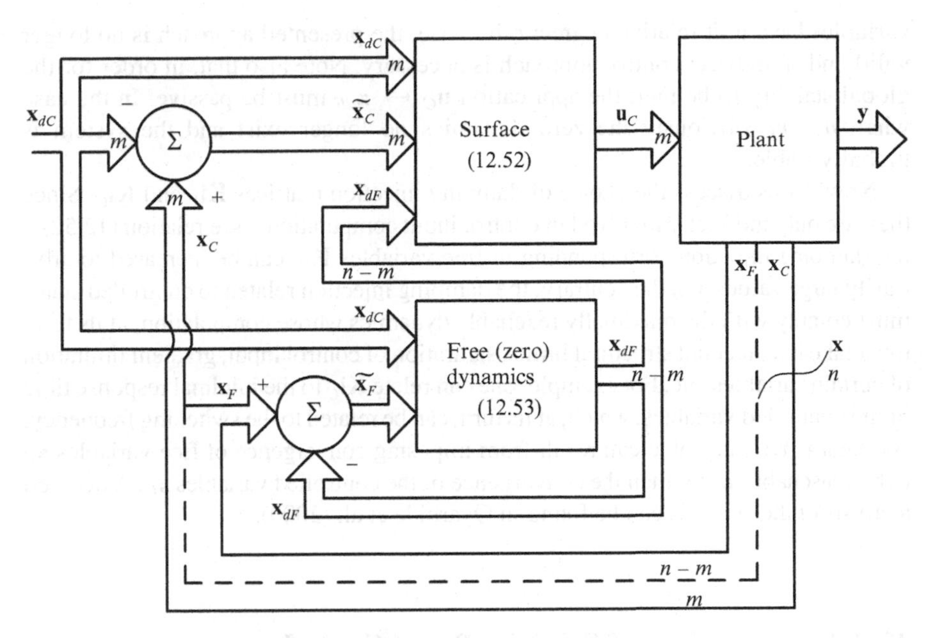

Fig. 12.8 Block diagram implementing computation of passivity-based control input based upon computation of desired free-state dynamics (Oyarbide et al. 2000)

Let $\mathbf{x}_{dC}$ be the desired trajectory that corresponds to controlled states and $\mathbf{x}_{dF}$ its counterpart for the free states. Let $\widetilde{\mathbf{x}_C} = \mathbf{x}_C - \mathbf{x}_{dC}$ and $\widetilde{\mathbf{x}_F} = \mathbf{x}_F - \mathbf{x}_{dF}$ be the corresponding state errors. Equation (12.50) allowing computation of the control input can also be decomposed as

$$\begin{cases} \mathbf{G}_C(\mathbf{u}_C, \mathbf{E}, \mathbf{I}) - \{\mathbf{H}_C \cdot \dot{\mathbf{x}}_{dC} + \mathbf{F}_C(\mathbf{u}_C) \cdot \mathbf{x}_{dC} + \mathbf{K}_C(\mathbf{u}_C) \cdot \mathbf{x}_{dC} + \mathbf{F}_{CF}(\mathbf{u}_C) \cdot \mathbf{x}_{dF}\} \\ \qquad + \mathbf{Ki}_C \cdot \widetilde{\mathbf{x}_C} = 0 \\ \mathbf{G}_F(\mathbf{u}_C, \mathbf{E}, \mathbf{I}) - \{\mathbf{H}_F \cdot \dot{\mathbf{x}}_{dF} + \mathbf{F}_F(\mathbf{u}_C) \cdot \mathbf{x}_{dF} + \mathbf{K}_F(\mathbf{u}_C) \cdot \mathbf{x}_{dF} + \mathbf{F}_{FC}(\mathbf{u}_C) \cdot \mathbf{x}_{dC}\} \\ \qquad + \mathbf{Ki}_F \cdot \widetilde{\mathbf{x}_F} = 0. \end{cases} \tag{12.51}$$

Note that for regulation purposes $\dot{\mathbf{x}}_{dC} = 0$ holds. Relations (12.51) represent a dynamic system that has $\mathbf{x}_{dC}$ as inputs and allows computation of control input $\mathbf{u}_C$ and of the free mode dynamics, the so-called zero dynamics, $\mathbf{x}_{dF}$. This is expressed formally as

$$\mathbf{u}_C = function(\mathbf{x}_{dC}, \mathbf{x}_{dF}, \mathbf{E}, \mathbf{I}, \mathbf{Ki}_C\widetilde{\mathbf{x}_C}), \tag{12.52}$$

$$\dot{\mathbf{x}}_{dF} = \mathbf{H}_F^{-1} \cdot [\mathbf{G}_F(\mathbf{u}_C, \mathbf{E}, \mathbf{I}) - \mathbf{F}_F(\mathbf{u}_C)\mathbf{x}_{dF} - \mathbf{K}_F(\mathbf{u}_C)\mathbf{x}_{dF} - \mathbf{F}_{FC}(\mathbf{u}_C)\mathbf{x}_{dC} + \mathbf{Ki}_F\widetilde{\mathbf{x}_F}] \tag{12.53}$$

and can be represented as a block diagram like that in Fig. 12.8. Equation (12.52) allows control input $\mathbf{u}_C$ being computed in a direct manner if the controlled variables have unit relative degree; otherwise, the presented approach is no longer valid and an indirect control approach is necessary. Note also that, in order for the global stability to be met, the application $\mathbf{u}_C \mapsto \mathbf{x}_{dF}$ must be passive. In the case when $n = m$, obviously the zero dynamics no longer exist and the system is globally stable.

Now, let us discuss the choice of damping injection matrices $\mathbf{Ki}_F$ and $\mathbf{Ki}_C$. Since they are only indirectly involved in control input computation – see relation (12.52) – the damping injection corresponding to free variables $\mathbf{Ki}_F$ can be increased to arbitrarily large values. On the contrary, the damping injection related to controlled states must comply with the maximally reachable dynamics whose computation, at its turn, must take into account structural limits, saturation of control input, gradient limitation of certain variables, etc. For example, one can relate $\mathbf{Ki}_C$ to the minimal response time of the controlled variables, which, at its turn, can be related to the switching frequency. As regards $\mathbf{Ki}_F$, its value can result from imposing convergence of free variables $\mathbf{x}_F$ to be reasonably faster than the convergence of the controlled variables $\mathbf{x}_C$. A detailed analysis of these issues can be found in Oyarbide et al. (2000).

12.4.2 Limitations of Passivity-Based Control

In the case of non-unit relative degree – when zero dynamics, or free modes $\mathbf{x}_{dF}$ exist – the passivity-based control approach loses its remarkable property of being intrinsically stable. Thus, the requirement of having a passive application $\mathbf{u}_C \mapsto \mathbf{x}_{dF}$ is limiting because it requires a local stability analysis, and a general analysis approach cannot be applied. Depending on each particular structure and constraint, a particular means of ensuring the zero dynamic stability can be found.

Another limitation is related to the strong dependence of the control structure on system parameters, which negatively affects its robustness. Indeed, control input computation according to (12.52) shows the necessity of measuring or estimating some of the system's parameters. In the case of DC-DC converters, for example, these are typically voltage source value E and load value R. In the case of variables that are difficult to measure, estimators must be embedded in the control structure, the latter becomes adaptive in this way.

12.4.3 Parameter Estimation: Adaptive Passivity-Based Control

Unlike classical control solutions based upon estimation, controller design and observer design are unified in the case of adaptive passivity-based control. This is possible if one can ensure the system is linear in relation to the estimated parameters (Ortega and Spong 1989). If it is, the Lyapunov stability of the controller–observer pair can be ensured.

Let $\mathbf{P}$ and $\mathbf{P}_{\text{est}}$ be vectors of real and estimated parameters, respectively. The dynamics of the difference between them can be written by using relation (12.48):

$$\Delta\mathbf{G}(\mathbf{u}_C, \mathbf{E}, \mathbf{I}) - \{\Delta\mathbf{H}\cdot\dot{\mathbf{x}}_d + \Delta\mathbf{F}(\mathbf{u}_C)\cdot\mathbf{x}_d + \Delta\mathbf{K}(\mathbf{u}_C)\cdot\mathbf{x}_d\}, \tag{12.54}$$

where notation $\Delta(\cdot) = (\cdot)|_{\mathbf{P}} - (\cdot)|_{\mathbf{P}_{\text{est}}}$ denotes parameterization error. Assuming it is always possible to perform a linear parameterization – eventually by using variable changes – then it is also possible to rearrange expression (12.54) to have the parameter estimation error $\Delta\mathbf{P} = \mathbf{P} - \mathbf{P}_{est}$ as a factor:

$$\Delta\mathbf{G}(\mathbf{u}_C, \mathbf{E}, \mathbf{I}) - \{\Delta\mathbf{H}\cdot\dot{\mathbf{x}}_d + \Delta\mathbf{F}(\mathbf{u}_C)\cdot\mathbf{x}_d + \Delta\mathbf{K}(\mathbf{u}_C)\cdot\mathbf{x}_d\} = \mathbf{Y}(\dot{\mathbf{x}}_d, \mathbf{x}_d, \mathbf{u}_C)\cdot\Delta\mathbf{P}. \tag{12.55}$$

In order to have both regulation error and estimation error decreasing to zero, it is sufficient to implement a parameter estimation law of the form (Bacha et al. 1997; Oyarbide-Usabiaga 1998):

$$\dot{\mathbf{P}}_{\text{est}} = \Gamma\cdot\mathbf{Y}(\dot{\mathbf{x}}_d, \mathbf{x}_d, \mathbf{u}_C)\cdot\widetilde{\mathbf{x}}, \tag{12.56}$$

where the diagonal positive semidefinite matrix Γ imposes estimation convergence speed, commonly larger than the imposed closed-loop dynamics corresponding to the previously chosen damping injection matrices. Elements of Γ must be chosen as a trade-off between convergence speed and sensitivity to high-frequency exogenous variations. Numerical simulation is useful to determine the best trade-off in each particular case.

12.4.4 Passivity-Based Control Design Algorithm

The main steps of implementing a passivity-based control approach for a given power electronic converter are listed below.

12.4.5 Example: Passivity-Based Control of a Boost DC-DC Converter

Let us consider the case of a boost converter supplying a resistive load R, whose electrical circuit has been presented in Fig. 3.5 of Chap. 3. The switched model of this converter – originally presented in Eq. (3.11) of Chap. 3 – is shown again here:

$$\begin{cases} L\dot{i}_L = -(1-u)v_C + E \\ C\dot{v}_C = (1-u)i_L - v_C/R, \end{cases} \tag{12.57}$$

where inductor current i_L and capacitor voltage v_C are the two states, and E is the voltage source value representing the external energy effort applied to the system. Parameters chosen are $L = 5$ mH, $C = 470$ μF, $R = 10$ Ω, $E = 15$ V, switching frequency $f = 20$ kHz.

12.4.5.1 Basic Control Design

The control goal of regulating output voltage will be solved next by passivity-based control, following the steps of Algorithm 12.3. To this end, model (12.57) can be expressed in the Euler–Lagrange formalism (12.42) and allows one to identify the matrices involved:

$$\underbrace{\begin{bmatrix} L & 0 \\ 0 & C \end{bmatrix}}_{\mathbf{H}} \cdot \underbrace{\begin{bmatrix} i_L \\ v_C \end{bmatrix}^{\cdot}}_{\dot{\mathbf{x}}} + \underbrace{\begin{bmatrix} 0 & 1-u \\ -(1-u) & 0 \end{bmatrix}}_{\mathbf{F(u)}} \cdot \underbrace{\begin{bmatrix} i_L \\ v_C \end{bmatrix}}_{\mathbf{x}} + \underbrace{\begin{bmatrix} 0 & 0 \\ 0 & 1/R \end{bmatrix}}_{\mathbf{K}} \cdot \underbrace{\begin{bmatrix} i_L \\ v_C \end{bmatrix}}_{\mathbf{x}} = \underbrace{\begin{bmatrix} E \\ 0 \end{bmatrix}}_{\mathbf{G(E)}}. \tag{12.58}$$

Algorithm 12.3

Steps of passivity-based control design for power electronic converters

#1. Write down the switched model of the converter, rearrange in E-L formalism and identify corresponding components.
#2. Conclude as to existence of zero dynamics (free modes). In the case of positive answer, identify controlled state subvector $\mathbf{x}_C$ and remainder free state subvector $\mathbf{x}_F$; deduce their sizes.
#3. Verify whether relative degrees of controlled variables $\mathbf{x}_C$ are equal to size of control input vector $\mathbf{u}$. If so, then passivity-based control is achievable.
#4. Choose desired reference values of controlled variables $\mathbf{x}_{dC}$.
#5. Choose values of damping injection matrices $\mathbf{Ki}_C$ and $\mathbf{Ki}_F$ in relation to maximally reachable dynamics and other operating constraints.
#6. Compute control input $\mathbf{u}_C$ by using Eq. (12.52) and reinject this value in Eq. (12.53) describing dynamics of "desired" free states $\mathbf{x}_{dF}$.
#7. Perform small-signal stability analysis – around equilibrium operating point corresponding to reference $\mathbf{x}_{dC}$ – of resulting closed-loop system. In this way, one can be sure that convergence of free variables $\mathbf{x}_F$ to their "desired" values $\mathbf{x}_{dF}$ is guaranteed.

(continued)

#8. Simulate closed-loop system numerically, eventually reiterating choice of damping injection matrices to improve dynamic performance.
#9. Analyze robustness of obtained control to parameter variations. Design parameter estimators for most variable parameters – e.g., by using relation (12.56) – to be embedded in the closed-loop system.

Like every DC-DC converter having boost capabilities, the considered converter exhibits nonminimum-phase behavior between control input u and output voltage v_C. This is the reason an indirect control is aimed at, namely the regulation of the inductor current i_L to the value i_L^* corresponding to the imposed output voltage v_C^*. The relation between these reference values characterizes an equilibrium point and is obtained by using model (12.57). This is a switched model, which in this case is the same as the averaged one, except for the nature of the control input and state variables. Thus, the averaged model results by substituting into (12.57) the switching function u by its averaged value, α:

$$\begin{cases} L\dot{i_L} = -(1-\alpha)v_C + E \\ C\dot{v_C} = (1-\alpha)i_L - v_C/R. \end{cases} \tag{12.59}$$

Further, equilibrium is characterized by zeroing state derivatives. By zeroing the derivative of i_L, in the first equation of (12.59), the equilibrium value α_e can be deduced as $\alpha_e = 1 - E/v_C^*$. This latter relation is then substituted into the second equation of (12.59), provided that the derivative of v_C also is zeroed. As has been shown in this chapter, in the example illustrating the design of a stabilizing control law for a boost DC-DC converter (see Sect. 12.2.3), the following relation connecting the two reference values results:

$$i_L^* = \frac{v_C^{*2}}{ER}. \tag{12.60}$$

Therefore, the controlled variable is $\mathbf{x}_C \equiv i_L$ and $\mathbf{x}_F \equiv v_C$ is the free variable of unit relative degree, equal to the control vector size. Passivity-based control is therefore feasible. The reference (desired) values are $\mathbf{x}_{dC} \equiv i_L^*$ and $\mathbf{x}_{dF} \equiv v_{dC}$, respectively. The error variables are $\widetilde{\mathbf{x}_C} = i_L - i_L^*$ and $\widetilde{\mathbf{x}_F} = v_C - v_{dC}$. A difference is noted between output voltage reference v_C^*, which is constant in the case of a regulation task, and the "desired" output voltage value v_{dC}, which exhibits dynamics depending on the control input. Finally, note that if v_C^* is constant, then $\mathbf{x}_{dC} \equiv i_L^*$ is constant, therefore $\dot{\mathbf{x}_{dC}} \equiv \dot{i_L^*} = 0$.

Having identified all these formal elements and supposing that damping injection matrices $\mathbf{Ki}_C$ and $\mathbf{Ki}_F$ have also been chosen, one can now proceed with particularizing relations that allow passivity-based control input computation. Thus, Eq. (12.51) become in this case:

$$\begin{cases} \mathbf{G}_C(u_C, \mathbf{E}) - \{\mathbf{F}_C(u_C) \cdot i_L^* + \mathbf{K}_C(u_C) \cdot i_L^* + \mathbf{F}_{CF}(u_C) \cdot v_{dC}\} + \mathbf{Ki}_C \cdot (i_L - i_L^*) = 0 \\ \mathbf{G}_F(u_C, \mathbf{E}) - \{\mathbf{H}_F \cdot \dot{v}_{dC} + \mathbf{F}_F(u_C) \cdot v_{dC} + \mathbf{K}_F(u_C) \cdot v_{dC} + \mathbf{F}_{FC}(u_C) \cdot i_L^*\} \\ \qquad + \mathbf{Ki}_F \cdot (v_C - v_{dC}) = 0, \end{cases} \tag{12.61}$$

where

$$\mathbf{G}_C(u_C, E) = E, \quad \mathbf{F}_C(u_C) = 0, \quad \mathbf{K}_C(u_C) = 0, \quad \mathbf{F}_{CF}(u_C) = 1 - u_C, \tag{12.62}$$

$$\begin{cases} \mathbf{G}_F(u_C, E) = 0, \quad \mathbf{H}_F = C, \quad \mathbf{F}_F(u_C) = 0 \\ \mathbf{K}_F(u_C) = 1/R, \quad \mathbf{F}_{FC}(u_C) = -(1 - u_C). \end{cases} \tag{12.63}$$

Note that matrices $\mathbf{Ki}_C$ and $\mathbf{Ki}_F$ are scalars in this case. By replacing the elements given in (12.62) in the first relation of (12.61), one obtains

$$E - (1 - u_C)v_{dC} + Ki_C(i_L - i_L^*) = 0,$$

which further allows the passivity-based control input being computed as

$$u_C \equiv \alpha = 1 - \frac{E + Ki_C(i_L - i_L^*)}{v_{dC}}. \tag{12.64}$$

Next, by replacing the elements given in (12.63) in the second relation of (12.61), one obtains

$$-C\dot{v}_{dC} - 1/R \cdot v_{dC} + (1 - u_C)i_L^* + Ki_F(v_C - v_{dC}) = 0, \tag{12.65}$$

which allows computation of the dynamics of the desired free state value v_{dC} being computed based on the computed control input u_C. The schematics of the closed-loop system are depicted in Fig. 12.9 that particularizes Fig. 12.8 to the case of the boost DC-DC converter.

12.4.5.2 Bounds of Damping Injection Coefficients

As regards the choice of damping injection values, one can start from switching frequency f and consider that it gives a lower bound of the time constant for the controlled variable, i.e., of the inductor current i_L; therefore

$$\tau_C \geq 1/f, \tag{12.66}$$

where frequency f is measured in rad/s.

As regards the dynamic of the free variable v_C, one can make some general remarks that hold in all cases when zero dynamics exist. The "free" dynamics are composed of two components: the first one can be tuned by choosing Ki_F, whereas

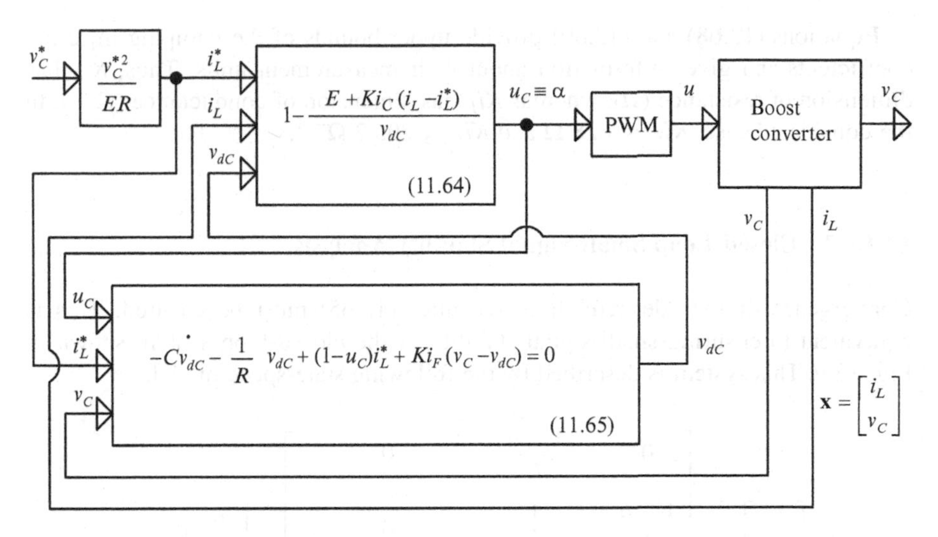

Fig. 12.9 Block diagram of passive-controlled boost DC-DC converter

the second one represents in fact the dynamic of the controlled variable, which is at its turn tuned by means of Ki_C. In general, the adjustable dynamic needs to be faster than that of the controlled variable. This means that, after some initial fast variations, the free variable will follow the controlled variable's variation. In conclusion, the convergence of the free variable v_C cannot be rendered faster than the controlled one.

Let τ_F be the time constant corresponding to the adjustable part of the free variable's dynamic. Following the above remarks, one usually imposes that τ_F be smaller than τ_C; for example, one-fifth the size consequently gives

$$\tau_F \geq 1/(5f). \tag{12.67}$$

Now, the closed-loop time constants τ_C and τ_F must be related to circuit element values. Note first that the closed-loop time constant of the controlled variable can be expressed as $\tau_C = L/Ki_C$, according to relation (12.64) being replaced in the first relation of (12.59). Taking account of (12.66) results in

$$Ki_C \leq L \cdot f. \tag{12.68}$$

As regards the time constant τ_F characterizing the adjustable part of the free variable's dynamic, its expression can be deduced from relation (12.65) if one supposes that the free and the dynamics of the controlled variables are decoupled, that is, if u_C would be constant. In this way, one obtains

$$\tau_F = \frac{1}{1/(RC) + Ki_F/C},$$

which, combined with (12.67), gives

$$Ki_F \leq 5Cf - 1/R. \tag{12.69}$$

Equations (12.68) and (12.69) provide upper bounds of the damping injection coefficients and gives information about their measurement units. Thus, Ki_C has dimension of resistance (Ω), whereas Ki_F has dimension of conductance (Ω^{-1}). In the considered case, $Ki_C \leq 628\ \Omega$ and $Ki_F \leq 295.2\ \Omega^{-1}$.

12.4.5.3 Closed-Loop Small-Signal Stability Analysis

Convergence of the "desired" free dynamics (12.65) must be ensured. This is equivalent to ensuring small-signal stability of the closed-loop system shown in Fig. 12.9. This system is described by the following state-space model:

$$\frac{d}{dt}\begin{bmatrix} i_L \\ v_C \\ v_{dC} \end{bmatrix} = \begin{bmatrix} 0 & -\frac{1-\alpha}{L} & 0 \\ \frac{1-\alpha}{C} & -\frac{1}{RC} & 0 \\ 0 & \frac{Ki_F}{C} & -\frac{1}{C}\left(\frac{1}{R}+Ki_F\right) \end{bmatrix} \cdot \begin{bmatrix} i_L \\ v_C \\ v_{dC} \end{bmatrix} + \begin{bmatrix} 0 \\ 0 \\ \frac{1-\alpha}{C} \end{bmatrix} \cdot i_L^* + \begin{bmatrix} \frac{E}{L} \\ 0 \\ 0 \end{bmatrix}; \tag{12.70}$$

so it involves a bilinear system as the state matrix depends on the control input α, which in turn depends on the system's state according to (12.64). The small-signal model of system (12.70) is deduced by linearization around the equilibrium point given by the equilibrium value of system's input i_{Le}^* Letting $\mathbf{x}_{SS} = [i_L \quad v_C \quad v_{dC}]^T$ be the state of system (12.70) gives $\mathbf{x}_{SSe} = [i_{Le} \quad v_{Ce} \quad v_{dCe}]^T$ as its equilibrium value and $\widetilde{\mathbf{x}_{SS}} = \mathbf{x}_{SS} - \mathbf{x}_{SSe}$ as state vector variation around the steady state $\mathbf{x}_{SSe}$. Letting $\widetilde{i_L^*} = i_L^* - i_{Le}^*$ denote the input's variation around its steady-state value i_{Le}^*, and α_e the associated control input steady-state value. The small-signal model describes the dynamics of variations around the quiescent operating point

$$\dot{\widetilde{\mathbf{x}_{SS}}} = \mathbf{A}_{SS} \cdot \widetilde{\mathbf{x}_{SS}} + \mathbf{B}_{SS} \cdot \widetilde{i_L^*},$$

where the involved matrices are

$$\mathbf{A}_{SS} = \left.\frac{\partial(\dot{\mathbf{x}}_{SS})}{\partial \mathbf{x}_{SS}}\right|_{\substack{\alpha=\alpha_e \\ \mathbf{x}_{SS}=\mathbf{x}_{SSe} \\ i_L^*=i_{Le}^*}}, \quad \mathbf{B}_{SS} = \left.\frac{\partial(\dot{\mathbf{x}}_{SS})}{\partial i_L^*}\right|_{\substack{\alpha=\alpha_e \\ \mathbf{x}_{SS}=\mathbf{x}_{SSe} \\ i_L^*=i_{Le}^*}}.$$

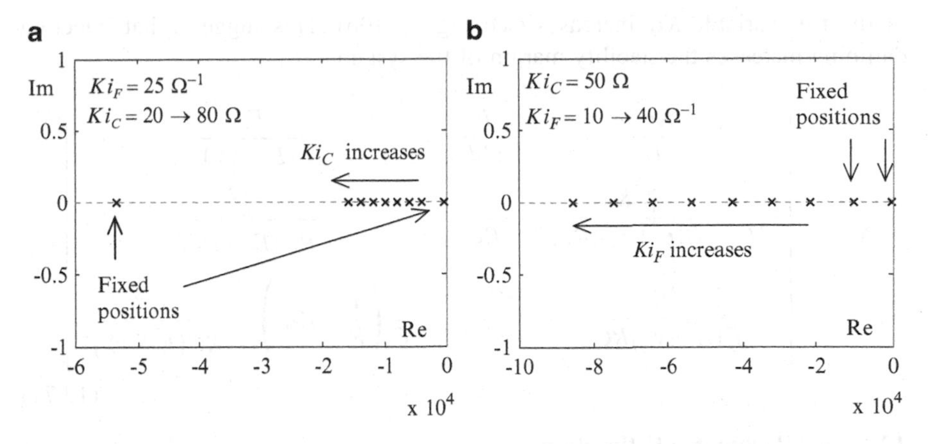

Fig. 12.10 Closed-loop small-signal stability analysis of passivity-based-controlled boost converter ($E = 15$ V, $v_C^* = 45$ V): (**a**) migration of poles in the complex plane as damping injection of current (controlled variable) increases; (**b**) migration of poles in the complex plane as damping injection of voltage (free variable) increases

From the stability viewpoint, only the state matrix $\mathbf{A}_{SS}$ is of interest. After some algebra, one obtains its analytical expression

$$\mathbf{A}_{SS} = \left. \begin{bmatrix} -\dfrac{Ki_C(E+v_C)}{L(E+v_{dC})} & -\dfrac{(1-\alpha)}{L} & \dfrac{(E+v_C)\left(E+Ki_C\left(i_L-i_L^*\right)\right)}{L(E+v_{dC})^2} \\ \dfrac{1-\alpha}{C}+\dfrac{i_L Ki_C}{C(E+v_{dC})} & -\dfrac{1}{RC} & -\dfrac{i_L\left(E+Ki_C\left(i_L-i_L^*\right)\right)}{C(E+v_{dC})^2} \\ \dfrac{i_L^* Ki_C}{C(E+v_{dC})} & \dfrac{Ki_F}{C} & -\dfrac{1}{C}\left(\dfrac{1}{R}+Ki_F\right)-\dfrac{i_L^*\left(E+Ki_C\left(i_L-i_L^*\right)\right)}{C(E+v_{dC})^2} \end{bmatrix} \right|_{\substack{\alpha=\alpha_e \\ \mathbf{x}_{SS}=\mathbf{x}_{SSe} \\ i_L^*=i_{Le}^*}} .$$

The eigenvalues of matrix $\mathbf{A}_{SS}$ are analyzed numerically for different values of damping injection coefficients Ki_C and Ki_F, assuming constant and known values of source voltage E and load R. Noting that output voltage reference v_C^* – assumed constant for sake of simplicity – is related to i_{Le}^* through relation (12.60), it is easy to verify that

$$v_{Ce} = v_{dCe} = v_C^*, \quad i_{Le} = i_{Le}^* = v_C^{*2}/(ER), \quad \alpha_e = 1 - E/v_C^*,$$

which allows putting matrix $\mathbf{A}_{SS}$ in the form (12.71).

Numerical analysis provides stable eigenvalue sets for large variation domains of Ki_C and Ki_F when the output voltage setpoint is $v_C^* = 45$ V; therefore, the free dynamics are stable. Figure 12.10a, b show the excursion of the three eigenvalues as Ki_C and, respectively, Ki_F varies. In each case all three eigenvalues are real and a single one that changes its position. One can note that the mobile eigenvalue gets farther from the imaginary axis in both cases, that is, as the damping injection of the controlled variable Ki_C increases (see Fig. 12.10a) and also as the damping injection of the free variable Ki_F increases (see Fig. 12.10b). This suggests that injecting damping increases the stability margin of the system.

$$\mathbf{A}_{SS} = \begin{bmatrix} -\frac{Ki_C}{L} & -\frac{E}{v_C^* L} & \frac{E}{L(E+v_C^*)} \\ \frac{E}{v_C^* C} + \frac{v_C^{*2} Ki_C}{E(E+v_C^*)RC} & -\frac{1}{RC} & -\frac{v_C^{*2}}{RC(E+v_C^*)^2} \\ \frac{v_C^{*2} Ki_C}{E(E+v_C^*)RC} & \frac{Ki_F}{C} & -\frac{1}{C}\left(\frac{1}{R} + Ki_F\right) - \frac{v_C^{*2}}{RC(E+v_C^*)^2} \end{bmatrix}. \tag{12.71}$$

12.4.5.4 Parameter Estimation

The final step of the control design procedure concerns its sensitivity analysis to parameter variations. Usually, the most variable parameters are E and R, and control law performance depends on how precisely these parameters are known. A parameter estimator can be designed following guidelines given in Sect. 12.4.3 and relation (12.56). Note first that estimation of parameter $Y = 1/R$ (load admittance instead of load resistance) is easier for reasons of linearity. Therefore, the parameter vector is $\mathbf{P} = [E \quad Y]^T$.

After having adopted notations $\Delta E = E - E_{est}$ and $\Delta Y = Y - Y_{est}$ for parameter estimation errors, the next step is to determine variations of matrices involved in the Euler–Lagrange representation (12.58) due to ΔE and ΔY. According to relation (12.58), one obtains

$$\Delta\mathbf{H} = 0, \quad \Delta\mathbf{F} = 0, \quad \Delta\mathbf{K} = \begin{bmatrix} 0 & 0 \\ 0 & \Delta Y \end{bmatrix}, \quad \Delta\mathbf{G} = \begin{bmatrix} \Delta E \\ 0 \end{bmatrix},$$

which allows Eq. (12.55) being particularized as follows:

$$\Delta\mathbf{G} - \Delta\mathbf{K} \cdot \begin{bmatrix} i_L^* \\ v_{dC} \end{bmatrix} = \begin{bmatrix} \Delta E \\ -\Delta Y \cdot v_{dC} \end{bmatrix} = \mathbf{Y}(\dot{\mathbf{x}}_d, \mathbf{x}_d, \mathbf{u}_C) \cdot \begin{bmatrix} \Delta E \\ \Delta Y \end{bmatrix},$$

taking into account that $\mathbf{x}_d = [i_L^* \quad v_{dC}]^T$ and $\Delta\mathbf{P} = [\Delta E \quad \Delta Y]^T$. From the last relation one deduces that

$$\mathbf{Y}(\dot{\mathbf{x}}_d, \mathbf{x}_d, \mathbf{u}_C) = \begin{bmatrix} 1 & 0 \\ 0 & -v_{dC} \end{bmatrix}.$$

Finally, Eq. (12.56) giving the parameter estimation dynamics appears as

$$\frac{d}{dt}\begin{bmatrix} E_{est} \\ Y_{est} \end{bmatrix} = \begin{bmatrix} \gamma_1 & 0 \\ 0 & \gamma_2 \end{bmatrix} \cdot \begin{bmatrix} 1 & 0 \\ 0 & -v_{dC} \end{bmatrix} \cdot \begin{bmatrix} i_L - i_L^* \\ v_C - v_C^* \end{bmatrix}, \tag{12.72}$$

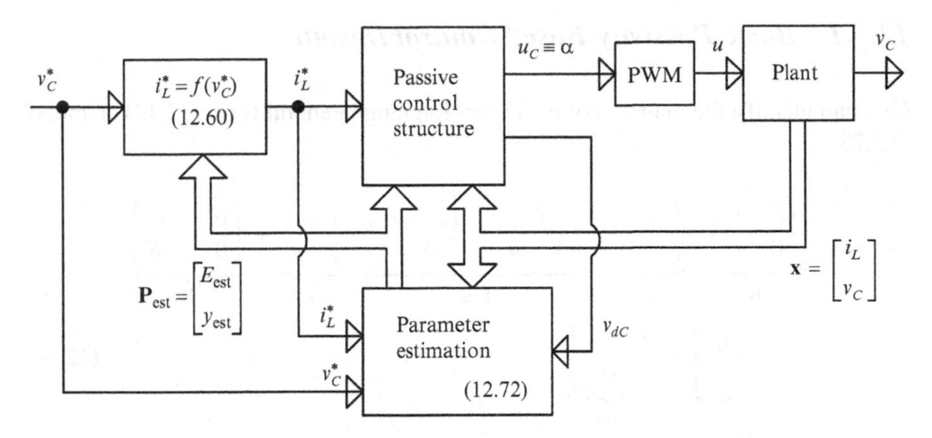

Fig. 12.11 Adaptive passivity-based control structure of boost converter based upon current tracking and output voltage regulation

where γ_1 and γ_2 are positive scalars which define the convergence speed of the estimation error to zero. Numerical simulation can be used for fine tuning these two coefficients for best performance, as well as for verifying that the adaptive control structure does not render the tracking system – i.e., the current loop – unstable. To conclude, the passivity-based control structure based upon parameter estimation is shown in Fig. 12.11.

12.5 Case Study: Passivity-Based Control of a Buck-Boost DC-DC Converter

This case study concerns the passivity-based control design for a buck-boost converter supplying a resistive load R, whose electrical circuit was presented in Fig. 3.26 in the text of Problem 3.5 in Chap. 3. The control goal is to regulate the output voltage at a setpoint denoted by v_C^*. The design will follow the steps of Algorithm 12.3. The switched model of this converter operating in continuous-conduction mode (ccm) – previously presented in the case study from Sect. 4.6 in Chap. 4 and repeated in relation (12.46) from this chapter – is given below (inductor resistance r being neglected):

$$\begin{cases} L\dot{i_L} = Eu + v_C(1-u) \\ C\dot{v_C} = -i_L(1-u) - v_C/R, \end{cases} \tag{12.73}$$

where the inductor current i_L and the capacitor voltage v_C are the two states and E is the source voltage representing the external energy effort applied to the system. Note that voltage values here are negative. The parameters are $L = 2.5$ mH, $C = 220$ μF, switching frequency $f = 20$ kHz, source voltage E varying between 5 and 45 V, rated $R = 10\ \Omega$ and voltage setpoint $v_C^* = -20$ V.

12.5.1 Basic Passivity-Based Control Design

One can identify the matrices of the Euler–Lagrange general form (12.42) on model (12.73):

$$\underbrace{\begin{bmatrix} L & 0 \\ 0 & C \end{bmatrix}}_{\mathbf{H}} \cdot \underbrace{\begin{bmatrix} \dot{i_L} \\ \dot{v_C} \end{bmatrix}}_{\dot{\mathbf{x}}} + \underbrace{\begin{bmatrix} 0 & -(1-u) \\ 1-u & 0 \end{bmatrix}}_{\mathbf{F(u)}} \cdot \underbrace{\begin{bmatrix} i_L \\ v_C \end{bmatrix}}_{\mathbf{x}} + \underbrace{\begin{bmatrix} 0 & 0 \\ 0 & 1/R \end{bmatrix}}_{\mathbf{K}} \cdot \underbrace{\begin{bmatrix} i_L \\ v_C \end{bmatrix}}_{\mathbf{x}} = \underbrace{\begin{bmatrix} E \cdot u \\ 0 \end{bmatrix}}_{\mathbf{G(u,E)}}. \tag{12.74}$$

Being a DC-DC converter with boost capabilities, the converter considered exhibits nonminimum-phase behavior between control input u and output voltage v_C. Similarly, as in the case of boost converters, an indirect control is recommended. Thus, inductor current i_L is to be regulated at the value i_L^* corresponding to the imposed output voltage v_C^*. The relation between these reference values is determined based upon the averaged model. Since in this case the switched and averaged models have the same form, it is sufficient to substitute into (12.74) the switching function u by its averaged value, α:

$$\begin{cases} L\dot{i_L} = (1-\alpha)v_C + E\alpha \\ C\dot{v_C} = -(1-\alpha)i_L - v_C/R. \end{cases} \tag{12.75}$$

Equilibrium values are further obtained by zeroing state derivatives in (12.75). Thus, after simple algebra, the following equilibrium values result:

$$\alpha_e = \frac{v_C^*}{v_C^* - E}, \quad i_L^* = \frac{v_C^*(v_C^* - E)}{E \cdot R}. \tag{12.76}$$

Hence, the controlled variable is $\mathbf{x}_C \equiv i_L$ and $\mathbf{x}_F \equiv v_C$ is the free variable of unit relative degree, equal to the control vector size; this renders the passivity-based control possible. The reference (desired) values are $\mathbf{x}_{dC} \equiv i_L^*$ and $\mathbf{x}_{dF} \equiv v_{dC}$, respectively, with $\widetilde{\mathbf{x}_C} = i_L - i_L^*$ and $\widetilde{\mathbf{x}_F} = v_C - v_{dC}$ being the error variables. Variable v_{dC} exhibits dynamics depending on the control input. Note that for a regulation task, v_C^* is constant, then $\mathbf{x}_{dC} \equiv i_L^*$ is constant, therefore $\dot{\mathbf{x}_{dC}} \equiv \dot{i_L^*} = 0$.

Once the formal elements have been identified and supposing that damping injection matrices $\mathbf{Ki}_C$ and $\mathbf{Ki}_F$ have also been chosen, one can particularize relations (12.51) that allow the passivity-based control input computation. In this case,

$$\begin{cases} \mathbf{G}_C(u_C,\mathbf{E}) - \left\{\mathbf{F}_C(u_C)\cdot \dot{i}_L^* + \mathbf{K}_C(u_C)\cdot i_L^* + \mathbf{F}_{CF}(u_C)\cdot v_{dC}\right\} + \mathbf{Ki}_C\cdot\left(i_L - i_L^*\right) = 0 \\ \mathbf{G}_F(u_C,\mathbf{E}) - \left\{\mathbf{H}_F\cdot \dot{v}_{dC} + \mathbf{F}_F(u_C)\cdot v_{dC} + \mathbf{K}_F(u_C)\cdot v_{dC} + \mathbf{F}_{FC}(u_C)\cdot i_L^*\right\} \\ \qquad\qquad + \mathbf{Ki}_F\cdot\left(v_C - v_{dC}\right) = 0, \end{cases} \tag{12.77}$$

where

$$\mathbf{G}_C(u_C,E) = Eu_C, \quad \mathbf{F}_C(u_C) = 0, \quad \mathbf{K}_C(u_C) = 0, \quad \mathbf{F}_{CF}(u_C) = -(1-u_C), \tag{12.78}$$

$$\begin{cases} \mathbf{G}_F(u_C,E) = 0, \quad \mathbf{H}_F = C, \quad \mathbf{F}_F(u_C) = 0 \\ \mathbf{K}_F(u_C) = 1/R, \quad \mathbf{F}_{FC}(u_C) = 1 - u_C. \end{cases} \tag{12.79}$$

Note that matrices $\mathbf{Ki}_C$ and $\mathbf{Ki}_F$ are scalars in this case. By replacing the elements given in (12.78) in the first relation of (12.77), one obtains

$$Eu_C + (1-u_C)v_{dC} + Ki_C\left(i_L - i_L^*\right) = 0,$$

which leads to the expression of the passivity-based control input as

$$u_C \equiv \alpha = \frac{v_{dC} + Ki_C\left(i_L - i_L^*\right)}{v_{dC} - E}. \tag{12.80}$$

Finally, by replacing the elements given in (12.79) in the second relation of (12.77), one obtains

$$-C\dot{v}_{dC} - 1/R\cdot v_{dC} - (1-u_C)i_L^* + Ki_F(v_C - v_{dC}) = 0, \tag{12.81}$$

which describes the dynamics of the desired free state value v_{dC} as depending on the computed control input u_C. Figure 12.12 presents the block diagram of the closed-loop system as a particularization of the general diagram given in Fig. 12.8 to the case of the buck-boost DC-DC converter.

12.5.2 Damping Injection Tuning

As explained in Sect. 12.4.1 and illustrated in Sect. 12.4.5 of this chapter, damping injection values are chosen in relation to switching frequency *f*, supposing that the latter gives a lower bound of the time constant for the controlled variable, i.e., of inductor current i_L. The result is

$$\tau_C \geq 1/f, \tag{12.82}$$

where frequency *f* is measured in rad/s. Regarding convergence of the free variable v_C, what is true for the case of the boost converter is true for the buck-boost converter as well. Thus, the "free" dynamics comprise two components, one of

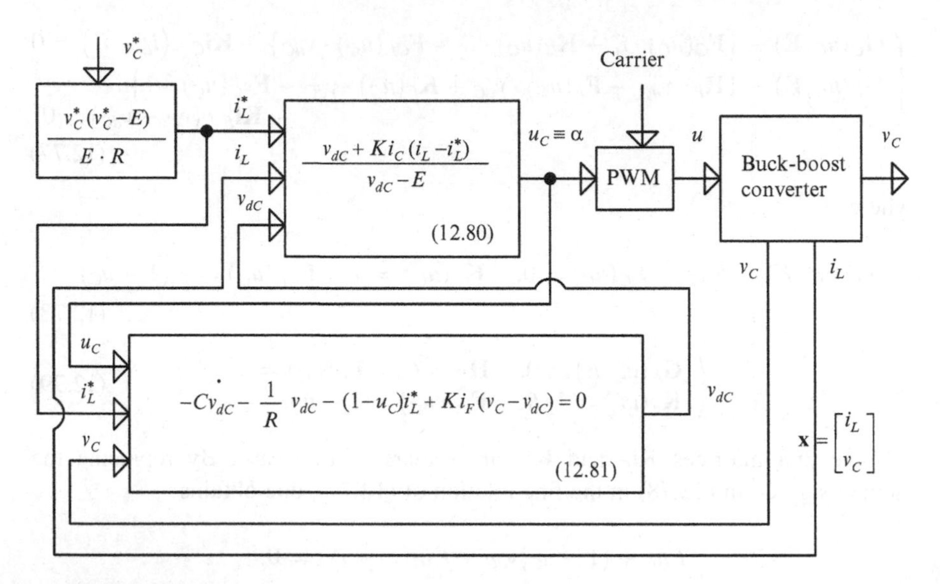

Fig. 12.12 Schematics of passive-controlled buck-boost DC-DC converter

which is practically speaking the dynamic of the controlled variable, and the other, which can be tuned by means of Ki_F. Imposing that this latter be five times faster than the controlled variable's dynamic, one obtains

$$\tau_F \geq 1/(5f). \tag{12.83}$$

On the other hand, the closed-loop time constant of the controlled variable can be expressed as $\tau_C = L/Ki_C$, according to expression (12.80) of the control input being replaced in the first relation of converter's dynamics (12.75). Taking into account (12.82) gives

$$Ki_C \leq Lf. \tag{12.84}$$

As for the time constant characterizing the adjustable part of the free variable's dynamic, its expression can be deduced from relation (12.81) of the free state desired dynamic by supposing u_C to be constant, namely,

$$\tau_F = \frac{1}{1/(RC) + Ki_F/C},$$

which, combined with (12.83), gives further

$$Ki_F \leq 5Cf - 1/R. \tag{12.85}$$

Relations (12.84) and (12.85) provide upper bounds on the damping injection coefficients. One notes that the same analytical relations as seen in the case of the boost converter have resulted (see Eqs. 12.68 and 12.69 respectively). In the considered buck-boost case $Ki_C \leq 314\ \Omega$ and $Ki_F \leq 138.13\ \Omega^{-1}$.

12.5.3 Study of Closed-Loop Small-Signal Stability

In order to prove the convergence of the "desired" free dynamics (12.81), the small-signal stability of the closed-loop system presented in Fig. 12.12 must be performed. This involves a bilinear system with the following state-space description:

$$\frac{d}{dt}\begin{bmatrix} i_L \\ v_C \\ v_{dC} \end{bmatrix} = \begin{bmatrix} 0 & \frac{1-\alpha}{L} & 0 \\ -\frac{1-\alpha}{C} & -\frac{1}{RC} & 0 \\ 0 & \frac{Ki_F}{C} & -\frac{1}{C}\left(\frac{1}{R}+Ki_F\right) \end{bmatrix} \cdot \begin{bmatrix} i_L \\ v_C \\ v_{dC} \end{bmatrix} + \begin{bmatrix} 0 \\ 0 \\ -\frac{1-\alpha}{C} \end{bmatrix} \cdot i_L^* + \begin{bmatrix} \frac{E}{L}\alpha \\ 0 \\ 0 \end{bmatrix}, \tag{12.86}$$

where α is the passivity-based control input computed according to (12.80). After introducing expression (12.80) of α into Eq. (12.86), one deduces the small-signal model of the closed-loop system by linearization around the equilibrium point corresponding to the equilibrium value of system input i_{Le}^*. It is the state matrix of the linearized system, $\mathbf{A}_{SS}$, that is interesting from the point of view of stability. One follows steps of the small-signal stability analysis detailed in Sect. 12.4.5 – concerning an application example to the case of a boost converter – and adopts analogous notations. In the case of a buck-boost converter, after some obvious manipulations, it turns out that this matrix has the expression

$$\mathbf{A}_{SS} = \left.\begin{bmatrix} \frac{Ki_C(E-v_C)}{L(v_{dC}-E)} & \frac{1-\alpha}{L} & \frac{(E-v_C)(i_L^*Ki_C-E)}{L(v_{dC}-E)^2} \\ -\frac{(1-\alpha)}{C}+\frac{i_LKi_C}{C(v_{dC}-E)} & -\frac{1}{RC} & -\frac{i_L(E+Ki_C(i_L-i_L^*))}{C(v_{dC}-E)^2} \\ \frac{i_L^*Ki_C}{C(v_{dC}-E)} & \frac{Ki_F}{C} & -\frac{1}{C}\left(\frac{1}{R}+Ki_F\right)-\frac{i_L^*(E+Ki_C(i_L-i_L^*))}{C(v_{dC}-E)^2} \end{bmatrix}\right|_{\substack{\alpha=\alpha_e \\ \mathbf{x}_{SS}=\mathbf{x}_{SSe} \\ i_L^*=i_{Le}^*}}$$

The eigenvalues of $\mathbf{A}_{SS}$ are computed numerically for different values of the damping injection coefficients Ki_C and Ki_F – placed within the limits previously determined, that is, $Ki_C \leq 314\ \Omega$ and $Ki_F \leq 138.13\ \Omega^{-1}$ – supposing in the first place that source voltage E and load R are constant. To ease computation, the relations (see also Eq. 12.76)

$$v_{Ce} = v_{dCe} = v_C^*, \quad i_{Le} = i_{Le}^* = \frac{v_C^*(v_C^* - E)}{E \cdot R}, \quad \alpha_e = \frac{v_C^*}{v_C^* - E},$$

are replaced in the expression of $\mathbf{A}_{SS}$, to allow rewriting it as

$$\mathbf{A}_{SS} = \begin{bmatrix} -\dfrac{Ki_C}{L} & -\dfrac{E}{L(v_C^* - E)} & -\dfrac{v_C^* Ki_C}{ERL} + \dfrac{E}{L(v_C^* - E)} \\ \dfrac{v_C^* Ki_C}{ERC} + \dfrac{E}{C(v_C^* - E)} & -\dfrac{1}{RC} & -\dfrac{v_C^*}{RC(v_C^* - E)} \\ \dfrac{v_C^* Ki_C}{ERC} & \dfrac{Ki_F}{C} & -\dfrac{1}{C}\left(\dfrac{1}{R} + Ki_F\right) - \dfrac{v_C^*}{RC(v_C^* - E)} \end{bmatrix}. \tag{12.87}$$

Numerical analysis was first performed for a boost operation, namely for the control goal $v_C^* = -20$ V and source voltage $E = 15$ V. This provided stable eigenvalue sets for large variation domains of Ki_C and Ki_F, allowing the conclusion that the free dynamics are stable. Thus, Fig. 12.13a shows the excursion of the three eigenvalues as Ki_C varies, with fixed Ki_F. This involves two complex-conjugated eigenvalues that move, the third being real and fixed. The figure shows that there is a value of Ki_C – in this case, this is around $Ki_C = 34\ \Omega$ – for which the two complex eigenvalues become equal and real. Increasing Ki_C further makes only one move, namely in the sense of increasing its absolute value.

Figure 12.13a contains information helpful for selecting a suitable value for the damping injection coefficient Ki_C. One can see none of the represented cases induce overshoot, practically speaking, because the damping coefficient is larger than 0.8. A suitable value can, for example, be one that corresponds to the two complex eigenvalues becoming real and quasi equal (i.e., $Ki_C = 34\ \Omega$ in this case).

Figure 12.13b shows the excursion of the three eigenvalues as Ki_F varies, Ki_C being fixed. Here also it is about two complex-conjugated eigenvalues and a third real one, all three moving. One notes there is a value of Ki_F – in this case, around $Ki_F = 13\ \Omega^{-1}$ – for which the two complex eigenvalues practically no longer move; only the third keeps moving in the sense of increasing its absolute value. In all the represented cases the pair of complex eigenvalues do not induce overshoot in the dynamic response, since its damping coefficient is larger than 0.9.

The above remarks suggest that both damping injections – in the controlled and in the free dynamic – improve the stability margin of the system when in boost operation.

Numerical analysis has also been performed for buck operation, namely for control goal $v_C^* = -10$ V and source voltage $E = 15$ V. The free dynamic is

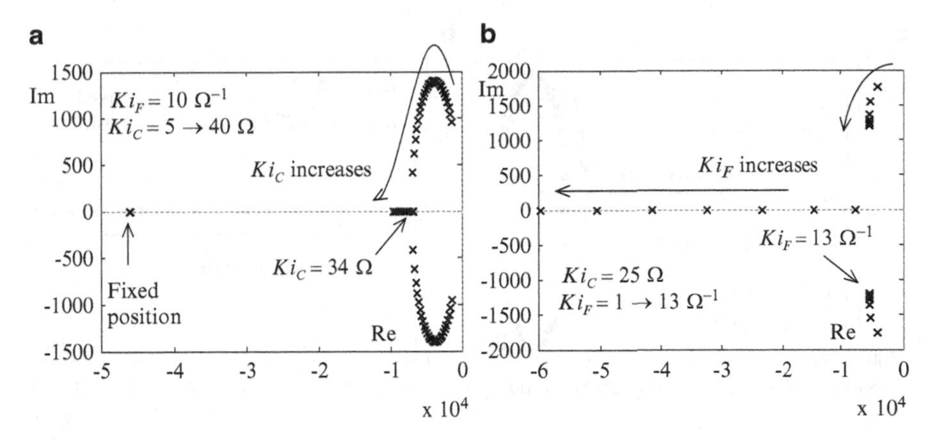

Fig. 12.13 Closed-loop small-signal stability analysis of a passivity-based-controlled buck-boost converter operating in boost mode ($E = 15$ V, $v_C^* = -20$ V): (**a**) migration of poles in the complex plane as damping injection of current (controlled variable) increases; (**b**) migration of poles as damping injection of voltage (free variable) increases

convergent also in this case for Ki_C and Ki_F varying within the limits previously determined, that is, $Ki_C \leq 314\ \Omega$ and $Ki_F \leq 138.13\ \Omega^{-1}$, as Figs. 12.14a, b show.

Figure 12.14a shows the excursion of the three eigenvalues as Ki_C varies, Ki_F being fixed. The three eigenvalues follow a similar motion scenario to those in the boost operation case, i.e., there are two moving complex-conjugated and one real, fixed. In this case the value of Ki_C for which the two complex eigenvalues become equal and real is around $Ki_C = 5.8\ \Omega$. Increasing Ki_C further makes one continue moving in the sense of increasing its absolute value, whereas the other moves in the opposite direction until reaching a fixed position.

Figure 12.14b shows the excursion of the three eigenvalues as Ki_F varies, Ki_C being fixed. Now all three eigenvalues are real, one of which moves while the other two keep approximately the same position. The moving one is getting farther from the imaginary axis. This suggests that the stability margin of the system increases as damping is injected in the dynamic of the free variable.

The analysis of the closed-loop linearized system's eigenvalues can further be resumed in the case of parameter variations, more precisely, by taking fixed values for Ki_C and Ki_F and allowing E and R to vary. The results of such analysis for $Ki_C = 25\ \Omega$, $Ki_F = 10\ \Omega^{-1}$ and $v_C^* = -15$ V are shown in Figs. 12.15a, b.

Figure 12.15a shows the influence of varying the source voltage E on the closed-loop small-signal stability, with the nature of two of eigenvalues changing from complex-conjugated to real as the operation changes from boost to buck (i.e., as E increases from 5 to 45 V). Figure 12.15b shows that the load variation has practically no influence on the closed-loop small-signal stability, as the position of the three eigenvalues, all real, varies very little as load R varies between 10 and 1000 Ω.

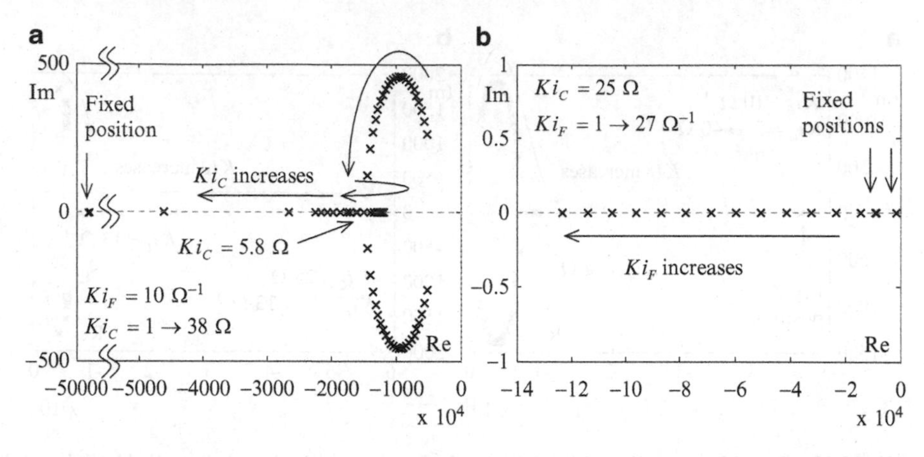

Fig. 12.14 Closed-loop small-signal stability analysis of a passivity-based-controlled buck-boost converter operating in buck mode ($E = 15$ V, $v_C^* = -10$ V): (**a**) migration of poles in the complex plane as damping injection of current (controlled variable) increases; (**b**) migration of poles in the complex plane as damping injection of voltage (free variable) increases

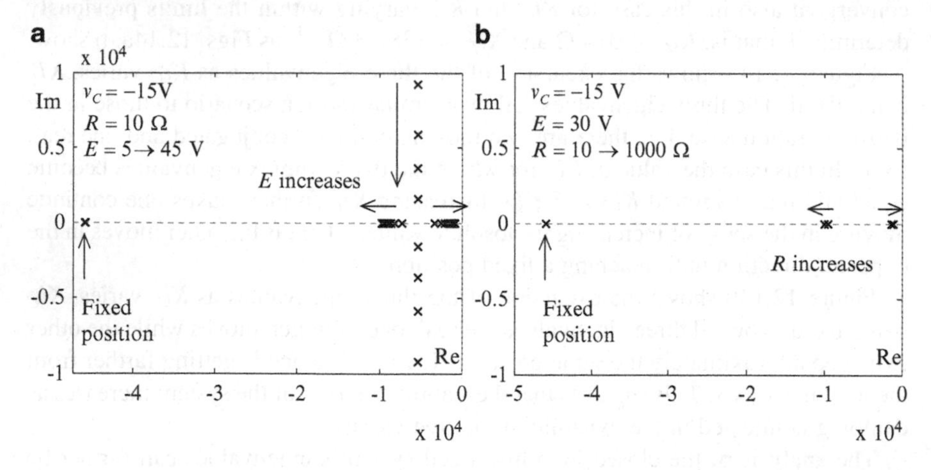

Fig. 12.15 Closed-loop small-signal stability analysis of a passivity-based-controlled buck-boost converter subject to parameter variations ($v_C^* = -15$ V): (**a**) migration of poles in the complex plane as source voltage E varies and load R is constant; (**b**) migration of poles in the complex plane as load R varies and source voltage E is constant

12.5.4 Adaptive Passivity-Based Control Design

The previous section ended with an analysis of sensitivity to parameter variations from the small-signal stability viewpoint. A parameter estimator for the most variable parameters E and R can be designed following guidelines given in

Sect. 12.4.3 and relation (12.56), where parameter $Y = 1/R$ (load admittance instead of load resistance) is estimated for reasons of linearity.

The parameter vector is denoted by $\mathbf{P} = [E \quad Y]^T$. $\Delta E = E - E_{\text{est}}$ and $\Delta Y = Y - Y_{\text{est}}$ are notations for the parameter estimation errors. The variations of matrices involved in the buck-boost converter's Euler–Lagrange representation (12.74) due to ΔE and ΔY are determined as

$$\Delta \mathbf{H} = 0, \quad \Delta \mathbf{F} = 0, \quad \Delta \mathbf{K} = \begin{bmatrix} 0 & 0 \\ 0 & \Delta Y \end{bmatrix}, \quad \Delta \mathbf{G} = \begin{bmatrix} \Delta E \cdot u_C \\ 0 \end{bmatrix}.$$

Taking into account that $\mathbf{x}_d = [i_L^* \quad v_{dC}]^T$ and $\Delta \mathbf{P} = [\Delta E \quad \Delta Y]^T$, the last expressions are substituted into relation (12.55), leading to

$$\Delta \mathbf{G} - \Delta \mathbf{K} \cdot \begin{bmatrix} i_L^* \\ v_{dC} \end{bmatrix} = \begin{bmatrix} \Delta E \cdot u_C \\ -\Delta Y \cdot v_{dC} \end{bmatrix} = \mathbf{Y}(\dot{\mathbf{x}}_d, \mathbf{x}_d, \mathbf{u}_C) \cdot \begin{bmatrix} \Delta E \\ \Delta Y \end{bmatrix},$$

and further, to identifying the expression of matrix $\mathbf{Y}$:

$$\mathbf{Y}(\dot{\mathbf{x}}_d, \mathbf{x}_d, \mathbf{u}_C) = \begin{bmatrix} u_C & 0 \\ 0 & -v_{dC} \end{bmatrix}.$$

Finally, Eq. (12.56) giving the parameter estimation dynamics appears as

$$\frac{d}{dt}\begin{bmatrix} E_{\text{est}} \\ Y_{\text{est}} \end{bmatrix} = \begin{bmatrix} \gamma_1 & 0 \\ 0 & \gamma_2 \end{bmatrix} \cdot \begin{bmatrix} u_C & 0 \\ 0 & -v_{dC} \end{bmatrix} \cdot \begin{bmatrix} i_L - i_L^* \\ v_C - v_C^* \end{bmatrix}, \tag{12.88}$$

with γ_1 and γ_2 being positive scalars responsible for the convergence speed of the estimation error to zero. Fine tuning of these two coefficients can be done by numerical simulation. Note that here – unlike in the case of the boost converter – the parameter estimation scheme needs the value of the passivity-based control input u_C. Also, one must verify that the adaptive control structure does not render the current loop unstable. Figure 12.16 presents the obtained passivity-based control structure based upon parameter estimation.

12.5.5 Numerical Simulation Results

This section is dedicated to illustrating the dynamic performance of the considered buck-boost converter controlled with the previously designed passivity-based control law. The nonlinear averaged model has been used for this purpose. Operation around a boost-mode steady-state point has been studied, namely, for the output voltage setpoint $v_C^* = -20$ V and source voltage $E < 20$ V, in the first place constant, then variable. This type of operation is more challenging in the sense of exhibiting nonminimum-phase behavior and, more important, nonlinear phenomena that lead to larger tracking errors.

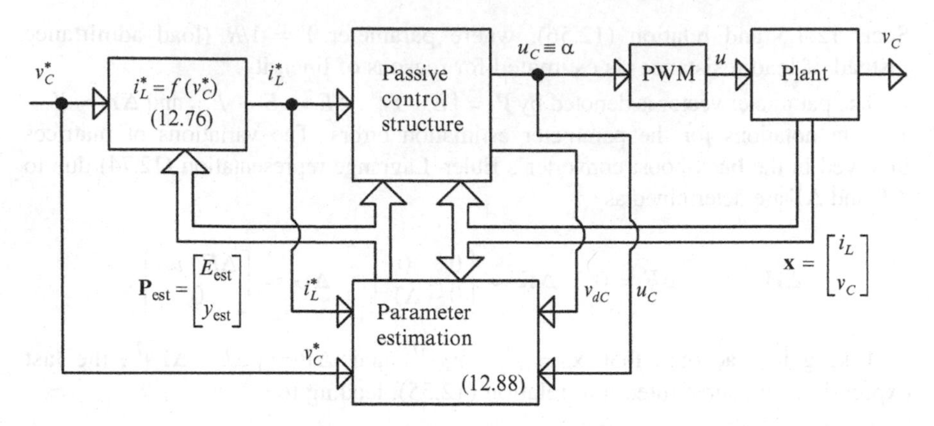

Fig. 12.16 Adaptive passivity-based control structure of a buck-boost converter based upon current tracking and output voltage regulation

Figure 12.17 illustrates the dynamic performance of the basic passivity-based control law presented in Fig. 12.12 when the parameters – load R and source voltage E – are assumed constant and known, that is, without embedding the parameter estimator in the closed-loop system. Figure 12.17a, b allow assessing the closed-loop dynamic response of the free and the controlled variable, respectively. Figure 12.17c presents the evolution of the duty ratio u_C as it reflects the control input effort.

The closed-loop time constant of the controlled variable τ_C can be evaluated from Fig. 12.17b by measuring the settling time of this variable, i_L, for a given value of the associated damping injection coefficient, Ki_C. For example, for $Ki_C = 5\ \Omega$ the settling time is around 2 ms, which gives $\tau_C \approx 0.5$ ms, which is coherent with evaluating the expression $\tau_C = L/Ki_C$. The time constant of the free variable, τ_F, can be evaluated from Fig. 12.17d by measuring the settling time of the error $v_C - v_{dC}$ for a given value of the corresponding damping injection coefficient Ki_F. For example, for $Ki_F = 2\ \Omega^{-1}$ the settling time is around 2.5 ms. Taking account that the free variable follows the controlled variable's dynamic, one obtains a settling time of $2.5 - 2 = 0.5$ ms for the "adjustable" free dynamic. Further, this means a time constant $\tau_F \approx 0.1$ ms that is coherent with evaluating the expression $\tau_F = 1/(1/(RC) + Ki_F/C)$.

Figure 12.18 contains a set of simulation results that emphasizes the influence of different choices of the damping injection coefficients Ki_C and Ki_F. Figure 12.18a is for a fixed value of Ki_F and different values of Ki_C; the response time of the output voltage decreases as Ki_C increases, but so does the nonminimum-phase effect. Figure 12.18b concerns the case of a fixed value of Ki_C and different values of Ki_F; the response time of the free variable error $v_C - v_{dC}$ decreases as Ki_F increases and so does the error amplitude.

Figure 12.19 presents the evolution of the variables of interest of the buck-boost converter controlled with the basic passivity-based control law in the presence of parameter variations. Figure 12.19a refers to load R variations and Fig. 12.19b

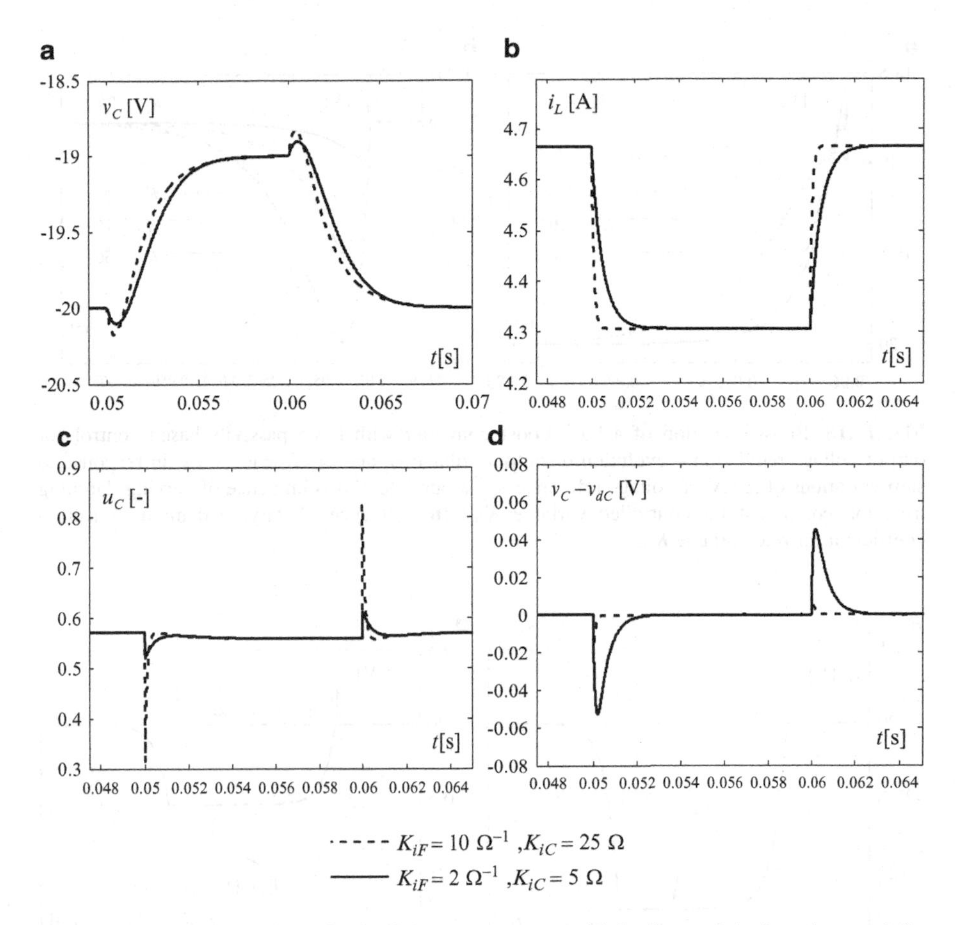

Fig. 12.17 Boost operation of a buck-boost converter with basic passivity-based control for output voltage regulation – evolution of variables of interest in response to step variations of ± 1 V of the voltage reference v_C^* around 20 V for two pairs of damping injection coefficients, Ki_C and Ki_F: (**a**) output voltage; (**b**) inductor current; (**c**) duty ratio (control input); (**d**) error between output voltage and its "desired" value, v_{dC}

regards the source voltage E variations. Both figures indicate that steady-state error occurs, which is larger in the case of load variations; hence, implementation of adaptive control law that embeds parameter estimation is necessary.

Figures 12.20 and 12.21 contain simulation results of implementing the adaptive passivity-based control structure given in Fig. 12.16. Suitable values of parameter estimator's coefficients $\gamma_1 = 40000$ and $\gamma_2 = 0.12$ – responsible for the estimation convergence speed –resulted from the simulation.

Figure 12.20a shows that the control goal is achieved in spite of load variations, with reasonable dynamic errors. The load variation also induces deviations of the estimated value of the second parameter E, as shown in Fig. 12.20d. Therefore, the parameter estimation works well and one can see that the estimation convergence speed depends on the operating point, as suggested in Fig. 12.20c. The steady-state

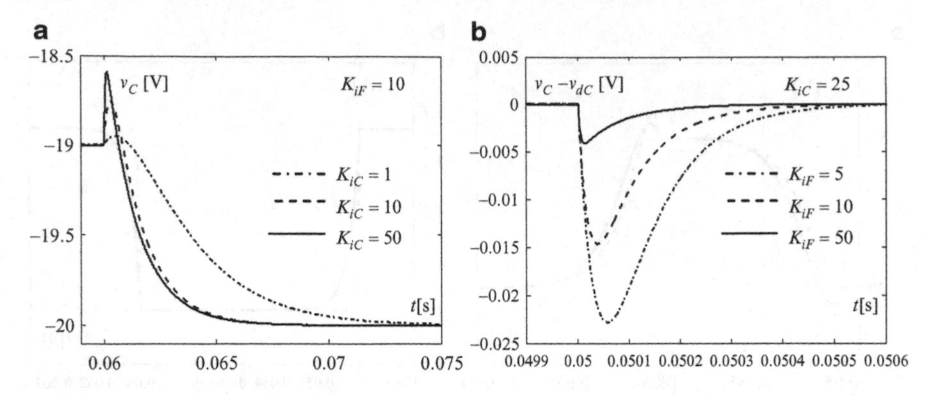

Fig. 12.18 Boost operation of a buck-boost converter with basic passivity-based control for output voltage regulation – evolution of output voltage v_C and of error $v_C - v_{dC}$ in response to step variations of ± 1 V of voltage reference v_C^* around 20 V: (**a**) influence of varying damping injection coefficient for controlled variable Ki_C; (**b**) influence of varying damping injection coefficient for free variable Ki_F

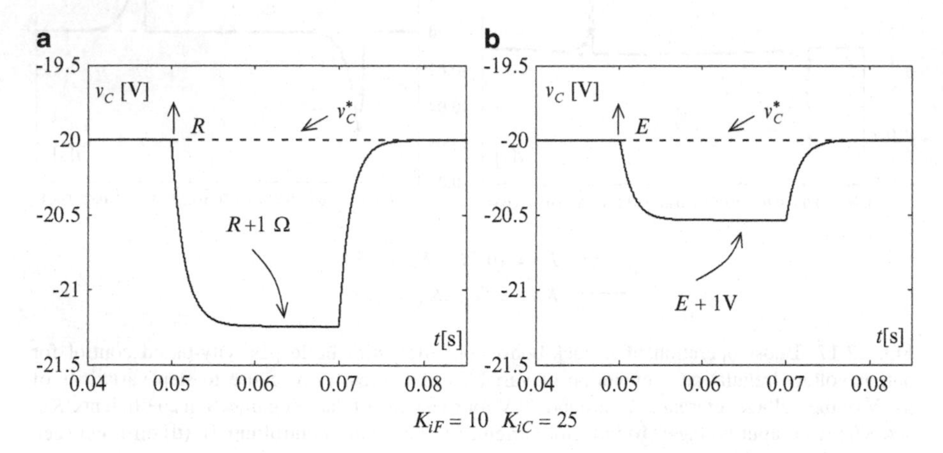

Fig. 12.19 Boost operation of a buck-boost converter with basic passivity-based control for output voltage regulation – influence of parameter variations (load R and source voltage E) on accomplishing control goal: (**a**) steady-state error due to load 1-Ω variation; (**b**) steady-state error due to source voltage 1-V variation

value of the controlled variable i_L changes in response to parameter change, as predicted by relation (12.76) (see Fig. 12.20b).

Figure 12.21a shows how the control goal is achieved in spite of source voltage variations. Reasonable dynamic errors also occur in this case. Voltage source variation determines deviations of the estimated value of the first parameter, R, as one can see in Fig. 12.21c. As in the previous case, the parameter estimation also works well in this case, with the dynamic depending on the operating point, as

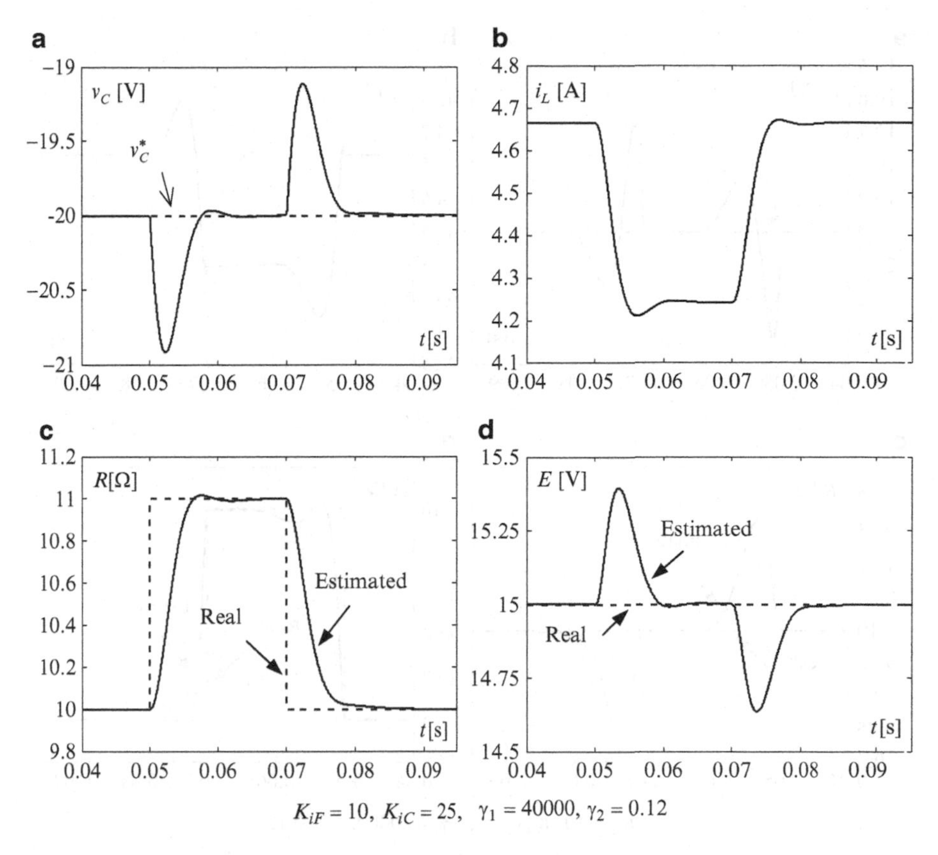

Fig. 12.20 Performance of parameter estimation in boost operation of a buck-boost converter with adaptive passivity-based control for output voltage regulation – evolution of variables of interest, real and estimated parameters in response to ±1-Ω load variation around the value $R = 10\ \Omega$ for which the control was initially designed

suggested in Fig. 12.21d. Figure 12.21b shows that the steady-state value of the controlled variable i_L changes in response to parameter change.

One can see that it is more important to obtain satisfactory load estimation than source voltage estimation, because it is highly probable that the latter could be measured, whereas this is not the case of the former. Besides, source voltage variations do not happen very abruptly in general, but load variations can occur suddenly.

Finally, Fig. 12.22 allows a comparison to be made between adaptive passivity-based control applied in the boost mode and in the buck mode of the considered buck-boost converter, when source voltage varies. Figures 12.22a, b show that, when the parameter estimated value is used, the output voltage setpoint is reached more slowly and with larger dynamic error than would happen if the real

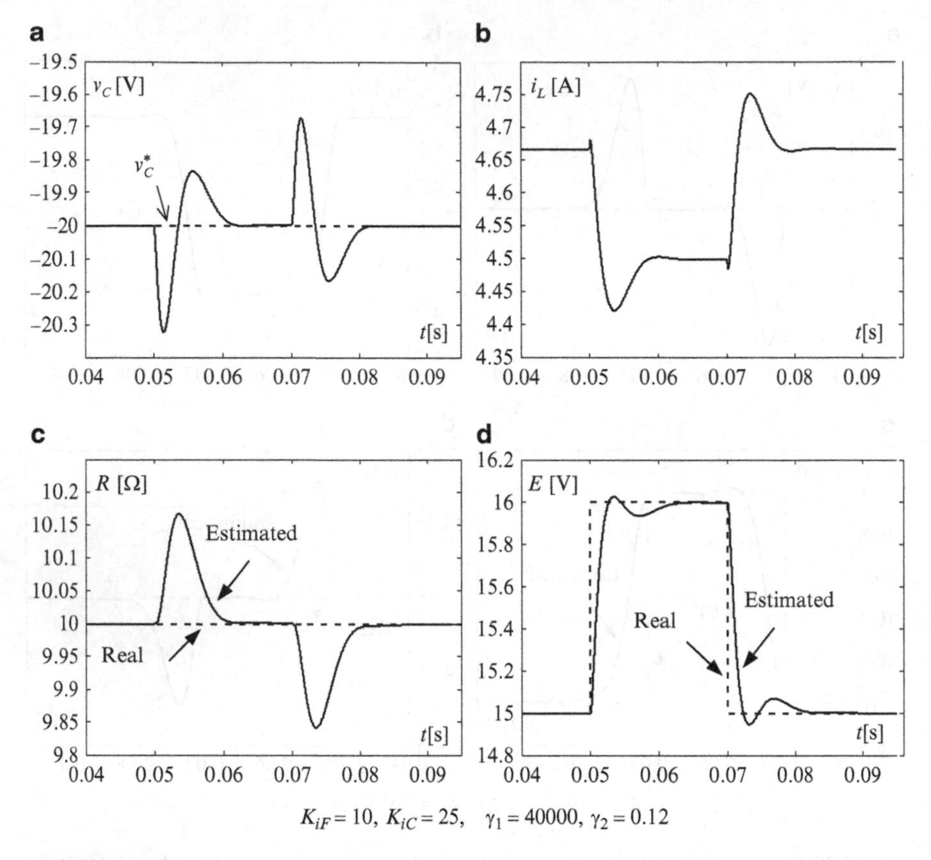

Fig. 12.21 Performance of parameter estimation in boost operation of a buck-boost converter with adaptive passivity-based control for output voltage regulation – evolution of variables of interest, real and estimated parameters in response to ±1-V variation of source voltage around value $E = 15$ V for which control was initially designed

(measured) value were used (supposing that such measure is available). Such errors are intuitively obvious; they can be estimated with the help of simulation results. Note that – even if the basic passivity-based control law was designed for $E = 15$ V, i.e., for boost-mode operation – dynamic errors due to variations of E are larger in the boost-mode operation; consequently, control input (duty ratio) variations are larger in this mode compared to the buck mode (see Fig. 12.22b compared with Fig. 12.22d).

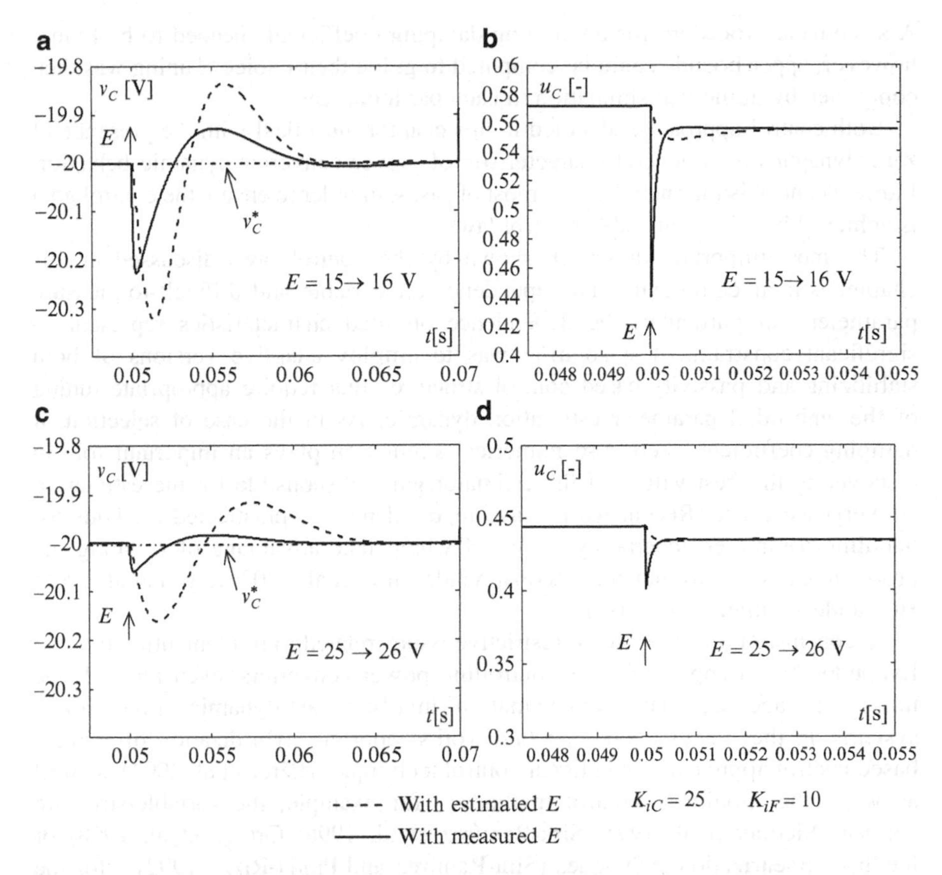

Fig. 12.22 Passivity-based control dynamic performance when source voltage varies with 1 V around typical value, comparatively for boost operation (**a** and **b**) and for buck operation (**c** and **d**), each case representing the use of real (measured) values (with *solid line*) vs. estimated values (with *dashed line*)

12.6 Conclusion

This chapter looked at two control techniques that exploit general features of power electronic converters related to energy processing. It was then natural for both approaches – stabilizing control and passivity-based control – to be theoretically developed with the help of Lyapunov methods, which supposed the use of energy functions. The property of power converters to dissipate a part of the received energy was emphasized; this property was further related with convergence to stable operating points and used for control purposes. More precisely, convergence speed could be controlled by means of the so-called damping injection. Technically speaking, both the stabilizing and the passivity-based control rely on conveniently selecting some damping injection coefficients, whose closed-loop action is equivalent to that of a state feedback that changes the poles of a linear system.

A systematic procedure for fine tuning damping coefficients needed to be found; however, upper bounds could be computed to guide their choice. Tuning was then completed by numerical simulation in each particular case.

Both control approaches detailed in this chapter must deal with the presence of zero dynamics as a general characteristic of power converter dynamic behavior. Indirect control is thus necessary in most of cases, in order to ensure the control goal is achieved by means of stable control laws.

The most important drawback shown by the control laws discussed in this chapter is their complexity and dependence on variable and difficult-to-measure parameters; in particular, the dependence on load characteristics represents a significant constraint. The solution was to employ adaptive versions of both stabilizing and passivity-based control structures that require appropriate tuning of the embedded parameter estimation dynamic. As in the case of selection of damping coefficients, here also numerical simulation plays an important role in discovering the best values of the estimator gains responsible for the estimation convergence speed. Recent works have reported more sophisticated methods for handling parameter uncertainty, many of which take advantage of restating the problem in a linear formalism (Escobar Valderrama et al. 2003; Leyva et al. 2006; Hernandez-Gomez et al. 2010).

In conclusion, there are some restrictive issues related to implementing the two Lyapunov-based approaches for controlling power converters, even more if one takes into account that implementation involves fast-dynamic, time-critical systems. In the literature one can find works reporting hybridization of energy-based control approaches with linear control techniques (Pérez et al. 2004), as well as with other nonlinear control methods – for example, the variable-structure control (Nicolas et al. 1995; Sira-Ramírez et al. 1996; Ortega et al. 1998) or feedback-linearization techniques (Sira-Ramirez and Prada-Rizzo 1992) – for the purpose of improving simplicity and closed-loop robustness and dynamic performance. Extensive details on the application of variable-structure control – and in particular, of sliding-mode control – to power electronic converters are presented in the next chapter.

Problems

Problem 12.1. Stabilizing control of a single-phase rectifier

Figure 12.23 presents the electrical circuit of a single-phase rectifier, where ω is the grid pulsation, E is the amplitude of the sinusoidal grid voltage, C is the DC-link capacitor's capacity, L is the grid-side inductor's inductance and the other notations have their usual meaning. R represents the variable DC load. Design a stabilizing control law aimed at regulating the output voltage at a setpoint v_0^*; this is the primary goal, with a secondary one being the requirement of zero reactive power. The rectifier averaged model in the synchronously rotating dq frame will be used to this end.

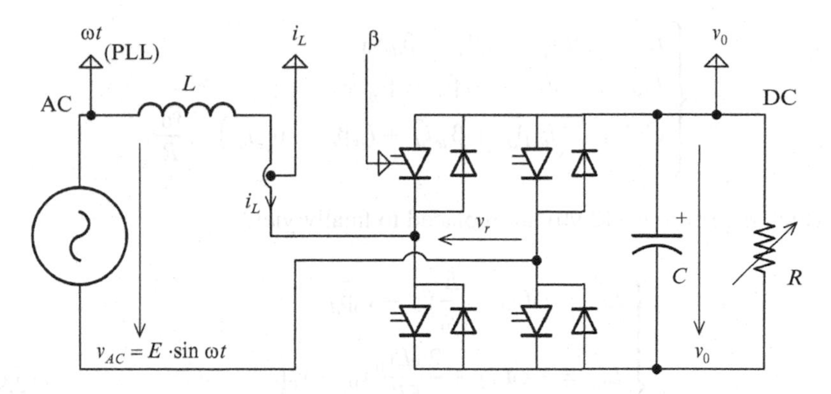

Fig. 12.23 Electrical circuit of single-phase rectifier

Solution Recall that the general averaged model (GAM) of this circuit expressed in the synchronously rotating dq frame is given by (see relation (5.63) in Chap. 5)

$$\begin{cases} L\dot{i_d} = \omega L i_q + E - v_0\beta_d \\ L\dot{i_q} = -\omega L i_d - v_0\beta_q \\ C\dot{v_0} = \frac{1}{2}\left(i_d\beta_d + i_q\beta_q\right) - \frac{v_0}{R}, \end{cases} \tag{12.89}$$

with $\mathbf{u} = \begin{bmatrix} \beta_d & \beta_q \end{bmatrix}^T$ being the input vector – composed of the d-component and the q-component of the duty ratio β – and $\mathbf{x} = \begin{bmatrix} i_d & i_q & v_0 \end{bmatrix}^T$ being the state vector. Equation (12.89) allows the considered AC-DC converter to be treated from the control viewpoint like a DC-DC one. Note also that the dq modeling has transformed the initial single-input system into a multi-input one. A stabilizing control input will be computed according to the steps of Algorithm 12.1 given in Sect. 12.2.3, in order to achieve the stated control goal.

Model (12.89) is linearized around the desired operating point, that is, the one corresponding to the output voltage setpoint v_0^* to the load value R, supposed in the first place known and constant, and to the steady-state value of the i_q current being zero (since the reactive power is imposed to be zero). The other values characterizing this operating point – denoted next by subscripts e – are found by zeroing derivatives in Eq. (12.89). It is easy to verify that

$$i_{de} = \frac{2v_0^{*2}}{ER}, \quad i_{qe} = 0, \quad \beta_{de} = \frac{E}{v_0^*}, \quad \beta_{qe} = -\frac{2\omega L v_0^*}{ER}. \tag{12.90}$$

The next step is to obtain the linearized version of model (12.89) around the operating point described by (12.90). By adopting notation $\tilde{\cdot}$ to denote variations around the chosen steady-state point, one then writes

$$\begin{cases} L\dot{\widetilde{i_d}} = \omega L\widetilde{i_q} - v_0^*\widetilde{\beta_d} - \beta_{de}\widetilde{v_0} \\ L\dot{\widetilde{i_q}} = -\omega L\widetilde{i_d} - v_0^*\widetilde{\beta_q} - \beta_{qe}\widetilde{v_0} \\ C\dot{\widetilde{v_0}} = \frac{1}{2}\left(i_{de}\widetilde{\beta_d} + \beta_{de}\widetilde{i_d} + i_{qe}\widetilde{\beta_q} + \beta_{qe}\widetilde{i_q}\right) - \frac{\widetilde{v_0}}{R}, \end{cases}$$

in which expressions (12.90) are replaced to finally yield

$$\begin{cases} L\dot{\widetilde{i_d}} = \omega L\widetilde{i_q} - \frac{E}{v_0^*}\widetilde{v_0} - v_0^*\widetilde{\beta_d} \\ L\dot{\widetilde{i_q}} = -\omega L\widetilde{i_d} + \frac{2\omega L v_0^*}{ER}\widetilde{v_0} - v_0^*\widetilde{\beta_q} \\ C\dot{\widetilde{v_0}} = \frac{E}{2v_0^*}\widetilde{i_d} - \frac{\omega L v_0^*}{ER}\widetilde{i_q} - \frac{1}{R}\widetilde{v_0} + \frac{v_0^{*2}}{ER}\widetilde{\beta_d}\,. \end{cases} \tag{12.91}$$

Expression (12.91) of the linearized model in variations is further written in the more compact form

$$\dot{\widetilde{\mathbf{x}}} = \mathbf{A}\cdot\widetilde{\mathbf{x}} + \mathbf{B}\cdot\widetilde{\mathbf{u}}, \tag{12.92}$$

with $\widetilde{\mathbf{x}} = \begin{bmatrix}\widetilde{i_d} & \widetilde{i_q} & \widetilde{v_0}\end{bmatrix}^T$ being the state vector, $\widetilde{\mathbf{u}} = \begin{bmatrix}\widetilde{\beta_d} & \widetilde{\beta_q}\end{bmatrix}^T$ being the input vector and state and input matrices being determined as follows:

$$\mathbf{A} = \begin{bmatrix} 0 & \omega & -\frac{E}{v_0^* L} \\ -\omega & 0 & \frac{2\omega v_0^*}{ER} \\ \frac{E}{2v_0^* C} & -\frac{\omega L v_0^*}{ERC} & -\frac{1}{RC} \end{bmatrix}, \quad \mathbf{B} = \begin{bmatrix} -\frac{v_0^*}{L} & 0 \\ 0 & -\frac{v_0^*}{L} \\ \frac{v_0^{*2}}{ERC} & 0 \end{bmatrix}. \tag{12.93}$$

The Lyapunov candidate function

$$V(\widetilde{\mathbf{x}}) = \frac{1}{2}\widetilde{\mathbf{x}}^T\mathbf{Q}\widetilde{\mathbf{x}} \tag{12.94}$$

is defined, where matrix $\mathbf{Q}$ is chosen such that function $\mathbf{V}$ represents the energy in the increment; therefore:

$$\mathbf{Q} = \begin{bmatrix} L & 0 & 0 \\ 0 & L & 0 \\ 0 & 0 & C \end{bmatrix}. \tag{12.95}$$

One can easily verify that $\mathbf{A}^T\mathbf{Q} + \mathbf{QA}$ is symmetric and negative definite. The stabilizing control law is effectively computed according to relation (12.20), after supposing that the positive scalar λ is conveniently chosen:

$$\widetilde{\mathbf{u}} = -\lambda \cdot \mathbf{B}^T\mathbf{Q} \cdot \widetilde{\mathbf{x}},$$

which, by taking into account expressions of matrices $\mathbf{B}$ and $\mathbf{Q}$ given by (12.93) and (12.95), respectively, leads in our case to the following state-feedback form of the stabilizing control input variation:

$$\widetilde{\mathbf{u}} = -\lambda \cdot \begin{bmatrix} -v_0^* & 0 & \dfrac{v_0^{*2}}{ER} \\ 0 & -v_0^* & 0 \end{bmatrix} \cdot \widetilde{\mathbf{x}}. \tag{12.96}$$

By replacing $\widetilde{\mathbf{u}} = \begin{bmatrix} \widetilde{\beta_d} & \widetilde{\beta_q} \end{bmatrix}^T$ and $\widetilde{\mathbf{x}} = \begin{bmatrix} \widetilde{i_d} & \widetilde{i_q} & \widetilde{v_0} \end{bmatrix}^T$ in expression (12.96), this latter can be written component wise, thus emphasizing the expressions of d- and q-components of the duty ratio:

$$\begin{cases} \widetilde{\beta_d} = \lambda v_0^* \widetilde{i_d} - \lambda \dfrac{v_0^{*2}}{ER} \widetilde{v_0} \\ \widetilde{\beta_q} = \lambda v_0^* \widetilde{i_q}\,. \end{cases} \tag{12.97}$$

One can see that scalar λ is measured in $A^{-1}V^{-1}$; its effective selection results from a numerical analysis of the closed-loop system's poles in the complex plane, as shown in the example detailed in Sect. 12.2.4. Expressions of duty ratio components obviously result as $\beta_d = \beta_{de} + \widetilde{\beta_d}$ and $\beta_q = \beta_{qe} + \widetilde{\beta_q}$, respectively, where steady-state values β_{de} and β_{qe} are given by Eq. (12.90) and variation values $\widetilde{\beta_d}$ and $\widetilde{\beta_q}$ are given by Eq. (12.97):

$$\begin{cases} \beta_d = \dfrac{E}{v_0^*} + \lambda v_0^* \widetilde{i_d} - \lambda \dfrac{v_0^{*2}}{ER} \widetilde{v_0} \\ \beta_q = -\dfrac{2\omega L v_0^*}{ER} + \lambda v_0^* \widetilde{i_q}\,. \end{cases} \tag{12.98}$$

Equation (12.98) shows that both obtained stabilizing control expressions depend on source voltage E and also on load value R. If the former might be not constant or it might be unknown in some applications, the latter habitually varies in an unpredictable manner in most applications. This is a quite important limitation of control law effectiveness because its practical use must rely upon parameter estimators.

The following problems are left to the reader to solve.

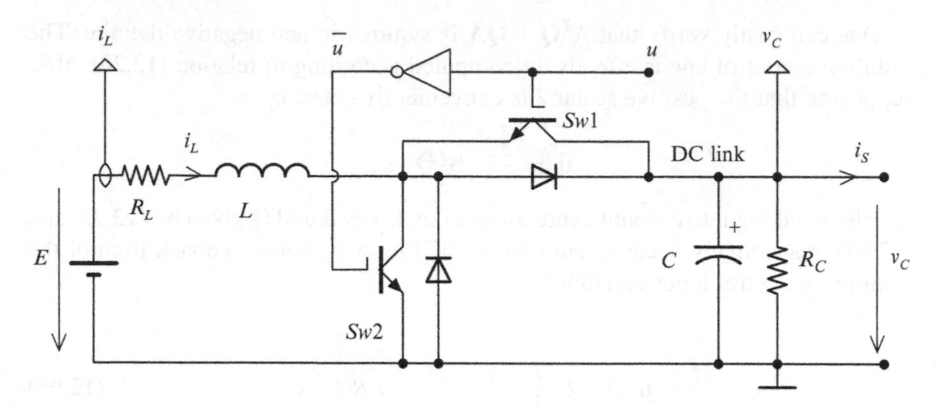

Fig. 12.24 Bidirectional-current DC-DC converter, $E < v_C$

Problems 12.2. Bidirectional-current DC-DC converter passivity-based control
Figure 12.24 presents the electrical circuit of a two-quadrant DC-DC converter – previously presented in Fig. 8.8 of Chap. 8 – where E is the source voltage of constant amplitude, R_C represents the variable DC load and the other notations have their usual meaning. Inductor losses are here neglected (i.e., $R_L = 0$).

It is required to obtain the Euler–Lagrange model of the converter – in the general form (12.38) given in Sect. 12.3 – and then use it to compute the expression of a passivity-based control law aimed at regulating the output voltage at a setpoint v_C^*.

Problems 12.3. Buck power stage stabilizing control
Figure 12.25 shows the electrical circuit of a buck DC-DC converter, which was previously used in Problem 3.5 of Chap. 3 (see Fig. 3.26). Notations have their usual meaning, with R representing the DC load. E and R are assumed known and constant. Values of parameters are $E = 12$ V, $L = 1$ mH, $C = 680$ μF and $R = 3$ Ω.

Address the following points.

(a) Obtain the averaged model of the converter. What can be said about its linearized model?
(b) Use the converter's linearized model to obtain the expression of a stabilizing control law aimed at regulating the output voltage at a setpoint v_C^*. To this end follow steps of Algorithm 12.1, choosing matrix $\mathbf{Q}$ as in (12.7).
(c) For $v_C^* = 9$ V perform a numerical analysis of the closed-loop system's poles in order to justify the choice of coefficient λ in the expression of the above computed stabilizing control law.

Problems 12.4. Stabilizing control of Ćuk converter
Let us consider the electrical circuit of a Ćuk converter presented in Fig. 12.26 (repeated from Fig. 3.17 of Problem 3.1 of Chap. 3). Meanings of notations are clear from the context, with v_{C2} being output voltage and R representing load. Source voltage E and R are assumed known and constant. Values of parameters are $L_1 = L_2 = 5$ mH, $C_1 = 470$ μF, $C_2 = 220$ μF, $E = 12$ V and $R = 15$ Ω.

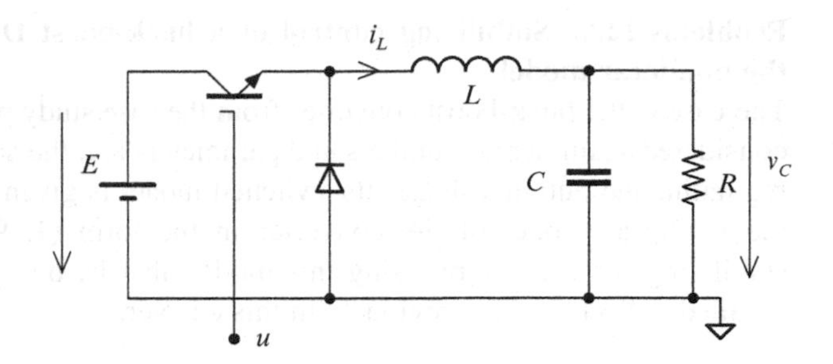

Fig. 12.25 Buck DC-DC converter

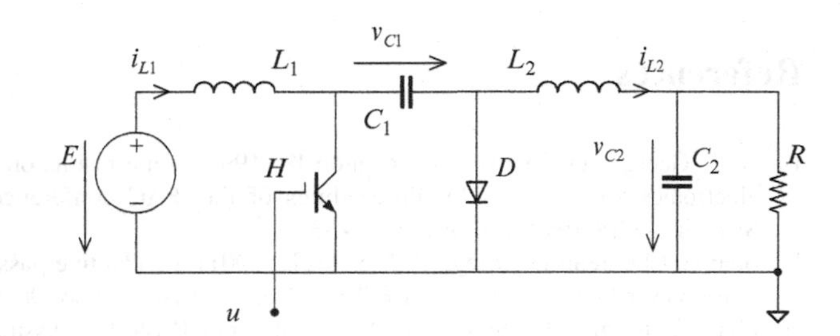

Fig. 12.26 Electrical circuit of Ćuk DC-DC converter

The following points must be addressed.

(a) Obtain the averaged model of the converter and then its linearized model.
(b) Use the converter's linearized model to compute the expression of a stabilizing control law aiming at regulating the output voltage at a setpoint v_{C2}^*. To this end follow steps of Algorithm 12.1, with matrix **Q** chosen as in (12.7).
(c) For $v_{C2}^* = 15$ V perform a numerical analysis of the closed-loop linearized system's poles and choose the value of coefficient λ in the expression of the above computed stabilizing control law so as to ensure a second-order dominant dynamic with 0.7 as damping coefficient.
(d) Build the numerical simulation diagram of the nonlinear closed-loop system in Simulink®. Perform simulations in order to validate the imposed dynamic performance of the system around the desired operating point – one corresponding to the setpoint v_{C2}^* – by exciting the system with step variations of voltage reference.

Problems 12.5. Adaptive passivity-based control of a boost DC-DC converter
Let us return to the case of the boost converter supplying a resistive load R from the example detailed in Sect. 12.4.5. Its parameters are $L = 5$ mH, $C = 470$ μF, $R = 10\ \Omega$, $E = 15$ V and the switching frequency $f = 20$ kHz. Its electrical circuit was presented in Fig. 3.5 of Chap. 3 and its switched model can be found in relation (12.57). The control task is to regulate the output voltage at a value v_C^*.

It is required to design a parameter estimator based upon Eq. (12.72) and to implement numerically the adaptive passivity-based control scheme presented in Fig. 12.11 by using developments given in the example. Perform an analysis of the estimator's dynamic performance depending on the coefficients γ_1 and γ_2.

Problems 12.6. Stabilizing control of a buck-boost DC-DC converter using the nonlinear model
The case of the buck-boost converter from the case study presented in Sect. 12.5 is considered again here, with the same parameters and the same control objective of regulating the output voltage. Its switched model is given by (12.73). First obtain the nonlinear model of the converter in the form (12.9) and then compute a stabilizing control law by using this model, that is, by applying Algorithm 12.2 detailed in Sect. 12.2.3 previously in this chapter.

References

Bacha S, Georges D, Oyarbide E, Rognon JP (1997) Some results on nonlinear control in power electronics applications. In: Proceedings of the IFAC conference on Control of Industrial Systems – CIS 1997, Belfort, pp 75–83

Escobar G, Chevreau D, Ortega R, Mendes E (2001) An adaptive passivity-based controller for a unity power factor rectifier. IEEE Trans Control Syst Technol 9(4):637–644

Escobar Valderrama G, Stanković AM, Mattavelli P (2003) Dissipativity-based adaptive and robust control of UPS in unbalanced operation. IEEE Trans Power Electron 18(4):1056–1062

Hernandez-Gomez M, Ortega R, Lamnabhi-Lagarrigue F, Escobar G (2010) Adaptive PI stabilization of switched power converters. IEEE Trans Control Syst Technol 18(3):688–698

Komurcugil H, Kukrer O (1998) Lyapunov-based control of three-phase PWM AC/DC voltage-source converters. IEEE Trans Power Electron 13(5):801–813

Kwasinski A, Krein T (2007) Passivity-based control of buck converters with constant-power loads. In: Proceedings of the Power Electronics Specialists Conference – PESC 2007, Orlando, pp 259–265.

Leyva R, Cid-Pastor A, Alonso C, Queinnec I, Tarbouriech S, Martinez-Salamero L (2006) Passivity-based integral control of a boost converter for large-signal stability. IEE Proc Control Theor Appl 153(2):139–146

Liserre M (2006) Passivity-based control of single-phase multilevel grid connected active rectifiers. Bull Pol Acad Sci Tech Sci 54(3):341–346

Maschke B, Ortega R, van der Schaft A (2000) Energy-based Lyapunov functions for forced Hamiltonian systems with dissipation. IEEE Trans Autom Control 45(8):1498–1502

Mattavelli P, Escobar G, Stanković AM (2001) Dissipativity-based adaptive and robust control of UPS. IEEE Trans Ind Electron 48(2):334–343

Nicolas B, Fadel M, Cheron Y (1995) Sliding mode control of DC-DC converters with input filter based on the Lyapunov-function approach. In: Proceedings of the 6th European conference on Power Electronics and Applications – EPE 1995, Sevilla, pp 1338–1343

Noriega-Pineda D, Espinosa-Perez G (2007) Passivity-based control of multilevel cascade inverters: high-performance with reduced switching frequency. In: Proceedings of the 2007 I.E. International Symposium on Industrial Electronics – ISIE 2007, Mexico, pp 3403–3408

Ortega R, Espinosa G (1993) Torque regulation of induction motors. Automatica 29(3):621–633

Ortega R, Garcia-Canseco E (2004) Interconnection and damping assignment passivity-based control: a survey. Euro J Control 10:432–450

Ortega R, Spong M (1989) Adaptive motion control of rigid robots: a tutorial. Automatica 25(6):877–888

Ortega R, Loria A, Nicklasson PJ, Sira-Ramirez H (1998) Passivity-based control of Euler-Lagrange systems. Springer, London

Ortega R, van der Schaft AJ, Mareels I, Maschke B (2001) Putting energy back in control. IEEE Control Syst Mag 21(2):18–33

Ortega R, van der Schaft A, Maschke B, Escobar G (2002) Interconnection and damping assignment passivity–based control of port–controlled Hamiltonian systems. Automatica 38(4):585–596

Oyarbide E, Bacha S (1999) Experimental passivity-based adaptive control of a three-phase voltage source inverter. In: Proceedings of the 8th European conference on Power Electronics and Applications – EPE 1999. Lausanne, Suisse

Oyarbide E, Bacha S (2000) Passivity-based control of power electronics structures. Part I: Generalization of structural properties associated to Euler-Lagrange formalism (in French: Commande passive des structures de l'électronique de puissance. Partie I: Généralisation des propriétés structurelles associées au formalisme d'Euler-Lagrange). Revue Internationale de Génie Électrique 3(1):39–57

Oyarbide E, Bacha S, Georges D (2000) Passivity-based control of power electronics structures. Part II: Application to three-phase voltage inverter (in French: Commande passive des structures de l'électronique de puissance. Partie II: Application à l'onduleur de tension triphasé). Revue Internationale de Génie Électrique 3(1):59–80

Oyarbide-Usabiaga E (1998) Passivity-based control of power electronics structures (in French: "Commande passive des structures de l'électronique de puissance"). Ph.D. thesis, Grenoble Institute of Technology, France

Penfield P, Spence R, Duinker S (1970) Telleghen's theorem and electrical networks, Research monograph 58. The M.I.T. Press, Cambridge

Perez M, Ortega R, Espinoza JR (2004) Passivity-based PI control of switched power converters. IEEE Trans Control Syst Technol 12(6):881–890

Rodriguez H, Ortega R, Escobar G (1999) A robustly stable output feedback saturated controller for the boost DC-to-DC converter. In: Proceedings of the 38th conference on decision and control – CDC 1999. Phoenix, Arizona, USA, pp 2100–2105

Sanders SR (1989) Nonlinear control of switching power converters. Ph.D. thesis, Massachusetts Institute of Technology

Sanders SR, Verghese GC (1992) Lyapunov-based control for switched power converters. IEEE Trans Power Electron 7(1):17–24

Sira-Ramírez H, Prada-Rizzo MT (1992) Nonlinear feedback regulator design for the Ćuk converter. IEEE Trans Autom Control 37(8):1173–1180

Sira-Ramírez H, Escobar G, Ortega R (1996) On passivity-based sliding mode control of switched DC-to-DC power converters. In: Proceedings of the 35th conference on decision and control – CDC 1996. Kobe, Japan, pp 2525–2526

Sira-Ramírez H, Pérez-Moreno RA, Ortega R, Garcia-Esteban M (1997) Passivity-based controllers for the stabilization of DC-to-DC power converters. Automatica 33(4):499–513

Stanković AM, Escobar G, Ortega R, Sanders SR (2001) Energy-based control in power electronics. In: Banerjee S, Verghese GC (eds) Nonlinear phenomena in power electronics: attractors, bifurcations, chaos and nonlinear control. IEEE Press, Piscataway, pp 25–37

Takegaki M, Arimoto S (1981) A new feedback method for dynamic control of manipulators. ASME J Dynam Syst Meas Control 102:119–125

van der Schaft A (1996) L_2-Gain and passivity techniques in nonlinear control, Lecture notes in control and information sciences 218. Springer, London

Chapter 13
Variable-Structure Control of Power Electronic Converters

Interest in variable-structure control is justified by the necessity of robustly controlling systems whose structure switches between several configurations. Power electronic converters are such class of systems because they can be described by differential equations with discontinuous right-hand sides (i.e., discontinuous inputs). Moreover, they exhibit nonlinear behavior which can in some applications render unsuitable standard linear control approaches. Good control performance such as large bandwidth is ensured because the switching solution is obtained directly without any other form of supplementary modulation (PWM, sigma-delta modulation).

This chapter first introduces some basic concepts specific to variable-structure control by relying upon some examples commonly used in the power electronics community. Next, mathematical developments are outlined in order to support the general algorithm containing the steps of a variable-structure control design procedure. Buck and boost DC-DC converters serve as benchmarks. Two case studies illustrate the variable-structure control approach: first, a single-phase power-factor-correction converter (PFCC) and second, a three-phase voltage-source converter used as PFCC, which is treated as a multi-input–multi-output (MIMO) system. Connections between variable-structure control and other nonlinear control approaches are emphasized. This chapter ends with two solved problems and several posed problems.

13.1 Introduction

Variable-structure control theory has its roots in early works of Filippov (1960) and Emelyanov (1967) on nonlinear systems described by differential equations with discontinuous right-hand side that may exhibit sliding modes. This theory has been further developed in many other works, like Utkin (1972) and Itkis (1976). As this approach has matured (Utkin 1977; Hung et al. 1993; Sira-Ramírez 1993; Young et al. 1999; Levant 2007; Sabanovic et al. 2004), its robustness has spread its application into many engineering and technology areas; thus, variable-structure

S. Bacha et al., *Power Electronic Converters Modeling and Control: with Case Studies*, Advanced Textbooks in Control and Signal Processing, DOI 10.1007/978-1-4471-5478-5_13,

control is particularly suitable for nonlinear and/or variant systems such as robotic manipulators (Slotine and Sastry 1983), motion control and electric drives (Utkin 1993; Šabanovic 2011) and renewable energy systems (Battista et al. 2000).

The relevant literature is particularly rich in the power electronic converter control area. Articles like Venkataramanan et al. (1985), Sira-Ramírez (1987, 1988), Malesani et al. (1995), Spiazzi et al. (1995), Carpita and Marchesoni (1996), Mattavelli et al. (1997), Carrasco et al. 1997; Martínez-Salamero et al. (1998), Guffon et al. (1998), Sira-Ramírez (2003) and Tan et al. (2005) are just few of the bibliographical references that the literature provides. There are also some notable textbooks like Sira-Ramírez and Silva-Ortigoza (2006) and Tan et al. (2011) that provide useful theoretical and practical insights in developing variable-structure control structures for power electronic converters.

13.2 Sliding Surface

The concept of sliding surface is one of the basic concepts related to variable-structure control; it will be introduced by means of the boost converter given in Fig. 13.1. Let us state a control goal for this configuration, namely to regulate the current i_L at the reference value $i_L{}^*$. This can be achieved by acting on switch H in the following way:

$$H = \begin{cases} 1, & i_L^* \geq i_L \\ 0, & i_L^* < i_L. \end{cases} \tag{13.1}$$

One can see that the circuit switches between two configurations – it changes its structure – according to the sign of the expression issued from (13.1) (see Fig. 13.2b, c):

$$s(x) = i_L^* - i_L. \tag{13.2}$$

It is this change of structure that gives such a system the name “variable-structure control”. Hysteresis has also been taken into account; it is due to noninstantaneous switching. This aspect will be discussed later.

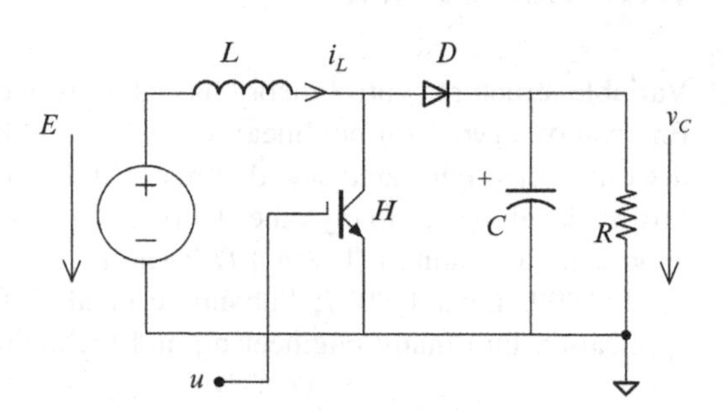

Fig. 13.1 Electrical circuit of a boost converter

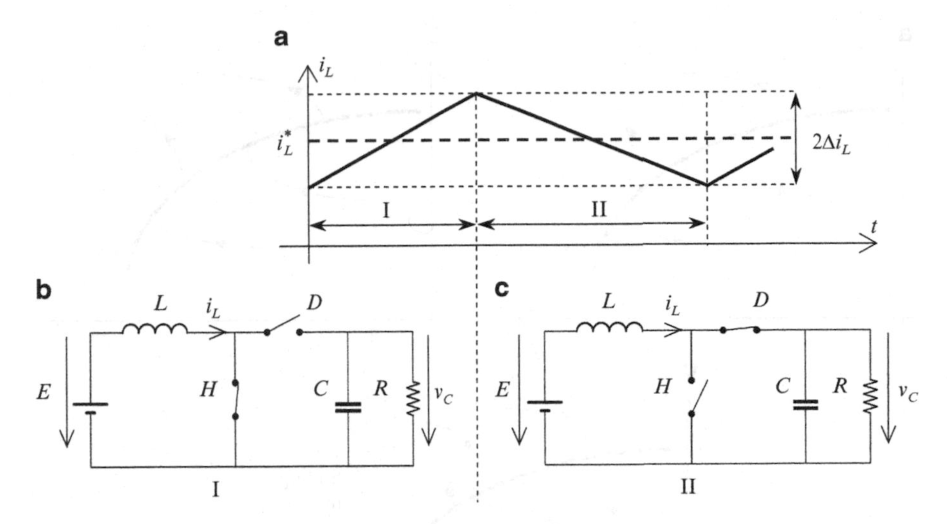

Fig. 13.2 Principle of variable-structure control illustrated on current regulation of boost converter (where Δi_L represents the hysteresis)

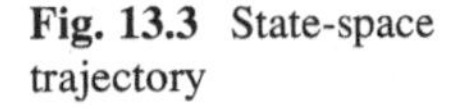
Fig. 13.3 State-space trajectory

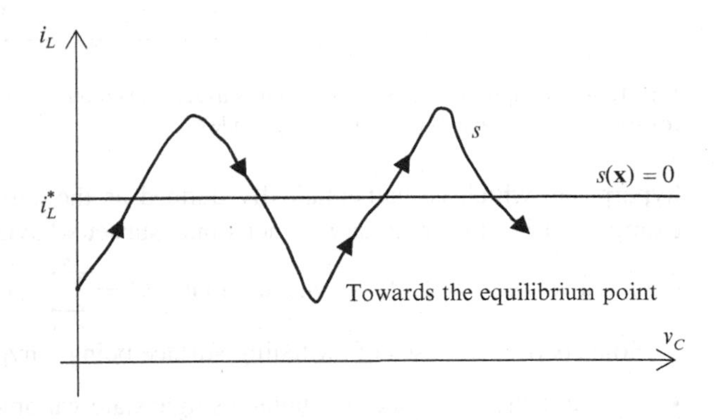

If now one is interested in the (v_C,i_L) state-space behavior, one observes that current i_L tends to switch between the two regions marked by the line described by $i_L = i_L^*$ (Fig. 13.3), which is called *switching surface*. In the general linear case, it is about the kernel of a linear application, i.e., a hyperplane.

In Fig. 13.3 the switching frequency is finite; in the theoretical case, this frequency is arbitrarily large (infinite). Thus, supposing that the current i_L switches instantaneously, then the state-space trajectory would slide on the surface defined by $i_L = i_L^*$. The dynamics on this surface are known as the associated *sliding modes*. The corresponding switching surface is called *sliding surface* (Utkin 1972).

Figure 13.4 shows several types of switching surface, along with the associated state-space trajectories.

In the single-output case, the switching surface dimension is (n – 1) for an n-dimensional state space. Thus, in the linear case the switching surface is a

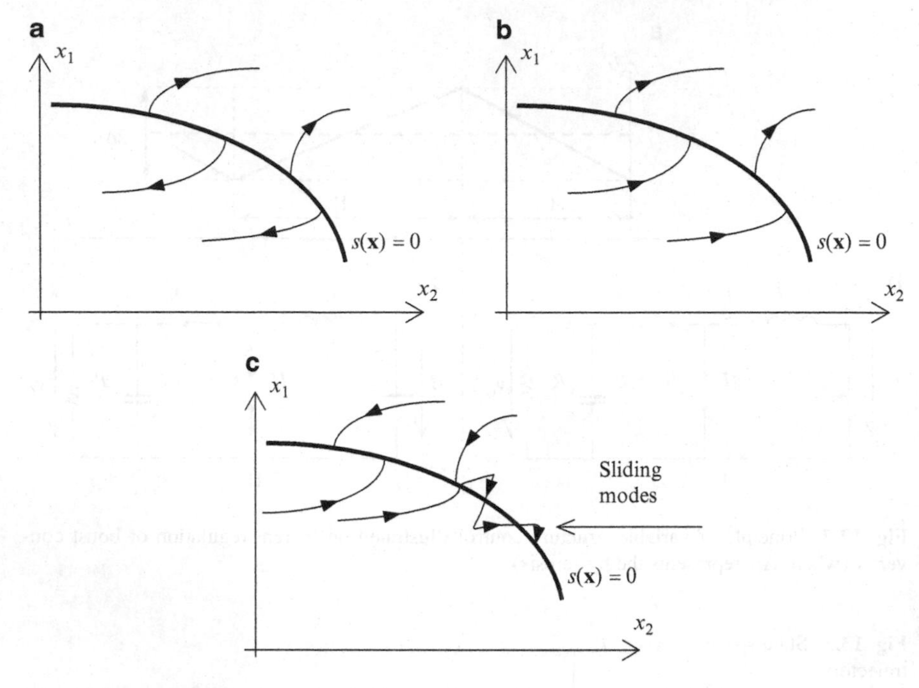

Fig. 13.4 Examples of types of switching surfaces: (**a**) repulsive surface; (**b**) refractive surface; (**c**) sliding surface and associated sliding modes

hyperplane, which is mathematically defined as the kernel of a linear form. For example, in $\mathbb{R}^n$ it is about an n-dimensional subset $\mathbf{x}$ having the property that there exist scalars a_k, $k = 1, 2, \cdots, n$, such that $s(\mathbf{x}) = \sum_{k=1}^{n} a_k x_k = 0$.

Some particular cases of switching surface being a hyperplane are as follows:

- a point if the state space contains a single state variable;
- a line in the case of a two-dimensional state space;
- a plane when there are three state variables;
- a volume for a four-dimensional state space.

In the case of the above boost converter the switching surface is a line because the state space is in fact a state plane, namely the (v_C, i_L) -plane. Note that in the general nonlinear case the switching surface is a variety (or manifold) of a nonlinear form.

13.3 General Theoretical Results

For the sake of simplicity, let us first approach the single-input–single-output (SISO) case and a time-invariant function $s : \mathbb{R}^n \rightarrow \mathbb{R}$. Let $\mathbf{S}$ be the set defined as

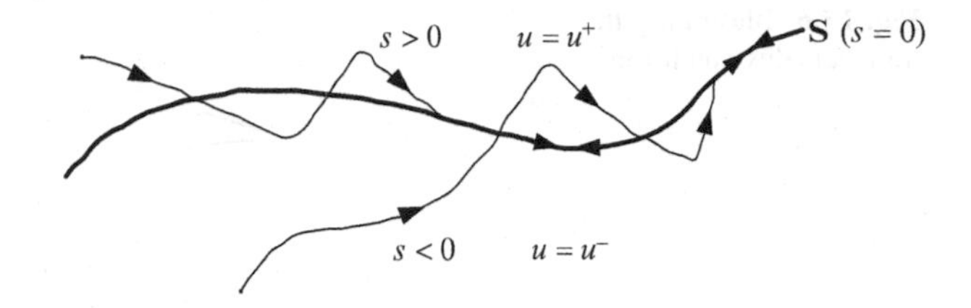

Fig. 13.5 Intuitive planar representation of reachability of surface **S** in finite time

$$\mathbf{S} = \{\mathbf{x} \in \mathbb{R}^n \text{ having property } s(\mathbf{x}) = 0\}, \tag{13.3}$$

which therefore defines a time-invariant switching surface. Suppose also that the state-space system is affine in its input, that is, it can be written as

$$\frac{d\mathbf{x}}{dt} = \mathbf{f}(\mathbf{x}) + \mathbf{g}(\mathbf{x}) \cdot u. \tag{13.4}$$

The goal of the variable-structure control can be stated as to bring and maintain the system state on the chosen switching surface while preserving the stable behavior on it.

13.3.1 Reachability of the Sliding Surface: Transversality Condition

The fact that the surface **S** is reached is written mathematically as $s(\mathbf{x}) = 0$. The result is that in a neighborhood,

- if $s(\mathbf{x}) > 0$, then one must apply u^+ such that $\frac{ds(\mathbf{x})}{dt} < 0$;
- if $s(\mathbf{x}) < 0$, then one must apply u^- such that $\frac{ds(\mathbf{x})}{dt} > 0$.

If the control law meets the above conditions, the reachability of surface **S** is ensured, but the manner of reaching is not specified. It could be reached either asymptotically or in finite time, as illustrated in Fig. 13.5. But in order to obtain sliding behavior on this surface, it must be reached in finite time.

Asymptotic convergence means that surface **S** is reached in infinite time, which is translated as

$$\lim_{\mathbf{s} \to 0^+} \frac{ds(\mathbf{x})}{dt} = 0 \text{ and } \lim_{\mathbf{s} \to 0^-} \frac{ds(\mathbf{x})}{dt} = 0. \tag{13.5}$$

An example of asymptotic convergence is that of a first-order system: its equilibrium point is reached asymptotically and in no case is there overshooting.

For a stronger condition, that is, for the surface to be reached in finite time, the time derivative $\frac{ds(\mathbf{x})}{dt}$ must not become zero in a neighborhood of **S**. In this neighborhood, locally, condition (13.5) becomes

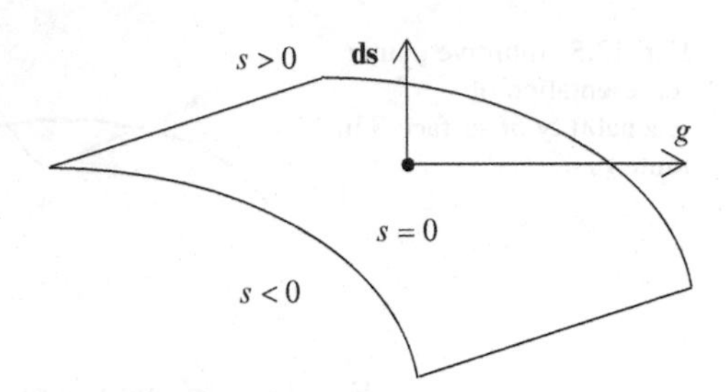

Fig. 13.6 Illustrating the transversality condition

$$\lim_{s \to 0^+} \frac{ds(\mathbf{x})}{dt} < 0 \text{ and } \lim_{s \to 0^-} \frac{ds(\mathbf{x})}{dt} > 0. \tag{13.6}$$

By using Lie derivatives defined in Chap. 10 and taking account of the system dynamics, condition (13.6) becomes

$$\lim_{s \to 0^+} \left(\mathrm{L}_{\mathbf{f}} s + u \cdot \mathrm{L}_{\mathbf{g}} s\right) < 0 \text{ and } \lim_{s \to 0^+} \left(\mathrm{L}_{\mathbf{f}} s + u \cdot \mathrm{L}_{\mathbf{g}} s\right) > 0. \tag{13.7}$$

By appropriately choosing the control law such that

$$u = \begin{cases} u^+ & \text{if } s(\mathbf{x}) > 0 \\ u^- & \text{if } s(\mathbf{x}) < 0 \end{cases}, \tag{13.8}$$

then condition (13.7) can be rewritten as

$$\left(\mathrm{L}_{\mathbf{f}} s + u^+ \cdot \mathrm{L}_{\mathbf{g}} s\right) < 0 \text{ and } \left(\mathrm{L}_{\mathbf{f}} s + u^- \cdot \mathrm{L}_{\mathbf{g}} s\right) > 0. \tag{13.9}$$

One observes that the sign change of $\frac{ds(\mathbf{x})}{dt}$ takes place due to the vector field $\mathbf{g}(\mathbf{x})$ being switched by the switching function u. Condition (13.9) cannot be ensured if $\mathrm{L}_{\mathbf{g}} s = 0$. A necessary, but not sufficient, condition for the switching surface being reached is the so-called *transversality* (or *hitting*) *condition*,

$$\mathrm{L}_{\mathbf{g}} s \neq 0, \tag{13.10}$$

which is intuitively illustrated in Fig. 13.6. Indeed, posing $\mathrm{L}_{\mathbf{g}} s = 0$ is equivalent to posing $\mathbf{ds} \bullet \mathbf{g} = 0$; the nullity of the latter scalar product indicates orthogonality of vector $\mathbf{g}$ and vector $\mathbf{ds}$, normal to surface $\mathbf{S}$. In other words, $\mathbf{g}$ has no component normal to $\mathbf{S}$. The system state cannot therefore be rendered on this surface, meaning that $\mathbf{S}$ is not reachable.

13.3.2 Equivalent Control

Invariance of the switching surface means that the state remains on this surface. If this is the case, the state will then have an equivalent averaged behavior defined by

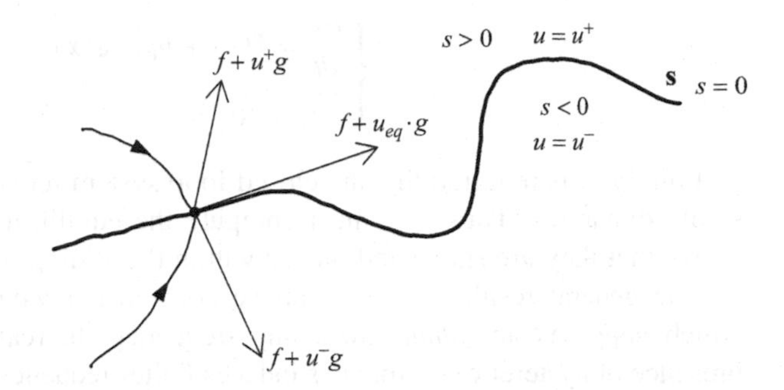

Fig. 13.7 Equivalent averaged behavior on surface **S** according to Filippov (1960)

the so-called *equivalent control* u_{eq} (Utkin 1972) and corresponding vector fields, as shown in Fig. 13.7. The above requirement can be translated as

$$s(\mathbf{x}) = 0 \text{ and } \frac{ds(\mathbf{x})}{dt} = 0, \tag{13.11}$$

which is further equivalent to

$$\mathrm{L}_{\mathbf{f}}s + u_{eq} \cdot \mathrm{L}_{\mathbf{g}}s = 0. \tag{13.12}$$

Equation (13.12) allows us to get the expression of the equivalent control:

$$u_{eq} = -\frac{\mathrm{L}_{\mathbf{f}}s}{\mathrm{L}_{\mathbf{g}}s} = -\frac{\mathbf{ds} \bullet \mathbf{f}}{\mathbf{ds} \bullet \mathbf{g}}, \tag{13.13}$$

which shows that the existence of the equivalent control is subject to the transversality condition being met, i.e., $\mathrm{L}_{\mathbf{g}}s \neq 0$. Note that in the multivariable case, Eq. (13.13) becomes

$$\mathbf{u}_{eq} = \left(\frac{\partial \mathbf{s}}{\partial \mathbf{x}} \cdot \mathbf{g}(\mathbf{x})\right)^{-1} \cdot \left(\frac{\partial \mathbf{s}}{\partial \mathbf{x}} \cdot \mathbf{f}(\mathbf{x})\right). \tag{13.14}$$

A necessary and sufficient condition for the existence of sliding modes is expressed as (the proof can be found in Sira-Ramírez 1988)

$$\min\{u^-, u^+\} < u_{eq} < \max\{u^-, u^+\}. \tag{13.15}$$

To condition (13.15) one can add the requirement that the region **R** where condition (13.15) is met have a nonvoid intersection with surface **S**, i.e., $\mathbf{R} \cap \mathbf{S} \neq \varnothing$, where set **S** is defined by (13.3).

13.3.3 Dynamics on the Sliding Surface

The equivalent control u_{eq} indicates sliding-mode behavior on switching surface **S**. The system behaves as its dynamics would be given by

$$\begin{cases} \dfrac{d\mathbf{x}}{dt} = \mathbf{f}(\mathbf{x}) + u_{eq} \cdot \mathbf{g}(\mathbf{x}) \\ s(\mathbf{x}) = 0. \end{cases} \tag{13.16}$$

Finally, it is required that the closed-loop system remain on surface **S** and have stable dynamics. Thus, one must compute the equilibrium points of (13.16) and verify that they are stable and placed within the sliding mode existence region.

The general results presented above concern the *ideal* sliding-mode behavior, which supposes an *infinite* switching frequency. In real world applications, the presence of hysteretic control laws induces finite-frequency switching. This does not impose theoretical foundations be rebuilt, but one should consider so-called *chattering* phenomena, i.e., oscillatory behavior around the sliding surface (Levant 2010).

13.4 Variable-Structure Control Design

13.4.1 General Algorithm

The following algorithm describes the steps for building the variable-structure control law. It is applicable to a SISO system; nevertheless the methodology is identical for the multivariable decoupled structures. An example will support this algorithm by repeating step-by-step its various stages.

13.4.2 Application Example

Let us consider the buck DC-DC power stage given in Fig. 13.8. One aims at regulating inductor current i_L in order to build a current-controlled generator. To this end, the following notations are adopted: x_1 is inductor current i_L; x_2 is output voltage v_C. Next, the previously-stated algorithm is applied.

Step #1. The system is described as

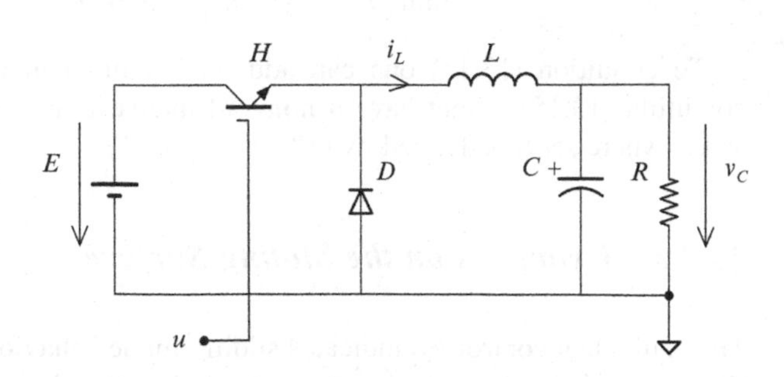

Fig. 13.8 DC-DC buck power stage diagram

$$\frac{d\mathbf{x}}{dt} = \mathbf{f}(\mathbf{x}) + \mathbf{g}(\mathbf{x}) \cdot u,$$

where

$$\mathbf{f}(\mathbf{x}) = \begin{bmatrix} -\frac{v_C}{L} & \frac{i_L}{C} - \frac{v_C}{RC} \end{bmatrix}^T, \mathbf{g}(\mathbf{x}) = \begin{bmatrix} \frac{E}{L} & 0 \end{bmatrix}^T,$$

and u gives the controlled switch H state: H is open if $u = 0$, and H is closed if $u = 1$.

Step #2. The current should be regulated to a certain value denoted by ${i_L}^*$, which leads to a switching surface **S**, defined by

$$\mathbf{S} = \left\{(x_1, x_2) \text{ having property } s(x_1, x_2) = x_1 - i_L^* = 0\right\}.$$

Step #3. The computation of **ds** gives $\mathbf{ds} = [1 \quad 0]$.
Computation of the scalar product $\mathbf{ds} \bullet \mathbf{g}$ gives

$$\mathbf{ds} \bullet \mathbf{g} = L_{\mathbf{g}} s = [1 \quad 0] \cdot [E/L \quad 0]^T = E/L \neq 0,$$

which verifies the transversality condition.

Step #4. The time derivative of s equals

$$\mathrm{L}_{\mathbf{f}} s + u \cdot \mathrm{L}_{\mathbf{g}} s = \mathbf{ds} \bullet \mathbf{f} + u \cdot (\mathbf{ds} \bullet \mathbf{g}).$$

Developing this relation one obtains

$$\frac{ds}{dt} = \frac{1}{L} \cdot (u \cdot E - x_2).$$

With the knowledge that in a buck converter input voltage E is always smaller than output voltage v_C (i.e., x_2), control input u is chosen as follows:

$$u = \begin{cases} 0 & \text{if } s > 0 \left(x_1 > i_L^*\right), \text{which gives } ds/dt < 0 \\ 1 & \text{if } s < 0 \left(x_1 < i_L^*\right), \text{which gives } ds/dt > 0. \end{cases} \tag{13.17}$$

Equation (13.17) defines the sliding-mode control law.

Step #5. The equivalent control input results as

$$u_{eq} = -\frac{\mathrm{L}_{\mathbf{f}} s}{\mathrm{L}_{\mathbf{g}} s} = \frac{x_2}{E}.$$

One notes that the equivalent control input u_{eq} is represented by the duty ratio. This illustrates that the closed-loop system averaged behavior on switching surface **S** results as the consequence of applying the equivalent control input.

Step #6. The region in which the sliding modes exist, i.e.,

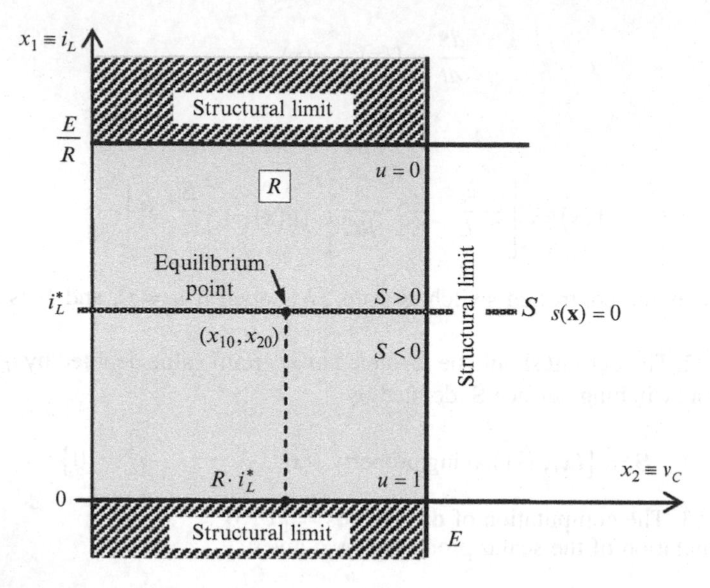

Fig. 13.9 Representation of **R** and **S** in (x_1, x_2) -plane

$$\min\{u^-, u^+\} < u_{eq} < \max\{u^-, u^+\},$$

is determined.

In this case $u^+ = 0$ and $u^- = 1$, the computation of u_{eq} gives that $0 < \frac{x_2}{E} < 1$. Knowing that the voltages x_2 and E are positive, u_{eq} is always positive. Also, the buck converter always delivers a voltage x_2 smaller than the input voltage E. To conclude, sliding modes exist on the entire operating region of the buck converter, defined as

$$\mathbf{R} = \{(x_1, x_2) \text{ having property } 0 < x_2 < E\}.$$

By studying the region **R** representation in the (x_1, x_2) -plane from Fig. 13.9, one can certify that $\mathbf{R} \cap \mathbf{S} \neq \varnothing$.

Step #7. This step concerns the stability assessment on the surface **S**.

The system dynamics on the surface **S** is given by Eq. (13.12), which, by particularization to the considered case, gives

$$x_1 = i_L^* \text{ and} \frac{d\mathbf{x}}{dt} = \mathbf{f}(\mathbf{x}) + u_{eq} \cdot \mathbf{g}(\mathbf{x}),$$

which leads to

$$\frac{dx_2}{dt} = \frac{i_L^*}{C} - \frac{x_2}{RC}. \tag{13.18}$$

The equilibrium point is given by the pair $x_{10} = i_L{}^*$ and $\dot{x_2} = 0$, which leads to $x_{20} = R \cdot i_L{}^*$. The dynamics on the surface **S** is defined by Eq. (13.18) and is always stable as it describes a first-order system with a time constant equal to RC.

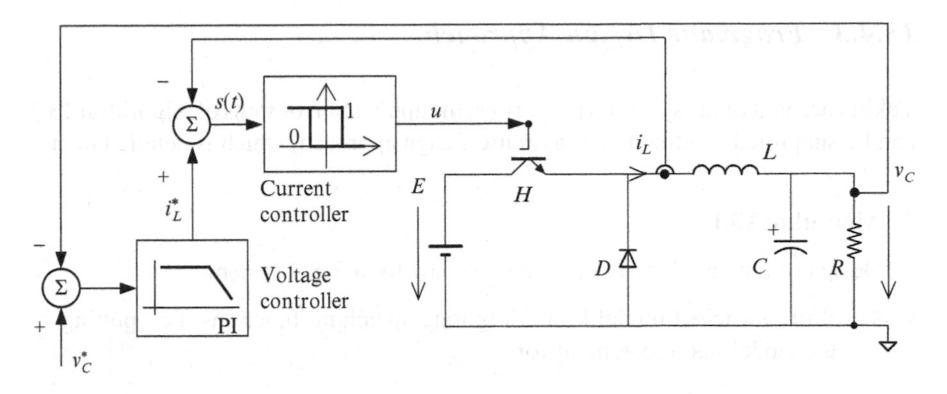

Fig. 13.10 Variable-structure current control employed by a voltage-regulation structure

Remark, This equilibrium point is not possible for any value of imposed inductor current $i_L{}^*$. For example, if $x_{20} = R \cdot i_L{}^* > E$, then one goes outside the existence zone of the sliding modes. In any event, current reference must be chosen in order to respect the converter structural limits:

- $i_L{}^*$ must be smaller than E/R;
- $i_L{}^*$ must be positive (x_{20} also must be positive), which is a trivial condition.

Figure 13.9 shows the structural limits: the operating zone in the state space is the rectangle defined by $0 < x_1 < E/R$ and by $0 < x_2 < E$.

At the end of this stage, one can consider the problem solved.

A supplementary problem arises if one considers the possibility of controlling in the same manner the output voltage v_C (i.e., x_2). If the switching surface **S** is defined as $s(x) = v_C{}^* - x_2$, one can note that the transversality condition is not respected because $\mathbf{ds} = [0 \quad 1]$ and, given that $\mathbf{g} = [E/L \quad 0]^T$, then $\mathbf{ds} \bullet \mathbf{g} = 0$. Therefore, it is not possible to regulate this voltage by means of variable-structure control design. Nevertheless, some solutions may arise, either

(a) by changing the surface **S** so the transversality condition holds (and its expression still contains the reference value $v_C{}^*$); or
(b) by using the previously studied inductor current control as the inner loop in a v_C voltage control structure.

The solution (b) displayed in Fig. 13.10 shows the implementation simplicity of the variable-structure control design. Besides the developed control laws, one can remark that the current control law given by Eq. (13.17) has been modified in the sense of including a hysteresis block; this latter limits the switching frequency of device *H*. The outer control loop built on a PI controller with the plant given by Eq. (13.18) is used for voltage regulation purpose.

13.4.3 Pragmatic Design Approach

Taking account of the system's properties, the application of steps of Algorithm 13.1 can be simplified, leading to a pragmatic design approach, which is detailed next.

Algorithm 13.1

Design of the variable-structure control law for a SISO system

#1. Write switched model by highlighting switching functions, i.e., putting the model into the general form

$$\frac{d\mathbf{x}}{dt} = \mathbf{f}(\mathbf{x}) + \mathbf{g}(\mathbf{x}) \cdot u.$$

#2. Define switching surface **S** (based on reference requirements).
#3. Verify transversality condition:

$$\mathrm{L}_{\mathbf{g}}s = \mathbf{ds} \bullet \mathbf{g} \neq 0.$$

#4. Define control input u that satisfies following relations:

$$\left(\mathrm{L}_{\mathbf{f}}s + u^{+} \cdot \mathrm{L}_{\mathbf{g}}s\right) < 0 \text{ and } \left(\mathrm{L}_{\mathbf{f}}s + u^{-} \cdot \mathrm{L}_{\mathbf{g}}s\right) > 0.$$

#5. Compute equivalent control input:

$$u_{eq} = -\frac{\mathrm{L}_{\mathbf{f}}s}{\mathrm{L}_{\mathbf{g}}s} = -\frac{\mathbf{ds} \bullet \mathbf{f}}{\mathbf{ds} \bullet \mathbf{g}}.$$

#6. Define domain **R** where $\min\{u^{-}, u^{+}\} < u_{eq} < \max\{u^{-}, u^{+}\}$ and verify that it has nonvoid intersection with switching surface **S**.
#7. Assess equivalent dynamics on surface **S** and verify:

- system stability on **S**,
- that equilibrium points are indeed situated within intersection $\mathbf{R} \cap \mathbf{S}$.

In the case that the equivalent dynamics on the surface **S** are unstable, find a stabilization solution by choosing another switching surface.

13.4.3.1 Transversality Condition

The time derivative of the switching surface has the expression

$$\frac{ds(\mathbf{x})}{dt} = \mathrm{L}_{\mathbf{f}(\mathbf{x})}s(\mathbf{x}) + \mathrm{L}_{\mathbf{g}(\mathbf{x})}s(\mathbf{x}) \cdot u \tag{13.19}$$

(see Eq. (10.7) in Chap. 10). In the analyzed case, if control input u appears explicitly in the expression of ds/dt, then obviously $L_g s \neq 0$, which verifies the transversality condition.

Hence, in the previous example $s = x_1 - {i_L}^*$ and $ds/dt = \dot{x}_1 = -v_C/L + u \cdot E/L$. Note that u appears explicitly; therefore the transversality condition is met.

13.4.3.2 Equivalent Control

From Eqs. (13.13) and (13.19) it is found that the equivalent control is the solution of the equation $ds/dt = 0$.

Hence, in the previous example $u_{eq} = v_C/E$, which is the same as the result provided by Step #5.

13.4.3.3 Dynamics on the Sliding Surface

On the sliding surface the following relations hold: $u = u_{eq}$ and $s = 0$. If these conditions are replaced in the systems dynamics $d\mathbf{x}/dt$ given by Eq. (13.4), then a new state vector emerges, whose dynamics $\dot{\mathbf{z}} = f(\mathbf{z}, u_{eq})$ describe the system behavior on the sliding surface.

In the previous example, putting $u = u_{eq}$ cancels the dynamics of the first state (current i_L), $s = 0$ leads to $i_L = {i_L}^*$, which is replaced in the second equation of $d\mathbf{x}/dt$. The behavior on the sliding surface is given by the voltage equation uniquely:

$$\frac{dv_C}{dt} = \frac{i_L^*}{C} - \frac{v_C}{RC},$$

which is identical with Eq. (13.18).

13.5 Supplementary Issues

This subsection is dedicated to how the steps of the variable-structure control design algorithm are particularized for somehow more complex cases that deserve a particular study. Namely, focus will be put on:

- how to deal with the case of a time-varying switching surface, for example, one issued from a sinusoidal reference;
- how to design a switching surface that is attractive;
- how to choose switching functions when the choice is not obvious;
- how to ensure finite switching frequency.

13.5.1 Case of Time-Varying Switching Surfaces

An invariant dynamical system is a system whose dynamics do not depend explicitly on time. A linear system may not be invariant, for example, the buck converter as described by its switched model: its structure changes within the operating period; on the contrary, the averaged model of the same converter presents a linear time-invariant structure.

In power electronics, one may propose time-dependent switching surfaces in the case of power-factor-correction PWM rectifiers or of active filters, for instance. In these cases, the reference to track is a sinusoidal signal or the difference between a harmonic-polluted current and its fundamental, respectively.

It is possible that the expression of $s(t)$ no longer includes a reference, but instead a model of the reference or a tracking model. In both of these cases, the switching surface depends on time.

To be able to use the above stated methodology and algorithm, one must

- assume a constant reference and thus study the various scenarios corresponding to these values of the reference,
- add a new state variable in the system which has constant derivative: time; with this artifice, the time variable can be eliminated from the surface expression.

A PWM rectifier ensuring unit power factor will be presented as the first case study in Sect. 13.6 in order to illustrate these approaches.

13.5.2 Choice of the Switching Surface

Conditions that determine the switching surface reaching strongly depend on the choice of the surface itself. Three methods of obtaining the switching surface are described next, namely: the direct method, the Lyapunov approach and reaching the switching surface with imposed reaching dynamic. The results that follow are also valid for the multivariable case.

The *direct method* derives from the transversality condition (summarized in the fourth step of Algorithm 13.1); it is the one chosen for the previously illustrated example concerning the buck converter. This very simple method is based on

$$\begin{cases} u = u^- \text{ with } ds/dt > 0 \text{ if } s(\mathbf{x}) < 0 \\ u = u^+ \text{ with } ds/dt < 0 \text{ if } s(\mathbf{x}) > 0. \end{cases} \tag{13.20}$$

If written more compactly, Eq. (13.20) becomes

$$s(\mathbf{x}) \cdot \frac{ds(\mathbf{x})}{dt} < 0. \tag{13.21}$$

One must remember that the condition expressed by (13.20) or (13.21) is global, but it does not guarantee surface reachability in a finite time. In the multivariable case, one again uses (13.20) or (13.21) by particularizing them with respect to the various switching surfaces s_i composing $\mathbf{s} = \begin{bmatrix} s_1 & s_2 & \cdots & s_p \end{bmatrix}^T$.

The Lyapunov stability theorem stated in Chap. 10 can be applied to finding the switching surface of a variable-structure control law. Thus, by choosing $V(\mathbf{x}, t) = \mathbf{s}^T\mathbf{s}$ in the general (multivariable) case – which gives $V(\mathbf{x}, t) = s^2$ in the scalar case – one can verify that V is a Lyapunov function. The computation of the time derivative $\dot{V}(\mathbf{x}, t)$ gives $2\mathbf{s}^T\dot{\mathbf{s}}$ in the first case and $2s\dot{s}$ in the second case. It is then sufficient to choose the control input $\mathbf{u}$ (or u) such that

$$\frac{d}{dt}V(\mathbf{x}, t) \leq -K \cdot \|\mathbf{x}\| \; \forall \, \mathbf{x}, t,$$

where K is a positive gain, in order to ensure finite-time convergence. Note that if the state space origin is taken as the equilibrium point, the results remain general, as it is sufficient to make the adequate change of state variable.

A third method of choosing the switching surface technique supposes not only the reaching of the surface **S** in finite time but also the manner in which this surface is reached. This is defined by the so-called *convergence law*; this is why this method can be called the *controlled-convergence method*. In this case the surface **S** can be defined by a noninvariant function **s(x)**.

As stated by Hung et al. (1993), the convergence law can be

(a) at constant speed:

$$\frac{d}{dt}\mathbf{s}(\mathbf{x}) = -\mathbf{K} \cdot \mathbf{sgn}(\mathbf{s}(\mathbf{x})), \tag{13.22}$$

where the **sgn** function is defined by

$$\mathbf{sgn}(s(\mathbf{x})) = \begin{bmatrix} \mathrm{sgn}(s_1(\mathbf{x})) & \mathrm{sgn}(s_2(\mathbf{x})) & \cdots & \mathrm{sgn}(s_p(\mathbf{x})) \end{bmatrix}^T$$

– with function sgn taking value 1 if its argument is positive and value -1 otherwise – and **K** being a $p \times p$ diagonal matrix with positive elements;

(b) at variable speed (depending on the distance from **S**):

$$\frac{d}{dt}\mathbf{s}(\mathbf{x}) = -\mathbf{K} \cdot \mathrm{sgn}(\mathbf{s}(\mathbf{x})) - \lambda \cdot \mathbf{s}(\mathbf{x}); \tag{13.23}$$

(c) at controlled speed expressed in the form of powers of $s_i(\mathbf{x})$:

$$\frac{d}{dt}s_i(\mathbf{x}) = -k_i \cdot |s_i(\mathbf{x})^r| \cdot \mathrm{sgn}(s_i(\mathbf{x})) - \lambda \cdot s_i(\mathbf{x}), i = 1, 2, \cdots, p, 0 < r < 1. \tag{13.24}$$

Remarks. Equation (13.22) is a particular case of Eq. (13.23). Taking **K** as the zero matrix in Eq. (13.23) does not ensure a finite-time convergence towards **S** and therefore no sliding modes are obtained. The case of the direct approach – Eqs. (13.20) or (13.21) – is a particular case of Eq. (13.22).

To conclude, a good choice of the switching surface is crucial for variable-structure control performance. Its expression must embed the control goal, e.g., the output setpoint, but it also must ensure the transversality condition by implementing simple nonlinear operation such as the relay function. Usually, in practice the switching surface results as a function of measurable states (Buhler 1986; Malesani et al. 1995; Tan et al. 2011); it may be altered in order to maintain some internal state variables (e.g., inductor currents) within reasonable limits.

13.5.3 *Choice of the Switching Functions*

Once the expression of **s** has been established, one must find the switching function (s) that allow(s) its implementation. In the previously studied example this is simple, but in the multivariable case this choice may be far more difficult.

Suppose that the switching surface has resulted by applying the controlled-convergence method at variable speed; therefore, it is defined as

$$\frac{d}{dt}\mathbf{s}(\mathbf{x}) = -\mathbf{K}\cdot \operatorname{sgn}(\mathbf{s}(\mathbf{x})) - \lambda \cdot \mathbf{s}(\mathbf{x}),$$

and knowing the system dynamics on one hand and the time surface derivative on the other hand

$$\frac{d\mathbf{s}(\mathbf{x})}{dt} = \frac{\partial \mathbf{s}(\mathbf{x})}{\partial \mathbf{x}}\cdot \mathbf{f}(\mathbf{x}) + \left(\frac{\partial \mathbf{s}(\mathbf{x})}{\partial \mathbf{x}}\cdot \mathbf{g}(\mathbf{x})\right)\cdot \mathbf{u},$$

the control input **u** is then chosen as

$$\mathbf{u}(\mathbf{x}) = -\left(\frac{\partial \mathbf{s}(\mathbf{x})}{\partial \mathbf{x}}\cdot \mathbf{g}(\mathbf{x})\right)^{-1}\cdot\left(\frac{\partial \mathbf{s}(\mathbf{x})}{\partial \mathbf{x}}\cdot \mathbf{f}(\mathbf{x}) + \mathbf{K}\cdot \operatorname{sgn}(\mathbf{s}(\mathbf{x})) + \lambda \cdot \mathbf{s}(\mathbf{x})\right). \quad (13.25)$$

If the Lyapunov approach is used, the surface defined by $\mathbf{s}(\mathbf{x}) = 0$ contributes at building the Lyapunov function $V(\mathbf{x},t) = 2\mathbf{s}^T\dot{\mathbf{s}}$; **u** is chosen such that to obtain

$$\frac{d}{dt}V(\mathbf{x},t) \le -\gamma \cdot \|\mathbf{x}\| \;\; \forall\, \mathbf{x}, t,$$

or else

$$2\cdot \mathbf{s}^T \cdot \left(\frac{\partial \mathbf{s}(\mathbf{x})}{\partial \mathbf{x}}\cdot \mathbf{f}(\mathbf{x}) + \frac{\partial \mathbf{s}(\mathbf{x})}{\partial \mathbf{x}}\cdot \mathbf{g}(\mathbf{x})\cdot \mathbf{u}\right) \le -\gamma \cdot \|\mathbf{x}\|, \;\; \forall\, \mathbf{x}, t. \quad (13.26)$$

Remarks. If the system to be controlled indeed has variable structure, meaning that the switching functions take discrete values, Eq. (13.26) is easier to comprehend.

This is not the case for the Eq. (13.25), which provides a continuous value of the control input vector **u**. Only in this latter case is the reasoning done based upon a continuous-time system where **u** can take infinitely many values, which is appropriate for a bang-bang-type approach.

One must therefore adapt the reasoning to a classical variable-structure control problem. For example, this is the case of a decoupled system with variable-gain relays where the problem is solved by taking the components $u_i(\mathbf{x})$ of **u** as

$$u_i(\mathbf{x}) = \begin{cases} k_i^+ & \text{if} \quad s_i(\mathbf{x}) > 0 \\ k_i^- & \text{if} \quad s_i(\mathbf{x}) < 0, \end{cases} \tag{13.27}$$

the relay gains $k_i^+(\mathbf{x})$ and $k_i^-(\mathbf{x})$ being chosen to satisfy Eq. (13.25).

The "unnatural" approach of controlling a continuous-time system by means of variable-structure control design can be built "locally" by choosing a control input of the form $\mathbf{u} = \mathbf{u}_{eq} + \Delta\mathbf{u}$, which brings the reasoning back to the classical methodology and algorithm already presented (DeCarlo et al. 2011).

In the multivariable case a $(n - p)$-dimensional switching surface results, which is the intersection between different surface components s_i.

In the case where the system is naturally decoupled, the variable-structure control problem is reducible to the scalar case: the p various surfaces are independently reached. A four-wire three-phase inverter with capacitor divider in the DC circuit is an example illustrating this situation.

If, on the contrary, the system is not decoupled, one must either

- make a variable change so any switching function affects only a single switching function; or
- find a preliminary decoupling control law $\mathbf{u} = F(\mathbf{x},\mathbf{u}')$ so the functions u_i' affect only their respectively corresponding surfaces s_i.

A case study will be considered for illustrating the nondecoupled case: a three-phase voltage-source inverter (without neutral wiring).

The surfaces s_i can be reached one by one; therefore, a single control input u_i switches each time. The approach can also be global. In this case the various functions u_i are chosen so the resulting vector field brings the state directly to the resultant switching surface, i.e., the intersection of the components s_i.

13.5.4 Limiting of the Switching Frequency

The results presented above suppose an infinite-frequency switching function; therefore, vectors switch instantaneously from one side to the other of the surface. In practice the real systems are not infinitely fast and, even if the power electronic switches become increasingly better-performing, their operation will always introduce delays that will alter more or less the state-space system trajectories.

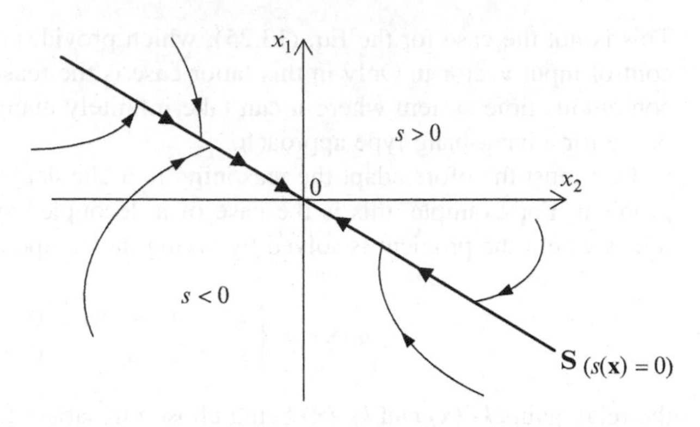

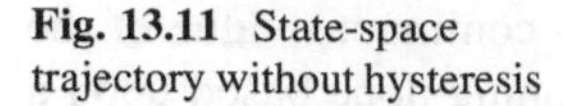
Fig. 13.11 State-space trajectory without hysteresis

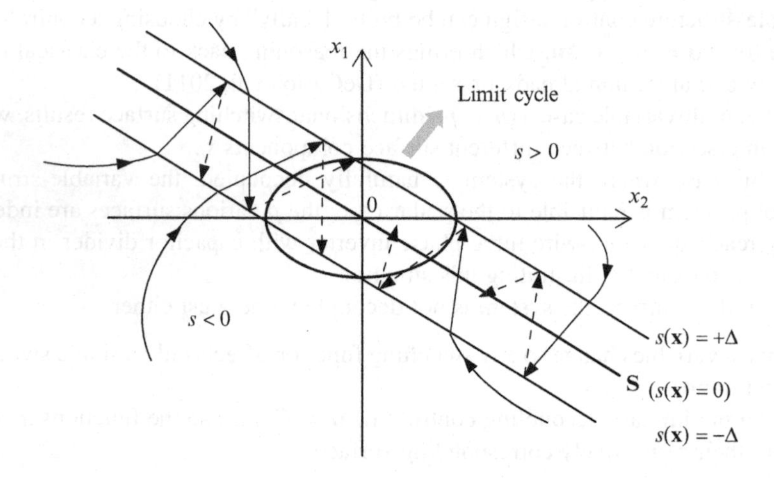

Fig. 13.12 State-space trajectory with hysteresis

In Fig. 13.11 is represented an ideal sliding mode trajectory in the state plane of a second-order system. One can remark that the vector fields bring the state trajectories onto the surface **S** and then towards the equilibrium point (this has been arbitrarily taken as the origin).

Multiple techniques are available that reduce switching frequency to a suitable value; among such techniques, two are presented next as examples.

Figure 13.12 shows the effect of a hysteresis of imposed width Δ. The switching series are done in an interval Δ around the surface **S**. A phenomenon appears around the equilibrium point which is never reached; namely, the state-plane trajectory gravitates around this point describing a limit cycle. This is the kind of undesirable phenomenon that must be reduced; it is called *chattering*. However, the faster the converter switchings are, the more reduced the surface of the limit cycle is, and so too the oscillations around the surface **S**.

In power electronics variable-structure control practice, some admissible current and voltage variation magnitudes must be allowed, thus allowing reasonably high switching frequency.

Suppose, for example, the case when the sliding surface is obtained as a linear combination of states that employ the feedback vector $\mathbf{k}^T$. Therefore,

$$\dot{s}(\mathbf{x}) = -\mathbf{k}^T\dot{\mathbf{x}}.$$

When the switching frequency is finite, the surface $s(\mathbf{x})$ has a triangular-like shape with a magnitude of Δ, and $\dot{s}(\mathbf{x})$ differs slightly from zero:

$$\dot{s}(\mathbf{x}) = -\mathbf{k}^T(f(\mathbf{x}) + g(\mathbf{x}) \cdot u_{\text{lim}}), \tag{13.28}$$

where u_{lim} is equal either to u^+ or to u^-. As result, the state trajectory $\mathbf{x}$ gets farther or closer to the switching hyperplane characterized by $s(\mathbf{x}) = 0$ and $\dot{s}(\mathbf{x}) = 0$ with a speed that represents the time derivative of the switching surface, $\dot{s}(\mathbf{x})$. Let $\mathbf{x}_0$ be that state vector value corresponding to the dynamics on the switching hyperplane, that is,

$$\dot{s}(\mathbf{x}_0) = 0 = -\mathbf{k}^T\left(f(\mathbf{x}_0) + g(\mathbf{x}_0) \cdot u_{eq}\right). \tag{13.29}$$

It follows that $\mathbf{x}$ varies slightly around $\mathbf{x}_0$, so $f(\mathbf{x}) \cong f(\mathbf{x}_0)$, $g(\mathbf{x}) \cong g(\mathbf{x}_0)$. By combining Eqs. (13.28) and (13.29) one obtains (Buhler 1986)

$$\dot{s}(\mathbf{x}) = -\mathbf{k}^T g(\mathbf{x}_0)\left(u_{\text{lim}} - u_{eq}\right). \tag{13.30}$$

Equation (13.30) is useful for computing the values of subinterval durations T_{on} and T_{off} by replacing u_{lim} by u^+ and u^-, respectively. Therefore, introducing a hysteresis of width Δ in the control law is a way of limiting the switching frequency to (Buhler 1986):

$$f_{\text{max}} = \frac{k^T g(\mathbf{x}_0)}{8\Delta}(u^+ - u^-), \tag{13.31}$$

showing that the switching frequency can be influenced by choosing either the hysteresis value or the vector of the state feedback, $\mathbf{k}^T$. Note that in the general case $s(\mathbf{x})$ is a nonlinear function and f_{max} depends proportionally on $\mathrm{L}_g s(\mathbf{x}_0)$.

Another way of limiting the switching frequency is by superposing an auxiliary signal, for example, a triangle-wave, over the original surface, $s(\mathbf{x})$. In this case, the resulting switching surface will have a ripple of 2Δ around an average value of a nonzero value, s_{av} (see Fig. 13.13).

Suppose that the switching surface in obtained as a linear state feedback, as it was in the previous case. Hence, $\dot{s}(\mathbf{x})$ results from Eq. (13.30) and is used to compute the subintervals T_{on} and T_{off} and then the switching frequency f. Imposing an upper bound of the switching frequency, f_{max}, will determine a lower limit of the hysteresis width, as these variables are related by the following relation (Buhler 1986):

$$\Delta = \mathbf{k}^T g(\mathbf{x}_0) \cdot \frac{\left(u^+ - u_{eq}\right)\left(u_{eq} - u^-\right)}{u^+ - u^-} \cdot \frac{1}{2f}.$$

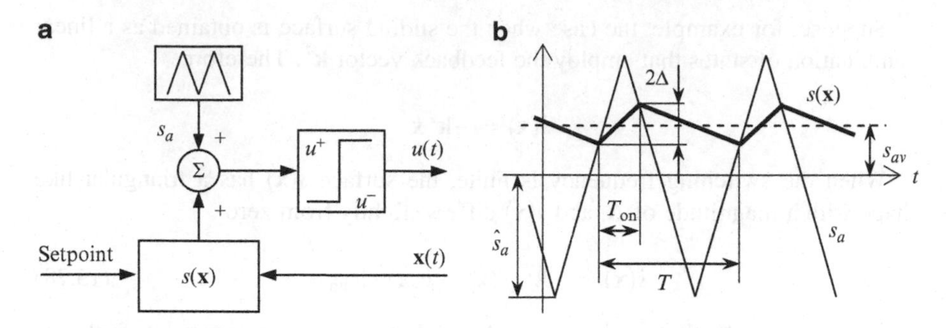

Fig. 13.13 Superposing an auxiliary signal over switching surface: (**a**) block diagram; (**b**) associated waveforms

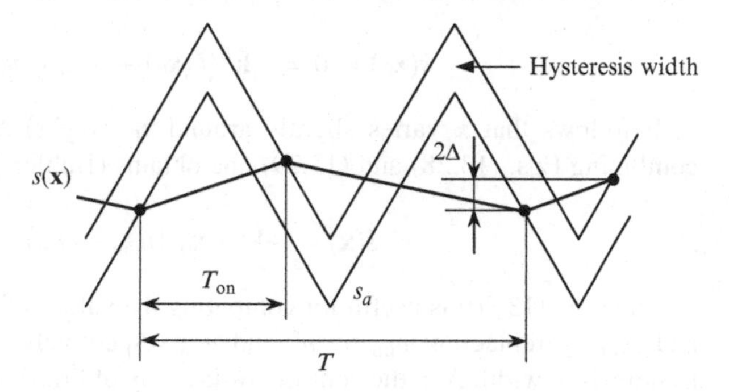

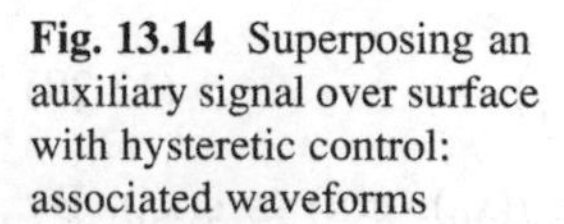

Fig. 13.14 Superposing an auxiliary signal over surface with hysteretic control: associated waveforms

Depending of the actual operating point, the average value of the switching surface, s_{av}, may take values between $-\hat{s}_a$ and $+\hat{s}_a$ (see Fig. 13.13b). Once the switching frequency has been chosen, one must take into account the fact that, in order to ensure proper operation, the slope of the auxiliary signal s_a must be higher than that of the switching surface $s(\mathbf{x})$ in the most unfavorable conditions. After some algebra this condition is translated into a relation between the switching frequency and the amplitude of the triangle wave, $\hat{s}_a$ (Buhler 1986):

$$\hat{s}_a > \frac{\mathbf{k}^T g(\mathbf{x}_0)(u^+ - u^-)}{4f_{\max}}.$$

In this way the auxiliary signal parameters can be effectively designed.

Another version of the diagram from Fig. 13.13a results by replacing the sign function with a hysteresis; the associated waveforms are shown in Fig. 13.14 (Spiazzi et al. 1995). The constraints on the switching frequency are less restrictive in this case.

Note that the list of above mentioned techniques for limiting the switching frequency domain is not exhaustive. For example, a PLL can also be used to control hysteresis width (Malesani et al. 1996; Guffon 2000).

13.6 Case Studies

Power electronic converters offer rich application field for variable-structure control. Increases of operating frequencies and of switched power are suitable for various applications, among which are power-factor-correction rectifying, active filtering, reactive power compensation, renewable energy conversion systems control. The main benefits are the improvement of the closed-loop response time and robustness.

The following case studies dealt have been chosen because they are representative from an instructional perspective.

13.6.1 Variable-Structure Control of a Single-Phase Boost Power-Factor-Correction Converter

The considered converter is shown in Fig. 13.15. Its aim is to output a constant DC voltage while reducing the harmonic content of the current absorbed from the grid and ensuring operation at unit power factor.

The main control objective is to regulate output voltage irrespective of load (in an acceptable power range). A constraint is to maintain AC current i_a as sinusoidal, in phase with grid voltage v_a. A single control input is available toward achievement of these two requirements: the turn-on/turn-off control signal u of the switch within the boost converter (see Fig. 13.15).

In order to fulfill the above stated goal, the switched models of each sub-circuit must be obtained by identifying the switching functions of:

- the boost converter by taking a constant positive voltage E;
- the diode rectifier by taking the voltage v_a as input and voltage E as output;
- the entire circuit by making the coupling occur between various variables.

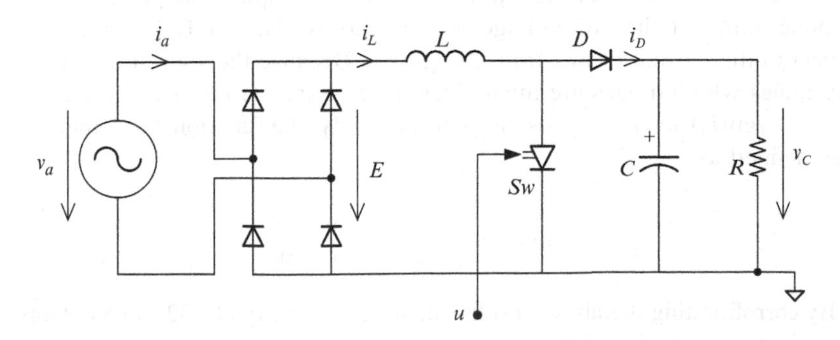

Fig. 13.15 General circuit description of boost power-factor-correction converter ($v_a = V_M \cdot \sin \omega t$ is AC power grid voltage)

After having established the averaged model of the rectifier – according to the methodology presented in Chap. 4 – one must render the model back to the boost converter switching time scale to obtain the complete circuit averaged model.

Variable-structure control design is based upon a cascade control structure, comprising an inner inductor current loop – which is computed by supposing voltage E constant – and an outer output voltage control loop. In this way the boost converter is decoupled from the rectifier. For the current control a static switching surface of the form $(i_L - i_L{}^*)$ is first chosen, then a dynamical switching surface with $i_L{}^* = I_M \cdot |\sin \omega t|$ – where v_a is taken as phase origin – will be chosen, thus allowing a sinusoidal input current to be drawn from the grid. The averaged behavior of the output voltage when the switching surface is reached can next be deduced. Finally, one can suggest a solution for the outer voltage loop.

The developments associated to the above issues will next be organized into three parts, dedicated respectively to modeling, sliding-mode control with constant switching surface and sliding-mode control with variable switching surface dedicated to operation at unit power factor.

(a) *Modeling*

The switched model of the boost power converter can be obtained following the steps indicated in Chap. 3:

$$\frac{d}{dt}\mathbf{x} = \mathbf{f}(\mathbf{x}) + \mathbf{g}(\mathbf{x}) \cdot u,$$

with $\mathbf{x} = \begin{bmatrix} i_L & v_C \end{bmatrix}^T$ being the state vector, where

$$\mathbf{f}(\mathbf{x}) = \begin{bmatrix} 0 & -\dfrac{1}{L} \\ \dfrac{1}{C} & -\dfrac{1}{RC} \end{bmatrix} \cdot \begin{bmatrix} x_1 \\ x_2 \end{bmatrix} + \begin{bmatrix} \dfrac{E}{L} \\ 0 \end{bmatrix}, \mathbf{g}(\mathbf{x}) = \begin{bmatrix} \dfrac{x_2}{L} \\ -\dfrac{x_1}{C} \end{bmatrix}. \tag{13.32}$$

The diode rectifier has the property that its output voltage E equals the absolute value of the AC voltage v_a. This holds also for DC current i_L with respect to the current drawn from AC grid i_a. Because the sign of the current i_a determines which diodes are in conduction, the expressions of DC variables are $E = v_a \cdot \text{sgn}(i_a)$ and $i_L = i_a \cdot \text{sgn}(i_a)$, respectively; the function "sgn" has already been defined as

$$\text{sgn}(x) = \begin{cases} 1 & \text{if} \quad x \geq 0 \\ -1 & \text{if} \quad x < 0. \end{cases}$$

By corroborating the above results and developing Eq. (13.32), one obtains

$$\begin{cases} \dot{x_1} = -\frac{1}{L}(V_M \cdot \sin(\omega t) \cdot \mathrm{sgn}(i_a) - x_2 \cdot u) \\ \dot{x_2} = \frac{1}{C}\left(x_1 - \frac{x_2}{R} - x_1 \cdot u\right), \end{cases} \tag{13.33}$$

which is the retained form for the global circuit switched model. Note that sgn $(i_a) = \mathrm{sgn}(\sin \omega t)$ due to the operation at unit power factor.

The diode rectifier averaged model has no dynamics as it includes no energy accumulation element. Therefore, one takes $E = |v_a|$ when its output current i_L is nonzero, also meaning that signals i_a and v_a are in phase. This means that the diode rectifier output average value on a (voltage grid) period is

$$\langle E \rangle_0 = \frac{2}{\pi} \cdot V_M.$$

The complete circuit averaged model rendered to the time scale of the boost converter may suppose in a first stage the voltage E as constant; hence, one takes the model given by (13.32) as the basis for the preliminary computations.

(b) *Case of a constant switching surface*
Variable-structure control design
Variable-structure control can be justified by several remarks:

- the control action must ensure a very accurate tracking of the current reference; this means that the closed-loop system bandwidth must be very large;
- the static switch *Sw* is operated by a binary ON/OFF signal;
- the boost power converter has nonlinear behavior depending on the operating point.

Note that controlling i_L means also controlling i_a, as it is a direct algebraic link between the two signals, that is, $i_a = i_L \cdot \mathrm{sgn}(i_a)$. In deducing the variable-structure control law, one can employ either a mathematical approach – that is, follow the steps of Algorithm 13.1 – or a pragmatic one.

Being given the switched model of the boost power stage and the switching surface, the *mathematical approach* begins by verifying the transversality condition ($\mathbf{L_g s} \neq 0$). Knowing that $s(\mathbf{x}) = x_1 - {i_L}^*$ and that $\mathbf{g}(\mathbf{x}) = [x_2/L \quad -x_1/C]^T$, one obtains

$$\mathbf{L_g s} = \mathbf{ds} \bullet \mathbf{g}(\mathbf{x}) = [1 \quad 0] \cdot [x_2/L \quad -x_1/C]^T = x_2/L.$$

Respecting the transversality condition implies that $x_2 \neq 0$ (nonzero output voltage), which concurs with the boost circuit operation.

Using the *pragmatic approach*, one can directly choose the control input u so as to ensure $s(\mathbf{x}) \cdot \dot{s(\mathbf{x})} < 0$ or, equivalently, $\mathrm{sgn}\left(\dot{s(\mathbf{x})}\right) = -\mathrm{sgn}(s(\mathbf{x}))$. Computations reveal the next statement:

- if $s(\mathbf{x}) > 0$ then u is chosen such that $E + (1 - u)x_2 < 0$;
- if $s(\mathbf{x}) < 0$ then u is chosen such that $E + (1 - u)x_2 > 0$.

Knowing that in normal operation the boost output voltage x_2 is larger than E, the choice of u becomes

- if $s(\mathbf{x}) > 0$ then $u = u^+ = 0$;
- if $s(\mathbf{x}) < 0$ then $u = u^- = 1$.

One can rapidly verify on the circuit diagram (Fig. 13.15) that, when the condition $i_L > i_L{}^*$ is fulfilled (equivalent with condition $s(\mathbf{x}) > 0$), if the switch is opened (meaning that $u = u^+ = 0$), the current i_L will begin to decrease, leading to $ds/dt < 0$. In the same way one can verify that the condition of $i_L < i_L{}^*$ leads eventually to an increase of the gradient of $s(\mathbf{x})$.

Equivalent control input
Computation of the equivalent control input u_{eq} gives:

$$u_{eq} = 1 - \frac{E}{x_2}.$$

One remarks that this variable is none other than the boost converter duty ratio.

The sliding mode existence domain is the geometrical locus described by the pair (x_1, x_2) defined by the relation $0 < u_{eq} < 1$, which is equivalent with writing $0 < E < x_2$. This latter inequality is satisfied when the boost converter operates normally. One remarks that the sliding domain presents a nonvoid intersection with the chosen switching surface $\mathbf{S}$.

Dynamics on the sliding surface
One can now analyze the dynamics on $\mathbf{S}$. As the surface $\mathbf{S}$ is reached, $x_1 = i_L{}^*$ and $u = u_{eq}$, the equivalent dynamic of the system is thus given by

$$\frac{d}{dt}x_2 = \frac{1 - u_{eq}}{C} \cdot i_L^* - \frac{x_2}{RC},$$

which after some development leads to

$$\frac{d}{dt}x_2 = \frac{E}{C} \cdot \frac{1}{x_2} \cdot i_L^* - \frac{x_2}{RC}. \tag{13.34}$$

Based on Eq. (13.34) one deduces equilibrium points as follows:

$$x_{20} = \pm\sqrt{R \cdot i_L^* \cdot E}.$$

Obviously, only the positive point $x_{20} = \sqrt{R \cdot i_L^* \cdot E}$ is taken into account.

As regards the stability assessment on the surface $\mathbf{S}$, it is possible to approach it in two ways, namely:

- by studying the pole of the transfer function of the small-signal system model around the equilibrium point x_{20};
- by studying the dynamic of the energy function $V(x_2) = \frac{1}{2}Cx_2^2$.

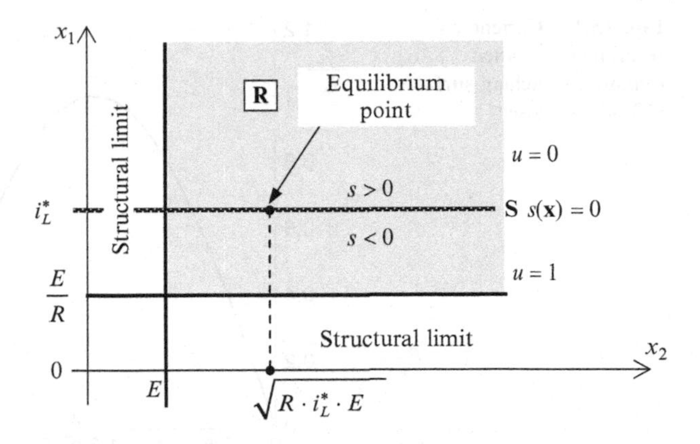

Fig. 13.16 Domain of sliding modes existence in case of boost power-factor-correction converter

If the first method is used, the small-signal model around the point x_{20} gives

$$\frac{d}{dt}\tilde{x}_2 = \frac{1}{C}\sqrt{\frac{E}{R \cdot I_L^*}} \cdot \tilde{i}_L^* - 2\frac{\tilde{x}_2}{RC}, \tag{13.35}$$

where $\tilde{i}_L^* = x_1 - i_L^*$ and $\tilde{x}_2 = x_2 - x_{20}$; one remarks that the transfer equivalent to $\tilde{x}_2 / \tilde{i}_L^*$ is described by a first-order stable transfer function with a time constant equal to $RC/2$ therefore the small-signal dynamic on surface **S** is stable.

The second method makes use of the energy function's derivative:

$$\frac{d}{dt}V = -2\frac{V}{RC} + E \cdot i_L^*. \tag{13.36}$$

Note that the large-signal model given by (13.36) is linear and stable, which confirms the previous result.

Domain of sliding modes existence

The existence domain of the sliding modes (previously deduced) is $0 < E < x_2$. The condition for which the equilibrium points are included in this domain is

$$x_{20} = \sqrt{R \cdot i_L^* \cdot E} > E,$$

which leads to a current reference that satisfies $i_L{}^* > E/R$. Figure 13.16 presents the existence domain of the sliding modes on surface **S**.

(c) *Choice of a dynamic switching surface for operation at unit power factor*

Next, a dynamic surface is taken into account, that is, i_L is no longer a constant but varies with time (see Fig. 13.17):

$$s(\mathbf{x}) = |I_M \sin \omega t| - x_1.$$

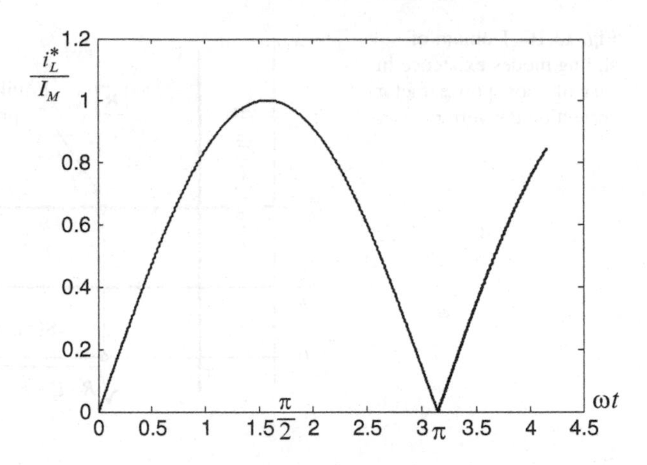

Fig. 13.17 Current reference $i_L{}^*$, when dynamic switching surface (13.38) is chosen

By introducing an artificial state, $x_3 \equiv \omega t$, Eq. (13.33) become

$$\begin{cases} \dot{x_1} = -\dfrac{1}{L}(V_a|\sin x_3| - x_2 + x_2 \cdot u) \\ \dot{x_2} = \dfrac{1}{C}\left(x_1 - \dfrac{x_2}{R} - x_1 \cdot u\right) \\ \dot{x_3} = \omega \end{cases} \tag{13.37}$$

and $s(\mathbf{x})$ can be written as

$$s(\mathbf{x}) = |I_M \sin x_3| - x_1. \tag{13.38}$$

This setup ensures that the current drawn from the AC grid, i_a, is sinusoidal and in phase with the grid voltage v_a.

One takes into account only the first semi-period of the AC voltage, v_a, for the computations that follow; this emphasizes the use of the absolute value function. It follows that x_3 varies within $[0, \pi)$ and therefore $|\sin x_3| = \sin x_3$. The results for the second semi-period are equivalent.

The switching function is chosen as in the time-invariant case; on the contrary, the equivalent control input computation is different. Using its definition relation and knowing that the system dynamics are given by Eq. (13.37), one obtains

$$\mathbf{ds} \bullet \mathbf{f} = [-1 \quad 0 \quad I_M \cos x_3] \cdot \begin{bmatrix} \dfrac{1}{L}(V_a \sin x_3 - x_2) \\ \dfrac{1}{C}\left(x_1 - \dfrac{x_2}{R}\right) \\ \omega \end{bmatrix},$$

$$\mathbf{ds} \bullet \mathbf{g} = \begin{bmatrix} -1 & 0 & I_M \cos x_3 \end{bmatrix} \cdot \begin{bmatrix} x_2/L \\ -x_1/C \\ 0 \end{bmatrix}.$$

Therefore, the equivalent control input is written as

$$u_{eq} = -\frac{\mathbf{ds} \bullet \mathbf{f}}{\mathbf{ds} \bullet \mathbf{g}} = 1 - \frac{V_a \sin x_3 - L\omega \cdot I_M \cos x_3}{x_2}. \tag{13.39}$$

Because the condition of existence of sliding modes has not changed, $0 < u_{eq} < 1$, and the variable x_2 is positive – by virtue of structural issues, the previous inequality can be rewritten as

$$x_2 > V_a \sin x_3 - L\omega \cdot I_M \cos x_3 > 0,$$

or, equivalently,

$$x_2 > V_a \left(\sin x_3 - \frac{L\omega \cdot I_M}{V_a} \cdot \cos x_3 \right) > 0. \tag{13.40}$$

As the quantity $L\omega \cdot I_M/V_a$ represents a real positive number, then there exists $\alpha \in [0, \pi/2)$ such that

$$\tan \alpha = \frac{L\omega \cdot I_M}{V_a}, \tag{13.41}$$

which leads condition (13.40) to become

$$x_2 > \frac{V_a}{\cos \alpha} \sin (x_3 - \alpha) > 0.$$

From notation (13.41) one deduces that $\cos \alpha = V_a / \sqrt{V_a^2 + (L\omega \cdot I_M)^2}$, which finally results in

$$x_2 > \sqrt{V_a^2 + (L\omega \cdot I_M)^2} \cdot \sin (x_3 - \alpha) > 0. \tag{13.42}$$

Equation (13.42) giving the sliding modes existence domain provides information on how to design power-factor-correction converter components; it also allows understanding of why the closed-loop circuit does not work at the beginning of each semi-period. Indeed, the condition $\sqrt{V_a^2 + (L\omega \cdot I_M)^2} \cdot \sin (x_3 - \alpha) > 0$ is valid for arguments of the sinus function placed between 0 and π. This is equivalent with saying that condition (13.42) is not satisfied for values of x_3 situated between 0 and α, so the sliding modes do not appear within the first α/ω seconds at the beginning of each period of v_a.

Besides this aspect, satisfying the inequality $x_2 > \sqrt{V_a^2 + (L\omega \cdot I_M)^2}$ leads to taking into account the variable $x_2 \equiv v_C$ (output voltage) desired in the electrical circuit design. For example, if v_C is too small, the circuit does not work, if one does not consequently reduce the value of V_a, which in its turn requires the reduction of the inductance L, and so on.

Equivalent dynamic on the switching surface
The equivalent dynamic on the chosen variable sliding surface is described by

$$\frac{dx_2}{dt} = \frac{x_1}{C} \cdot \left(1 - u_{eq}\right) - \frac{x_2}{RC},$$

where u_{eq} is given by (13.39) and x_1 by (13.38), where $s(\mathbf{x}) = 0$; this dynamic is therefore

$$\frac{dx_2}{dt} = \frac{V_a \cdot I_M}{x_2 C}(\sin x_3)^2 - \frac{L\omega \cdot I_M^2}{x_2 C} \sin x_3 \cos x_3 - \frac{x_2}{RC}. \tag{13.43}$$

Note that the primary objective of the rectifier is to offer a constant and noiseless output voltage (the capacitor C is chosen as a consequence); therefore, one of the variables of interest is the average value of x_2. If averaging Eq. (13.43) on one period of v_a (this means x_3 varies from 0 to π), one obtains

$$\frac{d\langle x_2\rangle_0}{dt} = \frac{V_a \cdot I_M}{2\langle x_2\rangle_0 C} - \frac{\langle x_2\rangle_0}{RC}. \tag{13.44}$$

As in the previous case of a time-invariant switching surface, the dynamic from Eq. (13.43) is stable and can be used to design the output voltage control.

Output voltage control
In order to propose a solution for the output voltage control, one notes the average behavior of x_2 on the surface $\mathbf{S}$ given by Eq. (13.43) is nonlinear, as it depends on the operating point. A classical solution would establish a controller based on the linearized (small-signal) model. This solution has an important drawback as it does not provide the desired performances when the output voltage reference varies.

Another quite elegant solution consists in tracking no longer the circuit output voltage but its squared value. Indeed, according to Eq. (13.44), the dynamics of x_2 averaged and squared is of first order:

$$\frac{d\langle x_2\rangle_0^2}{dt} = \frac{V_a}{C} \cdot I_M - 2\frac{\langle x_2\rangle_0^2}{RC},$$

which renders interesting the use of a classical PI controller with constant parameters, as presented in Fig. 13.18. However, one must acknowledge that the inherent measure

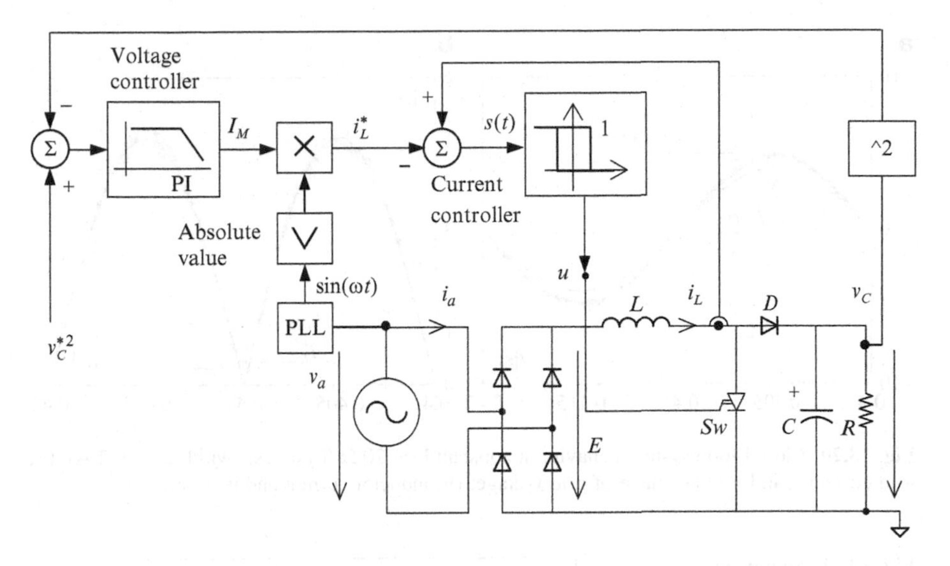

Fig. 13.18 Proposed complete control structure for power-factor-correction converter

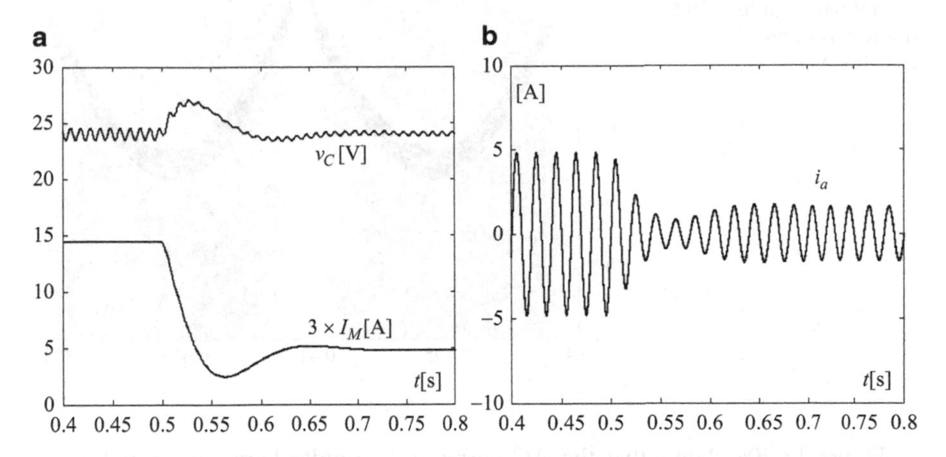

Fig. 13.19 Closed-loop behavior at load step variation from 20 to 60 Ω (hysteresis width $\Delta = 0.1$ A): **(a)** v_C and $3 \times I_M$; **(b)** grid current i_a

noise superposed on the output voltage may introduce problems in the control loop because the squaring operator significantly amplifies high-frequency noises.

Next, some illustrative simulation results for this case study will be presented. Circuit parameters are: $V_a = 12$ V, $\omega = 100\ \pi$ rad/s, $v_C^* = 24$ V, $L = 1$ mH, $C = 3600$ μF and rated power 75 W. Figure 13.19 shows the overall system performance in rejecting the load variations. Voltage v_C's evolution in Fig. 13.19a shows a transient regime lasting for about 0.1 s. The corresponding outer loop control input, I_M, is the AC current envelope (see Fig. 13.19b).

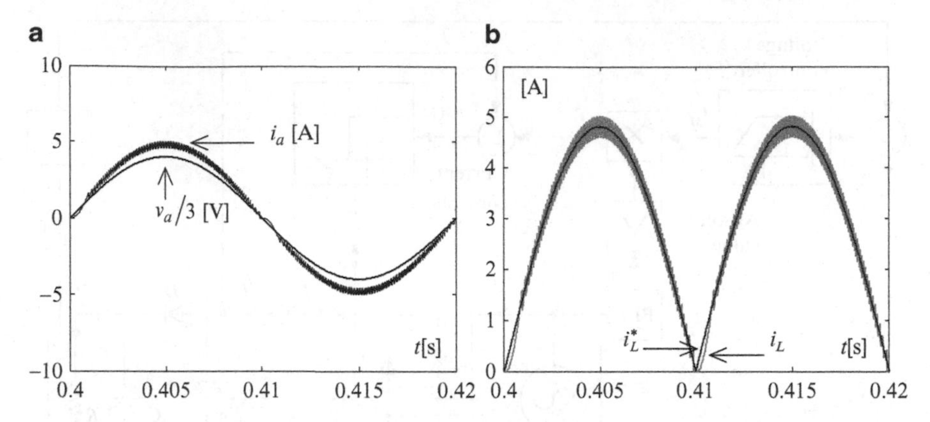

Fig. 13.20 Closed-loop system behavior at constant load 20 Ω (hysteresis width $\Delta = 0.2$ A): (**a**) grid current i_a and $v_a/3$ as image of grid voltage; (**b**) inductor current and its reference

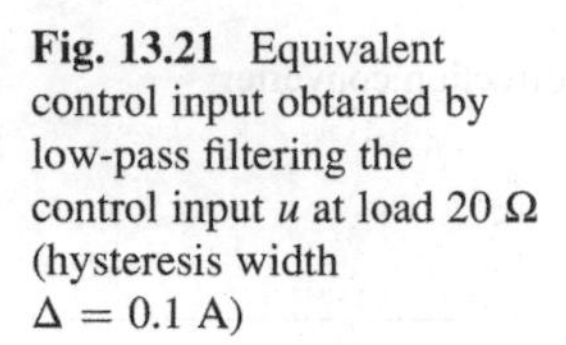

Fig. 13.21 Equivalent control input obtained by low-pass filtering the control input u at load 20 Ω (hysteresis width $\Delta = 0.1$ A)

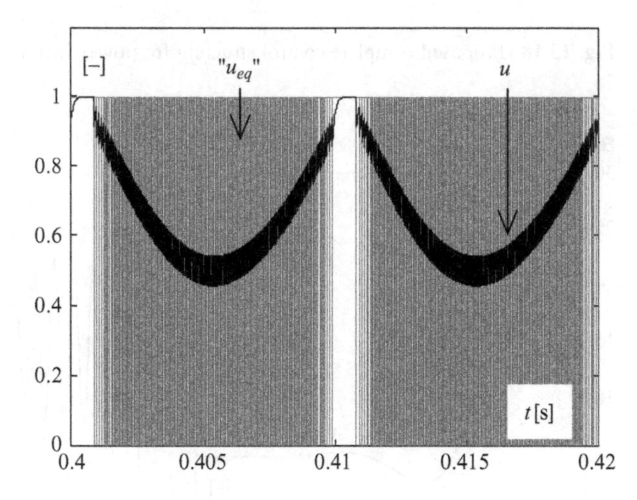

Figure 13.20a shows that the AC current's first-order harmonic is indeed in phase with the AC voltage. Figure 13.20b shows the tracking of the inductor current reference. Note that at the beginning of each voltage grid semi-period this tracking cannot be ensured as the sliding mode actually does not occur. This is also reflected in a nonsinusoidal AC current shape around the angle $\omega t = \pi$.

Preserving the same conditions, Fig. 13.21 shows the equivalent control input as a low-pass filtered version of the control variable u. Note that at around $\omega t = \pi$ the sliding mode does not occur as otherwise the equivalent control u_{eq} would have had values above one. Figure 13.22 shows the tracking of the current reference in somewhat extreme conditions, i.e., using a larger inductance L and also larger load corresponding to larger value of I_M. The closed-loop system performs worse as the sliding mode existence domain diminishes (see (13.41)).

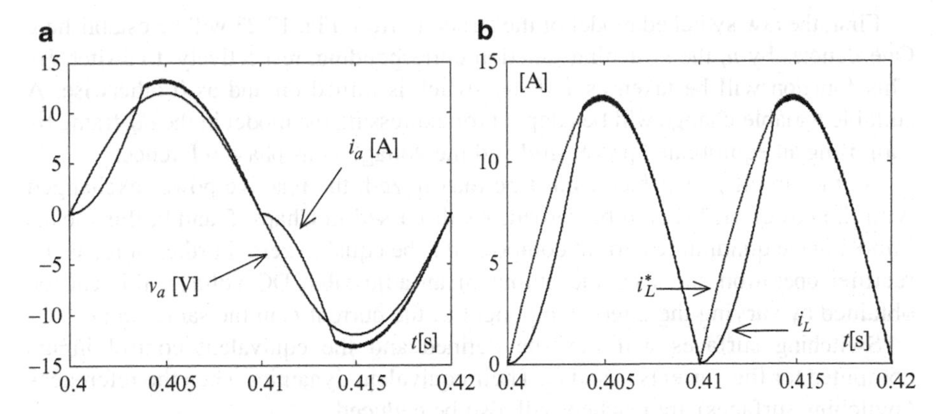

Fig. 13.22 Closed-loop system behavior at constant load 8 Ω (hysteresis width $\Delta = 0.1$ A): (**a**) grid current i_a and grid voltage v_a; (**b**) inductor current and its reference for $L = 2$ mH

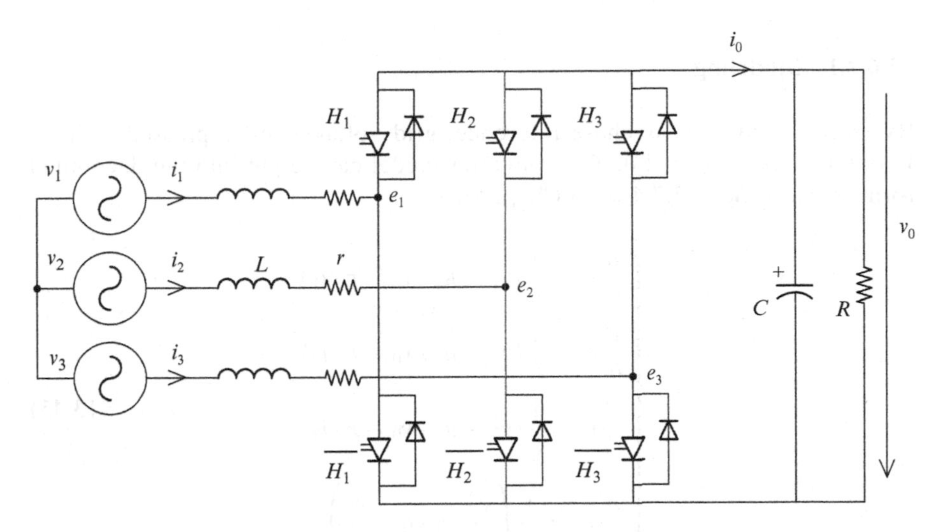

Fig. 13.23 Three-phase rectifier diagram

13.6.2 *Variable-Structure Control of a Three-Phase Rectifier as a MIMO System*

The voltage-source inverter (current rectifier) from Fig. 13.23 is a three-phase circuit characterized by a coupling between the various line currents. In this study it is proposed the circuit be forced to operate as a sinusoidal-input-current rectifier at unit power factor. The line currents must track a sinusoidal reference, having the same phase with the line voltage.

First, the raw switched model of the structure from Fig. 13.23 will be established. One denotes by u_i the switching function corresponding, respectively, to switch H_i. This function will be taken as 1 if the switch is turned on and as 0 otherwise. A suitable variable change will be adopted for expressing the model in the *dq*-frame by supposing an equilibrated power grid and the voltage v_1 as phase reference.

As the circuit power factor must be maximized, the reactive power exchanged with the power grid should be zeroed. As discussed in Chaps. 5 and 9, this means imposing the quadrature current component i_q be equal to zero. Furthermore, as the rectifier operation supposes the output of an adjustable DC voltage, this can be obtained by varying the direct component of the current i_d in the same frame.

Switching surfaces will next be defined and the equivalent control inputs computed for the currents i_d and i_q. Their equivalent dynamics when the references (switching surfaces) are reached will also be deduced.

Finally, a complete control diagram that includes both currents and voltage control loops will be proposed (see also Guffon 2000).

13.6.2.1 Modeling

By taking voltage v_1 as phase reference, grid voltages are expressed by $v_i = V \sin(\omega t - 2\pi/3 \cdot (i - 1))$. The converter model can be put into the decoupled form (see also Sect. 5.7.4 from Chap. 5):

$$\begin{cases} \frac{d}{dt} i_1 = \frac{1}{L}(v_1 - \widetilde{u_1} \cdot v_0 - r \cdot i_1) \\ \frac{d}{dt} i_2 = \frac{1}{L}(v_2 - \widetilde{u_2} \cdot v_0 - r \cdot i_2) \\ \frac{d}{dt} i_3 = \frac{1}{L}(v_3 - \widetilde{u_3} \cdot v_0 - r \cdot i_3) \\ \frac{d}{dt} v_0 = \frac{1}{C}\left(\sum_{i=1}^{3} \widetilde{u_i} \cdot i_i - \frac{v_0}{R}\right), \end{cases} \tag{13.45}$$

where

$$\begin{bmatrix} \widetilde{u_1} \\ \widetilde{u_2} \\ \widetilde{u_3} \end{bmatrix} = \frac{1}{3} \cdot \begin{bmatrix} 2 & -1 & -1 \\ -1 & 2 & -1 \\ -1 & -1 & 2 \end{bmatrix} \cdot \begin{bmatrix} u_1 \\ u_2 \\ u_3 \end{bmatrix}.$$

The system in (13.45) is decoupled, but one cannot obtain the control inputs u_i based upon the intermediary variables $\widetilde{u_i}$.

As detailed in Sect. 5.7.4 of Chap. 5, the switched model in the *dq* frame of the circuit of Fig. 13.23 is given by

$$\begin{cases} \frac{d}{dt} i_d = \frac{1}{L}\left(V - u_d \cdot v_0 - r \cdot i_d + \omega \cdot i_q\right) \\ \frac{d}{dt} i_q = \frac{1}{L}\left(-u_q \cdot v_0 - r \cdot i_q - \omega \cdot i_d\right) \\ \frac{d}{dt} v_0 = \frac{3}{2C}\left(u_d \cdot i_d + u_q \cdot i_q\right) - \frac{v_0}{R}. \end{cases} \tag{13.46}$$

The active and reactive power components exchanged with the electrical grid are respectively described by

$$P = 3/2 \cdot V \cdot i_d, \; Q = 3/2 \cdot V \cdot i_q. \tag{13.47}$$

Equation (13.47) shows that by controlling the current component i_q one controls in fact the reactive power drawn from the grid, while i_d determines the active power transfer and therefore influences the output voltage v_0 control.

13.6.2.2 Variable-Structure Control Design

By putting $\sigma_d = i_d - i_d{}^*$ and $\sigma_q = i_q - i_q{}^*$ one defines the corresponding switching surfaces $\mathbf{S}_d$ and $\mathbf{S}_q$. For deducing the equivalent control inputs in the *dq*-frame, one can use the model described by (13.46) in the above stated definitions by considering that

$$d\sigma_d/dt = di_d/dt = d\sigma_q/dt = di_q/dt = 0,$$

leading further to

$$\begin{cases} u_{deq} = 1/(L \cdot v_0) \cdot \left(V - r \cdot i_d + \omega \cdot i_q\right) \\ u_{qeq} = 1/(L \cdot v_0) \cdot \left(-r \cdot i_q + \omega \cdot i_d\right). \end{cases} \tag{13.48}$$

The equivalent dynamic is obtained while operating on the two surfaces $\mathbf{S}_d$ and $\mathbf{S}_q$, and therefore by considering that $i_d = i_d{}^*$ and $i_q = i_q{}^*$. By having two switching surfaces, the order of the system from (13.46) decreases by two. The dynamic that remains is the one of the DC voltage, v_0.

By replacing in (13.48) currents i_d and i_q by their references, one deduces the equivalent control input expressions; next, these latter are introduced in the model (13.46) to finally obtain the equivalent dynamic of v_0 on the intersection of the surfaces $\mathbf{S}_d$ and $\mathbf{S}_q$:

$$\frac{d}{dt} v_0 = \frac{3}{2C} \cdot \frac{1}{v_0}\left[V \cdot i_d^* - r\left(i_d^{*2} + i_q^{*2}\right) - \frac{v_0}{RC}\right]. \tag{13.49}$$

Table 13.1 Relation between the control inputs in different frames

(u_1,u_2,u_3)	$(\widetilde{u_1},\widetilde{u_2},\widetilde{u_3})$	(u_α,u_β)	Vector
(0,0,0)	(0,0,0)	(0,0)	**A**
(0,0,1)	(−1/3, − 1/3, 2/3)	$\left(-1/3,-1/\sqrt{3}\right)$	**B**
(0,1,0)	(−1/3, 2/3, − 1/3)	$\left(-1/3,1/\sqrt{3}\right)$	**C**
(0,1,1)	(−2/3, 1/3, 1/3)	(−2/3, 0)	**D**
(1,0,0)	(2/3, − 1/3, − 1/3)	(2/3, 0)	**E**
(1,0,1)	(1/3, − 2/3, 1/3)	$\left(1/3,-1/\sqrt{3}\right)$	**F**
(1,1,0)	(1/3, 1/3, − 2/3)	$\left(1/3,1/\sqrt{3}\right)$	**G**
(1,1,1)	(0,0,0)	(0,0)	**H**

If the voltage drop on the inductor resistance r is neglected in relation to voltage v_0, then one finds a form similar to the equivalent dynamic of the single-phase rectifier obtained in the previous case study, that is,

$$\frac{d}{dt}v_0 = \frac{3V \cdot i_d^*}{2C} \cdot \frac{1}{v_0} - \frac{v_0}{RC}.$$

Therefore, one may control the output voltage squared in which the equivalent dynamic is linear with respect to the control input $i_d = i_d{}^*$. Also, one may attempt to linearize the above system around a typical operating point and proceed with the design of a linear control loop.

The convergence condition towards the sliding surfaces is given by somewhat direct relations, $\sigma_d \cdot \dot{\sigma_d} < 0$ and $\sigma_q \cdot \dot{\sigma_q} < 0$. Taking into account Eq. (13.48) and the two first equations of (13.46), one obtains

$$\dot{\sigma_d} = v_0/L \cdot \left(u_{deq} - u_d\right), \dot{\sigma_q} = v_0/L \cdot \left(u_{qeq} - u_q\right).$$

Therefore, the convergence condition simply becomes:

$$\operatorname{sgn}\left(u_d - u_{deq}\right) = \operatorname{sgn}(\sigma_d),\ \operatorname{sgn}\left(u_q - u_{qeq}\right) = \operatorname{sgn}\left(\sigma_q\right). \tag{13.50}$$

The problem is now how to choose the vector fields (or switching orders u_i) in order to ensure the convergence condition given by (13.50). In this respect, one can use one of the three available representations: real, Concordia, Park. Once the dq control inputs u_d and u_q have been chosen, one should go backwards throughout these frames until the real control variables u_i are obtained. Table 13.1 gives the relations between these control inputs. The eight possible vectors – two of which are zero (A and H) – are classically represented in Fig. 13.24 in the $\alpha\beta$ -frame.

Control inputs u_d and u_q are not represented in this table, as they are defined in a rotating frame, so they are functions of ωt:

$$u_d = \cos\theta \cdot u_\alpha + \sin\theta \cdot u_\beta,\ u_\theta = -\sin\theta \cdot u_\alpha + \cos\theta \cdot u_\beta. \tag{13.51}$$

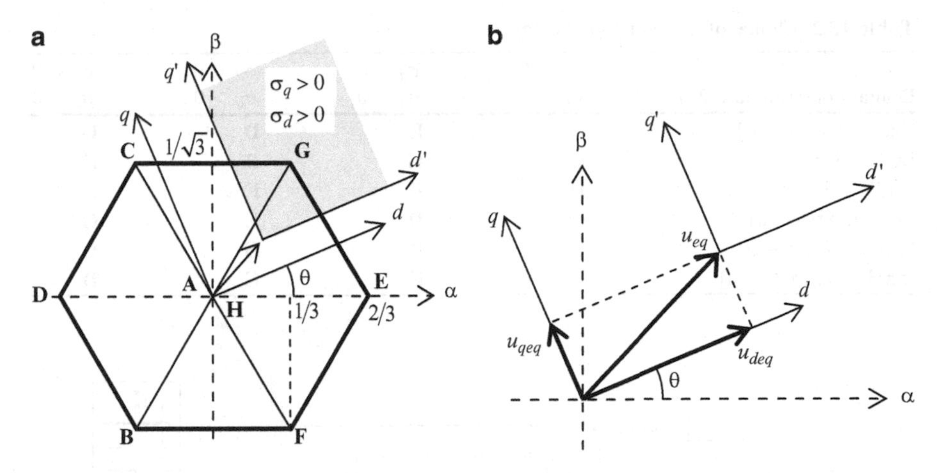

Fig. 13.24 (**a**) Representation of the eight positions possible for vector u_{eq} in the αβ -frame and example of choosing the control input vector; (**b**) detail of a

Taking into account the fact that the equivalent control inputs u_{deq} and u_{qeq} must be bounded by the maximal and minimal values of the controls u_d and u_q, respectively, from Eq. (13.51) one can extract the domain of existence of the sliding modes. The controls u_α and u_β being bounded at $\pm$ 2/3, the norm $\|\mathbf{u}_{eq}\| = \sqrt{u_{deq}^2 + u_{qeq}^2}$ must be smaller than the norm $\|\mathbf{u}_{\alpha\beta}\| = \sqrt{u_\alpha^2 + u_\beta^2}$, meaning that the vector $\mathbf{u}_{eq}$ is situated in the disc described by the hexagon in Fig. 13.24a. A sufficient condition imposes that the equivalent controls be bounded at $\pm$ 2/3.

The next stage is essential for the choice of the appropriate vector required for the convergence towards the desired sliding surfaces – see condition (13.50).

For representing the variables $(u_d - u_{deq})$ and $(u_q - u_{qeq})$ one must change the representation into another rotating frame $d'q'$ that is translated with respect to the initial frame dq with the value u_{deq} for the d-axis and with u_{qeq} for the q-axis, as shown in Fig. 13.24b. The signs of variables $(u_d - u_{deq})$ and $(u_q - u_{qeq})$ are directly deduced according to the quadrant in which the operating point is located.

One can better understand the approach when taking the example described in Fig. 13.24a. Suppose that the expressions σ_d and σ_q are positive, given the value of the angle θ (arbitrarily taken as an example). The convergence condition given by (13.50) needs the expressions $(u_d - u_{deq})$ and $(u_q - u_{qeq})$ to be positive, which is possible only in the first quadrant of the $d'q'$-frame. The sole vector that guarantees this condition is the vector **G**. Using the conversion described in Table 13.1, one finds from vector **G** the control input orders (u_1,u_2,u_3) for the corresponding switches.

To conclude, the choice of switching orders (u_1,u_2,u_3) depends on vectors (**B**,**C**, $\cdots$, **F**), which are chosen in turn as depending on the signs of σ_d and σ_q, the desired vector $\mathbf{u}_{eq}$, which is given by i_d^* and i_q^*, and on angle θ (i.e., ωt).

Table 13.2 Choice of the switching vector

Domain of θ (modulo 2π)	$\sigma_d > 0$, $\sigma_q > 0$	$\sigma_d > 0$, $\sigma_q < 0$	$\sigma_d < 0$, $\sigma_q > 0$	$\sigma_d < 0$, $\sigma_q < 0$
$[2\pi - \lambda, 2\pi/3 - \mu]$	**G**	**E**	**D**	**B**
$[\pi/3 - \lambda, \pi - \mu]$	**C**	**G**	**B**	**F**
$[2\pi/3 - \lambda, 4\pi/3 - \mu]$	**D**	**C**	**F**	**E**
$[\pi - \lambda, 5\pi/3 - \mu]$	**B**	**D**	**E**	**G**
$[4\pi/3 - \lambda, 2\pi - \mu]$	**F**	**B**	**G**	**C**
$[5\pi/3 - \lambda, 7\pi/3 - \mu]$	**E**	**F**	**C**	**D**

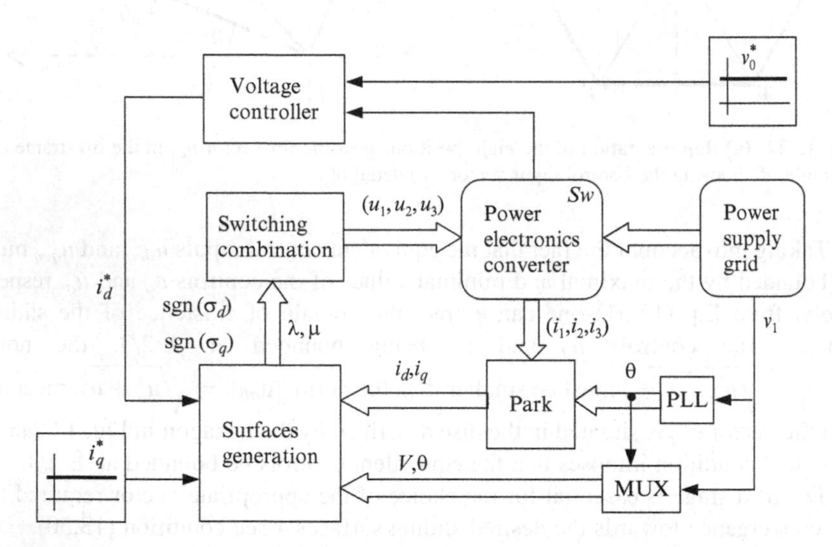

Fig. 13.25 Global control diagram of circuit of Fig. 13.23

By studying each possible combination of these three parameters and by posing $\lambda = \arcsin(3/2 \cdot u_{qeq})$ and $\mu = \arccos(3/2 \cdot u_{deq})$, one can conclude the choice of the suitable switching vector as shown in Table 13.2.

Once the choice of the switching vector has been made, all that remains is to deduce the corresponding switching orders (u_1, u_2, u_3) using Table 13.1.

13.6.2.3 Global Control Diagram

Next, a block diagram including the control loop of the continuous voltage squared is proposed (Fig. 13.25), where the innermost current control loop is the one described previously. A hysteresis is included within the block "switching combination" from Fig. 13.25. It renders more realistic the switching procedure, which cannot in fact be instantaneous.

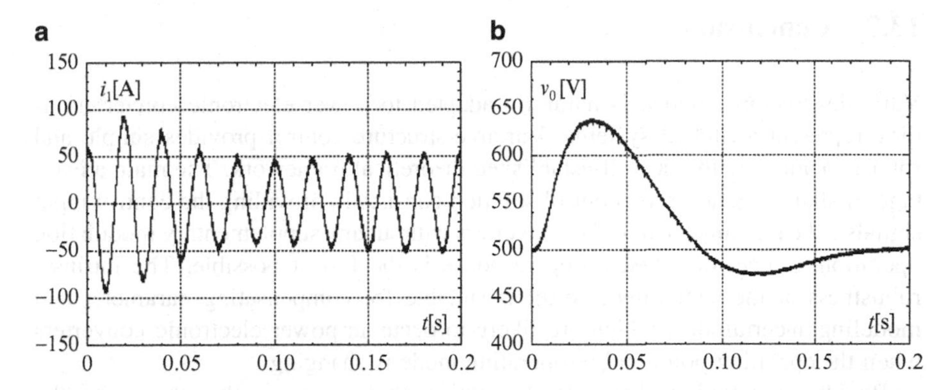

Fig. 13.26 Closed-loop behavior of converter in Fig. 13.23 when using hysteresis width of 3 A for both surfaces σ_d and σ_q: (**a**) evolution of AC current; (**b**) DC voltage time evolution

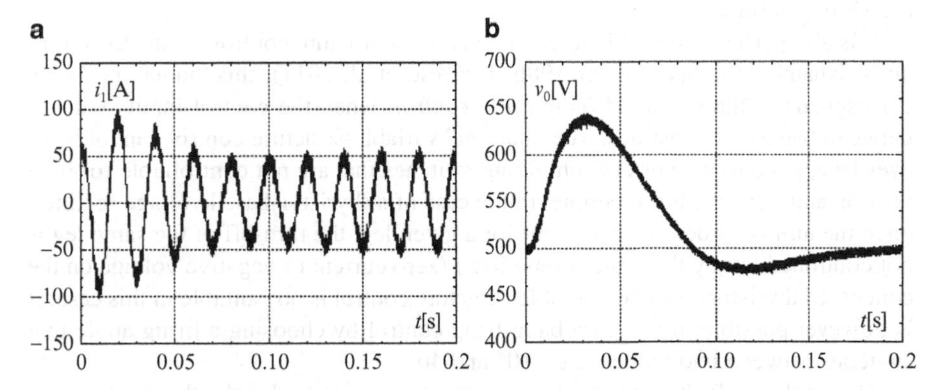

Fig. 13.27 Comparison with time evolutions given in Fig. 13.26 when hysteresis width increases to 10 A: (**a**) AC current time evolution; (**b**) DC voltage time evolution

The simulations presented have been performed based on the switched model of the current rectifier. Remember that the goal was to regulate the continuous voltage v_0 by drawing from the grid only sinusoidal currents. As a matter of fact, these currents are dominated by their fundamental component and contain also high-frequency harmonics that can be easily filtered.

Simulations have been performed for the following parameters: $L = 3$ mH, $r = 0\ \Omega$, $C = 1$ mF, $R = 30\ \Omega$ and $V = 100$ V. The chosen references are $i_q^* = 0$ (no reactive power) and $v_0^* = 500$ V.

In Fig. 13.26 are represented transients of the line current i_1 and of voltage v_0, with a hysteresis width of 3 A for σ_d and σ_q. A good tracking performance of the sinusoidal current reference by the line current is obtained, the total harmonic distortion (THD) being 6 %. Figure 13.27 gives the same case, but this time with a hysteresis width of 10 A, corresponding to THD 17 %.

13.7 Conclusion

Variable-structure control is naturally adapted to power electronic converters as they represent switched systems. Variable-structure control provides simple and robust solutions within a well-established theoretical framework. The main advantage of sliding-mode-based control solution consists in providing the control input signals to be fed directly into the converters without any supplementary modulation operation. Hence, the closed-loop response is the fastest possible. The intrinsic robustness of the switching control is suitable for compensating parameter and modeling uncertainties, which are likely to occur in power electronic converters when the operating point and/or operating mode is changing.

Besides developing the control solution that supposes the choice of the switching function, the variable-structure control framework also allows the control performance to be assessed in terms of delimiting the domain of sliding modes existence, deducing the equivalent control and obtaining the dynamics on the sliding surface.

It is also perfectly possible to extend variable-structure control results to continuous dynamic systems (Buhler 1986; DeCarlo et al. 2011); this "unnatural" case corresponds to the so-called *bang-bang control*, when the control input switches between two extreme values. The "natural" variable-structure control cannot however be applied to converters containing switches that are not controllable for both turn-on and turn-off. For example, in the case of a thyristor-based voltage rectifier, once the turn-on order has been sent for a given leg, the turn-off of the same leg is not controlled but by the system state itself (zero current or negative voltage on the concerned thyristor). Hence, variable-structure control is not suitable in this case. It is however possible to design a bang-bang control by choosing a firing angle that switches between two values, e.g., 10° and 40°.

The analysis dealt with in this chapter has concerned only the continuous-conduction mode – see the models used within the developments. However, in real applications the discontinuous-conduction mode is likely to occur (e.g., for weak loads). Control designed to perform well in the continuous-conduction case still ensures stable closed-loop operation, but obviously with altered performance.

Sliding mode control can be used in conjunction with other control-related purposes. For example, the estimation of nonmeasurable states can be done by employing a sliding-mode observer built upon a switching surface which is exactly the estimation error (Šabanovic et al. 2004). The sliding-mode approach can also be used in conjunction with other control techniques in order to enhance their performances (e.g., passivity-based or feedback-linearization control).

Without pretending to being exhaustive in what concerns the topic of the sliding-mode, this chapter has tried to give an overview of sliding-mode control basic issues and its methodology of application for power electronic converters. The reader is invited to go deeper in analyzing works from the relevant literature in order to get a more complete perspective (Tan et al. 2011).

Problems

Problem 13.1. Sliding-mode control of a flyback converter
Consider the flyback converter schematics given in Fig. 13.28 (whose modeling was performed in Problem 4.1 of Chap. 4).
Design a sliding-mode control law for regulating the output voltage v_C at the reference value $v_C{}^*$ by using the function u switching between $u^- = 0$ and $u^+ = 1$.

Solution. The switched model of the flyback converter is given below:

$$\begin{cases} L \cdot \dot{i_L} = -(1-u)\dfrac{v_C}{n} + u \cdot E \\ C \cdot \dot{v_C} = (1-u)\dfrac{i_L}{n} - \dfrac{v_C}{R}, \end{cases} \tag{13.52}$$

where the usual notations have been used and n is the transformer ratio. The state vector is $\mathbf{x} = [\,i_L \quad v_C\,]^T$. The surface defined by $s(\mathbf{x}) = 0$, where

$$s(\mathbf{x}) = v_C - v_C^* \tag{13.53}$$

with $v_C{}^*$ constant, is a suitable choice for beginning the design of a sliding-mode control aimed at regulating the output voltage as follows:

$$u = \begin{cases} u^+ = 1 & \text{if} \quad s(\mathbf{x}) < 0 \\ u^- = 0 & \text{if} \quad s(\mathbf{x}) > 0. \end{cases}$$

The first step of the design is to verify if the transversality condition is met. From a pragmatic viewpoint, this means to check if the factor multiplying u in the expression of $\dot{s}$ is nonzero. Taking into account the state Eq. (13.52), it then results that

$$\dot{s}(\mathbf{x}) = \dot{v_C} = (1-u)\frac{i_L}{nC} - \frac{v_C}{RC} = \frac{i_L}{nC} - \frac{v_C}{RC} - \frac{i_L}{nC} \cdot u, \tag{13.54}$$

in which the term $i_L/(nC)$ multiplying u is obviously nonzero because $i_L > 0$.

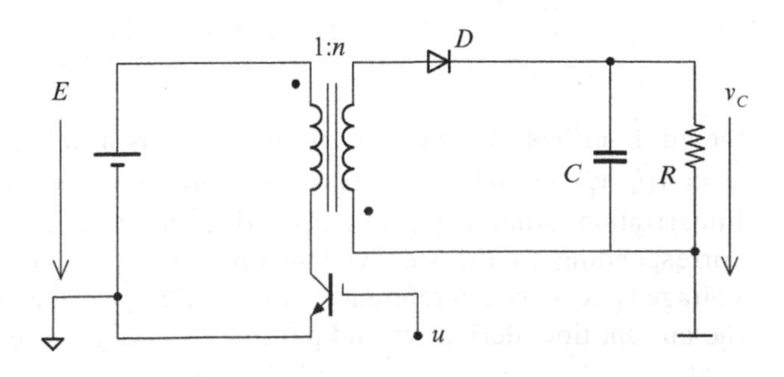

Fig. 13.28 Flyback converter schematics

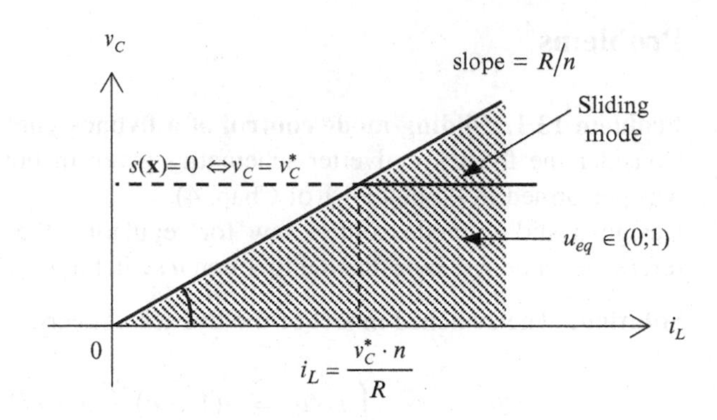

Fig. 13.29 Sliding modes existence domain in case of choosing switching surface $v_C = v_C{}^*$ for sliding-mode regulating output voltage of a flyback converter

In a second step one must compute the equivalent control. This is obtained by solving the equation $\dot{s}(\mathbf{x}) = 0$ for u. According to (13.54), this gives $(1 - u) \cdot i_L/n - v_C/R = 0$, hence:

$$u_{eq} = 1 - v_C \cdot n/(R \cdot i_L). \tag{13.55}$$

From imposing that $0 = u^- < u_{eq} < u^+ = 1$, one can deduce the domain of the sliding mode's existence. The condition $u_{eq} > 0$ gives $v_C \cdot n/(R \cdot i_L) > 0$, which is true because $v_C > 0$ and $i_L > 0$ in any operating point. The condition $u_{eq} < 1$ leads to $i_L > v_C \cdot n/R$, which bounds the hatched surface in Fig. 13.29. Note in this figure that the sliding modes existence domain is obtained as the nonvoid intersection between the domain of equivalent control existence and the switching surface defined by $s(\mathbf{x}) = 0$, i.e., the line $v_C = v_C{}^*$.

The third design step consists in deducing the equivalent dynamic when in sliding mode. To this end, Eq. (13.55) is substituted in the converter Eq. (13.52), meanwhile taking into account that $s(\mathbf{x}) = 0$ ($v_C = v_C{}^*$) and $\dot{s}(\mathbf{x}) = 0$. Thus, the equivalent dynamic is that of the inductor current:

$$L \cdot \dot{i}_L = -\left(1 - u_{eq}\right) \cdot v_C^*/n + u_{eq} \cdot E.$$

Using Eq. (13.55) one obtains after some simple manipulation:

$$L \cdot \dot{i}_L = \frac{v_C^*\left(nE - v_C^*\right)}{R \cdot i_L} + E, \tag{13.56}$$

which describes the dynamic of current i_L as a nonlinear function denoted by $\dot{i}_L = h\left(i_L, v_C^*\right)$. In order to assess the nature of this dynamic one can proceed by linearization around a given operating point. Let i_{eq0} be the current value corresponding to the steady-state operating point ensured by regulating the voltage v_C to a certain reference value $v_{C0}{}^*$; i_{eq0} is therefore obtained by zeroing the current time derivative and putting $v_C = v_{C0}{}^*$. The results is

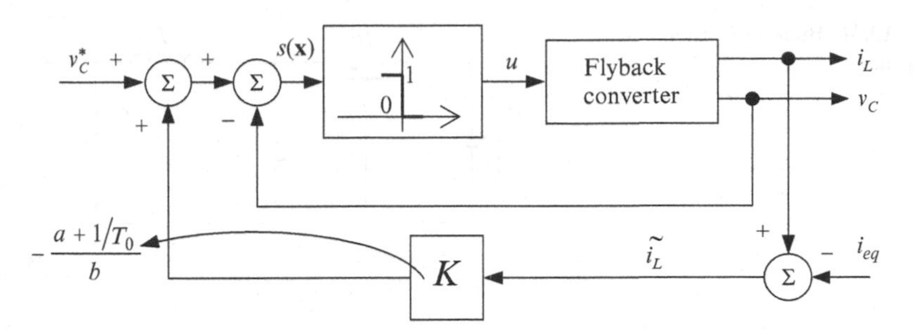

Fig. 13.30 Closed-loop diagram based on sliding-mode voltage control and feedback stabilization of equivalent dynamic for a flyback DC-DC converter

$$i_{eq0} = v_{C0}^*\left(v_{C0}^* - nE\right)/(ER). \tag{13.57}$$

Note that a positive value of i_{eq0} is ensured if and only if $v_{C0}^* > nE$. Let $\widetilde{i_L} = i_L - i_{eq0}$ and $\widetilde{v_C^*} = v_C^* - v_{C0}^*$ be small variations of the state variables around operating point $(i_{eq0}, v_{C0}{}^*)$. Linearization of system (13.56) around this point gives

$$\dot{\widetilde{i_L}} = \underbrace{\left(\frac{\partial h}{\partial i_L}\right)\Bigg|_{\left(i_{eq0}, v_{C0}^*\right)}}_{a} \cdot \widetilde{i_L} + \underbrace{\left(\frac{\partial h}{\partial v_C^*}\right)\Bigg|_{\left(i_{eq0}, v_{C0}^*\right)}}_{b} \cdot \widetilde{v_C^*},$$

where the values of a and b can be obtained after some algebra as

$$a = \frac{E^2}{L v_{C0}^*\left(v_{C0}^* - nE\right)}, \quad b = \frac{E}{L\left(v_{C0}^* - nE\right)}\left(\frac{nE}{v_{C0}^*} - 2\right).$$

Note that $a > 0$ because $v_{C0}{}^* > nE$, so the linearized dynamic of the inductor current i_L is unstable because the transfer function $\widetilde{V_C^*}(s)/\widetilde{I_L}(s)$ – with $\widetilde{V_C^*}(s)$ and $\widetilde{I_L}(s)$ being the Laplace transforms of $\widetilde{v_C^*}$ and $\widetilde{i_L}$, respectively – has a pole placed in the right-half-plane. In conclusion, the equivalent dynamic is unstable.

In order to stabilize the equivalent dynamic on the sliding surface, one can build a state feedback K such that the closed-loop dynamic is stable. Moreover, this dynamic can also be tuned to ensure desired performance, i.e., settling time. Thus, by imposing T_0 as desired stabilizing time constant, the value of K results as:

$$K = -\frac{a + 1/T_0}{b}.$$

The closed-loop sliding-mode control diagram embedding the stabilization of the equivalent dynamic is shown in Fig. 13.30. One can remark that this final form of the control law is equivalent with having chosen a switching surface that embeds a term depending on the current i_L: $s(\mathbf{x}) = v_C - v_C^* + K \cdot (i_L - i_{eq})$, where the

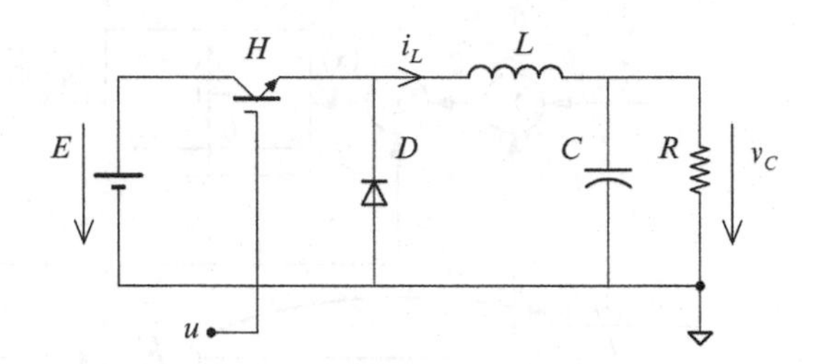

Fig. 13.31 Buck converter diagram

value i_{eq} is given by Eq. (13.57). Note that, according to this latter relation, the value i_{eq} depends on the load value R, which is usually not known. Therefore, implementing the control diagram in Fig. 13.30 requires a load estimator being designed based upon some reliable measures from circuit.

Problem 13.2. Sliding-mode control of a buck converter

Consider the buck converter schematics given in Fig. 13.31. Design a sliding-mode control law for regulating the output voltage v_C at the reference value $v_C{}^*$ by using the function u switching between $u^- = 0$ and $u^+ = 1$.

Solution The switched model of the buck converter is as follows:

$$\begin{cases} L \cdot \dot{i_L} = -v_C + u \cdot E \\ C \cdot \dot{v_C} = i_L - v_C/R, \end{cases} \tag{13.58}$$

where the usual notations have been used. The state vector is $\mathbf{x} = \begin{bmatrix} i_L & v_C \end{bmatrix}^T$. Note that choosing the switching surface $s(\mathbf{x}) = v_C - v_C{}^*$ would give $\mathrm{L}_{\mathbf{g}}s(\mathbf{x}) = 0$, therefore sliding-mode control is not possible. Let us put $s(\mathbf{x}) = 0$, where

$$s(\mathbf{x}) = v_C - v_C^* + \lambda\left(\dot{v_C} - \dot{v_C^*}\right) \tag{13.59}$$

– with $v_C{}^*$ constant and $\lambda > 0$ having a large value – is the switching surface depending on which control input is fed in order to regulate the output voltage:

$$u = \begin{cases} u^+ = 1 & \text{if} \quad s(\mathbf{x}) < 0 \\ u^- = 0 & \text{if} \quad s(\mathbf{x}) > 0. \end{cases}$$

Checking the transversality condition supposes the expression of $\dot{s}(\mathbf{x})$ being computed:

$$\dot{s}(\mathbf{x}) = \dot{v_C} - \dot{v_C^*} + \lambda\left(\ddot{v_C} - \ddot{v_C^*}\right), \tag{13.60}$$

in which one takes into account that $\dot{v_C^*} = \ddot{v_C^*} = 0$, replaces the expressions of $\dot{v_C}$ and $\ddot{v}_C$ based upon the converter model (13.58) and gets after some computation

$$\dot{s}(\mathbf{x}) = \left(1 - \frac{\lambda}{RC}\right) \cdot \left(\frac{i_L}{C} - \frac{v_C}{RC}\right) - \lambda \frac{v_C}{LC} + \lambda \frac{E}{LC} \cdot u, \tag{13.61}$$

where the control input u appears to be multiplied by a strictly positive value, $\lambda E/LC$. The conclusion is that the transversality condition is fulfilled.

Computation of the equivalent control also is based on Eq. (13.61); thus, u_{eq} results from solving $\dot{s}(\mathbf{x}) = 0$ for u and taking into account that $s(\mathbf{x}) = 0$ is also valid. Therefore, by combining results obtained by zeroing Eqs. (13.60) and (13.61), the equivalent control takes the following expression:

$$u_{eq} = \frac{v_C}{E} - \left(1 - \frac{\lambda}{RC}\right) \cdot \frac{(v_C - v_C^*)LC}{E\lambda^2}. \tag{13.62}$$

One must now ensure that $0 = u^- < u_{eq} < u^+ = 1$. First, because λ was assumed to take large values, it is reasonable to consider that $\lambda \gg RC$. Let us adopt the notation

$$d = \frac{LC}{\lambda^2}\left(\frac{\lambda}{RC} - 1\right) \tag{13.63}$$

and note that $d > 0$ because $\lambda \gg RC$; moreover, d has small values for the usual cases; the smaller they are, the larger λ is chosen ($d \ll 1$). Therefore, the condition $u_{eq} > 0$ becomes equivalent to

$$v_C \cdot (1 - d) > -v_C^* \cdot d,$$

which is met for any operating point, provided that v_C and v_C^* have positive values and d has small positive values. As regards the condition $u_{eq} < 1$, this leads to $v_C/E \cdot (1 - d) < 1 - v_C^*/E \cdot d$, or equivalently, given that $1 - d > 0$, to

$$v_C < \frac{E - v_C^* \cdot d}{1 - d}.$$

To conclude, equivalent control exists for output voltage meeting

$$0 < v_C < \frac{E - v_C^* \cdot d}{1 - d}. \tag{13.64}$$

Note that, according to (13.63), $d \to 0$ as $\lambda \to \infty$; therefore, according to (13.64), the larger λ is chosen, the upper limit of the equivalent control existence domain approaches E. The domain of equivalent control existence has been represented by the hatched surface in Fig. 13.32. In the same figure the surface (line) $s(\mathbf{x}) = 0$ with $s(\mathbf{x})$ given by (13.59) has also been shown for two values of λ. Note that the solid line segment representing the nonvoid intersection between the hatched surface and the line $s(\mathbf{x}) = 0$ has smaller slope as λ increases; for $\lambda = \infty$, it is superposed onto the abscissa segment $(0, E)$. This latter segment is the domain of the sliding modes existence.

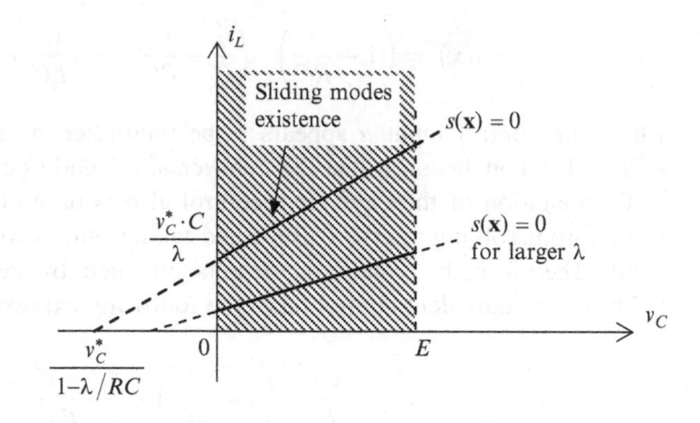

Fig. 13.32 Sliding modes existence domain in case of choosing switching surface $v_C = v_C^* - \lambda(\dot{v}_C - \dot{v}_C^*)$ for sliding-mode regulating output voltage of a buck converter

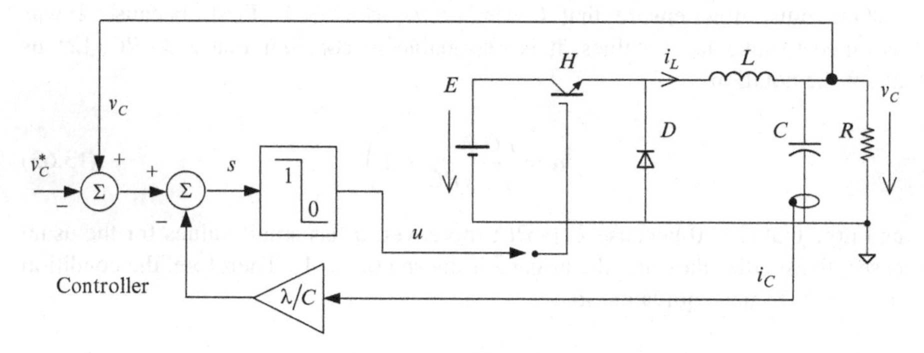

Fig. 13.33 Global sliding-mode-based control structure of a buck converter

One can conclude that a large value of λ practically ensures the existence of sliding modes for the whole operating range of the converter ($0 < v_C < E$).

The third step of the design concerns the computation of the equivalent dynamics, the ones on the sliding surface. These result by substituting Eq. (13.62) of the equivalent control into converter Eq. (13.58). Simple algebra finally leads to the following state equations:

$$\begin{cases} \dot{i}_L = -\left(\frac{\lambda}{RC} - 1\right)\frac{C}{\lambda} \cdot v_C + \left(\frac{\lambda}{RC} - 1\right)\frac{C}{\lambda} \cdot v_C^* \\ \dot{v}_C = \frac{i_L}{C} - \frac{1}{RC} \cdot v_C, \end{cases} \tag{13.65}$$

which describe stable linear dynamics under the assumption that $\lambda \gg RC$ (the proof of the state matrix poles having negative real parts is left to the reader).

To conclude, the variable-structure control structure is given in Fig. 13.33. Note that the output voltage derivative is computed based on a measure of the capacitor current, i_C.

The reader is invited to solve the following problems.

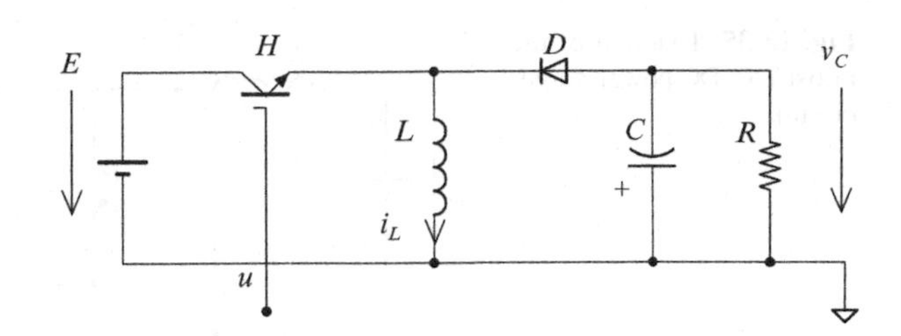

Fig. 13.34 Buck-boost DC-DC power stage diagram

Problem 13.3. Study of a variable-structure control for buck-boost power stage

The converter from Fig. 13.34 has $L = 0.5$ mH, $C = 1000$ μF, $E = 100$ V and a rated load value $R = 2\ \Omega$. The control goal is to maintain a constant output voltage, $v_C{}^* = -100$ V. The switching function u takes values in the discrete set {0;1}.

The following points must be addressed.

(a) The switching surface is $s_1(\text{x}) = v_C - v_C{}^*$, with $v_C{}^*$ being the output voltage setpoint.

(i) Verify the hitting condition and compute the equivalent control input u_{eq} corresponding to the above defined switching surface.
(ii) Show the sliding mode existence domain in the plane (i_L,v_C) and its intersection with the surface $s_1(\text{x})$.
(iii) Compute the equivalent dynamics. Explain the result by assessing system stability.

(b) The switching surface is $s_2(\text{x}) = i_L - i_L{}^*$, where $i_L{}^*$ is the current reference.

(i) Verify the hitting condition and compute the equivalent control input u_{eq} corresponding to the surface $s_2(\text{x})$.
(ii) Show the sliding mode existence domain in the plane (i_L,v_C) and its intersection with the surface $s_2(\text{x})$.
(iii) Compute the system equivalent dynamics. Assess the stability and evaluate its main time constant. Simulate the system using MATLAB®--Simulink® for different values of loads and verify the obtained theoretical results. Verify that the operation points outside the sliding mode existence domain cannot be reached. Find a solution to reduce the switching frequency to 100 kHz.

(c) Consider a controller $H_C(s) = \frac{K_C}{T_C s+1}$ within an outer voltage control loop.

(i) Find the controller parameters ensuring a voltage settling time of 0.1 s.
(ii) Compute the steady-state voltage error for the case when the circuit is loaded at rated value. Verify the result in simulation.
(iii) Study a solution that cancels the steady-state voltage error.

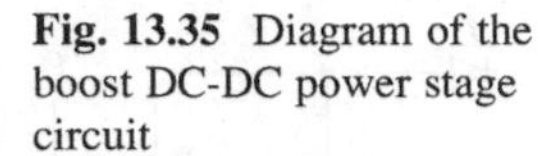

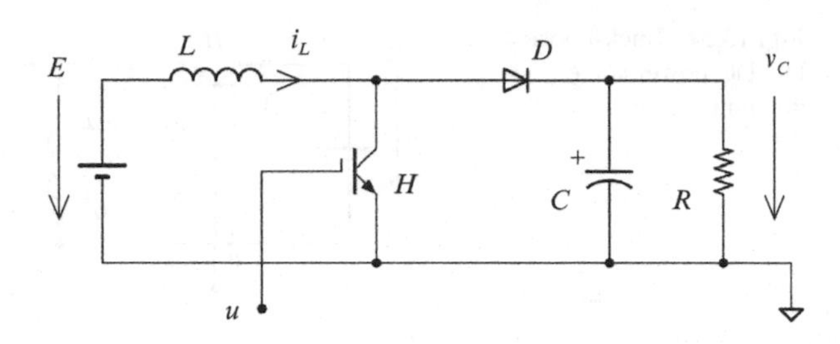

Fig. 13.35 Diagram of the boost DC-DC power stage circuit

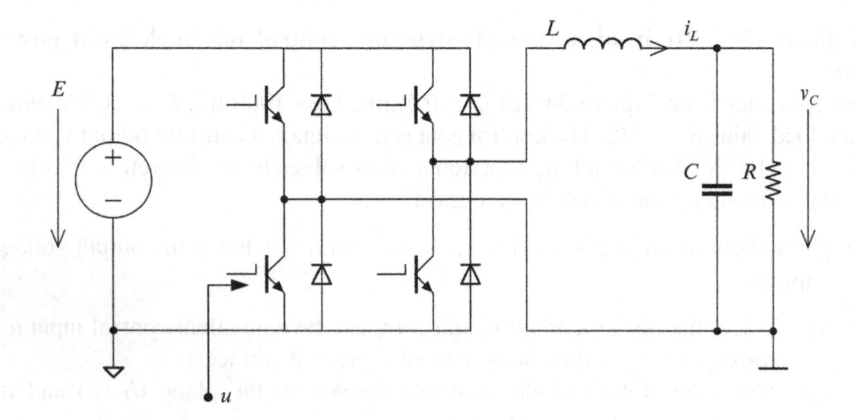

Fig. 13.36 Stand-alone voltage-source inverter diagram

Problem 13.4. Study of a voltage-regulated boost DC-DC power stage using variable-structure control
The converter from Fig. 13.35 has the following circuit parameters: $L = 0.5$ mH, $C = 1000$ μF, $E = 100$ V and a rated load value $R = 2\ \Omega$. The control scope is to maintain a desired constant output voltage, $v_C{}^* = 250$ V.

It is required to answer the same questions as in Problem 13.3.

Problem 13.5. Direct sliding-mode control of voltage-source inverter
The inverter in Fig. 13.36 serves at producing a sinusoidal voltage based upon the continuous voltage E.

Find a control solution of output voltage regulation via sliding modes irrespective of variable load R. The switching surface is chosen as $s(\mathbf{x}) = v_C - v_C^* + \lambda \cdot \left(\dot{v}_C - \dot{v}_C^*\right)$, where $v_C{}^* = V_{C\,\max} \cdot \sin \omega t$ and $\lambda > 0$ ($V_{C\,\max}$ and ω are both constant). Switching function u takes values in the discrete set $\{-1;\ 1\}$.

Problem 13.6. Indirect sliding-mode control of voltage-source inverter
Given the converter in Fig. 13.37, obtain a control solution of the output voltage via sliding modes. The scoped control structure is the so-called indirect control, where the switching surface is $s(\mathbf{x}) = i_L - i_L{}^*$. The voltage control loop has the reference $v_C{}^* = V_{C\,\max} \cdot \sin \omega t$, with fixed values of $V_{C\,\max}$ and ω. Switching function u takes values in the discrete set $\{-1;\ 1\}$.

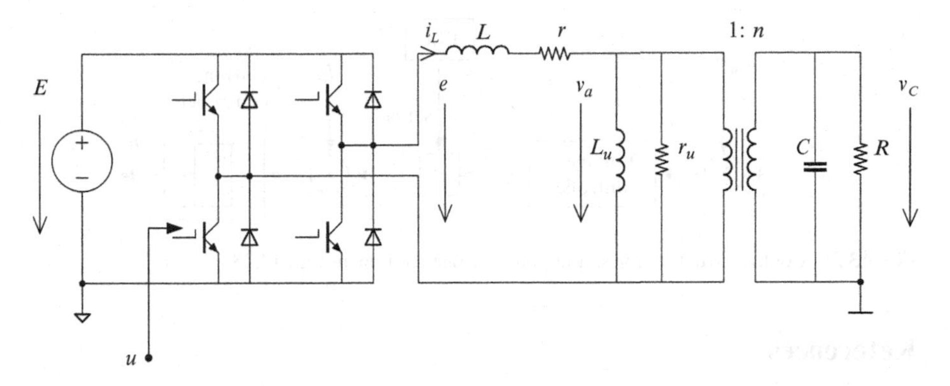

Fig. 13.37 Stand-alone insulated voltage-source inverter diagram

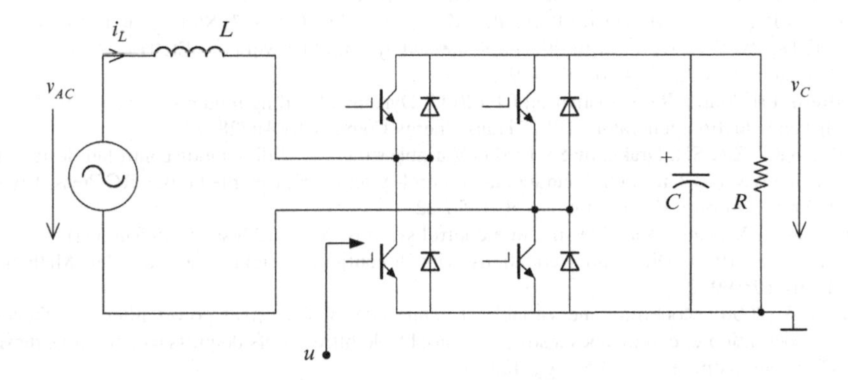

Fig. 13.38 Full-bridge rectifier schematics

Problem 13.7. Variable-structure control for power-factor-correction full-bridge single-phase rectifier

The AC-DC converter from Fig. 13.38 is controlled using variable-structure control technique in order to fulfill two objectives:

- to maintain a constant desired output voltage irrespective of load value (within the acceptable limits);
- to draw a sinusoidal current in phase with the grid voltage.

The switching function u takes values in the discrete set $\{-1; 1\}$ and the switching surface is chosen as $s = i_L - i_L{}^*$, where $i_L{}^* = I_{L\,\max} \sin \omega t$, with ωt being the grid phase.

It is required to design a sliding-mode current controller and a continuous voltage controller that match the control structure presented in Fig. 13.39.

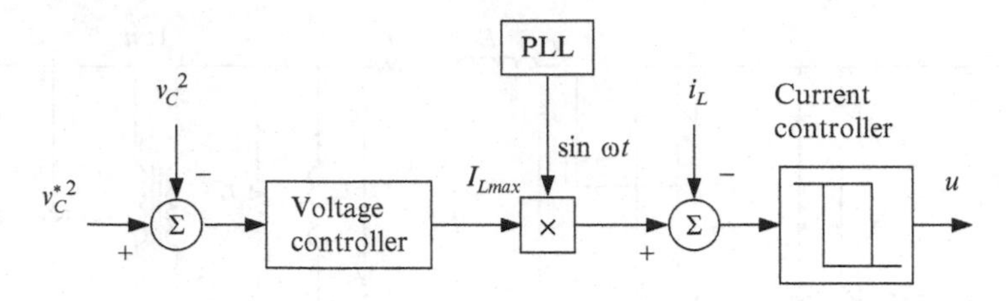

Fig. 13.39 Control structure for single-phase bridge rectifier in Fig. 13.38

References

Buhler H (1986) Sliding mode control (in French: Réglage par mode de glissement). Presses Polytechniques Romandes, Lausanne

Carpita M, Marchesoni M (1996) Experimental study of a power conditioning system using sliding mode control. IEEE Trans Power Electron 11(5):731–742

Carrasco JM, Quero JM, Ridao FP, Perales MA, Franquelo LG (1997) Sliding mode control of a DC/DC PWM converter with PFC implemented by neural networks. IEEE Trans Circuit Syst I Fundam Theor Appl 44(8):743–749

DeBattista H, Mantz RJ, Christiansen CF (2000) Dynamical sliding mode power control of wind driven induction generators. IEEE Trans Energy Convers 15(4):728–734

DeCarlo RA, Żak SH, Drakunov SV (2011) Variable structure, sliding mode controller design. In: Levine WS (ed) The control handbook—control system advanced methods. CRC Press, Taylor & Francis Group, Boca Raton, pp 50-1–50-22

Emelyanov SV (1967) Variable structure control systems. Nauka, Moscow (in Russian)

Filippov AF (1960) Differential equations with discontinuous right hand side. Am Math Soc Transl 62:199–231

Guffon S (2000) Modelling and variable structure control for active power filters (in French: "Modélisation et commandes à structure variable de filtres actifs de puissance"). Ph.D. thesis, Grenoble Institute of Technology, France

Guffon S, Toledo AS, Bacha S, Bornard G (1998) Indirect sliding mode control of a three-phase active power filter. In: Proceedings of the 29th annual IEEE Power Electronics Specialists Conference – PESC 1998. Kyushu Island, Japan, pp 1408–1414

Hung JY, Gao W, Hung JC (1993) Variable structure control: a survey. IEEE Trans Ind Electron 40(1):2–22

Itkis U (1976) Control systems of variable structure. Wiley, New York

Levant A (2007) Principles of 2-sliding mode design. Automatica 43(4):576–586

Levant A (2010) Chattering analysis. IEEE Trans Autom Control 55(6):1380–1389

Malesani L, Rossetto L, Spiazzi G, Tenti P (1995) Performance optimization of Ćuk converters by sliding-mode control. IEEE Trans Power Electron 10(3):302–309

Malesani L, Rossetto L, Spiazzi G, Zuccato A (1996) An AC power supply with sliding mode control. IEEE Ind Appl Mag 2(5):32–38

Martinez-Salamero L, Calvente J, Giral R, Poveda A, Fossas E (1998) Analysis of a bidirectional coupled-inductor Ćuk converter operating in sliding mode. IEEE Trans Circuit Syst I Fundam Theor Appl 45(4):355–363

Mattavelli P, Rossetto L, Spiazzi G (1997) Small-signal analysis of DC–DC converters with sliding mode control. IEEE Trans Power Electron 12(1):96–102

Šabanovic A (2011) Variable structure systems with sliding modes in motion control—a survey. IEEE Trans Ind Inform 7(2):212–223

Šabanovic A, Fridman L, Spurgeon S (2004) Variable structure systems: from principles to implementation, IEE Control Engineering Series. The Institution of Engineering and Technology, London

Sira-Ramírez H (1987) Sliding motions in bilinear switched networks. IEEE Trans Circuit Syst 34 (8):919–933

Sira-Ramírez H (1988) Sliding mode control on slow manifolds of DC to DC power converters. Int J Control 47(5):1323–1340

Sira-Ramírez H (1993) On the dynamical sliding mode control of nonlinear systems. Int J Control 57(5):1039–1061

Sira-Ramírez H (2003) On the generalized PI sliding mode control of DC-to-DC power converters: a tutorial. Int J Control 76(9/10):1018–1033

Sira-Ramírez H, Silva-Ortigoza R (2006) Control design techniques in power electronics devices. Springer, London

Slotine JJE, Sastry SS (1983) Tracking control of non-linear systems using sliding surface, with application to robot manipulators. Int J Control 38(2):465–492

Spiazzi G, Mattavelli P, Rossetto L, Malesani L (1995) Application of sliding mode control to switch-mode power supplies. J Circuit Syst Comput 5(3):337–354

Tan S-C, Lai YM, Cheung KHM, Tse C-K (2005) On the practical design of a sliding mode voltage controlled buck converter. IEEE Trans Power Electron 20(2):425–437

Tan S-C, Lai Y-M, Tse C-K (2011) Sliding mode control of switching power converters: techniques and implementation. CRC Press, Taylor & Francis Group, Boca Raton

Utkin VA (1972) Equations of sliding mode in discontinuous systems. Autom Remote Control 2 (2):211–219

Utkin VA (1977) Variable structure systems with sliding mode. IEEE Trans Autom Control 22 (2):212–222

Utkin V (1993) Sliding mode control design principles and applications to electric drives. IEEE Trans Ind Electron 40(1):23–36

Venkataramanan R, Šabanovic A, Ćuk S (1985) Sliding mode control of DC-to-DC converters. In: Proceedings of IEEE Industrial Electronics Conference – IECON 1985. San Francisco, California, USA, pp 251–258

Young KD, Utkin VI, Ozguner U (1999) A control engineer's guide to sliding mode control. IEEE Trans Control Syst Technol 7(3):328–342

General Conclusion

Nowadays power electronic converters play key roles within a variety of applications that fulfill important functions in modern society such as those of renewable energy conversion systems, electric vehicular applications or power delivery devices, to mention only a few. The operation of power converters in a very demanding context – requiring a performance set to be ensured along with hard real-time constraints – renders crucial the necessity of well-performing control structures.

This textbook has explored the most typical control approaches of power electronic converters employing both DC and AC power stages. To this end a formalization effort was detailed in the first part of the book – the first five chapters – aimed at providing a unified modeling framework, to be further employed in the control design approaches detailed in the second part of the book – the last seven chapters.

Modeling tools developed in the first part offer sufficient generality to be applied (possibly with minor adaptations) to any type of switching converter. Control approaches presented in the second part first provided insights on small-signal-model-based linear control, which is simple and intuitive, but relies upon approximations, hence it lacks robustness. A second class of nonlinear control approaches aims at alleviating this drawback; these mainly use the large-signal nonlinear models and result in quite complex control structures – such as feedback-linearization or passivity-based – or in less complex variable-structure sliding-mode controllers.

The discourse was addressed to students already in mastery of the basis of power electronic circuits and control systems theory. Insights provided may be used in order to implement control laws either in analog or in digital form and to analyze the behavior of open-loop and closed-loop power electronic converters. Although the discourse has been mainly intended for pedagogical goals, the issues considered indicate some still unexplored paths for research.

Typically a chapter within this book began with an introduction regarding its contents, followed by an algorithm reflecting a systematic way to solve the stated problem. The core of each chapter consisted of application examples and case

S. Bacha et al., *Power Electronic Converters Modeling and Control: with Case Studies*, Advanced Textbooks in Control and Signal Processing, DOI 10.1007/978-1-4471-5478-5,

studies which helped the reader clarify the main issues. Finally, the reader was invited to practice the acquired knowledge through a set of solved and unsolved problems provided at the end of the chapter.

Presentation and developments within this textbook did not aim at covering exhaustively the topics of power electronic converters control. Here below are mentioned some modeling and control approaches for possible further consideration.

Further study on modeling may be focused on going into a deeper study of the sampled-data models intended for digital use (Brown and Middlebrook 1981). Concerning control methods, direct design of digital controllers using discretized models of power electronic converters – resulting in discrete-time PID, RST or dead-beat controllers (Timbuş et al. 2009) – can be approached. Adaptive (Morroni et al. 2009), predictive (Rodriguez and Cortes 2012), flatness-based, fuzzy (Mattavelli et al. 1997) or model-free-based (Michel et al. 2010) control structures may be worthy of exploration.

The textbook approaches the current-mode operation of switching converters only in the averaged form; peak (or valley) current mode is equally interesting for further study (Middlebrook 1987).

As with many engineering systems, none of the existing control approaches for power converters can be declared the "best". Any chosen approach must be adapted to each application's initial data and will generally result from suitably trading-off cost (in terms of complexity) and performance (e.g., regulation quality and robustness).

References

Brown AR, Middlebrook RD (1981) Sampled-data modeling of switching regulators. In: Proceedings of the IEEE Power Electronics Specialists Conference – PESC 1981. Boulder, Colorado, USA, pp 349–369

Mattavelli P, Rossetto L, Spiazzi G, Tenti P (1997) General-purpose fuzzy controller for DC-DC converters. IEEE Trans Power Electron 12(1):79–86

Michel L, Join C, Fliess M, Sicard P, Chériti A (2010) Model-free control of dc/dc converters. In: Proceedings of the 12th IEEE workshop on Control and Modeling for Power Electronics – COMPEL 2010, CDROM, Boulder

Middlebrook RD (1987) Topics in multiple-loop regulators and current mode programming. IEEE Trans Power Electron 2(2):109–124

Morroni J, Zane R, Maksimović D (2009) Design and implementation of an adaptive tuning system based on desired phase margin for digitally controlled DC-DC converters. IEEE Trans Power Electron 24(2):559–564

Rodriguez J, Cortes P (2012) Predictive control of power converters and electrical drives. Wiley, New York

Timbuş A, Liserre M, Teodorescu R, Rodriguez P, Blaabjerg F (2009) Evaluation of current controllers for distributed power generation systems. IEEE Trans Power Electron 24(3):654–664

Index

S. Bacha et al., *Power Electronic Converters Modeling and Control: with Case Studies*, Advanced Textbooks in Control and Signal Processing, DOI 10.1007/978-1-4471-5478-5,